Ratio of Momentum Diffusivity to Thermal Diffusivity

This book provides a comprehensive introduction to the Prandtl number as well as measurements of thermo-physical parameters like viscosity, density, thermal conductivity, specific heat capacity, and thermal diffusivity. The textbook is intended for scientific researchers, fluid mechanics instructors, and heat and mass transfer professionals. More so, researchers on boundary layer flow, mechanical and chemical engineers, and physicists would find it very useful. It gives high-quality information to postgraduate students studying transport phenomena who require theoretical and empirical reviews on the impact of increasing the momentum diffusivity to thermal diffusivity ratio.

Features:

- Provides a systematic overview of the state of the art for understanding changes between dependent and independent variables.
- Highlights some theoretical and empirical reviews on the Prandtl number.
- Presents an in-depth analysis of various self-similar flows, emphasizing stretching-induced flows, nanofluid dynamics, suction, injection, free convection, mixed convection, and forced convection.
- Provides an insightful reports on thermal radiation, heat source, heat sink, energy flux due to concentration gradient, mass flux due to temperature gradient, thermo-capillary convection flow, Joule heating, viscous dissipation, thermal stratification, thermophoresis, and Brownian motion of particles.

Ratio of Momentum Diffusivity to Thermal Diffusivity

Introduction, Meta-analysis, and Scrutinization

Isaac Lare Animasaun
Nehad Ali Shah
Abderrahim Wakif
Basavarajappa Mahanthesh
Ramachandran Sivaraj
Olubode Kolade Koriko

CRC Press
Taylor & Francis Group
Boca Raton London New York

CRC Press is an imprint of the
Taylor & Francis Group, an **informa** business
A CHAPMAN & HALL BOOK

First edition published 2022
by CRC Press
6000 Broken Sound Parkway NW, Suite 300, Boca Raton, FL 33487-2742

and by CRC Press
4 Park Square, Milton Park, Abingdon, Oxon, OX14 4RN

CRC Press is an imprint of Taylor & Francis Group, LLC

Library of Congress Cataloging-in-Publication Data
Names: Animasaun, Isaac Lare, author. | Shah, Nehad Ali, author. |
Wakif, Abderrahim, author. | Mahanthesh, Basavarajappa, author. |
Sivaraj, Ramachandran, author. | Koríko, Olubode Kolade, author.
Title: Ratio of momentum diffusivity to thermal diffusivity : introduction,
meta-analysis, and scrutinization / Isaac Laare Animasaun, Nehad Ali
Shah, Abderrahim Wakif, Basavarajappa Mahanthesh, Ramachandran Sivaraj,
Olubode Kolade Koriko.
Description: First edition. | Boca Raton : Chapman & Hall/CRC Press, 2023. |
Includes bibliographical references and index.
Identifiers: LCCN 2022005066 (print) | LCCN 2022005067 (ebook) |
ISBN 9781032108520 (hbk) | ISBN 9781032310893 (pbk) | ISBN 9781003217374 (ebk)
Subjects: LCSH: Materials–Thermal properties. | Thermal diffusivity.
Classification: LCC TA418.52 .A55 2023 (print) | LCC TA418.52 (ebook) |
DDC 620.1/1296–dc23/eng/20220401
LC record available at https://lccn.loc.gov/2022005066
LC ebook record available at https://lccn.loc.gov/2022005067

ISBN: 9781032108520 (hbk)
ISBN: 9781032310893 (pbk)
ISBN: 9781003217374 (ebk)

DOI: 10.1201/9781003217374

Typeset in Palatino
by codeMantra

This textbook is dedicated to the Almighty God for blessing our collective minds with the objectives of this learning material. Also, this textbook is dedicated to the blessed memory of Late Professor Prandtl Ludwig, Late Racheal B. Animasaun, and Late Temitope Florence Benjamin.

[illegible]
[illegible]
[illegible]
[illegible]
[illegible]

Contents

Preface

In thermal engineering systems, restaurants, and industries, changes in the ratio of momentum diffusivity to thermal diffusivity is a yardstick suitable for influencing the end products, efficiency, and performance. Within the frame of heat transfer, thermal analysis, and fluid dynamics, this has led to many published reports on the increasing impact of the Prandtl number on various fluid flows through pipes, on wedges, over cones, on flat surfaces, and in arteries. This textbook presents the outcome of a broad scrutinization of the significance of increasing such a relative magnitude of two quantities on various transport phenomena to a global audience (i.e., professionals, scholars, and scientists within the research community of fluid dynamics who deal with dimensionless parameters). A number is needed to quantify the relationship between any two variables in a quantitative analysis, especially between dependent and independent variables. This textbook presents the theoretical and empirical reviews of over a hundred published reports on the effects of increasing the Prandtl number. After that, a slope linear regression through the data points was appraised as a tool suitable to quantify the Prandtl number's effect. The project's execution follows the procedure by Tawfik et al. [294], excluding protocol registration, which is not achievable. The steps include developing research questions; forming criteria; searching strategy; searching databases, titles, and abstracts; full- text screening; manual searching; extracting data; quality assessment; data checking; statistical analysis; double data checking; and manuscript writing. A combination of systematic review and scrutinization is shown in this textbook to be a feasible solution for helping experts in physical sciences and researchers to quantify any form of relationship between two variables dependent on one another. More so, such a relationship is shown to be helpful for the enhancement of discussion of results. The study's outcomes and conclusions are suitable for validating future observations and results related to the same dimensionless number.

Acknowledgements

It is crucial to recognize the contributions of Mr. Opeyemi Oladotun Olajimbiti, Late Miss. Temitope Florence Benjamin, Mr. Ayodele Moses Awe, Miss. Adejoke Esther Akinmusire, and Mr. Oluwafemi Samson Olagbami for their active participation in Fluid Dynamics and Survey Research Group. Together with Miss Adejoke Margaret Olotu, their supports towards the successful completion of the textbook is acknowledged.

Authors

Isaac Lare Animasaun is a Lecturer in the Department of Mathematical Sciences, the Federal University of Technology, Akure, Nigeria. His research interests are mathematical modeling, boundary layer analysis, heat and mass transfer, dynamics of Newtonian and non-Newtonian fluids, fluid flow through porous or non-porous media, approximate and analytical solutions of differential equations, and educational research (survey research). He has published more than one hundred articles through international and local refereed journals. The statistics show that Dr. Animasaun has published 115 reports indexed in Scopus (36 h-index) among which 43 articles are indexed in Web of Science (21 h-index). A few among the journals he has served as a reviewer and their respective numbers of reviews are the *Arabian Journal for Science and Engineering*, Springer (125), *Neural Computing and Applications*, Springer (61), *Heat Transfer – Asian Research*, Wiley (41), and *Journal of Thermal Analysis ad Calorimetry*, Springer (14). He won the 2019 Outstanding Reviewer Awards from Physica Scripta, IOP Publishing, and four awards from Publons to his credit. He was awarded the renowned Professor Aderemi Kuku Young Scientist Prize in Mathematical Sciences by the Nigerian Young Academy in 2020 (NYA). Recently, He was awarded Second place of the 2021 Mark Reed Young Research Award by Nanotechnology, IOP Publishing.

Dr. Nehad Ali Shah is an Assistant Professor in the Department of Mathematics, Lahore Leads University. His research interest is in fluid dynamics (i.e., Newtonian and non-Newtonian fluids, heat and mass transfer, viscoelastic models with memory, and fractional thermoelasticity). His research expertise was attracted by Ton Duc Thang University, Ho Chi Minh City, Vietnam, and led to his role as a part-time researcher in their Faculty of Mathematics & Statistics. He is the recipient of the Pre-PhD Quality Research Award from Abdus Salam School of Mathematical Sciences, GC University Lahore, Pakistan, in the year 2018. He was also a recipient of a research invitation by Prof. Haitao Qi, Vice Dean of School of Mathematics and Statistics, Shandong University of Weihai, China, and Prof. Shaowei Wang, Dean of Department of Engineering Mechanics, School of Civil Engineering, Shandong University, Jinan, P. R. China. He has published more than 70 research papers in different highly reputed journals. He has served as a reviewer of many manuscripts submitted to more than 20 international journals.

Dr. Abderrahim Wakif is a Lecturer in the Laboratory of Mechanics, Faculty of Sciences Aïn Chock, University Hassan II – Casablanca, B. P. 5366, Mâarif, Casablanca, Morocco. His research interests are peristaltic blood flow; nanofluids; boundary layer flows; natural, forced, and mixed convection; MHD/EHD flow in nanofluids; heat and mass transfer in mixtures; numerical, analytical, and semi-analytical methods; and linear and nonlinear stability analysis. Dr. Wakif is a board member of the *International Journal of Applied and Computational Mathematics*, *Journal of Computational Applied Mechanics*, *Mathematical Methods in the Applied Sciences*, *Progress in Computational Fluid Dynamics*, and *Thermal Science*. Some of the journals that have benefited from his review skills and the number of completed assignments are *Journal of Thermal Analysis and Calorimetry* (39), *Physica A: Statistical Mechanics and Its Applications* (19), *Scientia Iranica* (9), *International Journal of Applied and Computational Mathematics* (8), *International Communications in Heat and Mass Transfer* (5), *Journal of the Brazilian Society of Mechanical Sciences and Engineering* (5), and *Physica Scripta* (5). Seventeen of his publications were published through various journals indexed in Web of Science. He won the Top Reviewer Awards in Physics and Cross-Field by Publons Academy in the year 2019.

Dr. Basavarajappa Mahantheshis currently an Associate Professor at the Center for Mathematical Needs, Department of Mathematics, CHRIST (Deemed to be University), Bangalore-560029, India. His research covers various topics, including fluid mechanics, nanofluid dynamics, optimization techniques, response surface methodology, and sensitivity analysis. He is a member of the editorial board of renowned academic journals. He is an active reviewer for many renowned journals. He has edited the book *Mathematical Fluid Mechanics*, has four book chapters, and has published over 150 research articles in many renowned international journals. His scientific metrics according to (https://bit.ly/2Eod25I) Google Scholar show h-index = 42, citations index = 4,150, and i10-index = 96. He has organized several conferences and delivered key research papers in various countries and international conferences.

Dr. Ramachandran Sivaraj is working as a Senior Assistant Professor in the Department of Mathematics, School of Advanced Sciences, Vellore Institute of Technology, India. His area of research interest is fluid dynamics, and he works on non-Newtonian fluids, MHD, heat transfer, and porous media. He has published 34 research papers in highly reputed international journals. Under his guidance, two scholars received a PhD degree, and presently, four scholars are working. He received the VIT Research Award continuously in the years 2012–2018 and a travel grant from the Royal Society of London and CSIR, India. He worked at Guangdong University of Technology, Guangzhou, China, under the Faculty Exchange Programme. He visited the National Defense University of Malaysia, Kuala Lumpur, Malaysia, as a visiting scientist. He delivered lectures on fluid dynamics applications at the University of the West Indies, St. Augustine, Trinidad, West Indies. He delivered invited lectures in international conferences and seminars organized by the University of Botswana, the National University of Singapore, and Universiti Putra Malaysia. He has attended over 20 national and international conferences in India and abroad. He was invited to deliver lectures in several conferences, workshops, seminars, and guest lectures, and he has organized several international conferences, seminars, and workshops to promote the research activities. He served as a reviewer for 39 scientific journals. He served as one of the editors for the book *Applied Mathematics and Scientific Computing* published by Springer book series, *Trends in Mathematics, Advances in Fluid Dynamics* published by Springer book series, *Lecture Notes in Mechanical Engineering*, and the proceedings *Recent Trends in Pure and Applied Mathematics* published by AIP Conference Proceedings. He is the Joint Secretary of Academia for Advanced Research in Mathematics Society.

Olubode Kolade Koriko is a Professor in the Department of Mathematical Sciences, the Federal University of Technology, Akure, Nigeria. His research interests fall within the scope of mathematical modeling, boundary layer analysis, heat and mass transfer, dynamics of Newtonian and non-Newtonian fluids, flow through porous or non-porous media, and differential equations. He is a member of the Society for Industrial and Applied Mathematics, Nigeria Association of Mathematical Physics, and Nigeria Mathematical Society. Also, he has served in various journals as a reviewer. He had served as an external examiner of more than five PhD students. More so, he has produced two PhD students. The statistics show that he has published many papers indexed in Scopus and Web of Science. He was among the recipients of the Commonwealth Scholarship and Fulbright Junior Research Fellowship.

1

General Background Information

1.1 Introduction

Description of the changes in the dependent variables (i.e. unknown variables) with either one or more of the independent variables (i.e. known variables) is best achievable through exploration of the governing equations of its mathematical model. The governing equations that model welding, heating in various combustion devices, melting, and surface treatment have been shown in the literature to be something that can be parameterized and analyzed for exploration. Depending on the nature of the heat transfer, the parametrization of the governing equation may lead to dimensionless numbers such as an agglomerate number, Boltzmann number, Christensen number, Chvorinov cast mold system number, Chvorinov shrink formation number, Chvorinov time number, Craya–Curtet number, loss dissipation coefficient, melting efficiency, Niyam shrink formation number, Rosenthal number, solidification rate constant, and Tikhonov number. In this chapter, the general background information is divided into dimensionless numbers, thermo-physical properties of fluids, and slope linear regression through the data points.

1.2 Dimensionless Numbers

For a realistic simulation of any physical phenomenon, all parameters that emerge must be dimensionless. Kunes [168] listed more than 75 dimensionless numbers that often emerge in simulation of design and construction of machines (e.g., water pumps, fans, and mixers), including Addison shape number (Ad), advance ratio, bearing number, cavitation number, delivery number, discharge number, eccentricity of the bearing, Ekman number, gravitational-to-centrifugal acceleration ratio, homochronicity number, and Leroux number. For instance,

1. **Agglomerate number**: Adhesion forces, particle characteristics, impact rate, and particle shape are useful to quantify touch damage during agglomeration. According to physics, the energy required for a material to disintegrate is proportional to the accidental kinetic energy of the agglomeration. Surface energy, modulus of elasticity, particle diameter, particle density, and velocity play essential roles in the formation of an agglomerate number.
2. **Boltzmann number of combustion chamber**: Heat transmission in fireplaces is quantified by this dimensionless quantity (i.e., maximum heat transfer to the fireplace's surface through radiation).
3. **Christensen number**: Any research of dynamic thermal processes in electric arc welding calculates the ratio of motion velocity and surface to thermal diffusivity.

DOI: 10.1201/9781003217374-1

4. **Chvorinov cast mold system number**: According to Chvorinov's rule, areas of the casting object with a high volume-to-surface-area ratio should take a little while to become solidified. The parameter is determined by the Fourier number, volume parameter, and thermal conductivity.
5. **Craya-Curtet number**: The dimensionless number estimates not only the heat transfer but also the radiation flow in industrial furnaces. For example, the Craya-Curtet number, which is the square root of the ratio of the momentum flux of the coflowing stream to that of the central jet, is required to define the emerging boundary-layer approximation in the case of a steady narrow jet.

   Without variance in the emergent dimensionless quantities, fluid flow analysis is virtually trivial. The Mach number, Weber number, Euler number, Reynolds number, Grashof number, Froude number, and Prandtl number are some of the well-known dimensionless numbers in fluid mechanics. The ratio of the elastic force to the inertia force is known as the Mach number. Weber's number is the square root of the ratio of the surface tension force to the inertia force. The Euler number is the square root of the ratio of the inertia force to the pressure force. The Reynolds number is the ratio of the inertia force to the viscous force that causes fluid flow. The proportion of buoyant forces to viscous forces is known as the Grashof number. The ratio of the inertia force to the square root of the characteristic length and the gravitational force is known as Froude's number. The Chandrasekhar number is a dimensionless number highly useful to quantify the ratio of the Lorentz force to the viscosity of magnetic convection flow.

1.2.1 Categories of Dimensionless Numbers

Majority of the dimensionless numbers in the literature can be grouped into the following six categories:

1. **Surface gravity waves-related dimensionless numbers**: Ursell number and Iribarren number.
2. **Turbulence related dimensionless numbers**: Karlovitz number, Markstein number, Colburn J factor, and Dean number.
3. **Heat transfer-related dimensionless numbers**: Rayleigh number, Prandtl number, Biot number, Peclet number, Brinkman number, Grashof number, Eckert number, and Nusselt number.
4. **Mass transfer-related dimensionless numbers**: Lewis number, Schmidt number, and Sherwood number.
5. **Fluid dynamics-related dimensionless numbers**: Ohnesorge number, Gortler number, Laplace number, Keulegan-Carpenter number, Bejan number, Marangoni number, Galilei number, and Archimedes number.

In fluid flow, the dimensionless Froude number is the ratio of inertial to gravitational forces. This was ascribed to William Froude (father 1810–1879) and Robert Edmund Froude (son 1846–1924). The dimensionless Froude number is defined as

$$F_r = \frac{u}{\sqrt{gL}} = \frac{ms^{-1}}{\sqrt{ms^{-2}m}} = \frac{ms^{-1}}{m^{1/2}s^{-1}m^{1/2}} = 1 \tag{1.1}$$

where u is the velocity within the closed domain under investigation, g is the earth's gravity, and L is the characteristic length. Because of the insignificant impact of gravitational force, many theoretical studies of fluid dynamics ignore Froude's number, but inertial force is always substantial and is accounted for using a material derivative (i.e., unsteady and convective acceleration). However, naval architecture recognizes the importance of the Froude number since it may be used to investigate the resistance of moving objects through water that is partially submerged. Campbell [72] presented the Froude number as an important tool for determining the formation of the hydraulic jump, which is a problematic aspect of big runners' flow. The Froude number is especially useful in assessing scenarios when the surface's undulation causes its surface area to decrease, producing an entrainment condition.

1.3 Prandtl Number

The **Prandtl number** (P_r) is the ratio of momentum diffusivity to thermal diffusivity. The dimensionless number P_r comprises of thermal conductivity, heat capacity, dynamic viscosity, and density. Mathematically,

$$P_r = \frac{\vartheta}{\alpha} = \frac{\mu}{\rho} * \frac{\rho c_p}{\kappa} = \frac{\mu c_p}{\kappa}. \tag{1.2}$$

The turbulent Prandtl number and magnetic Prandtl number are the two types of variations of the Prandtl number.

1. **Turbulent Prandtl number**: This is defined as the ratio of momentum eddy diffusivity to heat transfer eddy diffusivity. This is mostly used to calculate heat flux in a turbulent flow.
2. **Magnetic Prandtl number**: For modeling an induced magnetic field, the development of a magnetic Prandtl number is unavoidable in fluid mechanics. This is the ratio of momentum diffusivity to magnetic diffusivity.

As estimated by Rapp [251], the Prandtl number of sodium at 100°C is 0.01, mercury at 25°C is 0.03, air at 30°C is 0.72, carbon disulfide at 25°C is 2.36, chloromethane at 25°C is 4.41, methanol at 25°C is 6.83, water at 25°C is 6.90, toluene at 25°C is 7.26, ethanol at 25°C is 18.05, argon at 30°C is 22.77, krypton at 30°C is 673.68, xenon at 30°C is 674.91, and glycerol at 25°C is 7,612.74. In another related study by Samee et al. [267], Prandtl number P_r of (i) liquid sodium is 0.005, (ii) sodium–potassium is 0.00753, (iii) lead is 0.02252, and (iv) helium is 0.666. Nehad et al. [211] estimated the Prandtl number P_r of methanol as 7.3786, water as 6.1723, blood as 22.9540, and ethylene glycol as 150.46.

1.3.1 Parameters Dependent on the Prandtl Number

The major laws that govern momentum and heat and mass transfer, most especially at the surface, are Newton's law (Eq. 1.3a), Fourier's law (Eq. 1.3b), and Fick's law (Eq. 1.3c).

$$\tau = \mu \frac{\partial u}{\partial y}, \quad q = -\kappa \frac{\partial T}{\partial y}, \quad n = -D_m \frac{\partial C}{\partial y}. \tag{1.3}$$

Numerous dimensionless numbers may be used to define the transport phenomena caused by the occurrence of momentum flow, heat flux, and diffusion flux, both theoretically and empirically. Some of these characteristics are affected by the ratio of momentum diffusivity to thermal diffusivity.

1. **Thermal Peclet number** (P_e): The passage of air with a differing temperature into the anticipated area on a synoptic scale is known as thermal advection. Meanwhile, a temperature gradient in a medium causes thermal diffusion, which is the relative mobility of a gaseous mixture's or solution's constituents. The ratio of the strengths of thermal advection to thermal diffusion is defined as

$$P_e = P_r \times R_e$$

where R_e is the Reynolds number.

2. **Rayleigh number** (R_a): This dimensionless number is equivalent to the product of the Grashof number and the Prandtl number. It is used to describe the transition from laminar to turbulence flow induced by natural convection. Mathematically, the Rayleigh number is defined as

$$R_a = G_r \times P_r.$$

3. **Brinkman number** (B_r): It is a dimensionless number used to calculate heat conduction from a wall to a moving viscous fluid. In other words, it is the ratio of heat produced through the dissipation of heat energy to heat transported by molecular conduction (i.e., external heating). Mathematically, the Brinkman number can be written as

$$B_r = P_r \times E_c.$$

4. **Lewis number** (L_e): This is the ratio of the Schmidt number to the Prandtl number. Mathematically, the Lewis number is expressed as

$$L_e = S_c / P_r.$$

5. **Graetz number** (G_z): The dimensionless number is most suitable to not only describe but also to characterize laminar flow through a conduit. It is defined as

$$G_z = \frac{D_H}{L} R_e P_r$$

In the above expression, D_H is the tubes diameter, L is the length of the conduit, and R_e is the Reynolds number.

6. **Stanton number** (S_t): This is the ratio of heat transferred into a liquid substance to the thermal capacity of the same fluid. Mathematically,

$$S_t = \frac{h}{\rho V c_p} = \frac{N_u}{R_e P_r}.$$

Here, the convection heat transfer coefficient is denoted by h, the density of the fluid is denoted by ρ, the specific heat of the fluid is denoted by c_p, the speed of the fluid is denoted by V, the Nusselt number is denoted by N_u, and the Reynolds number is denoted by R_e.

7. **Colburn j Factor** (j_H): It is a dimensionless number discovered by Prof. Colburn as an important factor to describe heat transfer dependent on local heat transfer coefficient h, mass velocity of fluid stream G, and specific heat of the fluid c_p. This is defined as

$$j_H = \frac{h}{c_p} G P_r^{2/3}.$$

One of the major connections between the heat and mass transfer is the relationship between the Prandtl number P_r and the Schmidt number S_c. Mathematically, the Prandtl number P_r and the Schmidt number S_c are defined as

$$P_r = \frac{\vartheta}{\alpha}, \qquad S_c = \frac{\vartheta}{D_m} \tag{1.4}$$

In Eq. (1.4), the kinematic viscosity ϑ implies momentum diffusivity, thermal diffusivity α, and mass diffusivity D_m implies the coefficient of mass diffusivity. Mathematically, Eq. (1.4) stands for

$$P_r = \frac{S_c D_m}{\alpha}. \tag{1.5}$$

Fick's first law is solely dependent on one parameter of the solute's interaction with the solvent, i.e., the diffusion coefficient. The diffusion coefficient determines the flux due to diffusion in dilute species movement. The most fundamental definition of the diffusion coefficient is the molar flux flow across a surface per unit concentration gradient out of plane.

1.4 Viscosity μ (Pa s = Ns m^{-2}) or kg m$^{-1}s^{-1}$

Viscosity is the amount of internal friction that holds the molecules of fluids together, whether gases or liquids. The reluctance of any fluid to flow from one place to another is called viscosity by rheologists. Internal friction is a phenomenon that can alter the velocity of a fluid's layers as they move with one another. This term is frequently used because the dynamic/absolute viscosity measures the mechanical friction between the whole molecules across the layers of fluids in motion. This explains why fluids with a high viscosity, such as motor oil, move slowly and fluids with low viscosity, such as kerosene and water, move quickly.

There is no agglomeration or clustering of any kind when viscosity is minuscule, as it is in the case of gases (i.e., air). However, whether the fluid is moving or not, adding additional nanoparticles to the base fluid causes agglomeration and clustering of particles, which is a factor in building stable systems. Yang et al. [325] concluded that high viscous fluids are preferable for forming stable nanofluids with little nanoparticle aggregation. However, it is essential to note that the same high viscous fluids are not suitable for heat transmission.

According to Timofeeva et al. [295], clustering nanometer-sized particles and agglomeration are the primary mechanisms responsible for the undesired increase in viscosity in nanofluids. The kinematic viscosity, often known as the ratio of dynamic viscosity to density, is another related characteristic. The viscosity of a fluid is defined as the ratio of shearing stress to velocity gradient, according to Sariyerli et al. [270]. Across the perpendicular layers (y-direction) of any fluid flow along horizontal surface (x-direction), the relation between molecular momentum and cohesion is most appropriate to classify fluids primarily as ideal fluids, Newtonian fluids, and non-Newtonian fluids.

> Viscosity is the degree of internal friction that holds the molecules that makeup fluids together.

1.4.1 Measurements of Viscosity

Some of the techniques for measuring the viscosity of fluids include the capillary tube, Zahn cup, falling sphere viscometer, vibrational viscometer, rotating viscometer, and VROC viscometer.

- **Capillary tube**: In the 1800s, time taken for a known volume of fluid to pass through a length of capillary tube was a typical methodology for measuring viscosity.
- **Zahn cup**: In the paint industry, this method of evaluating viscosity is commonly employed. The fluid material is placed in a cup with a small hole in the flat bottom. The time it takes for a the fluid material to flow through a the hole is the fluid's viscosity.

Because of the dynamics of simple fluids, particularly Newtonian fluids, any of the first three techniques stated are suitable. However, the nature of non-Newtonian fluids suggests that a rotational viscometer is best suited to determining the viscosity of these fluids because torque is required to cause an object of a specific weight to revolve in the fluid, particularly a high viscous fluid whose viscosity is to be determined. Furthermore, the amount of revolvement in various viscous fluids as a result of torque varies (Rajput [243]). A capillary viscometer is one of the best experimental devices for determining the viscosity of fluid materials. When gravity has a substantial impact, the functions of density become evident; the resistance of the fluid to flow is the kinematic viscosity; hence, it is the ratio of dynamic viscosity to density.

1.4.2 Review of Published Facts on Viscosity

The viscosity of liquids, particularly nanofluids, decreases with increasing temperature (Nguyen et al. [213]; Lee et al. [176]). In the case of alumina nanoparticles, cupric nanoparticles, and silicon dioxide nanoparticles distributed in ethylene glycol and water, the pattern of reduction in viscosity owing to temperature increase is exponential. The opposite is true for gas viscosity, which increases with temperature (Andrade [28]). Nasirzadehroshenin et al. [204] discovered that the viscosity of water-based titanium oxide and aluminum oxide (hybrid nanofluid) increased due to an increase in volume fraction and a reduction in nanoparticle size or temperature throughout the hybrid nanofluid. According to the findings of a meta-analysis conducted by Mohamoud Jama et al. [147], the size and shape of nanoparticles impact the viscosity of nanofluid. As the volume of nanoparticle concentration increases, the viscosity of nanofluid increases. Ubbelohde viscometers and Stabinger viscometers are two devices that can determine the viscosity of liquids at various temperatures (Sariyerli et al. [270]). Experimental measurements conducted by Wang et al. [311] shows a 130% increase in the viscosity of water-based nanofluids due to the addition of Fe_3O_4.

1.5 Classification of Fluids

The amount of molecules that make up a liquid per unit of space determines its viscosity. The viscosity is affected when the free-space decreases at a given volume of molecules. As a result, the relative free-space is the fractional increase in volume caused by expansion. Doolittle [92] defined "free-space" in a liquid as space that seems to arise from the liquid's entire thermal expansion without changing phase. As a result, relative free-space is the fractional increase in volume caused by expansion. The addition of copper nanoparticles to water, according to Saidi and Karimi [263], is a measure for increasing velocity (due to a decreased viscosity) and improving temperature distribution (due to increases of the thermal conductivity of the nanofluid). This section is limited to the classification of fluids into ideal, Newtonian, and non-Newtonian fluids.

- **Ideal fluids**: This category of incompressible fluids has zero shear stress ($\tau = 0$) at any level of velocity gradient $\frac{\partial u}{\partial y}$. In other words, whether the motion of such fluids is slow or fast, the shear stress remains zero. In an earlier report by Leonard Euler on fluid flow, it was proved that the viscosity of many less viscous fluids can be approximated to zero. This assumption once led to a group of experts to only deliberate on frictionless fluid (classical hydrodynamics).
- **Newtonian fluids**: This category of fluids demonstrate a linear relationship between shear stress and velocity gradient (shear strain rate). Mathematically,

$$\tau \propto \frac{\partial u}{\partial y}$$

$$\tau = \mu \frac{\partial u}{\partial y}. \tag{1.6}$$

  In other words, the ratio of shear stress to shear strain is known as viscosity. In Eq. (1.6), μ is the constant of proportionality or slope of the linear relationship between the two aforementioned variables. Typical examples are water, air (in the absence of vortex), and kerosene.
- **Non-Newtonian fluids**: This class of fluids has a nonlinear connection between velocity gradient and shear stress, which can be time-dependent or time-independent.

1.5.1 Category of Non-Newtonian Fluids

Because of variations in velocity gradient (shear rate) and shear stress, the viscosity of many fluids does not remain constant. This is true physically because, when subjected to force, the viscosity of any fluid may shift from more solid to more liquid.

1. **Shear thickening (dilatant)**: A higher shear stress causes an increase in viscosity (shear strain). The suspension of maize starch in water, for example, appears milky when swirled slowly, but becomes a highly viscous liquid when churned quickly. The second example is sand suspension in water.
2. **Shear thinning (pseudoplastic)**: The viscosity decreases due to a higher shear rate (e.g., nail polish and blood).
3. **Time-independent non-Newtonian fluids**: Non-Newtonian fluids with time-independent viscosity describe fluid viscosity that is not reliant on the history of fluid shear stress rate (e.g., shear thickening and shear thinning). Power-law fluid, Herschel–Bulkley fluid, Bingham fluid, Eyring–Powell fluid, and Casson fluid are a few examples (Ghassemi and Shahidian [108]).
4. **Time-dependent non-Newtonian fluids**: These fluids are affected by the strain rate and the duration of the applied shear because their viscosity varies while subjected to shear stress. Printer ink and gypsum plaster are two examples.

Other fluids affected by yield stress, shear stress, and velocity gradient are ideal Bingham plastic and thixotropic substances. Pressure and temperature, in particular, have a significant impact on viscosity. Higher viscous fluids, such as motor oil, require greater

TABLE 1.1
Viscosity of Some Liquid Substances mPa s

Liquid Substances	Viscosity	Liquid Substances	Viscosity
Air at 30°C	0.019	Blood at 38°C	2.69
Carbon disulfide at 25°C	0.352	Blood at 39°C	2.63
Chloromethane at 25°C	0.537	Blood at 40°C	2.55
Sodium at 100°C	0.542	Ethylene glycol at 25°C	16.1
Methanol at 25°C	0.544	Argon at 30°C	22.900
Toluene at 25°C	0.560	Xenon at 30°C	23.200
Methanol at 20°C	0.59	Krypton at 30°C	25.600
Water at 25°C	0.889999	Glycerol at 25°C	934
Ethanol at 25°C	1.074		
Mercury at 25°C	1.526		
Blood at 35°C	2.87		
Blood at 36°C	2.82		
Blood at 37°C	2.78		

force to produce flow, whereas lower viscous fluids, such as air and other gases, require little or no effort to induce flow. Table 1.1 presents the viscosity of some liquid substances as published by Lee et al. [176], Bergman et al. [58], Jahangiri et al. [146], Rapp [251], Wakif et al. [310], and Shell Chemicals [282].

1.6 Density ρ (kg m^{-3})

The density of any substance is the mass-to-volume ratio of the material. Because mass is involved, this characteristic is also known as mass density. Any material's density, whether in tiny or huge quantities, is always the same. As a result, density might be referred to as a material's distinctive attribute. According to some experts, density is a quantity that may be used to assess the compactness of the atoms that make up the substance. In other words, if the atoms in a substance are closely packed together, that material will have a higher density than another material made up of loosely packed atoms. Materials having a lower density may be categorically traced to the size of their mass, as illustrated in Table 1.2. This can be traced to the fact that the mass of a piece of wood is greater than that of water.

> Density is the number that quantifies the compactness of the atoms that form the material.

1.6.1 Description of Density

Density denotes denseness (i.e., the number of atoms per unit size). A higher density is achieved when atoms are densely packed in an item. In the meanwhile, the compression of these atoms serves as a meter for increasing density. The density of any object is independent of shape and size. As presented in Table 1.2, it is worth remarking that gases possess low density, followed by liquids and solids. This is due to the way their atoms are packed in nature. Furthermore, pressure and temperature can have a significant impact on

TABLE 1.2
Density of Some Materials

Material	Density (kg m^{-3})	Material	Density (kg m^{-3})
Diamond	3,500	Wood	700
Freshwater at temp. $< 4°C$	1,000	Liquid hydrogen	70
Ice at temp. $< 0°C$	916.7	Air at sea level	1.2
Potassium	860	Hydrogen	0.0898

density. Because of the looseness of particles or atoms, pressure and temperature on gases are significant. Another density-related term is specific gravity, defined as the ratio of a material's density to that of water (also known as relative density). For instance, if the relative density of a substance is less than one, then it is less dense than the reference.

To further explain the point mentioned above, palm oil floats on water because its density is less than water. In contrast, nails sink when thrown in water due to their density being higher than water. It is worth to noting that:

1. Densities are widely used to identify pure substances and to estimate the composition of any mixtures.

2. The analysis of chemical conversion and release of energy in liquids is possible through a change in density.

3. The charge in a storage battery can be described as an acidic solution.

As the battery discharges power, acids in the battery react with the lead to produce a new chemical, reducing the density of the material. This occurrence demonstrates the importance of utilizing the effects of temperature and pressure on the density of a material (i.e., the higher the pressure, the higher the density, while an increase in temperature leads to a reduction in density). It is worth noting that the density necessary for momentum diffusivity in the flow of compressible and incompressible fluids is the same as the density required for thermal diffusivity. The density of liquids is greater than that of gases. This disparity can be attributed to the high volume and mass of gases. Liquid atoms are more densely packed together than gas atoms (materials with loosely packed atoms). To further explain the aforementioned point, palm oil floats on water basically because its density is less than that of water. In contrast, nails sink when thrown in water due to its density higher than that of water. First, densities are widely used to identify pure substances and to estimate the composition of any kind of mixtures.

Second, the analysis of chemical conversion and release of energy in liquids is possible through a change in density. Third, the charge in a storage battery can be described as an acidic solution. As the battery discharges power, acids in the battery react with the lead to produce a new chemical, reducing the density of the material. This occurrence demonstrates the importance of utilizing the effects of temperature and pressure on the density of a material (i.e., the higher the pressure, the higher the density, while an increase in temperature leads to a reduction in density). It is worth noting that the density necessary for momentum diffusivity in the flow of compressible and incompressible fluid is the same as the density required for thermal diffusivity. The density of liquids is much higher than the density of gases. This disparity is because gases have a high volume and mass, but liquids have a large mass for each small volume of a particular material. Liquid atoms are more closely packed together than those in gases (materials with loosely packed atoms).

1.6.2 Review of Published Facts on Density

Hawkes [122] once observed that the rate of diffusion reduces with increasing density for polymers with weak intermolecular interactions (e.g., synthetic rubbers and hexadecane). In other words, the lower the rate of diffusion, the higher the density. One of the noteworthy findings reached by Vitello et al. [307] at the end of a correlation, Student's t test, and Welch's correction between 18 samples of blood aliquots and 18 distilled water is that the density of distilled water samples is comparable to the density of blood. In the instances of alumina-based nanofluid and $R141b$ refrigerant-based nanofluid, Vajjha and Das [305] and Mahbubul et al. [187] discovered that the density of the nanofluid increases with the concentration of nanoparticles. On the other hand, density diminishes when temperature rises (Mohamoud Jama et al. [147]). Elnaqeeb et al. [96] investigated the effect of increased suction and dual stretching on the dynamics of water colloidally mixed with three distinct nanoparticles of varying densities. The simulation shows that larger Nusselt numbers corresponding to heat transfer were observed in the mixture of water and high-density nanoparticles (i.e., copper oxide, copper, and silver) for all stretching ratios and suction levels.

1.6.3 Measurement of Density

The density of a substance may be determined by dividing its mass by its volume. The density of most substances may be measured using either the direct measurement method or the indirect measurement approach.

1. **Direct method:** This approach is employed when measuring the mass and volume of liquids and regularly shaped solids. In this example, beam balancing is used to determine the mass of the items (liquids or solids). The volume of liquids and solids is determined using a graduated cylinder and a ruler. The mensuration of the solid form is required when dealing with solids.
2. **Indirect method:** This approach is used to measure the mass of irregular solids using beam balance. The Archimedes principle is used to calculate the volume of irregularly shaped solids. A graduated cylinder was filled to a specific level with water. The amount of water in the cylinder varies as a solid (item) is dipped into it (displacement). The volume of the item is the difference between the new and old water levels. After obtaining the mass and volume of the substance, the density of the material is determined using $\rho = \frac{m}{V}$. A hydrometer and a pycnometer are two additional density measurements.

The fundamental physical property of substances known as density can be calculated mathematically as the ratio of mass (m) to volume (m^3) in a closed domain as

$$\rho = \frac{m}{v} = \frac{\text{kg}}{\text{m}^3} = \text{kgm}^{-3} \tag{1.7}$$

1.7 Thermal Conductivity κ ($\mathbf{Wm^{-1}K^{-1}}$)

Within the context of heat transfer, the notion of thermal conductivity is critical when the thermal energy of a homogeneous material flows from a higher to a lower position based solely on location and not time. The thermal conductivity may readily quantify the capacity of heat to pass through the aforementioned homogenous material. Any material that shows this property is referred to as a thermal conductor, whereas any material that does not exhibit this feature is referred to as a thermal insulator. An increase in a substance's thermal conductivity causes an increase in the rate at which heat travels through the material. On the other hand, thermal conductivity is essential, but insufficient to characterize the phenomenon when heat movement depends on both place and time. Thermal conductivity refers to a material's capacity to conduct heat.

> The capacity of heat to travel from a hot region to a region of lower temperature through a substance is known as thermal conductivity.

Another acceptable definition is the energy transfer caused by random molecule mobility in the direction of the temperature gradient. The heat flow in many solid objects is not necessarily parallel to the temperature gradient (i.e., thermal conduction is anisotropic). Heat is a term used to describe the movement of energy that occurs primarily due to thermal conduction. According to several studies, heat transmission happens at a slower pace in poor thermal conductivity materials (solids or liquids). As a result, these materials are suitable for use as thermal insulators. Gases (such as air and Styrofoam) and dense gases (such as xenon, dichlorodifluoromethane, sulfur hexafluoride, argon, and krypton) have a low heat conductivity in general. Surprisingly, sulfur hexafluoride has a high thermal conductivity due to its large heat capacity. Materials with high thermal conductivity, on the other hand, conduct heat at a faster pace. Metals (copper), natural diamonds, graphene, silver, gold, aluminum (pure), aluminum (alloys), lead, and stainless steel are examples of such materials.

1.7.1 Review of Published Facts on Thermal Conductivity

The notion of heat conductivity for liquids may also be described using molecular weight, which is a decreasing function of molecular weight. The symmetrical characteristic of some liquids and the creation of hydrogen bonding are exceptions to this rule (Palmer [223]). Due to their extremely high conductivities, alcohol and water were used to clarify this point further. In addition, phase transition (the conversion of one state of matter to another) is a factor that can dramatically alter the thermal conductivity. For example, when ice (solid) is exposed to air, it transforms into water (liquid). Thus, the thermal conductivity also reduces from 2.18 to 0.56 $Wm^{-1}K^{-1}$ (Ramires et al. [248]). The results of experimental research by Nasirzadehroshenin et al. [204] show that increasing the volume fraction or temperature improves the thermal conductivity of the hybrid nanofluid (water-based titanium oxide and aluminum oxide). A decrease in the size of suspended nanoparticles might also increase heat conductivity. Volume fraction, nanoparticle size, and temperature are all significant variables that have been shown to impact the thermal conductivity of nanofluids (Mohamoud Jama et al. [147]).

1. The thermal conductivity is an increasing property of volume fraction of nanoparticles; see Masuda et al. [192]; Lee et al. [175]; Said et al. [262].
2. Experimental study by Hemmat Esfe et al. [131] on (i) water conveying MWCNT-MgO and (ii) ethylene glycol conveying MWCNT-MgO forming two different hybrid nanofluids shows that the thermal conductivity increases with volume fraction $0 \leq \phi\% \leq 1$ in the temperature range for $30°C \leq T \leq 50°C$, with optimal thermal conductivity achievable at $T = 50°C$.
3. Also, the same thermal conductivity increases negligibly with temperature when solid volume fraction is small (i.e., $\phi = 0.015 vol.\%$).

4. At larger solid volume fractions (i.e., $0.04 \leq \phi vol.\% \leq 0.96$), the thermal conductivity increases significantly with temperature within the range of 30°C $\leq T \leq$ 50°C.
5. An experimental measurement conducted by Wang et al. [312] shows a 90% increase in the thermal conductivity of water-based nanofluids due to the addition of Fe_3O_4.

1.7.2 Measurement of Thermal Conductivity

The thermal conductivity of a block of material can be measured by placing it beside ice on one side and steam applied on the other side. The amount of ice that melts after sometimes is measured and is being converted to the amount of heat flux using enthalpy at fusion. The thermal conductivity of a material can also be measured using the steady-state method and transient method (non-steady method). The following properties must be considered before measuring the thermal conductivity

1. What type of material is to be measured?
2. What is the thermal property of the material?
3. What is the medium temperature of the material?

The estimation of Lee et al.[176] and Rapp [251] suggests that the thermal conductivity κ ($Wm^{-1}K^{-1}$) of sodium at 100°C is 60, mercury at 25°C is 8.250, air at 30°C is 0.026, carbon disulfide at 25°C is 0.149, chloromethane at 25°C is 0.117, methanol at 25°C is 0.200, water at 25°C is 0.607, toluene at 25°C is 0.131, ethanol at 25°C is 0.169, argon at 30°C is 0.018, krypton at 30°C is 0.010, xenon at 30°C is 0.006, and glycerol at 25°C is 0.292.

1.7.3 Steady-State Method

This method is used when there is no significant change in the measured temperature with respect to time. For this method to be used, it requires some experimental methods such as the divided bar method, Searle's bar method, and Lees' disc method.

1. **The divided bar method**: This technique employs two brass plates; the material is put vertically between the brass plates, with the hot brass plate on top and the cool brass plate underneath it. The heat is supplied from the top, and it moves downward. This is done to prevent any convection within the material. Measurement is taken after a constant heat is seen on the material. It can be used for both good and poor conductors of heat.
2. **Searle's bar method**: When the material is an excellent heat conductor, this approach is utilized.
3. **Lee's disc method**: When the material is a poor heat conductor, this approach is utilized.

1.7.4 Transient Method

When there is a change in the observed temperature of the substance over time, this approach is utilized. It is simple to set up and may be used to test the thermal conductivities of a variety of materials accurately. This approach's setup uses the transient hot wire technique, modified transient line source method, and laser flash method.

1. **Transient hot wire method**: This method uses a thin vertical wire with finite length as a step voltage. The wire is immersed in a fluid and acts as an electrical heating element and a resistant thermometer. The step voltage is applied to the vertical metal wire, and its transient temperature is measured. One of the products of Decagon Device, Inc., the USA is the KD2 Pro thermal property analyzer device used by Hemmat Esfe et al. [131] to determine the thermal conductivity of hybrid nanofluid. The stainless steel for KD2 Pro thermal property analyzer is a 60 mm KS^{-1} sensor with a 1.27 mm diameter.
2. **Modified transient line source method**: This approach is used to calculate the thermal conductivity of the earth's massive bulk for geothermal heat pump system design. This approach involves burying a pipe loop deep in the earth (in a wellbore), filling the annulus of the bore with a grout substance with known thermal characteristics, heating the fluid in the loop, and measuring the temperature decrease in the loop from the bore's input and return pipes. The line source approximation method estimates the ground thermal conductivity by drawing a straight line on the log of the observed thermal response. This technique needs a very steady heat source and pumping circuit.
3. **Laser flash method**: This method is used to measure the thermal diffusivity of a thin disk in the thickness direction. The method is based on measuring the temperature rise at the rear face of the thin-disk specimen produced by a short energy pulse on the front face. With a reference, sample-specific heat and density of the material can be determined. The thermal conductivity $\kappa = \alpha * C_p * \rho$, where α represents the thermal diffusivity, C_p stands for the specific heat capacity, and ρ represents the density.

1.8 Specific Heat Capacity C_p ($Jkg^{-1}K^{-1}$)

The amount of heat required to raise a body's temperature by one temperature unit is referred to as the heat capacity. The specific heat capacity is the amount of heat required to bring about a change in the temperature of a unit mass of the body by one unit of temperature (measured in $Jkg^{-1}K^{-1}$). The heat capacity is the heat required to raise the temperature of the whole mass of the body by one unit temperature (measured in JK^{-1}). The specific heat of a substance per unit of mass has the dimension $L^2T^{-2}\Theta^{-1}$. Therefore, the SI unit $Jkg^{-1}K^{-1}$ is equivalent to $m^2K^{-1}s^{-2}$. It is important to note that the SI unit of specific energy or energy density Jkg^{-1} is equivalent to m^2s^{-2}.

Definition 1: The amount of heat required to raise the temperature of a body by one unit of temperature is referred to as its heat capacity.

Definition 2: The quantity of heat required to change the temperature of a unit mass by one unit of temperature is known as the specific heat capacity.

1.8.1 Review of Published Facts on Specific Heat Capacity

Experimentally, it was confirmed by Denbigh [90] that the specific heat capacity in the expression of the Prandtl number does vary negligibly between the gaseous and liquid states of aggregation. The thermal radiation heating calorimetry technique was introduced by Morimoto et al. [197] as a better means for measuring specific heat capacity. The technique was used to measure the specific heat of MgO, and it was shown that the thermophysical property increases negligibly with increasing temperature in the range of 220°C $\leq T \leq$ 420°C. Mathematically, given the quantity of heat supplied Q and the corresponding temperature change $\Delta\Theta$, the heat capacity c_p of a body of mass m is given by $c_p = \frac{Q}{\Delta\Theta}$. Heat capacity is a physical property of a material that determines its tenacity when exposed to heat. While some compounds respond rapidly to temperature changes, others absorb heat to break up bonds or change the phase of some of their constituents. In other words, heat capacity varies depending on the condition, geometry (or molecular structure), and body size. A unique approach for evaluating the changes in specific heat capacity and heat capacity during weathering had been presented by Tamao [290]. It was discovered that specific gravity, density, and water content are required to derive the formulas for calculating specific heat capacity changes throughout the weathering process. In case a substance is made of different parts with each part having a known volumetric heat capacity, the addition of the volumetric heat capacities of the proportional parts is equivalent to volumetric heat capacity $\rho c_w = \rho_d c_s + \Delta\rho c_1$, where the specific heat capacity of solid at dry initial condition is c_s, the specific heat capacity of water is c_1, and the specific heat capacity of the wet weathered condition is c_w. The classical model for the heat capacity of nanofluid $(\rho C_p)_{nf}$ is

$$(\rho C_p)_{nf} = (1 - \phi)(\rho C_p)_{bf} + \phi(\rho C_p)_{sp}. \tag{1.8}$$

Objects in the gaseous state have the most significant heat capacity compared to those in the liquid and solid phases. This is not a far-fetched justification. In contrast to solids, where molecules are tightly packed, gaseous molecules are loosely packed. Furthermore, the larger the body's heat capacity, the heavier it is, indicating a direct relationship between mass and heat capacity. The specific heat capacity of a material (e.g., liquid) can be measured experimentally using a calorimeter. A calorimeter is an apparatus for measuring the heat generated or absorbed by a phase change or some other physical change. A calorimeter is made up of a dry inner chamber. The walls are made of a material with high thermal conductance, a stirrer for stirring the liquid during the experiment to allow uniformity, and a thermometer to detect the change in the temperature of the liquid considered. It is worth noting that the more well-insulated the calorimeter, the more precise the result. The estimation of Rapp [251] suggests that the heat capacity ρc_p ($\mathrm{Jkg^{-1}K^{-1}}$) of sodium at 100°C is 1.225, mercury at 25°C is 0.140, air at 30°C is 1.010, carbon disulfide at 25°C is 1, chloromethane at 25°C is 0.960, methanol at 25°C is 2.510, water at 25°C is 4.182, toluene at 25°C is 1.700, ethanol at 25°C is 2.840, argon at 30°C is 0.018, krypton at 30°C is 0.250, xenon at 30°C is 0.160, and glycerol at 25°C is 2.380.

1.8.2 Measurement of Specific Heat Capacity

A specific volume of water is put into the calorimeter up to a specific height, say 1 m, to measure the heat capacity and specific heat capacity of liquids (e.g., water). After that, a heat source (such as a hot metal) is dipped into the calorimeter's water. The stirrer is then used to agitate the water to ensure consistent heat transmission, while the thermometer monitors the temperature changes in the water. The heat capacity of a substance may be determined by measuring the temperature of the heater in the vicinity of the specimen and the temperature of the specimen in cooling and heating modes (Morimoto et al. [197]).

1.9 Thermal Diffusivity α $\left(\frac{\kappa}{\rho c_p} = \mathbf{Wm}^2 J^{-1}\right)$

Without understanding the function of thermal diffusivity, a description of a transitory process is inadequate. The thermal diffusivity is a measurement of how quickly heat moves from one part of a material to another due to a temperature differential. As a result, the thermal diffusivity may be described as a number that rises in part with the thermal conductivity and decreases in part with a material's heat storage capacity. It compares a material's capacity to transmit heat to the amount of heat it can store per unit volume. Experimentally, it has been established that the thermal diffusivity $\alpha(T)$ is equivalent to the ratio of thermal conductivity $\kappa(T)$ to heat capacity $\rho(T)C_p(T)$.

Definition: A thermal diffusivity is a number that quantifies how fast heat flows from one portion of a material to the other due to temperature differences.

1.9.1 Review of Published Facts on Thermal Diffusivity

Ahsan et al. [13] provided the description above as a thermo-physical characteristic that measures the degree of heat energy diffusion from a hot location to a location of low temperature. These concepts are linked to the thermal conductivity because the thermal conductivity is frequently used to assess the rate at which a substance placed in a hot or cold container gets hot or cold. This is not true as conductivity accounts for "how much" while thermal diffusivity accounts for "how quickly" heat flows. This misconception could be traced to the fact that good thermal conductors are usually thermal diffusers. However, this works perfectly only for solid and liquid states. It fails for the gaseous state because gases are bad conductors but diffusers at standard temperature and pressure (20°C, 1 atm). This can be traced to the fact that the thermal storage capacity per unit volume of gases lies within the range of 0.7×10^6 to 2×10^6 $\mathrm{Jm^{-3}K^{-1}}$ which is approximately half of solids and liquids. A typical example is an air with a low thermal conductivity of 0.026 compared to nickel and lead of 91 and 35 respectively, yet air diffuses as fast as the other two materials. It is also worth mentioning that the thermal properties of gases depend on pressure at a constant temperature. At standard temperature and pressure, all gases could be treated as ideal gases (i.e., pressure is not required to quantify thermo-physical properties such as specific heat capacities and the thermal conductivity). Hence, at constant thermal conductivity values and specific heat capacity, thermal diffusivity decreases with an increase in pressure for ideal gases.

The density and number of collisions among gas particles would be small whenever the pressure is low. This increases the average distance traveled by a moving particle between successive collisions, and it becomes easier for the particles to diffuse as fast as possible. Similarly, heat flows do not depend on the location when the material under consideration comprises two or more constituents. Such materials are referred to as composite or heterogeneous materials. The properties of the constituents of material are vital in quantifying the thermo-physical properties of the material. This gives rise to the principle of effective thermal properties, which is a function of the thermal properties of the constituents of a material. If the material constituents are good/bad conductors, the material would also be good/bad. Also, the effective thermal diffusivity of a material is a function of both its components' thermal conductivity and thermal diffusivity. Frittage or

TABLE 1.3
Thermal Diffusivity of Some Materials

Material	Thermal Diffusivity (mm^2/s)
Pure silver (99.9%)	165.63
Copper at 25°C	111
Aluminum	97
Aluminum 6061-T6 alloy	64
Air	19
Iron	23
Nitrogen (300 K, 1 atm)	22
Stainless steel 304 A at 27°C	4.2
Stainless steel 310 A at 25°C	3.352
Sandstone	1.15
Parafin at 25°C	0.081
Water at 25°C	0.143
Engine oil at 100°C	0.0738
Alcohol	0.07

sintering is the formation of mass in a solid form due to pressure or heat without melting. When the temperature is high, sintering causes the contact area between any close particles. This is the exact reason why heat diffuses rapidly at higher temperatures through contact. According to Brown (1958), Eckert [93], and Holman [133], the rate of heat transfer from the hot region to the cold region in some materials is presented as in Table 1.3.

1.9.2 Measurement of Thermal Diffusivity

When the temperature is high, analytical models are thought to be the best way to estimate thermal diffusivity. The thermal diffusivity α of the material of length L at the time it takes to reach half of the maximum temperature rise $(t_1/2)$ can be calculated as

$$\alpha = 0.1388 \times \frac{L^2}{t_1/2}, \quad \text{or} \quad \alpha = \frac{\kappa}{\rho c_\mathrm{p}} \tag{1.9}$$

1.10 Slope Linear Regression through Data Points S_{lp}

A discrete function is a function whose values are not connected (i.e., they have distinct and separate values). Absolute minimum, relative maximum, relative minimum, point of inflection, and absolute maximum are the terms used in advanced calculus to define discrete or continuous functions. For instance, every point on a perfect horizontal line is an absolute maximum and an absolute minimum (Wrede and Spiegel [317]). Geometrically, the interpretation of the slope of the tangent line to a curve at one of its points and representative of instantaneous velocity are facts that support practical applications of what is known as derivative and modeling of physical models. The rate of change between two physical quantities is used to quantify or describe the relationship between two variables. Mathematically, the slope quantifies the change in the dependent variable with respect to the independent variable as

$$\text{Slope} = m = \frac{\Delta y}{\Delta x} = \frac{y_2 - y_1}{x_2 - x_1}. \tag{1.10}$$

The formula in Eq. (1.10) requires the coordinates of the two points (i.e., the slope of a straight line between point $A = (x_1, y_1)$ and point $B = (x_2, y_2)$). The value of such slope (m) in Eq. (1.10) depicts the existing relationship between points A and B. In such a case, there are three possibilities of the nature of slope.

Remark on the Nature of Slope

1. $m > 0$ shows a positive or increasing relationship,
2. $m < 0$ shows a negative or decreasing relationship, and
3. $m = 0$ shows a constant relationship.

In the above case for points A and B, the slope is useful to discuss the changes in the discrete function $y(x)$ based on those two points. There are always more than two points involved in fluid mechanics and other areas of research. In applied mathematics, the line of best fit describes a line through a scatter plot of many data points (x_n, y_n) for $n > 2$ that best expresses the relationship between those points. For the line formed using many points (x_1, y_1), (x_2, y_2), (x_3, y_3), ..., (x_n, y_n), the least square method (LSM) mentioned above also provides the slope (m) of line of best fit given as $y = mx + c$, where c is the intercept as

$$m = \frac{\Sigma(x_{av} - x_i)(y_{av} - y_i)}{\Sigma(x_{av} - x_i)^2} \tag{1.11}$$

In Eq. (1.11), x_{av} and y_{av} are the average of the points in x and y, respectively. Also, x_i and y_i are the set of points in x and y, respectively.

For more explanation, check the video https://m.youtube.com/watch?v=C3DPNX6Hij8

1.10.1 Continuous Function $f(x) = x^3$

Let us consider a continuous function $f(x) = x^3$, where $-3 \leq x \leq 3$. As shown in Figure 1.1, the continuous function $f(x)$ increases as x ranges from -3 to 3. The major question needed to discuss the nature of function $f(x)$ is as follows:

- at what rate does the dependent variable $f(x)$ change with independent x?

$$m = \frac{\Sigma(x_{av} - x_i)(y_{av} - y_i)}{\Sigma(x_{av} - x_i)^2} = \frac{196}{28} = 7 \tag{1.12}$$

It is worth deducing from Tables 1.4 and 1.5 that the function $f(x)$ increases from -3 to 3 at the rate of 7. In order to use Microsoft Excel, it is advisable to follow the procedure below:

1. Open an Excel sheet.
2. In the first row, enter x and $f(x) = x^3$.
3. In Column A, enter all the variables for x.
4. In Column B, enter all the variables for $f(x)$. Or enter =A2$\hat{3}$ and right click + drag.
5. To compute the slope, type the formula `=SLOPE(B2:B8,A2:A8)` as shown in Figure 1.2 and press the enter button.

In other words, as shown in the `prtsc` of Microsoft Excel sheet in Figure 1.2, the rate of change in $f(x)$ with x is $S_{lp} = 7$.

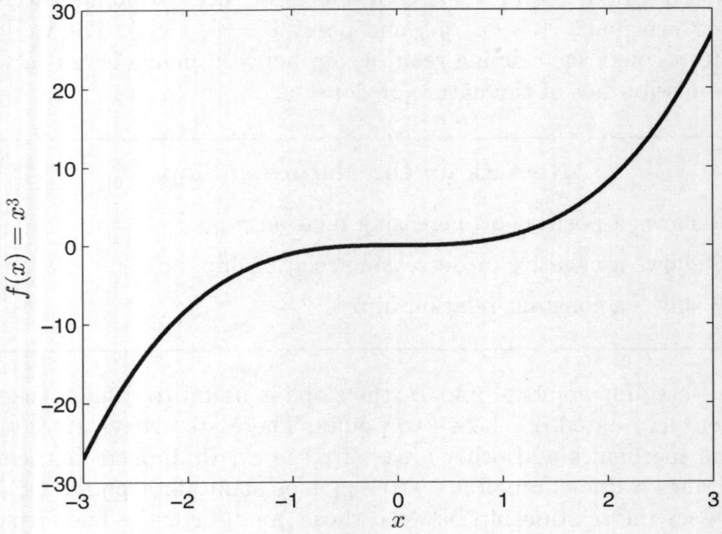

FIGURE 1.1
Graphical illustration of $f(x) = x^3$ for $-3 \leq x \leq 3$.

TABLE 1.4
Analysis of Least Square Method for $f(x) = x^3$ for $-3 \leq x \leq 3$: Part 1

x_i	y_i	$(x_{av} - x_i)$	$(y_{av} - y_i)$	$(x_{av} - x_i)^2$
−3	−27	−3	−27	9
−2	−8	−2	−8	4
−1	−1	−1	−1	1
0	0	0	0	0
1	1	1	1	1
2	8	2	8	4
3	27	3	27	9
$x_{av} = 0$	$y_{av} = 0$			$\Sigma(x_{av} - x_i)^2 = 28$

TABLE 1.5
Analysis of Least Square Method for $f(x) = x^3$ for $-3 \leq x \leq 3$: Part 2

$(x_{av} - x_i)(y_{av} - y_i)$
81
16
1
0
1
16
81
$\Sigma(x_{av} - x_i)(y_{av} - y_i) = 196$

FIGURE 1.2
`prtsc` of a Microsoft Excel sheet for $f(x) = x^3$ for $-3 \leq x \leq 3$.

TABLE 1.6
Relationship between the Independent Parameter K (Porosity Parameter) and Dependent Variable $Re^{0.5}C_r$ (Skin Friction Coefficients) as Presented by Shaw et al. [279]

$K(x_i)$	$Re^{0.5}C_r(y_i)$
1	3.08585135
3	3.57962179
8	4.55496146

1.10.2 Some Results Published by Shaw et al. [279]

Shaw et al. [279] published the data in Table 1.6 to describe the observed relationship between the independent parameter K (porosity parameter) and the dependent variable $e^{0.5}C_r$ (skin friction coefficients); see table two of the report.

From Tables 1.7 and 1.8,

$$m = \frac{\Sigma(x_{av} - x_i)(y_{av} - y_i)}{\Sigma(x_{av} - x_i)^2} = \frac{5.38267}{26} = 0.207025769 \qquad (1.13)$$

In order to use Microsoft Excel, follow the procedure below:

1. Open an Excel sheet.
2. In the first row, enter $K(x_i)$ and $Re^{0.5}C_r(y_i)$.

TABLE 1.7
Part 1 - Least Square Method (LSM) for Estimating the Relationship between the Independent Parameter K (Porosity Parameter) and Dependent Variable $Re^{0.5}C_r$ (Skin Friction Coefficients) as Presented by Shaw et al. [279]

$K(x_i)$	$Re^{0.5}C_r(y_i)$	$(x_{av} - x_i)$	$(y_{av} - y_i)$	$(x_{av} - x_i)^2$
1	3.085851	3	0.6542935	9
3	3.579621	1	0.1605230	1
8	4.554961	−4	−0.8148165	16
$x_{av} = 0$	$y_{av} = 0$			$\Sigma(x_{av} - x_i)^2 = 28$

TABLE 1.8
Part 2 - Least Square Method (LSM) for Estimating the Relationship between the Independent Parameter K (Porosity Parameter) and Dependent Variable $Re^{0.5}C_r$ (Skin Friction Coefficients) as Presented by Shaw et al. [279]

$(x_{av} - x_i)(y_{av} - y_i)$
1.962880
0.160523
3.259266
$\Sigma(x_{av} - x_i)(y_{av} - y_i) = 196$

3. In Column A, enter all the variables for $K(x_i)$.
4. In Column B, enter all the variables for $Re^{0.5}C_r(y_i)$.
5. To compute the slope, type the formula `=SLOPE(B2:B4,A2:A4)` as shown in Figure 1.3 and press the enter button.

In other words, as shown in the `prtsc` of Microsoft Excel sheet in Figure 1.3, the rate of change in $Re^{0.5}C_r(y_i)$ with $K(x_i)$ is $S_{lp} = 0.207025769$. The equation of the line of best fit is $y = mx + c$. To find the value of y-intercept c, it is necessary to use $x_{av} = 4$, $y_{av} = 3.740144867$, and $m = 0.207025769$. Then,

$$y_{av} = mx_{av} + c$$

$$3.740144867 = 4 \times 0.207025769 + c$$

$$c = 2.91204179 \tag{1.14}$$

Hence, the equation of the line of best fit for data presented in Table 1.8 above is

$$Re^{0.5}C_r = 0.207025769K + 2.91204179 \tag{1.15}$$

Based on the analysis above, it is worth concluding that the skin friction coefficients increase with a growth in the porosity parameter at the rate of 0.207025769. As the porosity across the domain of the fluid flow increases, the velocity of the transport phenomenon is enhanced,

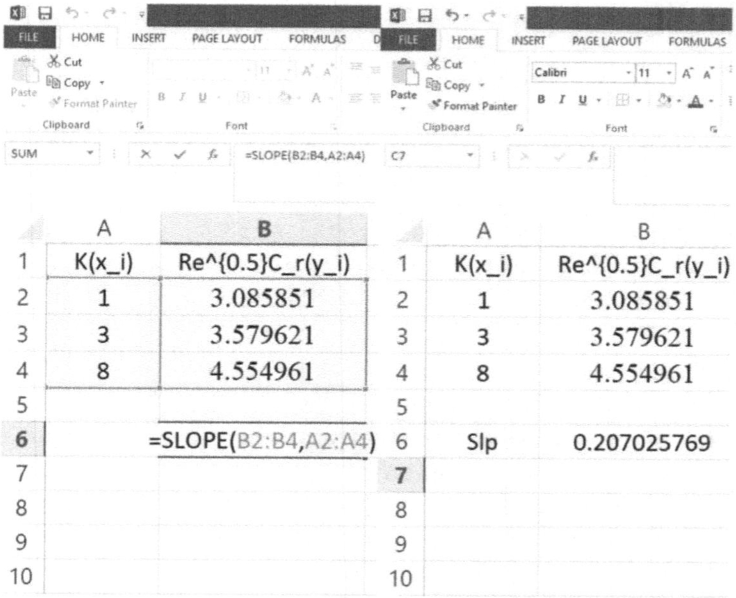

FIGURE 1.3
prtsc of a Microsoft Excel sheet for the data published by Shaw et al. [279].

TABLE 1.9
Reported Data by Nehad et al. [210] as Table Three When $S_r = 0.1$ and $D_f = 10$

d_p	**Local Skin Friction Coefficients**	**Heat Transfer Rate**	**Mass Transfer Rate**
0.5	−0.653806450627237	−8.856154914369313	1.471944690641082
2.5	−0.469235982235743	−8.531505838195230	1.483819380613837
4.5	−0.361939515178205	−8.330285252873948	1.492449292624451
6.5	−0.301600963981252	−8.217430056390326	1.498031241838610
8.5	−0.262740377859659	−8.146839475383260	1.501956963392201
S_{lp}	0.047488358	0.086635333	0.00371182

and this causes $Re^{0.5}C_r$ to increase. This is evident because when $K = 8$, the skin friction coefficient $Re^{0.5}C_r = 4.554961$. In other words, +0.207025769 is the rate of increase in the skin friction coefficient due to a rise in the porosity along the regression line $Re^{0.5}C_r = 0.207025769K + 2.91204179$.

1.10.3 Some Results Published by Nehad et al. [210]

In another study by Nehad et al. [210], an attempt was made to provide answer(s) to the following research question:

- When energy flux due to concentration gradient and mass flux due to temperature gradient are negligible, what is the significance of increasing radius of nanoparticles on the local skin friction coefficients, heat transfer rate, and mass transfer rate?

This led to the results presented in this learning material as Tables (1.9)–(1.11).

TABLE 1.10
Reported Data by Nehad et al. [210] as Table Four When $S_r = 10$ and $D_f = 0.1$

d_p	**Local skin friction coefficients**	**Heat transfer rate**	**Mass transfer rate**
0.5	−0.653806450627237	3.052651035544433	−3.874684483749367
2.5	−0.469235982235743	3.120598790031860	−3.954088416453056
4.5	−0.361939515178206	3.157453355416220	−3.988894824862508
6.5	−0.301600963981252	3.174211445520884	−3.998815684760363
8.5	−0.262740377859659	3.181969713124626	−3.998798514798066
S_{lp}	0.047488358	0.015612501	−0.014647767

TABLE 1.11
Reported Data by Nehad et al. [210] as Table Five when $S_r = 10$ and $D_f = 10$

d_p	**Local skin friction coefficients**	**Heat transfer rate**	**Mass transfer rate**
0.5	−0.653806450603498	0.792715106318922	0.495439010086529
2.5	−0.469235982226111	0.792495006570055	0.508203934009412
4.5	−0.361939515173113	0.794097753963524	0.517949200635406
6.5	−0.301600963978340	0.795989676393173	0.524492344750289
8.5	−0.262740377857890	0.797735022943715	0.529210531128244
S_{lp}	0.047488358	0.000676725	0.004191573

These results were further explored by estimating the slope linear regression through the data points to determine the nine relationships presented in Tables (1.9)–(1.11) in order to estimate the observed effects. It is worth concluding the following:

1. The observed changes in the local skin friction coefficients due to increasing the radius of nanoparticles (d_p) are negligible ($S_lp = 0.047488358$). The estimated slope using the LSM shows that neither energy flux due to concentration gradient nor mass flux due to temperature gradient is capable to significantly boost the local skin friction coefficients despite a growth in the radius of nanoparticles; see Tables (1.9)–(1.11).
2. When energy flux due to concentration gradient is small in magnitude but mass flux due to temperature gradient is large, a decrease in mass transfer rate due to enlargement in the radius of nanoparticles should be expected. This is true because the slope linear regression through the data points of radius of nanoparticles (d_p) and the mass transfer rate is $S_lp = -0.014647767$; see Table (1.10).
3. When energy flux due to concentration gradient and mass flux due to temperature gradient are large in magnitude, the heat transfer rate is very low. It is evident from Table 1.11 that the radius of nanoparticles (d_p) causes heat transfer rate to negligibly increase at the rate of $S_lp = 0.000676725$.

1.10.4 Variation in Concentration of a Fluid Flow

In the dynamics of a typical Newtonian fluid along a vertical porous surface due to free convection where the haphazard motion of tiny particles is negligible (for instance, $N_b = 0.1$) but the thermo-migration of tiny particles is highly significant (for instance, $N_t = 5$),

TABLE 1.12
Variation in the Concentration of Fluid Flow Across the Domain $0 \leq \eta \leq 5$

η	$\phi(\eta)$	η	$\phi(\eta)$	η	$\phi(\eta)$
0	1	2.1323	8.058	3.75	−13.1493
0.0852	1.3022	2.2264	7.4011	3.8068	−12.9596
0.1705	1.6284	2.3117	6.6773	3.892	−12.349
0.2557	1.9796	2.3969	5.8342	3.9773	−11.4624
0.3409	2.3564	2.4821	4.877	4.0467	−10.622
0.4261	2.7593	2.5673	3.812	4.1162	−9.7304
0.5114	3.1882	2.6615	2.5185	4.1414	−9.4007
0.5966	3.6424	2.7557	1.1138	4.154	−9.2354
0.6818	4.1203	2.8709	−0.7373	4.1667	−9.0701
0.7813	4.704	2.9861	−2.7083	4.2088	−8.5198
0.8807	5.3094	3.065	−4.1103	4.2677	−7.7568
0.9485	5.73	3.1439	−5.5388	4.3266	−7.0085
1.0164	6.152	3.1818	−6.2285	4.3834	−6.3047
1.1443	6.933	3.2365	−7.2224	4.4402	−5.6208
1.2721	7.6625	3.2912	−8.2044	4.4971	−4.9579
1.3999	8.296	3.3165	−8.6501	4.5539	−4.3166
1.5278	8.7872	3.3733	−9.6249	4.6107	−3.6968
1.6414	9.0687	3.4585	−10.971	4.6675	−3.0985
1.7551	9.1761	3.5438	−12.0866	4.7527	−2.2402
1.7929	**9.1694**	3.609	−12.7096	4.838	−1.4273
1.9155	8.9911	3.6742	−13.0773	4.919	−0.695
2.0381	8.5624	**3.7121**	**−13.1615**	5	0

the concentration of the fluid substance increases within the interval of $0 \leq \eta < 1.7929$. Within this interval, the optimal concentration is 9.1694. After the point $\eta = 1.7929$, the concentration decreases within $1.9155 \leq \eta < 3.7121$ where the most minimum concentration is −13.1615 (Table 1.12). It is seen that within $3.75 \leq \eta \leq 5$, the concentration increases significantly. As shown in Figure 1.4, it is pertinent to provide answers to such a question like, "what is the rate of change in $\phi(\eta)$ as $\eta \to 5$"

As shown in Table 1.13, there are sufficient data to describe the changes in the concentration across the domain as shown in Figure 1.4. It is worth providing the following answers to the question, "what is the rate of change in $\phi(\eta)$ as $\eta \to 5$?"

1. The concentration of the fluid increases within the layers of the fluid adjacent to the wall at the rate of 5.020780764.
2. Starting from a few distance away from the wall, $\eta = 1.9155$, the concentration decreases rapidly as $\eta \to 3.7121$ at the rate of −14.06935885.
3. The outcome of the slope linear regression through the data points reveals that the concentration increases at the rate of 11.28742381.

Overall, the linear regression slope across the data points shows that the fluid concentration diminishes at the rate of −3.372454663 as η changes from 0 to 5 (i.e., across the domain). Physically, the value indicates that the rate of decrease in the concentration dominates the observed increases at the interval of $0 \leq \eta < 1.7929$ and $3.75 \leq \eta \leq 5$. Sequel to this approach, it is now possible to quantify the changes in the dependent variable, the physical properties of fluid flow (i.e., velocity, temperature distribution, local skin friction coefficients, and Nusselt number), due to an increase or decrease in the independent variable, the magnitude of the Prandtl number.

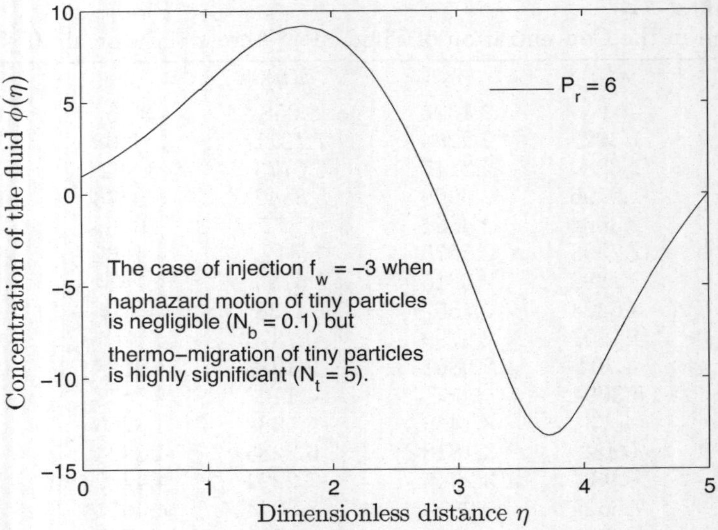

FIGURE 1.4
Variation in the Concentration of Fluid Flow Across the Domain $0 \leq \eta \leq 5$.

TABLE 1.13
Analysis of the Domain $0 \leq \eta \leq 5$

Dimensionless Distance	Description	Rate of Change in $\phi(\eta)$ with η
$0 \leq \eta < 1.7929$	Increases	5.020780764
$1.9155 \leq \eta < 3.7121$	Decreases	−14.06935885
$3.75 \leq \eta \leq 5$	Increases	11.28742381

1.10.5 Enhancement of the Discussion of Results

El-Omar [94] stated that reviewing a manuscript's discussion is one of the considerations evaluated by most Editors-in-Chief before assigning manuscripts to Associate Editors. It is necessary to describe the significant differences in the newly discovered results in most amazing published reports to improve the discussion of research presented by a manuscript. Meanwhile, another approach is to remind readers of the goals (i.e., preferably in a single sentence), refer back to the posed research questions, and refer to cited facts in the literature review under the discussion of results; see Koriko et al. [161], Song et al. [287], Oke et al. [216], and Nehad et al. [210]. It is no longer new that many authors find it challenging to discuss their results since the discussion section must sound both convincing and credible at the same time. In order to structure the discussion, Adrian [10] remarks that it is a must to provide answers to questions such as the following:

1. How do my findings compare with what others have found?
2. How consistent are they?

The authors would be more likely to reference another author's work directly relevant to the current findings, link the related published fact to the recent discoveries, and state how the latest results vary from related published facts if they respond to such a query. Without any doubt, the slope linear regression through data points may be instrumental in quantifying the relationship between two related variables. Such value is needed to improve the discussion of results.

1.11 Published Cases of Scrutinization

There are thousands of published facts within the scope of fluid dynamics, boundary layer analysis, and heat and mass transfer in the literature. However, there exists no meta-analysis or any form of scrutinization due to the unavailability of a technique and approach to quantify the observed results. One way to relate the weights of an individual to height is to adopt the linear regression model (i.e., to model the relationship between the two variables by fitting a linear equation using the data). Generally, regression is a well-known statistical method for determining the strength of the relationship between one or two independent variables and just a dependent variable. Simple linear regression and multiple linear regression are two basic types of regression. Recently, the body of knowledge has been updated with nonlinear regression for more detailed data and analysis. For simple linear regression, one independent variable is needed to predict the outcome of a dependent variable. Meanwhile, for multiple linear regression, two or more independent variables are needed to predict the outcome of a dependent variable. However, it is imperative to determine whether there is even such a relationship between the variables of interest. In such a case, a scatter plot is useful to determine the strength of the relationship between two variables. Following the step-by-step procedure of the meta-analysis by Neyeloff et al. [212], it may not be realistic to calculate the outcome (effect size), standard error (SE), variance (Var), individual study weights (w), each weighted effect size ($w \times es$), and Q test measures of heterogeneity and quantify the heterogeneity of the published reports on the increasing effects of the selected parameter (Prandtl number) on the property of any fluid flow. Next, the following are presented:

1. A meta-analysis of the effects of Grashof number on the flow of different fluids driven by convection over various surfaces.
2. A meta-analysis of the effects of the haphazard motion of tiny/nano-sized particles on the physical properties of some fluids.
3. A meta-analysis of the thermo-migration of tiny/nanosized particles in the motion of various fluids outlined using slope linear regression through the data points.

1.11.1 *Journal of Molecular Liquids*, 249, 980–990, 2018

Thermally generated convection is unavoidable in every scenario of fluid flow when the fluid and the surface are at different temperatures or when a solid surface is submerged in a fluid. As a result, a metric that measures the ratio of buoyancy to viscous forces is required. In fluid mechanics and heat transfer, the Grashof number is a dimensionless number that approximates the ratio of buoyant force to viscous force exerted on liquids (Shah et al. [276]). Because of this, there are many published observations on the significance of the Grashof number with no report on its systematic review. Thirty relevant articles were screened and reviewed for a study to scrutinize the effects of Grashof number on the flow of different fluids driven by convection over various surfaces (Shah et al. [276]). Each of the accepted articles was coded using study characteristics, and the corresponding effects of the parameter were scrutinized. The optimal level(s) of the parameter's effect(s) on the flow's physical properties was marked in each article, and the slope of the linear regression line across data points was calculated. The Grashof number and the mixed convection parameter on the flow of different fluids are examined using a systematic analysis method. Based on the outcome of the meta-analysis, it is worth concluding that:

1. A considerable increase in the velocity profile correlates to an increase in the magnitude of the buoyancy parameter. Physically, increasing the Grashof number

or another buoyancy-related parameter means increasing the wall temperature, which weakens the bond(s) between the fluids, reduces internal friction pressure, and increases gravity (i.e., causes a significant difference in specific weight between the fluid layers immediately next to the wall).

2. Buoyancy forces have only little influence on the local Nusselt number. This is particularly important when the Prandtl number is minimal, or the mandated wall heat flux (WHF) is considered.
3. The effect of the buoyancy parameter on fluid flow on a vertically moving cylinder can be considerably influenced by the defined surface temperature (PST) and prescribed WHF.
4. The Sherwood number, Bejan number, entropy generation, Stanton number, and pressure gradient are all growing properties of the buoyancy-related parameter. In contrast, concentration profiles, frictional force, and motile microbes are all lowering properties of the same dimensionless number.
5. The frictional force and fluid concentration are reducing properties of buoyancy forces.

The investigation of Brownian motion and thermophoresis of microscopic or nanoparticles was unraveled to be an open question at the end of the study.

1.11.2 *Chinese Journal of Physics*, 60, 676–687, 2019

A decline in the theoretical and empirical reviews of Brownian motion is worth noting, not only because of its importance in the field of mathematical physics, but also because statistical techniques are unavailable. The ongoing debate on the transport phenomenon and thermal performance of various fluids in the presence of haphazard motion of tiny particles as explained by Albert Einstein using kinetic theory and Robert Brown was further clinched by Animasaun et al. [35]. The report presents the outcome of detailed inspections of the significance of Brownian motion on the flow of various fluids as reported in 43 published articles using the method of slope linear regression through the data point. The technique of slope regression through the data points of each physical property of the flow and Brownian motion parameter was established and used to generate four forest plots. Based on the outcome of the meta-analysis, it is worth concluding that:

1. As the frequency of random particle motion rises, so does the internal strain on the tiny/nanosized particles. Collisions between particles are virtually certainly going to happen in such a circumstance.
2. According to Jang and Choi [330], the continuous and strong interactions between the molecules of the base fluid and nanosized/tiny particles result in nanoscale convection, which is predicated on previous translation into macroscopic conduction.
3. The Nusselt number, proportional to the heat transfer rate, is a decreasing feature of microscopic particle mixing motion.
4. In three-dimensional flow, Brownian motion can reach the maximum heat transfer rate in the presence of thermal radiation, thermal convection, and mass convection at the wall.
5. When Joule heating is essential, Brownian motion produces a reduction in mass transfer rate. Furthermore, with the movement of non-Newtonian nanofluids, Brownian motion ensures a decrease in mass transfer rate.

6. Increased Brownian motion causes a reduction in fluid content concentration, which is unavoidable. This is not conceivable in the case of high entropy generation and quartic autocatalytic chemical processes.

1.11.3 *Chinese Journal of Physics*, 68, 293–307, 2020

The relevance of thermophoresis and its function in particle movement has resulted in several publications (i.e., aims and objectives). However, no research has been done on the thermo-migration of tiny/nanosized particles in different fluids. Wakif et al. [308] considered a meta-analysis on the relevance of either nanosized or small particles exposed to thermophoretic force due to temperature differential during the dynamics of liquid substances. The method of slope linear regression across the data points was used to evaluate 60 published studies that discussed the impacts of thermophoresis (thermodiffusion) in the report. Based on the outcome of the meta-analysis, it is worth concluding that:

1. One of the main reasons concentration falls with thermophoresis is the development of a largely particle-free layer near the surface when small particles are blown away by thermophoresis.
2. The impact of thermophoresis should be ignored whenever energy flow generated by a composition gradient or mass flux produced by a temperature gradient is examined.
3. Different reactions to the force of a temperature gradient are adequate to improve the temperature distribution as thermophoresis increases. The effect of thermophoresis on Newtonian fluid flow concentration reduces, while it rises for non-Newtonian fluid flow.

In conclusion, a Microsoft Excel package for slope linear regression across data points seems to be a valuable tool for measuring the observed in either discrete or continuous functions or in any instance where there is a link between a dependent variable and an independent variable.

1.12 Four-Stage Lobatto IIIa Formula—`bvp5c`

The MATLAB package (`bvp5c`) was designed to integrate a system of ordinary differential equations of the form

$$\frac{dy_i}{dx} = f(x, y) \tag{1.16}$$

subject to two-point boundary value conditions `f0` and `finf` for $a \leq x \leq b$. The MATLAB package (bvp5c) is a finite difference code that implements the four-stage Lobatto IIIa formula. The collocation formula together with the collocation polynomial provides a C^1-continuous solution that is fifth-order accurate uniformly in $x\varepsilon[a, b]$. The selection of mesh and control of error is based on the residual of the continuous solution. An implicit Runge–Kutta integration scheme solves the algebraic equations directly (Kierzenka & Shampine [159]). Most experts chose this over `bvp4c` that uses analytical condensation, but handles the unknown parameters directly in Chapters 5–12 to solve all the emerged boundary value problems of ordinary differential equations.

1.13 Tutorial Questions

1. List at least five dimensionless properties that may be influenced by one or more of the following thermo-physical qualities: viscosity, density, thermal conductivity, specific heat capacity, and thermal diffusivity.
2. Explain the dimensionless Froude number. List at least three usefulness of the Froude number.
3. Mention one of the six (6) groups of dimensionless numbers that is of high interest to you.
4. Explain the major difference between momentum diffusivity and magnetic diffusivity.
5. Use suitable examples to explain the meaning of
   (a) Non-dimensionalization
   (b) Parametrization
   (c) Scrutinization.
6. Define viscosity with major emphasis on
   (a) Agglomeration and clustering of tiny particles
   (b) Molecules and free-space.
7. Use the concept of atoms to describe density.
8. Mathematically, use an indirect method to obtain the unit of density.
9. Explain the discrepancy between the higher density for liquid and smaller density for gases.
10. What is the significance of diffusion coefficient in heat and mass transfer?
11. Explain the effect of increasing temperature and volume concentration of nanoparticle on the density of nanofluid.
12. Mention three factors that can influence the thermal conductivity of nanofluids. Briefly discuss each factor.
13. What effect does rising temperature have on specific heat capacity?
14. Write a note on the dynamic viscosity (μ) and thermal conductivity (κ) of (a) nanofluid and (b) hybrid nanofluids?
15. Heat diffuses rapidly at higher temperatures through contact. Provide a scientific fact to either support or discredit the scientific fact?
16. What does it mean if slope linear regression through the data points is either positive or negative?
17. Mention some of the techniques suitable to improve the quality of discussion of results.
18. What are the factors that make four-stage Lobatto IIIa formula (`bvp5c`) for solving boundary value problem of ordinary differential equations better than `bvp4c`?

2

Conceptual and Empirical Reviews I

2.1 Background Information

German scientist Ludwig Prandtl, whose image appears in Figure 2.1, made several scientific contributions, most notably in the field of fluid mechanics. The Prandtl number was named after the amazing description of the Boundary layer principle as the ratio of momentum diffusivity to thermal diffusivity. The dimensionless number, Prandtl number, was later discovered to be appropriate for unraveling the relationship between conduction and convection in fluid flows. More specifically, the Prandtl number is important in studying rising currents of warm substances, the design of thermal hydraulic features, coolant collection, and the interpretation of momentum and heat transfer.

The momentum diffusivity of the fluid is the ability of the liquid substance to transport momentum. The inevitability of both momentum and thermal diffusivity in many transport phenomena has led to many reports on the effects of increasing Prandtl number (independent variable) on the following dependent variables: (i) vertical velocity, (ii) horizontal velocity, (iii) transfer of microorganisms, (iv) dust temperature and temperature distribution or temperature gradient, (v) Stanton number and Strouhal number, (vi) shear stress between two successive layers, (vii) ratio of Rayleigh number to critical Rayleigh number, (viii) local Nusselt number proportional to heat transfer rate, (ix) mean lift coefficient and magnetic field profile, (x) local skin friction coefficients, (xi) local Sherwood

FIGURE 2.1
German physicist Ludwig Prandtl (February 4, 1875–August 15, 1953).

DOI: 10.1201/9781003217374-2

number, (xii) centerline temperature, (xiii) spacing where the heat transfer rate is at maximum, (xiv) angular velocity, (xv) induced magnetic and average exit temperature, (xvi) concentration and concentration gradient, (xvii) displacement thickness, (xviii) drag force, and (xix) height of the capillary ridges.

2.2 Related Published Reports: 1946–2011

Literature reviews are utilized in a wide range of scholarly fields. As a result, it is feasible and necessary to employ a variety of strategies to plan and produce a literature review. Literature reviews include argumentative reviews, integrative reviews, historical reviews, methodological reviews, systematic reviews, and theoretical reviews. The focus of a methodological review is not always on what the author(s) of the paper stated, but rather on the study procedure. This method offers a structure for comprehending theories, substantive areas, research methods, data collection, and analysis techniques. Furthermore, researchers will be able to rely on a broad range of information, from philosophical to realistic papers, resulting from this. A theoretical review aims to look at a particular theory that can be used to investigate a problem, idea, theory, or phenomenon. The theoretical literature review aids in determining what theories already exist, their relationships, and the extent to which the existing theories have been studied and develop new hypotheses. Theoretical review is often used to identify a lack of relevant hypotheses or to show that the existing theories are insufficient to describe new or evolving research problems. On the other hand, a systematic review is a study that uses predefined and structured methods to identify and critically appraise relevant research and collect, publish, and analyze data from the studies included in the review. The papers mentioned in this subsection were chosen for analysis after a thorough review of all relevant published reports from 1946 to 2011, and the observed effects were estimated using slope linear regression through the data points

2.2.1 *Journal of the Society of Chemical Industry*, 65(2), 61–63, 1946

Phase changes of a substance are expected to cause rearrangement of its molecules, while the temperature remains fixed. The physical property of a substance known as the latent heat of vaporization is the amount of heat required to change one mole of such a substance, preferably liquid, at its boiling point under standard atmospheric pressure. The mathematical relation $(P_r)^{0.4} = 0.2\text{Exp}(0.18\Delta H/RT)$ was presented by Denbigh [90] as the most suitable model for estimating the significance of the Prandtl number on various kinds of fluid flows. According to the above-mentioned model, the latent heat of vaporization at the boiling point (ΔH), absolute temperature (T), and ideal gas constant (R) are the most beneficial in determining the Prandtl number for any liquid material.

1. Denbigh [90] deduced that when temperature decreases, $\Delta H/RT$ decreases for water.

2. It was further observed that with an increase in the temperature within the range $10°\text{C} \leq \text{Temp} \leq 80°\text{C}$, the Prandtl number for water decreases at the rate of -0.099.

3. For almost all the gases, the magnitude of the Prandtl number falls within the range of $0.7 \leq P_r \leq 1.0$.

2.2.2 *The Aeronautical Quarterly*, 15(04), 392–406, 1964

Smaller viscosity–temperature index (VI) values for a given fluid indicate that temperature variations are likely to impact viscosity significantly. Meanwhile, a higher VI indicates that the viscosity will stay constant even if the temperature changes. Young [326] discovered that the VI and the Prandtl number impact the friction between the fluid's last layer and the surface beneath the boundary layer flow of a laminar fluid.

1. It was confirmed by Young [326] that an increase in the magnitude of the Prandtl number corresponds to a change in the wall temperature T_w and viscosity due to the interaction between the pressure gradient (moderate, large, or adverse), and the density change near the wall.

It is worth noting that such a contact has a significant impact on skin friction. Since wall buoyancy is heavily affected by the same wall temperature, little is known about the effect of the same dimensionless parameter on (i) forced convection induced flow, (ii) free convection induced flow, and (iii) mixed convection induced flow; see Chapter 6.

2.2.3 *International Journal of Heat and Mass Transfer*, 22(10), 1401–1406, 1979

In Section 1.10.1, it was established that the buoyancy forces have a negligible positive effect on the local Nusselt number, but this is only significant when the magnitude of Prandtl number is small or the prescribed wall heat flux (WHF) is considered. Schneider [274] investigated the effect of buoyancy forces on the steady and laminar flow over a horizontal plate. In the article, an attempt was made to determine if the solution of the model depends on the Prandtl number when the buoyancy-related parameter $K = 0, K < 0$, and $K > 0$. Meanwhile, $K = 0$ implies no buoyancy, $K < 0$ implies that the wall temperature is smaller in magnitude than the free stream temperature, and $K > 0$ implies that teh free stream temperature is lower than the wall temperature.

1. Schneider [274] noticed that when $K = -0.0517, Pr = 0.5$; $K = -0.0787, P_r = 1.0$, and $K = -0.1099, P_r = 2.0$, the Prandtl number is strongly related to the buoyancy-related parameter. It is worth noticing that if $P_r(K)$, then the rate of decrease is -25.93483284. But, if $K(P_r)$, then the rate of decrease is -0.037714286.
2. The results further indicate that when $K = 0.5$, the local skin friction coefficient $f''(0)$ decreases with the Prandtl number at the rate of -0.128571429.
3. Also, when $K = 0.5$, the total heat transfer at the wall termed Stanton number $\sqrt{ReSt}$ decreases with the Prandtl number at the rate of -0.960285714.
4. In addition, the displacement thickness $\sqrt{U_\infty/\vartheta x}\delta^*$ increases with the Prandtl number at the rate of 0.642857143.

2.2.4 *International Journal of Heat and Mass Transfer*, 24(1), 125–131, 1981

The rectangular cavity is a hollow metallic tube with a rectangular cross section or a rectangular waveguide shorted at both ends. For example, eyelets as metallized holes create a rectangular cavity within the antenna substrate with an acceptable spacing distance between them. In order to generalize the result presented by Gill [110] for arbitrary values of the

Prandtl number, Graebel [113] was interested in determining the influence of increasing the Prandtl number on free convection in a rectangular cavity. Specifically, natural convection in a rectangular cavity was considered for a transport phenomenon where there is a difference in the temperature between two walls (one end is maintained cooled, while the other is heated).

1. In the study by Graebel [113], the effect of Prandtl number-dependent parameter $K = (P_r-1)/(P_r+1)$ and Prandtl number on the Nusselt number $Nu(RaL/H)^{1/4}$ was examined at various values of $Ra^{1/7}H/L$, where R_a is the Rayleigh number, H is the cavity height, and L is the cavity length.
2. In the same report, it is seen that when $Ra^{1/7}H/L = 1$, the Nusselt number increases with the Prandtl number at the rate of 0.349164499. More so, when $Ra^{1/7}H/L = 10,000$, the same property of the transport phenomena increases with the Prandtl number at a lesser rate of 0.133252574.

2.2.5 *Applied Scientific Research*, 40(4), 333–344, 1983

The physical meaning of the Boussinesq approximation for natural convection is based on the smallness of fluid flow accelerations instead of gravity acceleration. The effect of increasing the Prandtl number on natural convection in a horizontal porous medium was shown by Somerton [286] to be a critical explanation of the variation in heat transfer across the fluid flow. The observed changes in the Prandtl number led to the generalization of the influence of the dimensionless number. Somerton [286] concluded that the formation of porous layer convection is strongly dependent on the Prandtl number. Given the porous media Rayleigh number R_a and porous media Prandtl number Pr_m, the outcome of the correlation of data presented by Combarnous [81] shows the following:

1. Nusselt number N_u for $Pr_m = 0.18$ can be determined using
$$N_u = \frac{Ra^{0.68}}{12.63}.$$
2. Nusselt number N_u for $Pr_m = 5.3$ can be determined using
$$N_u = \frac{Ra^{0.77}}{14.39}.$$
3. Nusselt number N_u for $Pr_m > 11.8$ can be determined using
$$N_u = \frac{Ra^{0.81}}{14.12}.$$

The maximum deviation in the correlations, with respect to the experimental data, is ±12%.

2.2.6 *International Journal of Heat and Mass Transfer*, 33(11), 2565–2578, 1990

Buoyancy occurs naturally when the pressure on the fluid at the bottom is greater than the pressure on the fluid at the top or when the force on the fluid at the bottom pushes up while the force on the fluid at the top pushes down. Because the pressure rises with depth and the force is normal to the surface, the net force attributable to the fluid is directed upward. The impact of the Prandtl number on buoyancy-induced transport processes with and without solidification was investigated by Wei and Ming-Hsiung [315]. The study describes the transport methods by which high-Prandtl-number aqueous solutions attain the solidification properties of low-Prandtl-number metallic materials. The dependence of the Prandtl number on the buoyancy-induced transport phenomena was investigated using two-dimensional steady-state calculations for flow in a square enclosure at two different vertical wall temperatures.

1. With or without phase change, the transport characteristics are very insensitive to increasing Prandtl number when it is of large magnitude. However, the significance of P_r becomes more pronounced when it is lower than one (Wei and Ming-Hsiung [315]).
2. Also, as the Rayleigh number increases, a stronger impact of the Prandtl number on the convection strength is bound to occur. When the Rayleigh number is 10^4, the Nusselt number increases negligibly with the Prandtl number.
3. The increase in the Nusselt number with the Prandtl number is highly significant when the Rayleigh number is 10^5, most especially when $P_r = 10^{-2}, 10^{-1}$, and 10^0. This result was backed up by the fact that, at a constant value of Rayleigh number, the fluid's thermal diffusivity and kinematic viscosity simultaneously vary in the opposite directions and at the same rate due to an increase in the Prandtl number P_r.

2.2.7 *Metallurgical and Materials Transactions B*, 24(1), 197–200, 1993

Many engineers believe that circular ducts are more energy efficient and less expensive than straight ducts. Furthermore, circular ducts are more efficient in reducing hydraulic losses than square or rectangular ducts because they feature sharp edges at the corners where separation occurs. Das and Mohanty [87] looked at the entrance effects on low Prandtl number convection through a circular duct. It was discovered that convection in fluids with a low-Prandtl-number (such as liquid metals) has the feature of the thermal boundary layer being thicker than the hydrodynamic boundary layer.

1. When the axial distance $\xi = 0.0012$, the Nusselt number for UWT increases with the Prandtl number at the rate of 15.48884007, while the centerline temperature γ is zero throughout.
2. When $\xi = 0.062$, Nusselt number does not change, but centerline temperature decreases with the Prandtl number at the rate of -1.50958; see Das and Mohanty [87].
3. For the case of UHF, when the axial distance $\xi = 0.0012$, the Nusselt number increases with the Prandtl number at the rate of 26.21397, while γ is a constant function.
4. When the axial distance is a bit larger (i.e., $\xi = 0.058$), the Nusselt number appears to be a constant function, while the centerline temperature γ decreases with the Prandtl number at the rate of -1.97425; see Das and Mohanty [87].

2.2.8 *Astronomy and Astrophysics*, 286, 338–343, 1994

The energy transfer characteristic of turbulence is crucial to understanding the process of turbulent flows and producing turbulence (Hamba [119]). Heinrich [129] investigated the impact of turbulent energy transfer on narrow accretion disks approaching critical luminosity. Not only was the vertical structure of accretion disks surrounding permissive black holes inferred, but the non-laminar heat flow parameterized by the Prandtl number was also studied.

1. Heinrich [129] concluded that the structure of the innermost but dominated region of radiation pressure is a property strongly dependent on the ratio of momentum diffusivity to thermal diffusivity.
2. When a turbulent Prandtl number is infinity, density inversion occurs.
3. However, a finite value of a turbulent Prandtl number was recommended to model AGN accretion disks.

2.2.9 *Journal of Heat Transfer*, 116(2), 284–295, 1994

In the case of three-dimensional turbulence, eddy viscosity and eddy diffusivity are not feasible. As a result, investigating the turbulent Prandtl number is unrealistic. One of the questions raised during the 1992 Max Jakob Memorial Award Lecture was that "Turbulent Prandtl Number—Where are we?". In the report by Kays [150], the author focuses on the critical examination of experimental data on the effect of Prandtl number on the two-dimensional turbulent boundary layer. This was later extended to fully developed flow in a circular tube or a flat duct.

1. The methodology adopted by Kays [150] shows the existence of molecular Prandtl number $P_r = \frac{\mu c}{k}$, where the dynamic viscosity is denoted by μ and the specific heat at constant pressure c, and molecular thermal conductivity is denoted by κ.
2. Turbulent Peclet number $P_{et} = \frac{\epsilon_M P_r}{\vartheta}$, where eddy diffusivity for momentum is ϵ_M, and the kinematic viscosity coefficient $\vartheta = \frac{\mu}{\rho}$, and density ρ.
3. Turbulent Prandtl number $P_{rt} = \frac{\epsilon_M}{\epsilon_H}$, where the eddy diffusivity for heat is ϵ_H.

However, the relationship between the turbulent Prandtl number and the turbulent Peclet number is

$$P_{rt} = \frac{2.0}{P_e - t + 0.85}$$

Transpiration, either suction or blowing slightly affects turbulent Prandtl number P_{rt}.

2.2.10 *Journal of Heat Transfer*, 119(3), 467–473, 1997

Due to the existence of Dean vortices for the curved tubes, helically coiled water pipe is widely used in the industry, especially in the agriculture sector. In another attempt to illustrate the significance of geometric parameters and Prandtl number on the average convective and local heat transfer characteristics in the flow of air, water, and ethylene glycol through helical pipes, Xin and Ebadian [320] remarked that the appropriate Prandtl number for ethylene glycol is 120, that for water is 5, and that for air is 0.7.

1. The study reveals that the peripheral variation of the relative Nusselt number increases with the Prandtl number and Dean number. Dean number is inevitable in fluid dynamics in channels and curved pipes (Xin and Ebadian [320]).

2.2.11 *Journal of Fluid Mechanics*, 383, 55–73, 1999

The speed of the fluid at a given point in a turbulent flow is constantly changing in magnitude and direction. Verzicco and Camussi [306] utilized numerical experiments to study the on Prandtl number influence convective turbulent flow dynamics. In the study, the authors considered three experiments, two of which involve keeping the Prandtl number constant, while it was varied in the third experiment.

1. Verzicco and Camussi [306] remarked that the fluid dynamics for $P_r < 0.35$ is different. The temperature distribution across this fluid is highly diffusive to the extent that the generation of thermal plumes may be affected.

2. Such a result mentioned above is true because large-scale motion strongly affects temperature distribution and induces thermal and viscous boundary layers.

2.2.12 *International Journal of Thermal Sciences*, 40(6), 564–570, 2001

The propensity for heat and mass to migrate to higher surface tension within a liquid is known as Marangoni convection (something triggered by the change of surface tension with temperature along a surface). Christopher and Wang [79] examined the effects of the Prandtl number on Marangoni convection of a fluid substance over a flat surface. In the article, the solutions to the fluid flow problem are shown for surface velocity, total flow rate, and heat transfer for various temperature profiles and various Prandtl numbers.

1. According to Christopher and Wang [79], for $P_r = 0.04$ and $P_r = 0.1$, the temperature distribution decreases with the Prandtl number across a larger domain of $\eta \epsilon [0, 150]$ at the rate of -4.3333 which was estimated at $\eta = 25$.

2. When the temperature gradient exponent $k = 0.5, 0$, and 1, as the Prandtl number increases for $1E - 3 \leq P_r \leq 1,000$, the surface temperature gradient $-\theta'(0)$ increases at the rate of 0.052043097 for $k = -0.5$.

2.2.13 *Meccanica*, 37(6), 599–608, 2002

An alloy of the alkali metals sodium and potassium (i.e., sodium–potassium alloy) is one material that is highly reactive with water and may catch fire when exposed to air. Saravanan and Kandaswamy [269] presented the dynamics of two-dimensional laminar sodium–potassium alloy ($P_r = 0.054$), where the thermal conductivity of the liquid varies with temperature within the domain insulated for $0 \leq y \leq L$ when $x = 0$ and $x = H$. Also, the wall of the domain is cold for $0 \leq x \leq H$ when $y = 0$ and $y = L$.

1. For less viscous fluid, Saravanan and Kandaswamy [269] concluded that suppression of conduction of heat energy is substantial than that of convection at the hot wall.

2. Fluids with small magnitudes of Prandtl number are recommended for cooling materials.

2.2.14 *Physical Review E*, 65(6), 066306, 2002

Rayleigh–Benard convection describes the buoyancy-driven flow of a fluid heated from below and cooled from above. Lam et al. [171] studied the influence of the Prandtl number on the viscous boundary layer and the Reynolds numbers in the Rayleigh–Benard convection. The measurements of the viscous boundary layer and the Reynolds number in four different fluids over a wide range of the Prandtl and Rayleigh numbers were considered. In addition, all four different ways considered make use of a lone convection cell characterized by a unity aspect ratio:

1. The outcome of the experimental study for high-Prandtl-number turbulent Rayleigh–Bernard convection shows that the Reynolds number Re_{rms} can be quantified as

$$0.84R_a^{0.40\pm0.03}P_r^{-0.86\pm0.01}$$

   where the Rayleigh number is R_a and the Prandtl number is P_r (Lam et al., [171]).

2. In fact, the formation of the viscous boundary layer is also a function dependent on both dimensionless parameters. It was further remarked that the viscous layer δ_v can be estimated as

$$0.65R_a^{-0.16\pm0.02}P_r^{0.24\pm0.01}$$

   .

2.2.15 *Heat and Mass Transfer*, 40(3-4), 285–291, 2004

The combination of induced and natural convection is referred to as "mixed convection." According to science, mixed convection occurs when natural and forced convection processes collaborate to transport heat. Ali [23] examined the combined effects of buoyancy and Prandtl number on Newtonian fluid boundary layer flow along continuous surfaces owing to mixed convection. The influence of the various emerged fluid parameters such as Prandtl number, temperature exponent, and mixed convection parameter was presented accordingly.

1. Ali [23] discovered that with an increase in the Prandtl number for $-3.5 \leq f_w \leq -2.9$, the Nusselt number increases at the rate of 1.527795527 when $f_w = -2.9$.
2. However, when $f_w = -3.5$, the Nusselt number increases with the Prandtl number at the rate of 0.151821086.

2.2.16 *Progress in Natural Science*, 14(10), 922–926, 2004

Wang et al. [311] deliberated on the viscosity, thermal diffusivity, and Prandtl number of nanoparticle suspensions. In the article, the specific heat capacity and viscosity for copper(II) oxide nanoparticle suspensions and thermal diffusivity and Prandtl number were deduced.

1. Wang et al. [311] noticed that a simple but proportional relationship exists between the viscosity and thermal conductivity of fluid for which the Prandtl number is the constant of proportionality.
2. The outcome of the experimental study shows that if the Prandtl number of nanoparticle suspensions decreases, then there would be an increase in the nanofluid concentration.

2.2.17 *Physics of Fluids*, 18(12), 124103, 2006

As pressure and buoyant forces combine, this is known as mixed convection. The influence of Prandtl number on the stability of mixed convective flow through a vertical channel filled with a porous medium was studied by Bera and Khalili [57]. The Brinkman–Wooding-extended Darcy model was utilized to investigate the changing mechanisms of the basic flow and the influence of the Prandtl number on the fluid.

1. Bera and Khalili [57] showed that the possibility of buoyancy-opposed flows is strongly dependent on Prandtl number as this is associated with thermal and kinetic disturbances.
2. Fluid flow characteristics with a large magnitude of Prandtl number as in the case of oil (high viscous fluid $P_r = 70$) are most unstable (Bera and Khalili [57]).
3. Either for a less viscous fluid ($P_r = 0.7$) or for a high viscous fluid ($P_r = 70$), buoyancy force produces disturbance kinetic energy capacity to meet up with the dissipation of kinetic energy owing to surface drag.

2.2.18 *Numerical Heat Transfer, Part A: Applications*, 53(3), 273–294, 2007

Fourier's law governs heat conduction. As per Fourier's law, the quick pace of hotness transmission through a material corresponds to the negative inclination in temperature and space. The levels of weld pool morphology and molten metals can be largely determined by heat conduction for modest Prandtl values (Chakraborty [74]).

1. A low Prandtl number, in practice, indicates a higher magnitude of thermal conductivity, and such flow is diffusion-driven (Chakraborty [74]). The importance of convection at this level of the Prandtl number is negligible.

2.2.19 *Journal of Fluid Mechanics*, 592, 221–231, 2007

Lakkaraju and Alam [170] clearly explained how the Prandtl number affects the linear stability of a planar thermal plume using a quasi-parallel approximation. The consequences of changing the Prandtl number's magnitude were investigated.

1. In an attempt to investigate the instability mode in a plane thermal plume, at 10^{-3} (i.e. across the plume width), Lakkaraju and Alam [170] observed that the velocity of the flow decreases with the Prandtl number at the rate of -0.00034.

2.2.20 *Applied Mathematics and Computation*, 206(2), 832–840, 2008

Bataller [54] investigated the effects of increasing the Prandtl number, thermal radiation, and convective heating of the wall temperature (solid–fluid interface) beneath Blasius flow and Sakiadis flow.

1. Bataller [54] discovered that with an increase in Prandtl number, the wall temperature $\theta(0)$ in Blasius flow decreases at the rate of -0.027941259 and the wall temperature in Sakiadis flow decreases at the rate of -0.051197003.

2.2.21 *International Journal of Thermal Sciences*, 47(6), 758–765, 2008

Different scientific disciplines use mass transfer for different processes and mechanisms such as evaporation, membrane filtration, and adsorption. Alam et al. [17] considered the effects of variable suction and thermophoresis on steady MHD combined free–forced convective heat and mass transfer flow over a semi-infinite permeable inclined plate in the presence of thermal radiation.

1. Alam et al. [17] discovered that when thermal radiation is negligible ($R = 0.1$), with an increase in Prandtl number, the local skin friction coefficient $f''(0)$ decreases at the rate of -0.03659, the local Sherwood number $-\phi'(0)$ decreases at the rate of -0.01338, while the Nusselt number $-\theta'(0)$ increases at the rate of 0.070757.
2. When thermal radiation is more enhanced ($R = 0.5$), the local skin friction coefficient $f''(0)$ decreases at the rate of -0.03972, the local Sherwood number $-\phi'(0)$ decreases at the rate of -0.01338, while the Nusselt number $-\theta'(0)$ increases at the rate of 0.168 due to an increase in the magnitude of the Prandtl number.

2.2.22 *Nonlinear Analysis: Modeling and Control*, 13(4), 513–524, 2008

In heat conduction, microscopic particles such as molecules, atoms, and electrons can exist. The kinematic and potential energy of microscopic particles is included in internal energy. Rahman et al. [240] researched the influences of temperature-dependent thermal conductivity on magnetohydrodynamic (MHD) free convection flow along a vertical flat plate with heat conduction.

1. When the thermal conductivity of viscous fluid flow was assumed to be temperature-dependent, Rahman et al. [240] discovered that at the leading edge (i.e., $x = 0$), the Prandtl number does not affect skin friction and surface temperature.
2. At some distance away from the wall, it was discovered that both skin friction and surface temperature decrease Prandtl number functions at the rates of -0.3139 and -0.32476, respectively.
3. Furthermore, the velocity decreases with the Prandtl number at the rate of -0.17917. Meanwhile, the temperature distribution across the fluid flow decreases with the Prandtl number at the rate of -0.35542.

2.2.23 *Nuclear Engineering and Design*, 238(9), 2460–2467, 2008

The performance of a system called an artificial neural network is based on wall superheat, the ratio of the gap size to the diameter of the heated surface, Prandtl number, and Rayleigh number. Such a study led Zhao et al. [328] to explain an experimental and theoretical study on transition boiling concerning downward-facing horizontal surface in confined space.

1. Zhao et al. [328] noticed that the Nusselt number is a property that increases with a higher Prandtl number for gap size lesser than 4.0 mm.
2. Based on the fact that the formation of free convection is easier for a large gap size, the effect of the Prandtl number on the Nusselt number changes.
3. However, the wall temperature is a tool capable of influencing the effect of the Prandtl number on the Nusselt number.

2.2.24 *International Journal of Heat and Mass Transfer*, 52(15-16), 3790–3798, 2009

Wei et al. [314] investigated the motion of two-dimensional Marangoni convection caused by thermo-capillary force, where energy distribution and scanning speed were integrated using time-dependent incident flux.

1. In the hot region on the wavy weld boundary, the surface temperature was found to be a property that can be decreased due to an increase in the Marangoni and Prandtl numbers. Not only that, these dimensionless numbers greatly influence molten pool shapes and flow patterns in the flow along a horizontal surface (Wei et al. [314]).
2. In the hot region, the surface temperature is a decreasing property of higher Prandtl number (Wei et al. [314]).
3. The observed increase in the surface temperature due to a decrease in the Prandtl number for $P_r > 1$ was traced to a decrease in surface speed in order to satisfy energy conservation in the thermal boundary layer whose thickness is less than that of the momentum layer (Wei et al. [314]).
4. Meanwhile, for $P_r < 1$, it was due to a decrease in surface velocity and enhanced heat conduction through the momentum boundary layer to the pool bottom in the thick thermal boundary layer (Wei et al. [314]). Note that heat conduction transfers internal thermal energy within a body caused by microscopic particle collisions and electron motion.

2.2.25 *Heat and Mass Transfer*, 46(2), 147–151, 2009

In the presence of stretching at the free stream, the significance of Prandtl number and thermal radiation on the free convective flow of a Newtonian fluid along a plate with temperature $T_w \sim x^{-1/2}$, where $T_w = T_\infty + cx^{-1/2}$ and the variable x denotes the distance from the leading edge, was investigated by Ishak [143]. Boussinesq approximation of the form $g\beta\frac{\partial}{\partial x}\int_y^\infty(T - T_\infty)dy$ was incorporated into the momentum equation to model the buoyancy-induced flow.

1. In the study conducted by Ishak [143], the critical value λ_c decreases with the Prandtl number P_r at the rate of -0.0358 when thermal radiation is highly negligible.

2.2.26 *Journal of Applied Fluid Mechanics*, 2(1), 23–28, 2009

The heat sink for a certain case of fluid flow was modeled by Crepeau and Clarksean [83] as a function of thermal conductivity, Grashof number, and the difference between the wall and free stream temperatures. In the presence of these fluid properties, a better analysis of the effect of Prandtl number on free convection flow when thermal conductivity and dynamic viscosity are temperature-dependent was considered by Mahanti and Gaur [186].

1. The variation in a constant function $\theta'(0)$ with P_r was investigated by Mahanti and Gaur [186]. The study shows that with an increase in the magnitude assigned to the Prandtl number, $\theta'(0)$ decreases at the rate of -0.083211877. Table 3

presented in page 27 of the report shows the significance of the Prandtl number can be explored using the statistical approach in eight cases.

2. Firstly, $\gamma = -0.4$ implies that the variation in viscosity due to temperature is low. In other words, as presented in Eq. 2.5 of the report, $\gamma = -0.4$ implies that $(T - T_\infty) < 0$. Secondly, $\gamma = 0.4$ implies that $(T - T_\infty) > 0$ and intermolecular forces holding the molecules are not high. In the study, $\varepsilon = 0$ implies that the thermal conductivity is a constant function (Mahanti and Gaur [186]).

3. When $\gamma = -0.4$ and $\varepsilon = 0$, the local skin friction coefficients $f''(0)$ decreases with P_r at the rate of -2.684384615. Moreover, when $\gamma = 0.4$ and $\varepsilon = 0$, $f''(0)$ decreases with P_r at the rate of -2.740292308. When $\varepsilon = 0.3$, it is implies that the thermal conductivity changes with temperature laterally moderate. When $\gamma = -0.4$, it is seen that $f''(0)$ decreases with P_r at the rate of -2.441307692 (Mahanti and Gaur [186]). Furthermore, when $\gamma = 0.4$, $f''(0)$ decreases with P_r at the rate of -2.513538462.

4. Next is to explore the observed effects of the Prandtl number on the Nusselt number $-\theta'(0)$ proportional to the heat transfer rate. It was seen that when $\gamma = -0.4$, $-\theta'(0)$ increases with P_r at the rate of 3.000538462 for $\varepsilon = 0$ and it increases with P_r at the rate of 2.555307692 for $\varepsilon = 0.3$ (Mahanti and Gaur [186]).

5. Lastly, Mahanti and Gaur [186] observed that $-\theta'(0)$ increases with P_r at the rate of 2.933 when $\varepsilon = 0$ and increases with P_r at the rate of 1.641153846 when $\varepsilon = 0.3$. It was also remarked by the authors that the temperature distribution is a decreasing property of the Prandtl number at the rate of -0.376964417. The above-mentioned facts were estimated at a unit span away from the wall ($\eta = 1$).

2.2.27 *International Journal of Heat and Mass Transfer*, 53(11-12), 2477–2483, 2010

The Sherwood number proportional to the mass transfer rate is very significant in processes such as evaporation, membrane filtration, and adsorption. Khan and Pop [158] investigated the effect of Prandtl number on the heat and mass transfer during the two-dimensional laminar flow of a nanofluid along a flat sheet when Brownian motion and thermophoresis are negligible and highly significant.

1. First, in the absence of Brownian motion and thermophoresis, Khan and Pop [158] discovered that with an increase in the Prandtl number, the Nusselt number proportional to the heat transfer rate $Re_x^{-\frac{1}{2}} Nu = -\Theta'(0)$ increases at the rate of 0.087817875.

2. When Brownian motion and thermophoresis are highly negligible ($N_b = N_t = 0.1$), $Re_x^{-\frac{1}{2}} Nu$ increases due to an increase in the Prandtl number at the rate of -0.048888889. When $N_b = N_t = 0.5$, $Re_x^{-\frac{1}{2}} Nu$ increases with the Prandtl number at the rate of -0.0248888889.

3. The Sherwood number $Re_x^{-\frac{1}{2}} Sh$ proportional to the mass transfer rate increases with the Prandtl number at the rate of 0.026888889 when $N_b = N_t = 0.1$. When Brownian motion and thermophoresis is highly significant ($N_b = N_t = 0.5$), $Re_x^{-\frac{1}{2}} Sh$ increases with the Prandtl number at the rate of 0.028888889.

2.2.28 *Journal of Fluids Engineering*, 132(4), 044502, 2010

An attempt was made by Makinde and Olanrewaju [189] to explore the effect of increasing Prandtl number on skin friction coefficients and heat transfer rate in the flow of a Newtonian fluid along a convectively heated vertical sheet.

1. When the Biot number $Bi_x = 0.1$ and the Grashof number $Gr_x = 0.1$, with an increase in the Prandtl number, Makinde and Olanrewaju [189] discovered that the local skin friction coefficient decreases at the rate of -0.003799677 while the Nusselt number decreases at the rate of -0.001715545.
2. In addition, the study's outcome indicated that the temperature distribution decreases with the Prandtl number at the rate of -0.017605758.

2.2.29 *Heat and Mass Transfer*, 47(4), 419–425, 2011

The mass transfer is the net flow of mass from one location to another, usually in a stream, step, fraction, or part. Bakier and Gorla [51] investigated the importance of Prandtl number and heat radiation when thermophoresis strongly depends on constant, laminar Newtonian fluid flow within a semi-infinite vertical permeable sheet.

1. In the study of Bakier and Gorla [51], a modified Prandtl number defined as $P_{rm}^{-1} = \frac{2(3+4R)}{3P_r}$ is presented. The effects of P_{rm}^{-1} on the skin friction coefficient $f''(0)$, Nusselt number $-\theta'(0)$, and Sherwood number $-\phi'(0)$ when thermophoresis (quantified using the thermophoretic coefficient - κ) is ignored, moderately large, and highly significant are presented.
2. With an increase in the magnitude of P_r, in the absence of thermophoresis (i.e., when $\kappa = 0$), $f''(0)$ decreases at the rate of -0.042368819, $-\theta'(0)$ increases at the rate of 0.101332657, and $-\phi'(0)$ decreases at the rate of -0.002590221.
3. At the moderate level of thermophoresis (i.e., when $\kappa = 5$), $f''(0)$ decreases at the rate of -0.041597786, $-\theta'(0)$ increases at the rate of 0.101720111, and $-\phi'(0)$ decreases at the rate of -0.004745018.
4. When thermophoresis is highly significant ($\kappa = 10$), $f''(0)$ decreases at the rate of -0.040810886, $-\theta'(0)$ increases at the rate of 0.102117897, and $-\phi'(0)$ decreases at the rate of -0.006905166.

2.2.30 *International Journal of Advances in Science and Technology*, 2(4), 102–115, 2011

The outcomes of various kinds of fluid flow over a static surface (Blasius flow) or a moving flat wall (Sakiadis flow) have received considerable attention due to their usefulness in metallurgical processes and manufacturing of rubber/plastic sheets. Olanrewaju et al. [218] investigated the effect of Prandtl number on Blasius and Sakiadis flows over a convectively heated surface in the presence of thermal radiation and viscous dissipation.

1. The empirical review of the research conducted by Olanrewaju et al. [218] on the effect of Prandtl number on the skin friction coefficient $f''(0)$, wall temperature $\theta(0)$, and Nusselt number $-\theta'(0)$ during Sakiadis and Blasius flows is presented.
2. First, in the absence of thermal radiation and viscous dissipation, when the level of convective heating of the surface is small in quality (i.e., $a = 0.05$) during Blasius flow, the wall temperature $\theta(0)$ decreases with the Prandtl number at the

rate of −0.014566469. However, when $a = 20$, the rate of decrease in the property of the flow changes significantly. It was discovered that $\theta(0)$ decreases with P_r at the rate of −0.002547173.

3. Second, when the thermal radiation and the level of convective heating of the surface are more significant (i.e., $N_R = 0.7$ and $a = 1$), it was observed that $\theta(0)$ decreases with P_r at the rate of −0.00278 for Blasius flow and −0.00457 for Sakiadis flow.

4. Due to an increase in the magnitude of P_r, the wall temperature $\theta(0)$ decreases at the rate of −0.003 for Blasius flow and −0.00496 for Sakiadis flow. It was also observed that the skin friction coefficients decrease with the Prandtl number at the rate of −0.00093 for Blasius flow and −0.000902368 for Sakiadis flow.

5. Lastly, $-\theta'(0)$ varies with the Prandtl number at the rate of −0.00299 for Blasius flow and 0.004962 for Sakiadis flow. The above-mentioned results imply that the Nusselt number increases with the Prandtl number during Sakiadis flow but decreases during Blasius flow. However, near the free stream, a decrease in the temperature distribution due to the Prandtl number is inevitable.

2.2.31 *Meccanica*, 46(5), 1103–1112, 2011

Ali et al. [22] investigated the influence of the Prandtl number on the mixed convection boundary layer flow of a laminar MHD fluid in conjunction with a vertical flat sheet. The significance of Hall currents and magnetic fields was also considered.

1. Ali et al. [22] tested for the effect of Prandtl number on temperature. It was discovered that when the Lorentz force is highly significant (i.e., $M = 40$), a decrease in the Prandtl number resulted in an increase in the temperature profile at the rate of 0.01651. Meanwhile, as P_r increases, the temperature distribution decreases at the rate of −0.01651.

2. Furthermore, while testing for the effect of Prandtl number on the skin friction and the Nusselt number in the absence of the Lorentz force (i.e., $M = 0$), it was discovered that the local skin friction decreases with an increase in the Prandtl number at the rate of −0.10706. In contrast, the Nusselt number increases with the Prandtl number at the rate of 0.44822.

3. Moreover, when the Lorentz force is highly significant (i.e., $M = 25$), the local skin friction decreases with the Prandtl number at the rate of −0.01172. In contrast, the Nusselt number increases at the rate of 0.360608 with an increase in the Prandtl number.

2.3 Related Published Reports: 2012–2015

Following a study of all the relevant published literature from 2012 to 2015, the following reports were analyzed. The slope linear regression through the data points was used to approximate the observed results.

2.3.1 *AIP Conference Proceedings*, 1450(1), 183–189, 2012

It is noteworthy that thermophoresis and Brownian motion drive nanoparticle migration in nanofluids, and both have a major impact on nanofluids' thermo-physical characteristics. Uddin et al. [299] investigated the influence of Prandtl number on the boundary slip flow of MHD nanofluid in the presence of thermophoresis and Brownian motion. The fluid was assumed to be a chemically reactive substance flowing on a permeable plate.

1. In a study on the effect of Prandtl number on the Nusselt number as in the case of steady boundary layer flow of nanofluid on a stretching sheet in the presence of slips by Uddin et al. [299], it was discovered that an increase in the value of Prandtl number resulted in a corresponding increase in the Nusselt number at the rate of 0.081429.

2.3.2 *Applied Mathematical Modelling*, 36(5), 2056–2066, 2012

Rahman et al. [241] explained the mixed convective flow within a well-ventilated chamber containing a heat-generating, solid circular barrier in the middle. The relevance of the Prandtl number and the Reynolds number was incorporated in the mathematical formulation of the problem.

1. The Richardson number is a dimensionless number that expresses the ratio of buoyancy to shear flow. Rahman et al. [241] discovered that when Richardson's number is 5 (i.e., $R_i = 5$), an increment in the Prandtl number resulted in a decrement in the average fluid temperature in the cavity at the rate of -0.01077.

2. When the Richardson number is zero (i.e., $R_i = 0$), the Prandtl number does not affect the drag force, but when the Richardson number is 5, an increment in the Prandtl number leads to an increment in the same drag force at the rate of 0.40307.

3. When the Richardson number is zero (i.e., $R_i = 0$), an increase in the Prandtl number resulted in an increase in the average Nusselt number at the rate of 1.381095. Also, when the Richardson number is 5 (i.e. $R_i = 5$), an increment in the Prandtl number leads to an increment in the average Nusselt number at the rate of 1.59318.

2.3.3 *Applied Mathematics and Mechanics*, 33(6), 765–780, 2012

Every boundary layer flow of fluid subjected to shearing forces has a spectrum of velocities ranging from maximum to zero, as is well known. The boundary layer flow of a nanofluid via a flattened vertical sheet was studied by Anbuchezhian et al. [27]. With the impact of thermophoresis and Brownian motion owing to thermal stratification, the relevance of the Prandtl number was incorporated in the governing equation. The scaling group of transformations was utilized to get the symmetry groups from the governing equation.

1. In the test for the effect of various values of the Prandtl number on the Nusselt number, Anbuchezhian et al. [27] discovered that an increase in the Prandtl number resulted in a corresponding increase in the Nusselt number at the rate of 0.087819.

2.3.4 *Applied Mathematics*, 3(7), 685–698, 2012

Viscoelastic fluids are non-Newtonian fluids containing a viscous and elastic component. The boundary layer flow of a viscoelastic fluid within a vertical sheet under the effect of a transverse magnetic field was studied by Ezzat and Khatan [98]. The importance of thermal relaxation was also added to the mathematical formulation of the problem. The one-dimensional problem was solved using the state space approach, while the Laplace transform was adopted to solve the thermal stock reactive problem.

1. The influence of the Prandtl number on the temperature distribution was discussed by Ezzat and Khatan [98]. It was discovered that an increase in the value of the Prandtl number resulted in a decrease in the temperature distribution across the fluid at the rate of −0.12044.

2.3.5 *Engineering Computations*, 30(1), 97–116, 2012

The Lorentz force is the product of electromagnetic fields combining magnetic and electric forces on a point charge. A moving charge encounters a force perpendicular to both its velocity and the magnetic field in a magnetic field. The effect of Prandtl on the natural convective MHD flow under the influence of the magnetic field within an open enclosure was scrutinized by Kefayati et al. [151].

1. Kefayati et al. [151] discovered that in the absence of Lorentz force, the average Nusselt number increases with the Prandtl number at the rate of 0.452995, while when the Lorentz force is 150, the average Nusselt number increases with the Prandtl number at the rate of 0.049756.

2.3.6 *International Journal of Theoretical and Mathematical Physics*, 2(3), 33–36, 2012

The force of gravity is a force that attracts two mass objects. Due to the frequent occurrence of free convective flows under the influence of gravitational forces, many researchers have studied the problem of boundary layer flow over a semi-infinite vertical plate. The aim mentioned above prompted Abah et al. [2] to observe the effect of Prandtl number on the velocity, temperature, concentration, heat, and mass transfer.

1. The results of Abah et al. [2] showed that an increase in the Prandtl number resulted in a decrease in the temperature and velocity at the rates of −0.07311 and −0.00189, respectively.

2.3.7 *Journal of Aerospace Engineering*, 27(4), 04014006, 2012

It is a well-known fact that many fluid flows in industries are non-Newtonian fluids. This research aim was considered by Mustafa et al. [201] to investigate the effect of Prandtl number, thermophoresis, and Brownian motion on the flow of an electrically conducting non-Newtonian nanofluid along a flat plate.

1. The result of the research carried out by Mustafa et al. [201] showed that an increase in the Prandtl number led to an increase in the Sherwood number at the rate of 0.0205, but it led to a decrease in the temperature at the rate of −0.02287 and the Nusselt number at the rate of −0.0465.

2.3.8 *Journal of Fluids Engineering*, 134(8), 081203, 2012

There are a countable number of types of free convection in which the ratio of momentum diffusivity to thermal diffusivity is inevitable. However, studies have revealed that the suspension of nanoparticles in base fluid significantly changes the conventional base fluid's transport properties and heat transfer features. The research mentioned above propelled Ibrahim and Shankar [138] to observe the effects of convectively heated boundary layer flow of nanofluid, heat transfer, thermophoresis, Brownian motion of nanoparticles, and nanoparticle fraction over a vertical flat plate using Oberbeck–Boussinesq approximation. To obtain the dimensionless governing equation, the similarity solution was obtained using $\eta = \frac{y}{x}Ra_x^{\frac{1}{4}}$ and $\psi(x, y) = \alpha f(\eta)\frac{y}{x}Ra_x^{\frac{1}{4}}$. Sequel to this, the Prandtl number appears in the dimensionless momentum equation.

1. The result of the study carried out by Ibrahim and Shankar [138] revealed that the velocity of Newtonian fluids conveying tiny particles decreases with the Prandtl number at the rate of -0.327.
2. Within the fluid domain, the temperature distribution decreases with the Prandtl number at the rate of -0.00656.
3. In addition, due to an increase in the Prandtl number, the local Nusselt number increases at the rate of 0.0003, the local Sherwood number decreases at the rate of -0.000318367, while the surface temperature decreases at the rate of -0.002967347.

2.3.9 *Mathematical Problems in Engineering*, 2012, Article ID 934964, 2012

Uddin et al. [299] investigated the effect of the Prandtl number on the problem of convective heat transfer on an incompressible, electrically conducting nanofluid flow with slip boundary condition over a porous stretching plate under the influence of heat generation. The significance of thermophoresis, magnetic field, and Brownian motion was incorporated in the model of the problem.

1. When suction is small (i.e., $f_w = 0.1$), Uddin et al. [299] noticed that the wall temperature $\theta(0)$ decreases with the Prandtl number at the rate of -0.04. When suction is more enhanced (i.e., $f_w = 0.6$), the temperature distribution at the wall decreases with the Prandtl number at the rate of -0.03.
2. An attempt was further made to consider the effect of the Prandtl number on the flow when injection is minimum ($f_w = -0.1$) and maximum ($f_w = -0.6$). The results presented by Uddin et al. [299] showed that $\theta(0)$ decreases with P_r at the rate of -0.055 when $f_w = -0.1$ and $\theta(0)$ increases with P_r at the rate of 0.0065 when $f_w = -0.6$. More so, the Nusselt number $-\theta'(0)$ increases with the Prandtl number at the rate of 0.081461238.

2.3.10 *Thermal Science*, 16(1), 79–91, 2012

Sharma and Dhiman [278] attempted to examine the problem of forced convective heat transfer on an incompressible fluid flow about a rotating circular disk. The effects of the Reynold and Prandtl numbers on the Nusselt number at various levels of rotation.

1. The results of Sharma and Dhiman [278] showed that when the Reynolds number is 10, the Nusselt number increases with the Prandtl number at the rate of

0.068609. Also, it was observed that when the Reynolds number increases to 40, the Nusselt number increases with the Prandtl number at the rate of 0.144052.

2. Sharma and Dhiman [278] stated that when the Reynolds number (ratio of inertia forces to viscous forces) is small in magnitude and the cylinder is stationary, the Nusselt number increases with the Prandtl number at the rate of 0.024245. Also, it was discovered that when the Reynolds number is high ($R_e = 35$), and the cylinder remains stationary, the Nusselt number increases with the Prandtl number at the rate of 0.126799.
3. Furthermore, when the cylinder is rotating and the ratio of inertia forces to viscous forces is small, the Nusselt number increases at the rate of 0.012906. In contrast, the same Nusselt number increases with P_r at the rate of 0.000143 when $R_e = 35$ and the cylinder is rotating.

2.3.11 *Journal of Applied Mathematics*, 2012, 1–15, 2012

The movement of a Newtonian fluid with temperature-dependent viscosity through a porous medium in the presence of energy flow from a concentration gradient and mass flux from a temperature gradient was studied by Moorthy and Senthilvadivu [196] using $\psi = \alpha f(Ra_x)^{1/2}$ and $\eta = \frac{y}{x}$, where Ra_x is the Rayleigh number.

1. The analysis conducted by Moorthy and Senthilvadivu [196] showed that the velocity and temperature distribution increase negligibly with an increasing Prandtl number at the rates of 0.002192982 and 0.004473684 respectively, estimated at $\eta = 2$.
2. However, the fluid concentration decreases with the Prandtl number at the rate of 0.00119.

2.3.12 *Procedia Engineering*, 56, 54–62, 2013

The impacts of the Prandtl number on the movement of water–alumina nanofluid because of free convection inside a sunlight-based authority with a glass cover sheet at the top and a wavy safeguard sheet at the base were investigated by Nasrin et al. [205].

1. Due to higher Prandtl number, the heat transfer rate Nu increases at the rate of 1.25, the mean bulk temperature of the nanofluid θ_w decreases at the rate of -0.12, and the average sub-domain velocity of the nanofluid decreases at the rate of -0.023. These findings by Nasrin et al. [205] led to the conclusion that the Prandtl number has a substantial influence on isotherms and streamlines.

2.3.13 *Applied Nanoscience*, 4(7), 897–910, 2013

The effect of Prandtl number on the two-dimensional flow of a rheological fluid conveying tiny particles along a horizontal surface in the presence of Navier's partial slip, the Lorentz force, thermophoresis, and Brownian motion of those tiny particles was presented by Uddin et al. [298].

1. It is worth deducing from Table 1 presented by Uddin et al. [298] that when the Prandtl number increases, the Nusselt number also increases at the rate of 0.072214.
2. The further reported that when the fluid flow is Newtonian, the temperature distribution decreases with the Prandtl number at the rate of -0.11654.

3. However, the Prandtl number decreases the temperature distribution in the flow of shear-thinning fluids (pseudoplastic fluids) at the rate of −0.1176.

2.3.14 *Advances in Applied Science Research*, 4(2), 190–202, 2013

The problem of the natural convective incompressible flow of an electrically-conducting nanofluid under the influence of the transverse magnetic field, when the ratio of momentum diffusivity is non-zero, was scrutinized by Poornima and Reddy [235]. The significance of thermophoresis and Brownian motion was included in the model.

1. The research conducted by Poornima and Reddy [235] on the effect of Prandtl number on the velocity and temperature showed that an increase in the Prandtl number decreases both the temperature and the velocity of the nanoparticles at the rates of −0.17314 and −0.09455 respectively.

2.3.15 *Boundary Value Problems*, 2013(1), 136, 2013

Awad et al. [45] investigated the problem of thermophoresis on the laminar flow of a two-dimensional nanofluid over a horizontal surface due to stretching when the energy flux due to the concentration gradient, mass flux due to the temperature gradient and variation in the nanoparticle volume fraction along the x- and y-directions are non-zero.

1. The outcome of the study by Awad et al. [45] on the energy flux due to the temperature gradient and mass flux due to the concentration gradient in MHD nanofluid flow over a stretching sheet indicated that an increase in the Prandtl number resulted in a decrease in the temperature at the rate of −0.04421.
2. Also, when the thermophoresis parameter is small (i.e., $N_t = 0.1$), a higher Prandtl number causes the Nusselt number to increase at the rate of 0.07875, and when the thermophoresis parameter is large (i.e., $N_t = 0.4$), the Prandtl number causes the Nusselt number to increase at the rate of 0.0375.

2.3.16 *Brazilian Journal of Chemical Engineering*, 30(4), 897–908, 2013

Joule heating, also known as resistive, resistance, or Ohmic heating, is the process by which the passage of an electric current through a conductor produces heat. In the presence of Joule heating, viscous dissipation, thermal radiation, and thermophoresis, Shehzad et al. [280] explained the effect of Prandtl number on Jeffery's non-Newtonian fluid on a linearly stretched vertical surface.

1. At some distance away from the wall (i.e., $\eta = 2$) Shehzad et al. [280] discovered that the horizontal velocity in the flow of Jeffrey's fluid decreases with the Prandtl number at the rate of −0.038, the temperature distribution decreases at the rate of −0.152, the local skin friction increases at the rate of 0.0294, the local Nusselt number $-\theta'(0)$ increases at the rate of 0.283620833, and the local Sherwood number $-\phi'(0)$ decreases at the rate of −0.0616.

2.3.17 *Energy Procedia*, 36, 788–797, 2013

The influence of Rayleigh number and Prandtl number on natural convection between two horizontal confocal elliptic cylinders was examined and presented by Bouras et al. [65].

1. Bouras et al. [65] remarked that when the Rayleigh number was large (i.e., $R_a = 10,000$), an increase in the Prandtl number resulted in an increase in the average Nusselt number at the rate of 1.125.

2.3.18 *International Journal of Engineering and Innovative Technology*, 3(3), 225–234, 2013

A comparative analysis between the flow of fluid conveying nanoparticles on the porous surface in the presence of viscous dissipation and thermal radiation when there is stretching at the wall and the free stream was presented by Hady et al. [182].

1. Hady et al. [182] discussed the effect of Prandtl number on Blasius and Sakiadis flows at the wall and illustrated in their Table 3 that in the absence of viscous dissipation, the temperature distribution across Blasius and Sakiadis flows decreases with an increase in the Prandtl number at the rates of −0.00304 and −0.00504, respectively.

2.3.19 *International Journal of Engineering Mathematics*, 2013, Article ID 581507, 2013

Deep knowledge of Marangoni convection is very important in stabilizing soap films. For this reason, Sastry et al. [273] examined the effect of heat transfer on the boundary layer flow of an electrically conducting fluid conveying Cu, Al_2O_3, and TiO_2 nanoparticles due to Marangoni convection in which the ratio of momentum diffusivity to thermal diffusivity is non-zero. In the report, the surface tension is assumed to vary with the temperature. In addition, the wall temperature was assumed to vary along the x-direction of the flow. Meanwhile, the nanofluid's thermal diffusivity, density, heat capacity, thermal conductivity, and dynamic viscosity vary with the volume fraction.

1. In the study of Marangoni convection of water conveying three different nanoparticles in two-dimensions, Sastry et al. [273] remarked that Nusselt number increases with Prandtl number in the flow of water conveying Cu, Al_2O_3, and TiO_2 nanoparticles at the rate of 0.378, 0.399, and 0.342, respectively.
2. However, the temperature distribution across the flow of the three nanofluids decreases with the Prandtl number at the rates of −0.0625, −0.0575, and −0.0575, respectively.

2.3.20 *International Journal of Mechanical Sciences*, 70, 146–154, 2013

Solar energy is one of the renewable energy sources where the momentum diffusivity to thermal diffusivity ratio is not insignificant. As with biomass and hydroelectricity, the resultant radiation prompted Kandasamy et al. [149] to consider the problem of laminar fluid flow on a stretchy porous vertical surface susceptible to temperature stratification at the free stream using the Oberbeck–Boussinesq approximation. In addition, Brownian motion and thermophoresis of tiny particles in the flow were considered.

1. The outcome of the study by Kandasamy et al. [149] on the boundary layer flow, heat transfer, and nanoparticles volume fraction over a porous vertical stretched surface due to solar radiation in the presence of thermal stratification and conduction parameters revealed that an increase in the Prandtl number resulted in an increase in the Nusselt number at the rate of 0.085284.

2.3.21 *Journal of Mathematics*, 2013, 1–10, 2013

Pavithra and Gireesha [230] investigated the boundary layer flow and heat transfer of dusty fluids over an exponentially stretchable surface subject to viscous dissipation and internal heat generation/absorption due to heat transfer's importance in industries. It affects the quality of final products. The study's goal was to investigate the effect of increasing the Prandtl number when the heat transfer phenomena specified exponential order surface temperature (PEST) and prescribed exponential order heat flux (PEST/PEST) (PEHF).

1. The outcome of the study conducted by Pavithra and Gireesha [230] shows that the Nusselt number $-\theta'(0)$ increases with the Prandtl number at the rate of 1.101755337, while the wall temperature $\theta(0)$ decreases with the same dimensionless number at the rate of −0.186559778.

2.3.22 *Journal of Mechanics*, 29(3), 423–432, 2013

During the heat treatment of a material traveling between a feed roll and the extrusion of polymer, Mustafa et al. [200] pointed out the significance of Prandtl number on the transport phenomenon in which the significance of Brownian motion and thermophoresis of tiny particles is considered. The above-mentioned fact led to a theoretical study on the unsteady laminar boundary layer flow of nanofluids caused by a linearly stretchable surface.

1. Between the transition from the initial unsteady and final steady-state (i.e. $\tau = 0.5$), Mustafa et al. [200] discovered that the temperature distribution $\theta(\tau, \eta = 1)$ across the flow conveying tiny nanoparticle decreases with the Prandtl number at the rate of −0.18282.

2.3.23 *Journal of Scientific Research*, 5(1), 67–75, 2013

The problem of forced convection of water conveying CuO nanoparticles through an open cavity with porous wavy isothermal wall was considered by Parvin et al. [227] with emphasis on the effects of Prandtl number on streamlines, isotherms, rate of heat transfer, mean temperature, and mid-height U-velocity.

1. Parvin et al. [227] discovered that when the Prandtl number increases, the heat transfer increases at the rate of 0.226539. Also, an increment in the Prandtl number resulted in a decrease in the mean temperature and mid-height U-velocity at the rates of −0.00817 and −0.10955, respectively.

2.3.24 *PloS One*, 8(8), e69811, 2013

The effects of Prandtl number on the dynamics of Oldroyd-B nanofluid in the presence of thermophoresis and Brownian motion were presented by Nadeem et al. [203].

1. In the findings of Nadeem et al. [203] on Oldroyd-B nanofluid flow over a stretching sheet, it was ascertained that as the Prandtl number increased, the local Nusselt number increased at the rate of 0.081402. Also, the temperature distribution decreased with the Prandtl number at the rate of −0.212244898.
2. It is shown that the Prandtl number has a dual effect on the concentration of nanoparticles ϕ. Near the wall, $\phi(1)$ increases with P_r at the rate of 0.23877551. Near the free stream, $\phi(5)$ decreases with P_r at the rate of −0.126122449.

2.3.25 *Journal of Heat Transfer*, 135(5), 054501, 2013

The problem of stagnant flow of a forced convective nanofluid in the presence of thermophoresis and Brownian motion induced by stretching at the wall and the free stream was investigated by Bachok et al. [49].

1. The effect of the Prandtl number was investigated by Bachok et al. [49] on the Nusselt number, and it was discovered that the Nusselt number increases with the Prandtl number at the rate of 0.1667 when Brownian motion is small (i.e., $N_b = 0.1$). However, the Nusselt number increases with the Prandtl number at the rate of 0.3333 when $N_b = 0.5$.

2.3.26 *Journal of Heat Transfer*, 135(10), 102702, 2013

Ramesh and Gireesha [246] deliberated on the effect of Prandtl number on the motion of air conveying dust particles along a horizontal stretchable surface using Stokes drag.

1. The findings of Ramesh and Gireesha [246] showed that with an increase in the Prandtl number, the Nusselt number $-\theta'(0)$ increases for the case of prescribed power-law surface temperature at the rate of 0.23125 and the temperature at the wall $g(0)$ for the case of prescribed power-law heat flux decreases at the rate of -0.54892.

2.3.27 *Open Physics – Central European Journal of Physics*, 12(12), 862–871, 2014

Haq et al. [121] explained the effect of Prandtl number on the motion of Casson fluid due to exponentially stretchable convectively heated objects when the Lorentz force, thermophoresis, and Brownian motion of tiny particles are highly significant.

1. At $\eta = 0.4$, Haq et al. [121] pointed out that the concentration of Casson fluid as it flows decreases with the Prandtl number at the rate of -0.11.
2. At $\eta = 0.6$, the temperature distribution across the fluid flow decreases with the Prandtl number at the rate of -0.04. Further, they revealed that the local Nusselt number increases with the Prandtl number at the rate of 0.289272.

2.3.28 *IOSR Journal of Engineering*, 4(8), 18–32, 2014

Animasaun and Aluko [34] demonstrated the influence of the Prandtl number on the motion of a two-dimensional Newtonian fluid on a porous horizontal surface in the presence of thermal radiation, a heat source, and viscous dissipation, with the dynamic viscosity varying with temperature.

1. The velocity and temperature distribution across two-dimensional flow is decreasing the property of the Prandtl number. The linear regression slope through data points was used to measure the observed effect at $\eta = 5$ at the rate of -0.204561451 for temperature and -0.072929302 for velocity; see Animasaun and Aluko [34].

2.3.29 *Canadian Journal of Physics*, 93(7), 725–733, 2014

Ghalambaz et al. [106] studied the influence of Prandtl number on velocity, temperature, and convective profiles of the boundary layer flow of nanofluid over a horizontal stretching surface. Variable thermal conductivity, variable viscosity, and relative slip velocity between the nanoparticles and base fluid were considered.

1. Ghalambaz et al. [106] discovered that at $5.0E + 4$ value of Lewis number, the concentration of the base fluid decreases with an increase in the Prandtl number at the rate of -0.0295. It was further observed that at a lesser value of Lewis number $1.0E + 4$, the concentration also reduces with the Prandtl number at the rate of -0.0145.
2. Furthermore, it was revealed that when the thermal conductivity of the fluid is a constant function of temperature and the volume fraction of nanoparticles $N_c = 0$, the Nusselt number increases at the rate of 0.163.
3. Meanwhile, when the thermal conductivity of the nanofluid varies with the volume fraction of nanoparticles $N_c = 0.1$, the Nusselt number also increases with the Prandtl number at the rate of 0.161.
4. Moreover, Ghalambaz et al. [106] remarked that $-\theta'(0)$ increases with the Prandtl number at the rate of 0.081424.

2.3.30 *Journal of Heat and Mass Transfer Research*, 2(2), 63–78, 2015

Many scholars have published numerous theoretical investigations on two dimensional non-Newtonian incompressible fluid flows over a surface with stretching or shrinking qualities over the years due to their extensive relevance in engineering areas and businesses such as toothpaste and shampoo manufacture. Lare [174] described the effects of Prandtl number on the velocity, temperature, and temperature gradient of Casson fluid with variable viscosity and thermal conductivity within the boundary layer formed over a vertical surface embedded in a structured medium with suction and exponentially decaying surface-dependent internal heat generation in the same context.

1. From the result of the research carried out by Lare [174], it was noticed that an increase in the Prandtl number corresponds to an increase in the heat transfer $-\theta'(0)$ at the rates of 0.35055 and 0.2905 when the magnitude of thermal radiation is 0.5 and 1, respectively.

2.3.31 *Journal of Generalized Lie Theory and Applications*, 9(2), 1000232, 2015

Abd-el-Malek et al. [4] observed the low free convection laminar boundary layer flow across a vertical flat plate to respond to the influence of thermophoresis and heat generation while also taking into consideration the effect of Prandtl number on velocity, concentration, and temperature.

1. Abd-el-Malek et al. [4] ascertained that increasing the Prandtl number resulted in a decrease in both the horizontal velocity and temperature profile of the fluid flow at the rates of -0.062 and -0.6, respectively.

2.3.32 *International Journal of Mechanical Engineering and Technology*, 6(4), 87–100, 2015

Nanofluids are useful in heat transfer due to their high thermal conductivity and convective heat transfer coefficient. Because of this, Abel et al. [6] investigated the effect of a convective boundary condition and magnetic field on the boundary layer flow, heat, and nanoparticles fraction over a stretching surface in a nanofluid while also taking into account the effect of the Prandtl number on the temperature and the nanoparticle concentration distribution.

1. Abel et al. [6] discovered that an increment in the Prandtl number leads to decrement in the temperature distribution at some distance away from the wall (i.e. $\eta = 1$) at the rate of -0.0098.

2.3.33 *International Journal of Applied and Computational Mathematics*, 1(3), 427–448, 2015

Taking into account the effect of Prandtl number, Brownian motion, and thermophoresis, Buddakkagari and Kumar [71] investigated and analyzed the unsteady, two-dimensional laminar free convective boundary layer flow over a permeable vertical cone or plate.

1. The results of the research conducted by Buddakkagari and Kumar [71] on the effect of the Prandtl number in the presence of Brownian motion $N_b = 1$ over a cone showed that when $X = 0$ and $X = 0.1$ an increase in the Prandtl number leads to a decrease in the local skin friction at the rates of -0.035 and -0.056; an increase in the Nusselt number at the rates of 0.0159 and 0.0127; a decrease in the Sherwood number at the rates of -0.01878 and -0.153, respectively.
2. Over a plate when $X = 0$ and $X = 1$, the local skin friction decreases at the rates of -0.03204 and -0.05823, the local Nusselt number at the rate of 0.0026 and the Sherwood number decreases at the rate of -0.02292 when the Prandtl number is increases.

2.3.34 *Canadian Journal of Physics*, 93(10), 1131–1137, 2015

Khan et al. [154] examined the effect of Prandtl number on the temperature profile of a three-dimensional boundary layer flow of a Maxwell nanofluid along a bidirectional stretching surface. The influence of Brownian motion, thermophoresis, and heat source/sink was considered.

1. According to Khan et al. [154], the Nusselt number proportional to heat transfer rate increases with the Prandtl number at the rate of 0.026537331.
2. At two levels of Schmidt number, the temperature distribution decreases with the Prandtl number at the rate of -0.007815789 when $S_c = 50$ and at the rate of -0.156973684 when $S_c = 0.5$.
3. Sequel to an increase in the Prandtl number, the concentration of the fluid negligibly increases near the wall at the rate of 0.002, but decreases a sone distance away from the wall at the rate of -0.004868421.

2.3.35 *Applied Mathematics*, 6(8), 1362–1379, 2015

The influence of the Prandtl number on heat and mass transfer of an upper-convected Maxwell nanofluid flowing on a horizontal melting surface was studied by Adegbie et al. [9]. The effect of melting heat transfer, variable viscosity, thermal conductivity, solutal stratification, and thermal stratification was considered.

1. Adegbie et al. [9] made known the effect of increasing the Prandtl number on the three properties of boundary layer flow in the presence and absence of melting heat transfer. In the absence of melting heat transfer ($m = 0$), the local skin friction

coefficient $f''(0)$ increases with the Prandtl number at the rate of 0.001792033. When $m = 0.5$, the observed rate of increase in $f''(0)$ with P_r was estimated as 0.039599031.

2. Next is the estimation of decrease in the Nusselt number $-\theta'(0)$ due to higher magnitude of the Prandtl number at the rate of −0.035400889 for $m = 0$. When $m = 0.5$, $-\theta'(0)$ decreases with P_r at the rate of −0.039797923.

3. Lastly, it was shown in the report that when $m = 0$, the Sherwood number $-\phi'(0)$ increased due to an increase in P_r at the rate of 0.000292067. When melting heat transfer is more significant ($m = 0.5$), $-\phi'(0)$ decreases with P_r at the rate of −0.030371336.

2.3.36 *Frontiers in Heat and Mass Transfer (FHMT)*, 6(1), 3, 2015

In another study, the effects of Prandtl number on the temperature distribution, velocity, and heat transfer rate of the steady laminar flow of a fluid flowing over a vertically convectively heated stretchable sheet due to partial slip were examined by Daba and Subhashini [84] exclusively.

1. Sequel to the findings of Daba and Subhashini [84], in the absence of partial slip (i.e., stretching at the wall is linear, $K = 0$), the Nusselt number increased due to higher Prandtl number at the rate of 0.006966667. When $K = 0.5$, the Nusselt number increases with the Prandtl number at a lower rate of 0.003866667.

2. In the absence of convective heating at the wall ($\lambda = 0$), the Nusselt number increases with the Prandtl number at the rate of 0.023166667. When $\lambda = 1$, 0.019166667 was the estimated rate of decrease in the Nusselt number due to higher Prandtl number.

3. Also, at the wall, −0.027340714 was estimated as the amount of decrease in the temperature due to higher Prandtl number.

2.3.37 *Journal of Particle Science & Technology*, 1(4), 225–240, 2015

Observations were made by Ghadami Jadval Ghadam and Moradi [105] to explore the increasing effect of Prandtl number on the assisting and opposing flows of a fluid induced by pressure forces, the Lorentz force, thermal radiation, thermophoresis, and Brownian motion of certain unknown tiny particles.

1. The results obtained by Ghadami Jadval Ghadam and Moradi [105] showed that due to a higher ratio of momentum diffusivity to thermal diffusivity P_r, the local skin friction coefficients decrease at the rate of −0.011124007 for assisting flow, but increase at the rate of 0.011697619 for opposing flow.

2. The Nusselt number $NuRe_x^{-0.5}$ was found to be an increasing property of Prandtl number at the rate of 0.069133706 for assisting flow and at the rate of 0.064724787 for opposing flow.

3. The Sherwood number $ShRe_x^{-0.5}$ decreases with the Prandtl number at the rate of −0.00736454 for assisting flow and at the rate of −0.006995654 for opposing flow.

2.3.38 Master of Technology: Thesis, Federal University of Technology Akure, Nigeria, 2015

In the study of fluid flow, the importance of temperature, heat, and mass transfer rate cannot be neglected since the increase or decrease in the temperature brings a decrease or increase in the viscosity and molecule of fluid and vice versa. Animasaun [29] considered a dimensionless number called the Prandtl number as a factor that acts on the concentration profile of nanofluids and thereby causes a change in the magnitude of the temperature of the fluid. Results were presented with governing equations and graphs, which show the decrease in the velocity and temperature profile as the Prandtl number increases.

1. When the energy flux due to mass concentration gradient and mass flux due to temperature concentration is highly negligible (i.e., $D_f = S_r = 0$), Animasaun [29] remarked that the velocity of the flow decreases with the Prandtl number at the rate of -0.046. In that case, the temperature distribution across the flow decreases with the Prandtl number at the rate of -0.19. However, the concentration of the fluid increases with the Prandtl number at the rate of 0.015.
2. Furthermore, when $D_f = S_r = 3$, the Prandtl number increases within the range $0.2 \leq P_r \leq 0.8$, the velocity of the flow parallel to the vertical surface decreases at the rate of -0.0365, and the temperature distribution slightly increases near the wall at the rate of 0.0024, but decreases after that at the rate of -0.189. When $D_f = S_r = 3$, at $\eta = 5$, it is noteworthy that the concentration increases significantly with the Prandtl number at the rate of 0.14.
3. In addition, when $D_f = S_r = 0$, due to an increase in P_r, $f''(0)$ decreases at the rate of -0.0385, $-\theta'(0)$ increases at the rate of 0.05235, and $-\phi'(0)$ decreases at the rate of -0.01333.
4. When energy flux due to mass concentration gradient and mass flux due to temperature gradient are significant, due to an increase in P_r, $f''(0)$ decreases at the rate of -0.00576, $-\theta'(0)$ decreases at the rate of -0.033735211, and $-\phi'(0)$ increases at the rate of 0.051613.

2.3.39 *Numerical Algorithms*, 70(1), 43–59, 2015

The case of non-similarity solution of stretching-induced flow of fluid in the presence of Brownian motion and thermophoresis of tiny particles with the velocity $U_w(x) = \frac{x}{1+x}$ was studied by Farooq et al. [99].

1. At $\eta = 1$, Farooq et al. [99] pointed out that the temperature distribution diminishes due to higher Prandtl number at the rate of -0.106857143.

2.3.40 *Physica Scripta*, 90(3), 035208, 2015

The usefulness of wavy rolls solution is numerous in astrophysical and geophysical sciences. This influenced Dan et al. [85] to examine the significance of Prandtl number on Rayleigh–Benard convection. It was remarked that a chaotic regime is bound to occur in between two quasi-periodic regimes whenever the Prandtl number is small in magnitude.

1. The reduced Rayleigh number r is the ratio of Rayleigh number to critical Rayleigh number. Dan et al. [85] noticed that the changes in W_{101} with r is almost linear when $P_r = 0.025$, but changes rapidly when $P_r = 0.4$.

2. When $0 \leq W_{101} \leq 0.15$, $r \approx 1$ at all the values of P_r. But, when $W_{101} = 10$, the ratio of Rayleigh number to critical Rayleigh number r changes (i.e., increases) with P_r at the rate of 0.377435897.

2.3.41 *World Academy of Science, Engineering and Technology, International Journal of Mechanical, Aerospace, Industrial, Mechatronic and Manufacturing Engineering*, 9(1), 138–143, 2015

Considering the extensive applications of nanofluid flowing along a stretching cylinder in fiber technology, and piping and casting systems, Sarojamma and Vendabai [272] investigated the influence of Prandtl number on the temperature distribution and heat transfer of the steady boundary layer flow of a Casson nanofluid flowing on a vertically stretching cylinder. The effects of magnetic waves, Brownian motion, and thermophoresis on the fluid were examined.

1. At $\eta = 1.35$, the distribution of temperature decreases with the Prandtl number at the rate of -0.038213592 (Sarojamma and Vendabai [272]).

2.4 Tutorial Questions

1. What was the contribution of Ludwig Prandtl to the body of knowledge on fluid dynamics?
2. Describe thermal diffusivity and momentum diffusivity.
3. Explain the major significance of turbulent Prandtl number and magnetic number.
4. What is the exact relationship between the Nusselt number and heat transfer rate?
5. Using the appropriate phenomenon, explain the relationship between the Prandtl number and (a) thermal Peclet number P_e, (b) Rayleigh number R_a, (c) Brinkman number B_r, (d) Lewis number L_e, (e) Graetz number G_z, and (f) Stanton number S_t.
6. According to Saidi and Karimi [263], what is the significance of increasing copper nanoparticles on velocity and temperature distribution?
7. What are the major differences between theoretical and empirical reviews?
8. Consider Figure 17 published by Hayat, T., Hayat, F., Muhammad, T., & Alsaedi, A. (2020) Darcy–Forchheimer flow by rotating disk with partial slip. *Applied Mathematics and Mechanics, 41*(5), 741–752. doi:10.1007/s10483-020-2608–9, Ref. [272]. Determine the difference in the rate at which the local Nusselt number $-\theta'(0)$ varies with a growth in the thermo-migration of nanoparticles when the thermal slip parameter $\gamma_2 = 0$ and $\gamma_2 = 1.0$.
9. Estimate the changes in the Nusselt number $-\theta'(0)$ and Sherwood number $-\phi'(0)$ due to the upsurge in the Lorentz force associated with parameter M as reported in Table 2 by Khan, S. U., Shehzad, S. A., Rauf, A., & Abbas, Z. (2020). Thermally developed unsteady viscoelastic micropolar nanofluid with modified heat/mass fluxes: A generalized model. *Physica A: Statistical Mechanics and Its Applications, 550*, 123986. doi:10.1016/j.physa.2019.123986, Ref. [272].

[illegible]

[illegible]

[illegible]

[illegible]

[illegible]

[illegible]

[illegible]

[illegible]

[illegible]

[illegible]

[illegible]

[illegible]

[illegible]

[illegible]

3

Conceptual and Empirical Reviews II

3.1 Background Information

Differential equations, heat and mass transfer, and fluid dynamics are few among the fundamental subjects in applied mathematics. Considering the contributions of several published articles within the three major areas of interest mentioned above, it is noteworthy that validation of results, theoretical/conceptual review, and empirical review have decreased trends in quality. After 2015, the publication of information about the importance of increasing the Prandtl number continues. As a result, this chapter seeks and presents the conceptual and empirical analyses of the screened published studies from 2016 to 2021.

3.2 Related Published Reports: 2016–2018

After screening all the relevant published papers from 2016 to 2018, the reports mentioned in this subsection were chosen for analysis. The observed results were estimated using slope linear regression across the data points.

3.2.1 *Heat Transfer Engineering*, 37(18), 1521–1537, 2016

Reliable values of heat transfer coefficients are of great importance to industries to produce boilers and evaporators. Consequently, Baranwal and Chhabra [52] were prompted to investigate the effect of Prandtl number on the laminar natural convection heat transfer to Newtonian fluids in a square enclosure consisting of one hot and one cold circular cylinder.

1. Baranwal and Chhabra [52] discovered that when buoyancy is $Gr = 10$, the Prandtl number has no significant effect on the average Nusselt number. When buoyancy is considerably large ($Gr = 10$) for the case where the two cylinders are close to the bottom wall (i.e., $\delta = -0.25$), the average Nusselt number increases with P_r at the rate of 0.082591 for shear thinning ($n = 0.4$) and at the rate of 0.386171 for shear thickening ($n = 1.6$).
2. Furthermore, when the two cylinders are symmetric (i.e., $\delta = 0$), the average Nusselt number increases with P_r at the rate of 2.690854 for shear thinning ($n = 0.4$) and at the rate of 0.421477 for shear thickening ($n = 1.6$).
3. Finally, when the two cylinders are close to the top wall (i.e., $\delta = 0.25$), the average Nusselt number is an increasing property of P_r at the rate of 2.204528 for shear thinning ($n = 0.4$) and at the rate of 0.0310079 for shear thickening ($n = 1.6$).

DOI: 10.1201/9781003217374-3

3.2.2 *Powder Technology*, 301, 858–867, 2016

In view of the enhancement of thermal conductivity of fluids due to dispersion of nanoparticles into base fluids, Animasaun and Sandeep [37] deliberated on the effects of increasing Prandtl number on the motion of a nanofluid (water conveying alumina nanoparticles) within the boundary layer in the presence of buoyancy, thermal radiation, the Lorentz force, space-dependent internal heat, variable thermal conductivity, and viscosity over a horizontal surface of a paraboloid of revolution.

1. At $\eta = 1$, it is obvious in the work by Animasaun and Sandeep [37] that the distribution of temperature decreases with the Prandtl number at the rate of -0.02.

3.2.3 *Journal of Molecular Liquids*, 219, 703–708, 2016

Based on the unique thermo-physical properties of nanofluids in the field of engineering, Raza et al. [255] examined the effect of Prandtl number on such dynamics within the bounded surfaces $y = \pm h$ subject to the Lorentz force, Brownian motion, thermophoresis, and slip effects (velocity, thermal, and concentration) over a rotating channel. The thermal slip at the wall and free stream was modeled as

$$T - \gamma_2 \frac{\partial T}{\partial y} = T_w, \quad at \quad y = -h$$

$$T + \gamma_2 \frac{\partial T}{\partial y} = T_o, \quad at \quad y = +h$$

1. At $y = -h$, the temperature gradient $\theta'(-1)$ decreases with the Prandtl number at the rate of -0.055929003. At the other side of the wall, $y = +h$, reverse is the case as the temperature gradient $\theta'(+1)$ increases with the Prandtl number at the rate of 0.390582202.
2. At $y = -h$ and $y = +h$, the concentration gradients $\phi'(-1)$ and $\phi'(+1)$ increase with the Prandtl number at the rates of 0.059556423 and 0.505737766 respectively.

3.2.4 *Journal of the Nigerian Mathematical Society*, 35(1), 1–17, 2016

The study of magnetohydrodynamic (MHD) flow, emphasizing the heat transfer phenomenon, is an imperative case study due to its numerous industrial applications. Owing to this, Animasaun et al. [33] investigated the influence of Prandtl number on Casson fluid flow over a vertical exponential stretching sheet due to free convection when magnetic field, space-dependent heat source, thermal radiation, suction, and decaying heat are significant.

1. Animasaun et al. [33] discovered that both the velocity and temperature distribution are decreasing properties of Prandtl number. At $\eta = 5$, the horizontal velocity decreases with the Prandtl number at the rate of -0.070654628 while the temperature distribution decreases with Prandtl number at the higher rate of -0.230022573.

3.2.5 *Physics of Fluids*, 28(11), 113603, 2016

Taking into account the importance of flow through a cylinder in engineering applications such as heat exchange and chimney stack, Ajith Kumar et al. [15] explained the effect of

Prandtl number on laminar mixed convection flow ($R_e \leq 35$) due to buoyancy over a heated cylinder at constant temperature.

1. Flow separation α_{sep} and plume generation angles α_{plume} emerged in the study of Ajith Kumar et al. [15]. At all the levels of Reynolds number and Richardson number, it was found that α_{sep} and α_{plume} increase with the Prandtl number.

3.2.6 *Propulsion and Power Research*, 5(4), 326–337, 2016

In the presence of Ohmic heating, nonlinear thermal radiation, haphazard motion of tiny particles, and thermophoresis, the effect of Prandtl number on boundary layer flow due to nonlinear stretching (at $y = 0$, $u = ax^m$) was presented by Mishra et al. [195].

1. Mishra et al. [195] described the velocity of the flow parallel to the surface and temperature distribution as decreasing properties of Prandtl number. At $\eta = 0.5$ and $\eta = 1.0$, the velocity decreases at the rate of -0.030204589, while the temperature distribution decreases at the rate of -0.053936273.

3.2.7 *Results in Physics*, 6, 805–810, 2016

A report on the two-dimensional flow of non-Newtonian Sutterby fluid through a vertical symmetric channel with convective heat at the wall of both sides was presented by Hayat et al. [128]. At both walls $y = \eta$ and $y = -\eta$, the heat at the walls was modeled as

$$k\frac{\partial T}{\partial y} = -h(T - T_o), \quad k\frac{\partial T}{\partial y} = -h(T_o - T).$$

1. In the study of peristatic flow of incompressible Sutterby fluid through a vertical symmetric channel, Hayat et al. [128] discovered that an increase in the magnitude of Prandtl number leads to an increase in the temperature distribution across the fluid at the rate of 0.2. This was measured at the middle of the vertical symmetric channel ($y = 0$).

3.2.8 *Results in Physics*, 6, 1015–1023, 2016

When the wall temperature $T_w = T_\infty + A_1 \left(\frac{x}{l}\right)^m$, Mabood et al. [183] explained the effects of Prandtl number on the boundary layer flow of stagnant MHD fluid when Coriolis force, heat source, thermophoresis, Brownian motion, thermal radiation, viscous dissipation, and chemical reaction are significant.

1. For a limiting case, Mabood et al. [183] also discovered that the Nusselt number $-\theta'(0)$ increases with the Prandtl number at the rate of 0.08143913.

3.2.9 *Thermal Science*, 20(6), 1835–1845, 2016

In some configuration systems, the convective boundary condition is most suitable for the variation in the wall temperature and wall concentration. At $y = 0$, Hayat et al. [124] defined the condition for wall temperature and wall concentration as $-\kappa\frac{\partial T}{\partial y} = h(T_f - T)$ and $-D_m\frac{\partial C}{\partial y} = k_m(C_f - C)$, where h is the convective heat transfer coefficient, κ is the thermal conductivity, D_m is the molecular diffusivity of the species concentration, and k_m is the wall mass transfer coefficient. In such a case, the effect of the Prandtl number on the temperature distribution across the flow of MHD nanofluid in the presence of thermophoresis and Brownian motion was examined.

1. Hayat et al. [124] discovered that the temperature distribution decreases with the Prandtl number. But at $\eta = 1$, the rate of decrease due to higher P_r was estimated as -0.157627119.
2. Also, the concentration decreases with the Prandtl number. But at $\eta = 1$, the rate of decrease due to a growth in P_r was estimated as -0.120384615.

3.2.10 *International Journal of Engineering and Innovative Technology*, 3(3), 225–234, 2016

The significance of increasing the ratio of momentum diffusivity to thermal diffusivity on the motion of unsteady nanofluid along a vertical surface in the presence of gyrotactic microorganisms, viscous dissipation, and passively controlled nanoparticles was presented by Hady et al. [182].

1. The velocity of the flow decreases significantly with the Prandtl number within the domain $10^{-3} \leq \eta \leq 10^{0}$. At $\eta = 10^{-3}$, the rate of decrease in the velocity with P_r is -0.000325331. However, the Prandtl number does not affect the velocity within the domain $5 \leq \eta \leq 10^{2}$.
2. There is an instability in the fluid flow when the Prandtl number is of a larger magnitude, and this is associated with the generation of perturbation kinetic energy due to buoyancy force (Hady et al. [182]).

3.2.11 *Communications in Theoretical Physics*, 66(1), 133–142, 2016

The dynamics of a viscoelastic fluid conveying tiny particles through a porous medium due to free convection subject to thermophoresis, Brownian motion, Dufour effect, and Soret effect was presented by Ramzan et al. [250].

1. The ninth figure presented by Ramzan et al. [250] shows that the temperature distribution of the fluid decreases with the Prandtl number at the rate of -0.113456464.
2. A reduction in the nanoparticles concentration is observed by increasing the Prandtl number at the rate of -0.148548813.
3. In addition, the Nusselt number increases with P_r at the rate of 0.15857551 and the Sherwood number increases at the rate of 0.588446939.

3.2.12 *Zeitschrift fur Naturforschung A*, 71(9), 837–848, 2016

Different types of fluids, such as colloidal suspensions, are encountered in the industries. Azhar et al. [47] examined the transport phenomenon of a steady Sutterby fluid along a vertical surface subject to thermophoresis and Brownian motion of tiny particles due to dual stretching (i.e., at the wall and the free stream).

1. Azhar et al. [47] showed that the concentration increases but is highly significant at the middle of the domain. At $\eta = 0.6$, the concentration of the fluid substance increases with the Prandtl number at the rate of 0.0376.
2. The Nusselt number $Re_x^{-1/2}Nu_x$ decreases with the Prandtl number at the rate of -0.029094, while the Sherwood number $Re_x^{-1/2}Sh_x$ increases with the Prandtl number at the rate of 0.027696.

3.2.13 *Modelling, Measurement and Control B*, 86(1), 271–295, 2017

Owing to the development of hydrophobic surfaces, slip flows of nanofluids have attracted the attention of many scholars. Excited by the applications of nanomaterials in enrobing dynamics, Beg et al. [55] examined the steady-state transport phenomenon in Casson nanofluid flow over a vertical plate with partial slip.

1. Beg et al. [55] pointed out that the concentration increases with the Prandtl number P_r at the rate of 0.0786; this was estimated at $\eta = 0.9$.
2. The temperature distribution decreases with the Prandtl number at the rate of -0.054; this was estimated at $\eta = 1.4$.
3. The velocity of the fluid flow decreases with the Prandtl number significantly within the domain $0 \leq \eta \leq 3$. Thereafter, an increase in the velocity due to higher Prandtl number is seen. At $\eta = 0.85$, $f'(\eta)$ decreases with P_r at the rate of -0.01648.

3.2.14 *Defect and Diffusion Forum*, 377, 127–140, 2017

The analysis of local skin friction, transfer of microorganisms, and heat and mass transfer across the dynamics of fluids conveying tiny particles has been a topic of interest to many researchers in the past decades. The contribution of Avinash et al. [44] was to examine such fluid flow subject to bioconvection, thermophoresis, and Brownian motion effects by applying the magnetic field at different angles.

1. Avinash et al. [44] noticed that as the Prandtl number increases within the interval $0 \leq P_r \leq 2$, the local skin friction coefficient C_f increases at the rate of 0.001, the Nusselt number $Nu_x/Ra_x^{-0.25}$ increases at the rate of 0.00075, the Sherwood number $Sh_x/Ra_x^{-0.25}$ increases at the rate of 0.0025, and the transfer of microorganisms is at the increasing rate of 0.005.

3.2.15 *Chinese Journal of Physics*, 55(3), 963–976, 2017

In geothermal energy companies, the modeling of oil reservoir insulating process and porous media flows is very common among engineers and mathematicians. Muhammad et al. [199] were keen to study the features of MHD flow of a Maxwell fluid through Darcy–Forchheimer porous medium by employing the heat and zero nanoparticles mass flux conditions at the boundary by taking into account the effects of Brownian motion and thermophoresis in the temperature and concentration species expressions.

1. Muhammad et al. [199] revealed through Table 2 of their report that local the Nusselt number increases with the Prandtl number at the rate of 0.05555.
2. Temperature distribution practically decreases with the Prandtl number. At the wall ($\eta = 0$), the temperature distribution decreases with the Prandtl number at the rate of -0.1365.
3. Dual effects of the Prandtl number P_r on the concentration were reported by Muhammad et al. (2017). Near the wall ($0 \leq \eta \leq 1.4$), the concentration increases with P_r at the rate of 0.028; this was estimated at $\eta = 1$. But, at $\eta = 4$, the concentration decreases with the Prandtl number at the rate of -0.02.

3.2.16 *Global Journal of Pure and Applied Mathematics*, 13(7), 3083–3103, 2017

Considering the wide application areas such as cooling or drying of papers and textile materials fabricated by extrusion, the boundary layer behavior over a porous surface is important to different engineering processes. Furthermore, researchers have been drawn to nanofluids due to their enhancement of physical properties in heat transfer. Hence, Ittedi et al. [145] investigated the partial slip effect of MHD boundary layer flow of nanofluids and radiative heat transfer over a permeable stretching sheet while also taking into account the combined effects of the magnetic field, viscous dissipation, Brownian motion, and thermophoresis.

1. The results obtained by Ittedi et al. [145] indicated that the temperature distribution decreases with the Prandtl number. At $\eta = 2$, the rate of decrease in the temperature with P_r was estimated as -0.136.

3.2.17 *Multidiscipline Modeling in Materials and Structures*, 13(4), 628–647, 2017

Because of the significance of nanofluid in the crude oil extraction, food processing, and aerodynamics, Archana et al. [39] investigated the effect of Prandtl number on velocity, temperature, heat transfer, and concentration of a nonlinear radiating MHD Casson nanofluid moving over both a moving wedge and a static wedge.

1. At all the values of Prandtl number, the temperature distribution for the case of non-linear thermal radiation is greater than that of linear thermal radiation (Archana et al. [39]).

3.2.18 *Journal of the Egyptian Mathematical Society*, 25(1), 79–85, 2017

Considering the increased use of micropolar fluids in the production of polymers and insulators, Animasaun [30] analyzed the influence of Prandtl number on heat transfer, thermal conductivity, temperature, and velocity of a micropolar fluid on a horizontal linearly stretching melting sheet and experiencing mixed convection.

1. The temperature-dependent thermal conductivity parameter and Prandtl number affect the temperature distribution as Animasaun [30] observed.
2. When the velocity at the wall is less than unity ($sigma1$), the vertical velocity at the wall ($f(0)$) rises at the rate of 3.52791603 due to increasing Prandtl number. Meanwhile, when the velocity at the wall is greater than unity ($\sigma > 1$), $f(0)$ increases with the Prandtl number at the rate of 3.598260955.
3. The next stage is to quantify the observed effects of Prandtl number on the local skin friction coefficient $f''(0)$ when the wall's velocity is less than unity and greater than unity. For $\sigma < 1$, $f''(0)$ increases with P_r at the rate of 0.159065735. Reverse is the case for $\sigma > 1$; $f''(0)$ decreases with P_r at the rate of -0.161044985.
4. For $\sigma < 1$, $-\theta'(0)$ decreases with P_r at the rate of -0.35643077. In the case of $\sigma > 1$, $-\theta'(0)$ decreases with P_r at the rate of -0.427234755.
5. For $\sigma < 1$, $-\phi'(0)$ decreases with P_r at the rate of -0.68572596. In the case of $\sigma > 1$, $-\phi'(0)$ decreases with P_r at the rate of -0.731817285.

3.2.19 *International Journal of Chemical Sciences*, 15(3), 1–12., 2017

Heat transfer is the transmission of thermal energy between physical systems. The temperatures of the structures and the characteristics of the medium through which the heat is transported impact the rate of heat transfer. The three main types of heat transport are conduction, convection, and radiation. Hari Krishna et al. [323] investigated the effect of radiation and chemical reactions on MHD flow past an oscillating inclined porous plate with variable temperature and heat transfer, respectively, due to the importance of MHD flow in engineering disciplines such as geothermal energy extraction processes. Also, the effect of buoyancy forces to viscous force on the velocity of the fluid was examined.

1. Hari Krishna et al. [323] observed that the velocity and temperature distributions decrease with the Prandtl number. At $\eta = 1$, the velocity decreases at the rate of -0.059953039.
2. At $\eta = 1$, the temperature distribution decreases at the rate of -0.035598851. It was remarked that a rise in the magnitude of Prandtl number results in a significant reduction in the thermal energy and consequently in a lower temperature.

3.2.20 *International Journal of Current Research and Review*, 9(22), 5–12, 2017

The knowledge of MHD stagnation flow over a stretchable surface is very important in optimizing crude oil, tragedy core cooling system, and glass industries. This led Rayapole and Jakkula [254] to investigate the significance of Brownian motion and thermophoresis on the MHD stagnation point of a nanofluid boundary layer flow on a stretched surface with variable thickness, significant magnetic field on velocity, temperature, and nanoparticles was examined, respectively. Also, the effect of Prandtl number on skin friction was examined.

1. Rayapole and Jakkula [254] ascertained the fact that the local skin friction coefficient is a constant function of Prandtl number, while the Nusselt number $-\theta'(0)$ and Sherwood number $-\phi'(0)$ increase with the Prandtl number at the rates of 0.03485 and 0.01545 respectively.

3.2.21 *International Journal of Engineering Research in Africa*, 29, 10–20, 2017

Because the non-Newtonian fluids are very important in engineering and industrial processes such as heating and cooling of the batch tank and metallurgical processes, Koriko et al. [162] investigated the problem of non-Newtonian fluid (micropolar) flow over a horizontal melting surface in the presence of internal heat source and dual stretches (i.e.. at the wall and the free stream).

1. Koriko et al. [162] discovered that the induced magnetic profile denoted as $g'(\eta)$ decreases near the wall with magnetic Prandtl number. At $\eta = 4.7$, $g'(\eta)$ decreases with P_r at the rate of -0.1732.
2. Temperature gradient increases near the wall $0 \leq \eta \leq 2$, but decreases thereafter till the free stream due to higher magnitude of Prandtl number. At $\eta = 0.2$, the temperature gradient increases with P_r at the rate of 0.150346878. At $\eta = 2$, the temperature gradient decreases with P_r at the rate of -0.160307235.

3.2.22 *International Journal of Mathematics Trends and Technology*, 47(2), 113–127, 2017

Non-Newtonian fluid flow over a stretched surface is very important in industrial applications such as rubber sheets, glass blowing, plastic extrusion, drying/cooling of textile and paper, textile productions, crystal growing, and fiber spinning. This led Avinash and Reddy [42] to compare the effect of increasing Prandtl number on the concentration of liquid film of MHD Carreau fluid over a stretched sheet in the presence of Brownian motion and thermophoresis between the steady and unsteady flows.

1. Avinash and Reddy [42] showed clearly that the concentration ϕ of steady flow is higher than that of unsteady flow at all the values of Prandtl number. At $\eta = 1$ for a steady flow, ϕ increases with P_r at the rate of 0.055. Also, at $\eta = 1$ for an unsteady flow, ϕ increases with P_r at the rate of 0.0375.
2. At $\eta = 1$ for a steady flow, the temperature distribution θ decreases with P_r at the rate of -0.075. Also, at $\eta = 1$ for an unsteady flow, θ decreases with P_r at the rate of -0.05.
3. Due to an increase in the magnitude of Prandtl number, $f''(0)$ neither increases nor decreases, $-\theta'(0)$ increases at the rate of 0.21591, and $-\phi'(0)$ decreases at the rate of -0.2008915.

3.2.23 *International Journal of Mechanical Sciences*, 130, 31–40, 2017

Ramzan et al. [249] deliberated on the effect of thermal and solutal stratification on a mixed convective flow of Jeffery's magneto-nanofluid along a stretched cylinder in the presence of the Lorentz force, thermal radiation, Brownian motion, and thermophoresis.

1. Ramzan et al. [249] noticed that at all the levels of Lewis number, the Sherwood number $Sh_x Re_x^{-1/2}$ decreases with the Prandtl number.
2. When $L_e = 1$, $Sh_x Re_x^{-1/2}$ decreases with P_r at the rate of -0.245714286. But when $L_e = 4$, $Sh_x Re_x^{-1/2}$ decreases with P_r at the rate of -1.386571429.
3. The effect of increasing Prandtl number on the temperature distribution $\theta(\eta)$ as the fluid flow on a flat surface ($\gamma = 0$) and along a cylinder ($\gamma = 0.3$) was presented by Ramzan et al. [249] as Figure 13. For $\gamma = 0$, $\theta(\eta = 4.5)$ decreases with P_r at the rate of -0.06, while for $\gamma = 0.3$, $\theta(\eta = 4.5)$ decreases with P_r at the rate of -0.025.

3.2.24 *Powder Technology*, 318, 390–400, 2017

Considering the importance of nanoparticles in improving the thermal conductivity of the convectional fluid, Dogonchi and Ganji [91] investigated the flow of MHD nanofluid between non-parallel walls (convergent and divergent horizontal walls) subject to Brownian motion and thermophoresis with major emphasis on the effect of Prandtl number on heat and mass transfer.

1. Dogonchi and Ganji [91] noticed that the temperature distribution increases with the Prandtl number, but the concentration gradient decreases with Prandtl number for both fluid flows along convergent and divergent horizontal walls.
2. When the Prandtl number is very small ($P_r = 0.7$), the temperature distribution for convergent flow increases similar to that for divergent flow through horizontal

walls. At the wall $\eta = 0$, for convergent flow, the temperature distribution increases with P_r at the rate of 0.007534236. For divergent flow at $\eta = 0$, the temperature distribution increases with P_r at the rate of 0.00651331.

3. For convergent and divergent flows, the concentration of the fluid decreases with the Prandtl number, but at a different pattern. For divergent flow, the concentration decreases with the Prandtl number at the rate of −0.006522678. For convergent flow, the concentration decreases with the Prandtl number at the rate of −0.006976242.

3.2.25 *Multidiscipline Modeling in Materials and Structures*, 14(2), 261–283, 2018

The importance of non-Newtonian fluid in chemistry engineering, solar engineering, and polymer processing led Koriko et al. [163] to consider the effect of Prandtl number on velocity, temperature profile, and concentration of a three-dimensional nonlinear radiating Eyring–Powell aluminum–water nanofluid flowing on a stretchable surface and having a quantic autocatalytic homogenous and heterogeneous chemical reaction and electromagnetic radiation.

1. Koriko et al. [163] concluded on the fact that the temperature distribution across the motion of Eyring–Powell aluminum–water nanofluid is a decreasing property of Prandtl number. At $\eta = 3$, the property decreases at the rate of −0.0231.
2. Increasing Prandtl number has dual effects on the temperature gradient. At the wall ($\eta = 0$), $\theta'(\eta)$ increases with P_r at the rate of 0.0228. Meanwhile, at the free stream ($\eta = 8$), $\theta'(\eta)$ decreases with P_r at the rate of −0.007.
3. The concentration of the bulk fluid is greatly decreased with the Prandtl number at the wall. At the wall ($\eta = 0$), the property of the fluid decreases with P_r at the rate of −0.026.
4. The catalyst concentration at the wall is greatly increased with the Prandtl number at the wall. At the wall ($\eta = 0$), the property of the fluid increases with P_r at the rate of 0.0191.

3.2.26 *Physical Review Fluids*, 3(1), 013501, 2018

In the past decade, many studies on Rayleigh–Benard convection have been devoted to investigating the effects of variation of Prandtl number on heat transport and flow properties. Chong et al. [78] considered the effect of the Prandtl number on heat transport and flow structures of Rayleigh–Benard convective fluid through DNS within the interval $0.1 \leq P_r \leq 40$.

1. When the Prandtl number increases within $0.01 \leq P_r \leq 1$, there is a significant difference in the flow structure changes with the domain. However, the amount, existence, and enhancement of heat transport are solely dependent on the Prandtl number (Chong et al. [78]).
2. Enhancement in the heat transport becomes significant for $P_r \geq 0.5$. Meanwhile, this was invisible for $P_r = 0.1$ and $P_r = 0.2$.
3. Although the Rayleigh number plays a major role, as the Prandtl number increases for $0.5 \leq P_r \leq 40$, heat transport enhancement is seen to rise from 5.3% to 15.3%.

4. The motion of the fluid can be characterized as less turbulent when P_r is very large, while the Reynolds number decreases.

5. It had been established theoretically by Chong et al. [78] and experimentally by Xia et al. [319] that there exists weak dependence of Prandtl number on the Nusselt number, most especially when P_r is large. In other words, the Nusselt number is a constant function of the Prandtl number for $P_r \gg 10^0$.

3.2.27 *Heat Transfer - Asian Research*, 47(1), 203–230, 2018

Due to the growing utilization of magnetic nanofluids in medical engineering and industrial processing systems, Annasagaram et al. [38] examined the effect of Prandtl number on velocity, temperature, and concentration of an MHD Casson nanofluid flow across an isothermal vertical cone with partial slip.

1. Annasagaram et al. [38] discovered that the velocity of the flow decreases with a higher Prandtl number at the rate of -0.01632. This was estimated at $\eta = 1.25$. However, $f'(\eta)$ negligibly increases with P_r at the rate of 0.03 in the range of $3.75 \leq \eta \leq 7.5$.

2. For $0 \leq \eta \leq 4$, the temperature distribution diminishes as P_r grows large. At $\eta = 1$, the temperature distribution decreases with P_r at the rate of -0.0668.

3. The concentration of the Casson nanofluid increases with the Prandtl number. At $\eta = 0.75$, it was seen that the concentration decreases with the Prandtl number at the rate of 0.058.

3.2.28 *Heat Transfer – Asian Research*, 47(4), 603–619, 2018

Considering the increasing usage of nanofluids as coolants in nuclear reactors and the production of fuels, Samee AD et al. [267] studied the influence of Prandtl number on the mean exit temperature of nanofluids flowing between heat-generating vertical plates.

1. The results obtained by Samee AD et al. [267] presented as Table 3 show the effects of increasing B_{opt} spacing where the heat transfer rate is at maximum and Nu_H is the average Nusselt number.

2. When the Reynolds number $Re_H = 500$, B_{opt} decreases due to higher P_r at the rate of -1.968243061.

3. When $Re_H = 1,000$, B_{opt} decreases with P_r at a lower rate of -1.811032965. The lowest rate of decrease in B_{opt} with P_r is ascertained at -0.772024621 when $Re_H = 1,500$.

4. It is noteworthy from the same table that when the Reynolds number $Re_H = 500$, Nu_H increases due to higher P_r at the rate of 36.71995442.

5. When $Re_H = 1,000$, Nu_H increases with P_r at a higher rate of 53.27373814. The highest rate of increase in Nu_H with P_r is ascertained at 64.93835902 when $Re_H = 1,500$.

6. At all the numerical values of conduction–convection parameter N_{cc}, at $N_{cc} = 0.3$, average exit temperature (Q_{ae}) of the coolant decreases with the Prandtl number at the rate of -0.192716341.

3.2.29 *International Journal of Applied and Computational Mathematics*, 4(3), 85, 2018

Owing to the significance of non-Newtonian fluids as a heat transfer material in chemical engineering and plastic processing, Reddy et al. [256] analyzed the impact of Prandtl number on the temperature distribution and velocity of a Casson fluid moving along a vertical cylinder.

1. The results obtained by Reddy et al. [256] show that when $\beta = 1.0$, the local skin friction coefficients decrease with the Prandtl number at the rate of -0.044685075, while the Nusselt number increases with the Prandtl number at the rate of 0.052480883.
2. In the case of Newtonian fluids (when $\beta = \infty$), the local skin friction coefficients decrease with the Prandtl number at the rate of -0.031916936, while the Nusselt number increases with the Prandtl number at the rate of 0.062814659.

3.2.30 *AIP Advances*, 8(3), 035219, 2018

Considering the practical significance of non-Newtonian fluids in industrial applications and engineering such as the production of boilers, Atif et al. [41] were led to study the effect of Prandtl number and magnetic Prandtl number on the dynamics of a micropolar Carreau fluid conveying tiny articles along a vertically stretching surface subject to thermal radiation, internal heating, Ohmic and viscous dissipation, thermophoresis, and Brownian motion.

1. Based on the analysis by Atif et al. [41], the Nusselt number increases with the Prandtl number at the rate of 0.1251215. In addition, the Sherwood number was found to decline with the Prandtl number at the rate of -0.1253365.
2. The growth in the magnetic Prandtl number causes the angular velocity to be boosted negligibly at the rate of 0.037142857, measured at $\eta = 1.8$.
3. From the wall till the free stream, the temperature distribution decreases with a higher Prandtl number. At $\eta = 2$, the observed rate of decrease was estimated as -0.12440678.
4. In contrast, the concentration of the fluid increases with the Prandtl number at the rate of 0.08877551, measured at $\eta = 2$.
5. The magnetic field profile $h'(\eta)$ increases with magnetic Prandtl number. In fact, at the wall ($\eta = 0$), the observed rate of increase was as 10.9.

3.2.31 *Alexandria Engineering Journal*, 57(3), 1859–1865, 2018

Taking into account the importance of MHD nanofluid in the industrial production of heat exchangers and pumps, Kumaran et al. [167] examined the effect of Prandtl number on heat transfer behavior and Carreau and Casson fluid flows on the upper part of the paraboloid of revolution with an induced and space-dependent internal heat source or sink.

1. Kumaran et al. [167] showed that the temperature distribution across a non-Newtonian Carreau fluid is higher than that of a non-Newtonian Casson fluid at all the chosen values of Prandtl number. However, this is prevailing at smaller values of Prandtl number. At $\eta = 1$, the temperature distribution decreases with the Prandtl number at the rate of -0.077 across the flow of non-Newtonian

Carreau fluid and at the rate of -0.071285714 across the flow of non-Newtonian Casson fluid.

2. In the transport phenomenon of non-Newtonian Carreau fluid, increasing the Prandtl number corresponds to higher local skin friction coefficients at the rate of 0.001683 and the Nusselt number at the rate of 0.3186955.

3. Meanwhile, due to an increase in the magnitude of Prandtl number, the local skin friction coefficients for the transport phenomenon of non-Newtonian Casson increases at the rate of 0.0035125, while the Nusselt number increase at the rate of 0.31603.

3.2.32 *Advanced Engineering Forum*, 28, 33–46, 2018

In light of the ongoing studies on the motion of non-Newtonian fluid and its application process in industries, Avinash et al. [43] deliberated on the impact of the magnetic field, thermophoresis, Brownian motion, and viscous dissipation on Casson fluids flow over a stretching sheet. The effect of the Prandtl number on the steady and unsteady flows of Casson fluid was examined.

1. Avinash et al. [43] remarked that there exists a significant difference between the effect of increasing Prandtl number on the temperature distribution of the steady flow of Casson fluid and the unsteady flow of the same fluid. The slope linear regression through the data points was explored to estimate the differences at $\eta = 2$ as -0.12 for the steady flow of Casson fluid and as -0.0925 for the unsteady flow of the same fluid.

2. It was also discovered that with an increase in the ratio of momentum diffusivity to thermal diffusivity for the steady flow of Casson fluid, local skin friction coefficients decrease at the rate of -0.942944, the Nusselt number increases at the rate of 0.4595, and the Sherwood number decreases at the rate of -0.461703.

3. However, for the unsteady flow of Casson fluid, with an increase in the ratio of momentum diffusivity to thermal diffusivity for the steady flow of Casson fluid, local skin friction coefficients diminish at the rate of -0.900560, the Nusselt number increases at the rate of 0.2577095, and the Sherwood number decreases at the rate of -0.1957185.

3.2.33 *Physics Letters A*, 382(11), 749–760, 2018

Ever since the second law of thermodynamics, together with entropy generation minimization (EGM), was introduced by Professor A. Bejan as a reliable means of controlling accessible irreversibility in a procedure, many researchers have explained its significance. Sequel to this, Hayat et al. [125] investigated the effect of Prandtl number on the motion of a nanofluid over a convectively heated object with non-uniform thickness $A(x + b)^n$ in the presence of the Lorentz force, first-order chemical reaction, Brownian motion of tiny particles, zero mass flux at the wall, thermophoresis, and heat source.

1. The eleventh figure presented by Hayat et al. [125] shows that the temperature distribution across the fluid flow decreases with the Prandtl number. At $\eta = 1$, the rate of decrease in the temperature distribution was estimated as -0.015257143.

3.2.34 *Multidiscipline Modeling in Materials and Structures*, 14(4), 744–755, 2018

Over the years, mass and energy transport along a stretching domain has been a topic of widespread research due to their abundance of usage in polymer industries and metallurgical progressions. The gap in the literature prompted Sulochana and Ashwinkumar [288] to investigate the impact of thermophoresis and Brownian movement on the two-dimensional flow of nanofluids induced by forced convection on a permeable stretching sheet in the presence of thermal diffusion.

1. Based on the results presented by Sulochana and Ashwinkumar [288], Ishak [144], and Pramanik [236], it is worth concluding that the Nusselt number $-\theta'(0)$ increases with the Prandtl number at the rate of 0.379206429.

3.2.35 *Journal of Molecular Liquids*, 260, 436–446, 2018

Considering the increasing usage of nanofluids as a heat transport material in aerodynamics and nuclear reactors, Hamid and Khan [120] studied the impact of Prandtl number on heat transfer in the mixed convective flow of Williamson a nanofluid in a magnetic field with changing thermal conductivity.

1. Hamid and Khan [120] remarked that in the absence of thermophoresis ($N_t = 0$), the Nusselt number $Re^{-1/2}Nu$ increases with the Prandtl number at the rate of 0.2285.
2. However, when thermophoresis is significant ($N_t \approx 0.53$), the same property proportional to heat transfer rate increases with the Prandtl number at the rate of 0.0891.

3.2.36 *Microgravity Science and Technology*, 30(3), 265–275, 2018

In order to understand the effects of melting heat transfer in the flow of Newtonian fluids on the non-porous medium subject to zero mass flux for concentration at the wall, Hayat et al. [126] examined the effect of Prandtl number on the temperature, nanoparticle concentration, Brownian diffusion, and thermophoresis.

1. It is worth remarking that the Nusselt number decreases with the Prandtl number at the rate of -0.2992 (Hayat et al. [126]).
2. The temperature distribution $\theta(\eta)$ increases with the Prandtl number. It is seen that $\theta(\eta = 1.4)$ increases with the Prandtl number at the rate of 0.21.
3. The concentration of the fluid $\phi(\eta)$ decreases with the Prandtl number. It is noticed that at the wall, $\phi(\eta = 0)$ decreases with Prandtl number at the rate of -1.36.

3.2.37 *International Journal of Heat and Mass Transfer*, 122, 1255–1263, 2018

As a result of numerous applications of heat and mass transfer rate in nanofluids over a stretchable slant surface at an angle α, Usman et al. [303] studied the flow of non-Newtonian Casson nanofluids over an inclined stretching cylinder subject to thermo-migration of tiny/nanosized particles and haphazard motion of tiny/nano-sized particles.

1. When the Lewis number $Le = 1$, Usman et al. [303] discovered that the Sherwood number $-\phi'(0)$ increases with the Prandtl number at the rate of 0.331994598. When $Le = 0$, $-\phi'(0)$ is approximately a constant property of P_r.
2. When the curvature parameter $c = 0.5$, the Nusselt number $-\theta'(0)$ decreases with the Prandtl number at the rate of −0.070856695.
3. The concentration of the fluid substance ϕ is a decreasing property of Prandtl number. At $\eta = 1$, the concentration ϕ decreases at the rate of −0.051936046.

3.2.38 *Monthly Notices of the Royal Astronomical Society*, 479(2), 2827–2833, 2018

In another study to unravel the significance of magnetic and electric conducting fluids in astronomy, Brandenburg et al. [69] examined the effect of a small-scale dynamo with the ratio of viscosity to magnetic diffusivity and its dependence on forcing scale. In the study, the effect of the Prandtl number on astrophysical fluids was studied.

1. Brandenburg et al. [69] revealed that, "there exists a significant difference in the magnetic energy spectrum between Prandtl numbers $Pr_M = 0.1$ and for a large values of Pr_M."

3.2.39 *The European Physical Journal E*, 41, 37, 2018

Sequel to the usefulness of nanoparticles, Ur Rehman et al. [302] deliberated on the analysis of both Newtonian fluid flows as a result of the stretching over a vertically inclined cylindrical surface with physical effects such as magnetic field, mixed convection, heat generation, thermal radiation, Brownian motion, first-order chemical reaction, and stratification of concentration when the temperature at the wall is $T_w = T_0 + \frac{bx}{L}$ and free stream temperature is $T_\infty = T_0 + \frac{cx}{L}$. The study presented the effects of increasing the Prandtl number on the flow of non-Newtonian Casson nanofluids.

1. Ur Rehman et al. [302] discovered that the Prandtl number has no effect on the local skin friction coefficient; see their Table 1. It further shows that the Nusselt number increases with the Prandtl number at the rate of 0.4578. In addition, the local Sherwood number increases with the Prandtl number at the rate of 0.12975.

3.2.40 *Results in Physics*, 9, 1201–1214, 2018

Due to the significant applications of micropolar nanofluid in the purification of crude oil and aerodynamics, Shah et al. [277] examined the influence of Prandtl number on the temperature distribution and concentration of MHD micropolar nanofluid along a rotating parallel plate.

1. Shah et al. [277] discovered that the temperature distribution $\theta(\eta)$ and the concentration of the fluid flow $\phi(\eta)$ decrease with the Prandtl number. Using the proposed technique, $\theta(\eta = 0.4)$ and $\phi(\eta = 0.4)$ decrease at the rates of −0.0185 and −0.0674 respectively.

3.2.41 *Radiation Physics and Chemistry*, 144, 396–404, 2018

Based on the biochemical applications of non-Newtonian fluids, the heat and mass transfer in non-Newtonian (tangent hyperbolic) nanofluids with the influence of buoyancy, mixed convection, magnetic field, Brownian motion, thermophoresis, thermal radiation, heat absorption, and heat generation over a stretching surface was studied by Qayyum et al. [238]. In the research, the effect of the Prandtl number on the non-Newtonian nanofluids was presented.

1. In the research of Qayyum et al. [238] on the significance of magnetohydrodynamics on the motion of tangent hyperbolic nanofluid, it was observed that the Nusselt number increases due to an increase in the Prandtl number at the rate of 0.2304.

3.2.42 *Scientific Reports*, 8(1), 3709, 2018

The effect of the mixture of metallic particles submerged in base fluids that are of low thermal conductivity prompted Lu et al. [180] to investigate the motion of electrically conducting fluid where the significance of momentum diffusivity and thermal conductivity are non-zero. The effect of increasing the ratio of momentum diffusivity to thermal conductivity was also investigated in the temperature and concentration field.

1. According to the research carried out by Lu et al. [180] on the impact of nonlinear radiation and Prandtl number on the MHD Carreau nanofluid flow, it was observed that the temperature distribution decreases with the Prandtl number at the rate of −0.096.
2. However, the dual effects of the Prandtl number on the concentration of the fluid are worth deducing from their Figure 9. Near the wall, the concentration across the fluid increases with the Prandtl number at the rate of 0.02. Meanwhile, far away from the wall, the concentration decreases with the Prandtl number at the rate of −0.04.

3.2.43 *International Journal for Computational Methods in Engineering Science and Mechanics*, 19(2), 49–60, 2018

Due to the net mass movement from the high concentration region, the interaction between pressure forces and buoyant forces in manufacturing industries, the relationship between the diffuso-thermal and Prandtl number can be considered a factor suitable to improve performance. However, when thermophoresis, diffuso-thermal (Dufour effect), the Lorentz force, and non-uniform absorption and injection of heat source are inevitable, the two-dimensional flow of an incompressible electrically conducting fluid over an inclined vertical surface with suction and injection when the ratio of thermal conductivity to momentum diffusivity plays a major role was presented by Pal and Mondal [222].

1. The results of Pal and Mondal [222] showed that the velocity of the transport phenomenon at $\eta = 0.5$ decreases with P_r at the rate of -0.09178.
2. At some distance away from the wall (i.e., $\eta = 0.75$), it is seen that the temperature distribution across the flow also decreases with P_r at the rate of -0.11612.

3.2.44 *International Journal of Computing Science and Mathematics*, 9(5), 455–473, 2018

At the initial unsteady stage, in between the transition and final steady stage, Motsa and Animasaun [198] investigated the effects of Prandtl number on the flow of air conveying tiny microstructures with full potential to rotate in the presence of internal heat source on an object at absolute zero of temperature with the aid of bivariate quasi-linearization.

1. During the impulse-driven flow of non-Newtonian micropolar fluids, in the initial unsteady stage $\theta(\eta = 3, \xi = 0)$ decreases with the Prandtl number at the rate of -2.5; in between the transition from the initial unsteady stage to final steady stage, $\theta(\eta = 3, \xi = 0.5)$ decreases with the Prandtl number at the rate of -2.25; and in the final steady stage, $\theta(\eta = 3, \xi = 1)$ decreases with the Prandtl number at the rate of -0.75; see Motsa and Animasaun [198].
2. In other words, as time scale $\xi \to 1$, the rate of decrease in the temperature distribution is diminishing drastically (Motsa and Animasaun [198]).

3.2.45 *Defect and Diffusion Forum*, 387, 625–639, 2018

In the extraction of petroleum products from crude oil, food mixing, syrup drugs, the flow of plasma, and the flow of mercury amalgams, non-Newtonian fluids are very important with distinct features. Mahanthesh et al. [185] investigated the boundary layer flow and heat transfer in Casson fluids submerged with particles over three geometries (vertical cone, wedge, and plate) where the effect of magnetic field on velocity and the mass concentration of dust particles on velocities at a constant value of Prandtl number was examined.

1. Considering the outcome of the research conducted by Mahanthesh et al. [185] on the rate at which heat is being transferred in the flow of dusty Casson fluid over a plate, cone, and wedge, it was noticed that at every increase in the Prandtl number, the local skin friction decreases at the rate of -0.0609 over a plate, -0.0526 over a cone, and -0.0428 over a wedge.
2. The Nusselt number increases at the rate of 1.4745 over a plate, 0.3317 over a cone, and 0.3307 over a wedge.

3.2.46 *Applied Sciences*, 8(2), 160, 2018

Due to the usefulness of MHD nanofluids in engineering and industrial disciplines, Khan et al. [153] examined the effect of Prandtl number on the temperature of nanofluids in order to investigate the impact of MHD and electric fields on the unsteady Maxwell nanofluid flow over a stretching surface in the presence of variable heat and thermal radiation.

1. The outcome of the research conducted by Khan et al. [153] on the significance of the Prandtl number on the flow of nanofluids shows that the temperature distribution across the fluid flow decreases at the rate of -0.008 as the magnitude of Prandtl number increases.

3.2.47 *Defect and Diffusion Forum*, 389, 50–59, 2018

Another research study was designed with the aim to report the significance of electrically conducting nanofluid as applicable in biochemical engineering. Basha et al. [53] explained the fluid transport properties of a high order chemical reacting nanofluid with a non-uniform heat source/sink, Brownian motion, diffusivity, and thermophoretic diffusivity on a cone and a plate through a porous medium. The effect of Prandtl number on MHD nanofluids was studied.

1. Basha et al. [53] ascertained that an increase in the Prandtl number resulted in a decrease in the velocity profile of the Casson nanofluid over a plate and cone at the rate of -0.08393 and -0.06608, respectively.

2. Furthermore, an increase in the Prandtl number resulted in a decrease in the temperature over a cone at the rate of −0.09456 and over a plate at the rate of −0.09984.

3.2.48 *International Communications in Heat and Mass Transfer*, 91, 216–224, 2018

The importance of activation energy in nonlinear thermal radiation and heat transfer over a moving surface triggered Khan et al. [155] to investigate the flow of electrically conducting fluids such as Cross-nanofluid over a stretching sheet. Further explanation was made on the inverse relation between the Prandtl number and thermal diffusivity.

1. Khan et al. [155] remarked that the temperature distribution decays with the Prandtl number at the rate of −0.02875. This was estimated at $\eta = 2$.

3.3 Related Published Reports: 2019–2021

The papers mentioned in this subsection were chosen for analysis after a thorough review of all the relevant published reports from 2019 to 2021, and the observed effects were estimated using slope linear regression through the data points.

3.3.1 *Zeitschrift fur Naturforschung A*, 74(12), 1099–1108, 2019

In order to understand the dynamics of non-Newtonian fluids along an inclined plane, Patra and Panda [229] deliberated on the impact of Prandtl number on the flow of polymeric fluid downward an inclined locally heated plate at an angle within the range $(0 < \Omega \leq \frac{\pi}{2})$ using second-grade fluid's model. An attempt was made to account for the capillary ridges formed along the inclined plate in the study.

1. Patra and Panda [229] ascertained that as the Prandtl number increases, the height of the capillary ridges h_{max} in the flow of polymeric fluid increases at the rate of 0.003069.

3.3.2 *Zeitschrift fur Naturforschung A*, 74(10), 879–904, 2019

The motion of air conveying particles that are non-interacting and of different diameters on an object with variable thickness at various values of Prandtl number was investigated by Animasaun et al. [36] with the aid of Stokes' drag. An internal heat source across the domain on an upper horizontal surface of a paraboloid of revolution, nonlinear thermal radiation, and quartic autocatalysis chemical reaction were incorporated in the model for the transport phenomenon.

1. For a limiting case, Animasaun et al. [36], Ramesh and Gireesha [230], and Grubka and Bobba [117] ascertained that an increase in the Prandtl number leads to an increase in the Nusselt number at the rate of 0.392662343.
2. Animasaun et al. [36] further revealed that when the thickness of the surface in which the dusty fluid flows is small ($\chi = 0.1$), the velocity of the fluid parallel to the surface is found to be a decreasing function of the Prandtl number at the

rate of -0.09325. When $\chi = 1.5$, the velocity decreases with the Prandtl number at the rate of -0.085.

3. However, when $\chi = 0.1$ the shear stress between two successive layers of the fluid at $\varsigma = 0.6$ decreases with the Prandtl number at the rate of -0.10075. In addition, $F''(0.6)$ decreases with P_r at the rate of -0.04575 when $\chi = 1.5$.
4. In addition, the temperature of the fluid at the wall of an object with a smaller thickness (i.e., $\chi = 0.1$) decreases at the rate of -0.1175, while at a small distance away from the wall ($\varsigma = 2.5$), the same property of the fluid decreases at the rate of -0.125.
5. When $\chi = 1.5$, the temperature of the dusty fluid $\Theta(\varsigma)$ at the same point near the wall decreases at the rate of -0.11575, while $\Theta(\varsigma = 2.5)$ decreases with P_r at the rate of -0.12325.
6. Near the wall ($\varsigma = 0.1$) when the thickness of the surface is small ($\chi = 0.1$) and away from the wall $\varsigma = 1.1$, the concentration of the bulk fluid increases and decreases with the Prandtl number at the rates of 0.06075 and -0.27875 respectively. When the thickness of the surface is large $\chi = 1.5$, the concentration of the bulk fluid decreases with the Prandtl number at the rates of -0.019 and -0.17375 at $\varsigma = 0.1$ and $\varsigma = 1.1$, respectively.

3.3.3 *Journal of the Brazilian Society of Mechanical Sciences and Engineering*, 41(10), 439, 2019

In the presence of viscous dissipation, the Lorentz force, thermophoresis, and Brownian movement of tiny particles, variations in the characteristics of infinite shear rate viscosity and entropy generation in magneto-mixed convective flow due to increases in the ratio of momentum diffusivity to thermal diffusivity were presented by Sultan et al. [289].

1. Figure 7 presented by Sultan et al. [289] shows that the temperature distribution $\theta(\eta)$ is a decreasing property of Prandtl number P_r. At $\eta = 2$, it was deduced that $\theta(\eta)$ decreases with P_r at the rate of -0.1085. More so, the Nusselt number increases with the Prandtl number at the rate of 0.139905.

3.3.4 *Arabian Journal for Science and Engineering*, 44(9), 7799–7808, 2019

Taking into account the peristaltic (wavy motion) transport phenomenon and its utilization in engineering and biological sciences, Tariq et al. [293] investigated the impact of Prandtl number on the wavy motion of dusty Walter's B fluid through an inclined asymmetric medium with an inclined magnetic field and heat transfer considered.

1. The findings of Tariq et al. [293] showed that as the Prandtl number is increased, the fluid's temperature reduces at the rate of -0.035.

3.3.5 *Mathematical Problems in Engineering*, 2019, Article ID 3478037, 2019

Bioconvection is a major factor that can affect thermal diffusivity. In a note on the effects of Prandtl number on the motion of unsteady MHD Jeffery's nanofluid over a rotating vertical cone, the significance of gyrotactic microorganisms, haphazard motion of nanoparticles, and thermophoresis was established by Saleem et al. [260].

1. Saleem et al. [260] discovered that an increase in the Prandtl number increased the local Nusselt number and decreased the Sherwood number at the rates of 0.043565 and −0.00011, respectively.
2. Furthermore, an increment in the Prandtl number led to decrease in the temperature profile of the fluid flow at the rate of −0.343711

3.3.6 *Mathematical Problems in Engineering*, 2019, Article ID 4507852, 2019

In another related study, Ibrahim and Tulu [139] deliberated on the two-dimensional Blasius flow of a Newtonian fluid over a porous wedge due to forced convection where viscous dissipation, thermophoresis, and Brownian motion are highly significant.

1. Ibrahim and Tulu [139] discovered that as the Prandtl number increases, the skin friction coefficient remains constant, the local Nusselt number $-\theta'(0)$ decreases at the rate of −0.17373, and the local Sherwood number $-\phi'(0)$ increases at the rate of 0.551350777.
2. Also, at $\eta = 2.5$, the temperature distribution across the fluid flow decreases at the rate of −0.361904762, and the concentration of the fluid $\eta = 2$ decreases at the rate of −0.372108844.

3.3.7 Ph.D. Thesis submitted to Quaid-I-Azam University Islamabad, Pakistan, 2019

The significance of Prandtl number on the motion of non-Newtonian fluids (food products, i.e., ketchup, milk, alcoholic beverages, and mayonnaise) and (biological materials, i.e., blood and synovial fluid) led Qayyum [237] to explain its effects on the following:

1. the temperature distribution and Nusselt number on the flow of Jeffrey's fluid on a convectively heated inclined cylinder in the presence of heat source, and thermal radiation,
2. the temperature distribution and Nusselt number on the flow of Jeffrey's fluid on a radially nonlinear stretching surface, and
3. the temperature distribution across the flow of Jeffrey's fluid on a vertical convected heated surface with variable thickness.
4. The first case extracted from Qayyum [237] shows that the flow of Jeffrey's fluid on a convectively heated inclined cylinder in the presence of a heat source and the temperature distribution across the flow of Jeffrey's fluid is found to decrease with the Prandtl number at the rate of −0.060273973, while the Nusselt number increases with the Prandtl number at the rate of 0.078615385.
5. In the second case, the temperature distribution across Jeffrey's fluid flow due to a radially nonlinear stretching surface decreases with the Prandtl number at the rate of −0.0601085. The Nusselt number increases with the Prandtl number at the rate of 0.047311538.
6. The results for case three, the flow of Jeffrey's fluid on a vertical convected heated surface with variable thickness, show that the temperature distribution decreases with the Prandtl number at the rate of −0.028476085.

3.3.8 *Arabian Journal for Science and Engineering*, 44(2), 1269–1282, 2019

Nayak et al. [208] examined the effects of Prandtl number on the axial velocity, transverse velocity, and temperature of a free convective MHD nanofluid flowing on an exponentially stretching surface under the force of variable magnetic field and thermal radiation.

1. Nayak et al. [208] observed that the temperature reduces with the increasing Prandtl number as water, ethylene glycol 30%, and ethylene glycol 50% convey Cu nanoparticles at the rate of −0.20115.
2. The skin friction along both the X- and Y-axes was also observed to decrease with increasing Prandtl number at the rates of −0.03566 and −0.02603, respectively, for different solutes of Cu nanoparticles.

3.3.9 *SN Applied Sciences*, 1(7), 705, 2019

Kumar et al. [165] investigated the effect of Prandtl number on the heat transport behavior of two different non-Newtonian fluids (i.e., Williamson and Casson fluids) due to stretching at the wall $u_w = P_o x l_o^{-1}$ and at the free stream $u_e = P_o x l_o^{-1}$. In addition, the temperature at the wall and at the free stream are stratified (i.e., $T_w = T_o + m_1 x l_o^{-1}$ and $T_\infty = T_o + m_2 x l_o^{-1}$), respectively.

1. Kumar et al. [165] discovered that an increase in the Prandtl number resulted in a decrease in the local skin friction of both Casson and Williamson fluids at the rates of −0.1777 and −0.1938, while the Nusselt number of both fluids increases at the rates of 0.7735 and 0.7667, respectively.
2. Furthermore, it was discovered that an increase in the Prandtl number resulted in a decrease in the temperature across Casson and Williamson fluids at the rates of −0.0173 and −0.04189, respectively.
3. The velocity of Casson and Williamson fluid flows decreases with the Prandtl number at the rates of −0.07351 and −0.0328, respectively.

3.3.10 *Symmetry*, 11(10), 1282, 2019

The dynamics of the three-dimensional flow of a nano-liquid on a rotating disk in the presence of the Lorentz force, heat generation and absorption, Brownian motion of nanometer-sized particles, thermophoresis, Arrhenius activation energy, and binary chemical reaction when the temperature at the wall is thermally stratified was explored; see Asma et al. [40].

1. Asma et al. [40] discovered that the temperature distribution decreases with the Prandtl number at the rate of −0.179411765, this was estimated at $\eta = 2$.

3.3.11 Ph.D. Thesis Submitted to the Federal University of Technology Akure, Nigeria, 2019

A comprehensive report on the significance of quartic autocatalysis kind of chemical reaction on the flow of water conveying 29 nm CuO and 47 nm Al_2O_3 nanoparticles, and air conveying dust particles and microstructures on objects with variable thickness was presented in a dissertation by Animasaun [29]. The case of dusty fluid can be found in Animasaun et al. [29] and Animasaun [29].

1. Animasaun [29] discovered that when the thickness of the horizontal object is small ($\chi = 0.1$), the shear stress across the water flow conveying 29 nm CuO decreases with Prandtl number at the rate of -0.0133, meanwhile that across the water conveying 47 nm Al_2O_3 decreases at the rate of -0.0135.
2. The temperature distribution across the flow of water conveying 29 nm CuO diminishes with the Prandtl number at the rate of -0.0170 and at the rate of -0.0175 in the motion of water conveying 47 nm Al_2O_3 nanoparticles.
3. The temperature distribution in the flow of air conveying micro-structures decreases with the Prandtl number at the rate of -0.25075 when micro-rotation is negligible ($\aleph = 0.01$), but when micro-rotation is significant ($\aleph = 3$), the temperature distribution decreases with the Prandtl number at the rate of -0.2665.
4. The Nusselt number increases with P_r at the rate of 0.026 and 0.027 in the motion of water conveying 29 nm CuO and 47 nm Al_2O_3 nanoparticles, respectively.
5. When $\chi = 0.1$ and micro-rotation is negligible in the flow of a micropolar fluid, the Nusselt number $-\Theta'(\varsigma = 0)$ increases with P_r at the rate of 0.376. However, when micro-rotation is significant ($\aleph = 3$), the Nusselt number increases with P_r at the rate of 0.4225.
6. When the thickness of the paraboloid of revolution is small ($\chi = 0.1$), the negative temperature gradient at the free stream $-\Theta'(\varsigma = 4)$ decreases at the rates of -0.09925 and -0.08325 when $\aleph = 0.01$ and $\aleph = 3$, respectively. Surprisingly, when $\chi = 1.5$, $-\Theta'(\varsigma = 4)$ decreases at the rates of -0.07 and -0.0475 when $\aleph = 0.1$ and $\aleph = 3$, respectively.
7. The concentration of the homogeneous bulk fluid increases with the Prandtl number at the rates of 0.0222 and 0.024 in the dynamics of water conveying 29 nm CuO and 47 nm Al_2O_3 nanoparticles, respectively.

3.3.12 *Pramana*, 93(6), 86, 2019

Due to the significance of Carreau fluids in industries and engineering, Ramadevi et al. [245] examined the impact of the addition and removal of internal heat energy in the motion of three-dimensional MHD flow of a Carreau fluid over a stretched surface where the thermal conductivity and Joule heating were considered.

1. The results of Ramadevi et al. [245] showed that an increase in the Prandtl number resulted in a decrease in the temperature distribution at the rate of -0.041.

3.3.13 *Mathematical Modelling of Engineering Problems*, 6(3), 369–384, 2019

The quadratic convective flow of a non-Newtonian Casson fluid in a microchannel when induced magnetic field and exponential space-dependent heat source are significant was presented by Kunnegowda et al. [169]. The effect of magnetic Prandtl number on the velocity, induced magnetic field, and the volume flow rate was discussed.

1. Kunnegowda et al. [169] investigated the quadratic convective flow of a Casson fluid with an exponential heat source in two vertical plates, and there is symmetrical heating. An increase in magnetic Prandtl number does not affect

the Knudsen number, but when one side of the plate is heated while the other is not, an increase in the Prandtl number resulted in the increase in the Knudsen number at the rates of 0.0065 and 0.0066 for $\beta_v k_n = 0.05$ and $\beta_v k_n = 0.1$, respectively.

2. In addition, when the volume flow rate and symmetrical heating are significant, an increase in magnetic Prandtl number causes a decrease in the Knudsen number at the rates of -0.005 and 0.0057 for $\beta_v k_n = 0.05$ and $\beta_v k_n = 0.1$, respectively.

3.3.14 *Journal of Thermal Analysis and Calorimetry*, 138(2), 1311–1326, 2019

Considering the passage of an electric current through a conductor in which heat or thermal energy is being produced during such motion of liquid, Ramesh and Prakash [247] were led to focus on the motion of Sutterby nanofluids through a microfluidic vertical channel in the presence of electroosmosis. It was assumed that the flow was induced by sinusoidal peristaltic waves between $Y = -H$ and $Y = +H$ with constant speed c, where

$$H = \pm \left[a - bcos^2 \left(\frac{\pi X}{\lambda} - \frac{\pi ct}{\lambda} \right) \right],$$

Here, the half-width of the channel is a, the direction of the wave propagation is X, wave amplitude of the peristaltic microchannel is b, the time is denoted as t, and the wavelength is λ. In such a transport phenomenon where melting heat transfer occurs at both walls, an increase in the ratio of momentum diffusivity to thermal diffusivity was investigated.

1. Ramesh and Prakash [247] observed that an increase in the Prandtl number resulted in a dual effect on the axial velocity at the rates of -0.3929 and 0.37143 as it is evident in the parabolic nature of the graph.
2. Also, it was discovered that an increase in the Prandtl number resulted in a dual effect on the thermal temperature at the rates of -38.0 and 37.5, respectively. Furthermore, Ramesh and Prakash [247] discovered that an increase in the Prandtl number resulted in a dual effect on the concentration of the particles at the rates of -6.5 and 6.5.
3. Moreover, it was discovered that an increase in the Prandtl number resulted in a dual effect on the Nusselt number at the rates of -4.85 and 5.45. In addition, it was unveiled that an increase in the Prandtl number also resulted in a dual effect on the Sherwood number at the rates of -9.0 and 11.25, respectively.

3.3.15 *Journal of Applied Fluid Mechanics*, 12(1), 257– 269, 2019

The analysis of mixed convective flow forming vortices over a heated square body, especially in the construction of high-rise buildings, led Tanweer et al. [292] to investigate the effect of Prandtl number on a cross-buoyancy flow over a heated square cylinder in an unsteady laminar flow region. In addition, the heat transfer across the fluid flow and the formation of vortices due to buoyancy were presented.

1. Tanweer et al. [292] discovered that as the Prandtl number increased, the mean lift coefficient increased at the rate of 0.007707 and a variation of instantaneous drag coefficient decreased at the rate of -0.00012.
2. It was further shown that as the Prandtl number was increased, the Strouhal number increased at the rate of 0.000855 and the Nusselt number increased at the rate of 0.153413.

3. Also, the increasing effect in Prandtl number resulted in a dual effect of decrease and increase in the time-averaged stream-wise velocity at the rates of -0.00304 and 0.002692; see the next chapter for more details.

3.3.16 *Journal of Applied and Computational Mechanics*, 5(5), 849–860, 2019

The Hall effect and the creation of a voltage differential across an electric conductor on the MHD boundary layer movement and heat transfer across a stretchy surface are highlighted. Trivedi et al. [297] considered the effect of Prandtl number on the MHD flow, hydromagnetic, concentration boundary layer, and heat transfer of a Casson fluid..

1. With the aid of paired quasi-linearization method, Trivedi et al. [297] discovered that the temperature distribution decreases with the Prandtl number at the rate of -0.046. The analysis of results illustrated in Figure 20 was estimated at $\eta = 4$.
2. It was also discovered that the nanoparticle concentration $h(\eta, \chi)$ is a constant function of Prandtl number near the wall, but decreases within the interval $0.15 \leq \eta < 10$. At $\eta = 5$, $h(\eta, \chi = 0.5)$ decreases with P_r at the rate of -0.0005.

3.3.17 *Heliyon*, 5(4), e01555, 2019

Darcy's law was proposed as an equation describing the flow of a fluid through a porous medium based on the results of Henry Darcy's experiments on water dynamics through sand beds. Taking into account the numerous applications of non-Newtonian fluids in engineering and industries, Patel [228] studied the effect of cross-diffusion on the Casson fluid flow in the presence of a uniform magnetic field, thermal radiation, chemical reaction, heat generation, and heat absorption over an infinite plate with a porous medium. The effect of the Prandtl number was investigated on the fluid flow with other fluid characteristics.

1. Patel [228] discovered that the local skin friction and local Nusselt number are decreasing with an increase in the Prandtl number at the rates of -0.0000685 and -0.01197 respectively, while the Sherwood number increases with the Prandtl number at the rate of 0.003192.

3.3.18 *Heliyon*, 5(3), e01345, 2019

During liquid coating on photographic film, the velocity at the wall can be modeled as the combination of $C\text{Exp}\left(\frac{x}{L}\right)$ and $N\vartheta\frac{\partial u}{\partial y}$. In addition, the temperature at the wall is known as $T_\infty + T_o\text{Exp}\left(\frac{2x}{L}\right)$ and $K\vartheta\frac{\partial T}{\partial y}$ where C is the stretching rate per second, N is the velocity slip, K is the thermal slip, and l is the characteristic length of the sheet. Physically, rheologists believe that non-Newtonian Williamson fluids are suitable to model such liquids. This led Lund et al. [181] to investigate the dynamics of a Williamson fluid on a vertical exponential surface in the presence of thermal slip and velocity slip.

1. The methodology adopted by Lund et al. [181] showed that there are two solutions to the mathematical model of the motion of non-Newtonian Carreau fluid along a vertical surface due to partial slip and thermal jump.
2. The increase in the magnitude of the Prandtl number was observed as a factor suitable to diminish the thermal diffusivity, resulting in the decrease in the temperature distribution at the rate of -0.09.

3. Also, the velocity was also considered negligible at the first solution, while the second solution shows that the velocity decreases with the Prandtl number at the rate of −0.055.

3.3.19 *Applied Mathematics and Mechanics*, 40(6), 861–876, 2019

In cases of fluid flows through porous media, inertial effects are significant. Consequently, scientists used the Darcy–Forchheimer model to account for the pressure difference's nonlinear behavior, which leads to nonlinear high-speed flows. Because the motion of the fluid substance through a porous space at high Reynolds number is commonly encountered in mechanical processes, Bilal et al. [62] investigated the effects of thermal slip, velocity slip, the ratio of momentum diffusivity to thermal diffusivity on the Darcy–Forchheimer slip flow of a Sutterby fluid over a stretchable surface subject to magnetic and radiative heat fluxes.

1. Bilal et al. [62] noticed that the Nusselt number defined as $-\left(1+\frac{4}{3}N_r\right)\theta'(0)$ increases with the Prandtl number at the rate of 0.555.

3.3.20 *The European Physical Journal Special Topics*, 228(1), 35–53, 2019

Acknowledging the importance of convective heat transfer over a cone due to its importance in the design of geothermal reservoirs, Sivaraj et al. [284] observed the effect of Prandtl number on temperature as well as the significance of thermodiffusion (Soret effect) and diffuso-thermal (Dufour effect) subject to varying molecular diffusivity, thermal conductivity, and momentum diffusivity.

1. It is worth deducing in the report by Sivaraj et al. [284] that the fluids of higher Prandtl number have a decreasing heat transfer rate.
2. The temperature distribution changes at a rate of −0.0826 for the dynamics on the cone and −0.08021 for the dynamics on the plate as the Prandtl number grows.
3. The Sherwood number varies at a rate of −0.19395 for the dynamics on the cone and −0.19672 for the dynamics on the plate as the size of the Prandtl number increases.
4. As the magnitude of Prandtl number increases, the local skin friction coefficient changes at the rate of −0.0298 for the dynamics on the cone and at the rate of −0.03859 for the dynamics on the plate.

3.3.21 *Multidiscipline Modeling in Materials and Structures*, 15(2), 337–352, 2019

Owing to the problems arising in many industrial applications, including welding and magma solidification, Kumar et al. [166] studied the impact of Joule heating on boundary layer flow and melting heat transfer of a Prandtl fluid over a stretching sheet in the presence of fluid particles suspension.

1. Sequel to the analysis presented by Kumar et al. [166], it is worth remarking that in the case of melting surface at the wall, the temperature of the dust increases with the Prandtl number.
2. At the wall $\eta = 0$, the observed increase in $\theta_p(\eta)$ with P_r is at the rate of 0.051.

3. Within the domain, the increase in the temperature distribution due to a higher Prandtl number is at the rate of 0.019.

3.3.22 *Journal of Applied and Computational Mechanics*, 6(1), 77–89, 2020

In some cases, an electrically conducting fluid made up of two molecules reacts with a molecule of catalyst at the wall. In such cases, the thermal wave dissemination with normal speed due to thermal relaxation time is significant. In the presence of the Lorentz force, cubic autocatalysis kind of chemical reaction, nonlinear thermal radiation, and relaxation heat transfer, Sarojamma et al. [271] examined the effects of Prandtl number on the dynamics of a micropolar fluid along a stretchable surface.

1. The results of Sarojamma et al. [271] presented as Table 1 reveal that the Nusselt number $-\theta'(0)$ is an increasing property of Prandtl number at the rate of 0.06983.
2. It was established that the result corroborates with that of Grubka and Bobba [117], Ishak [142], and Keimanesh and Aghanajafi [152].

3.3.23 *Computer Methods and Programs in Biomedicine*, 183, 105061, 2020

For an isolated system, entropy never decays. This fact is true because the exchange of heat energy and work between any isolated system and the external environment is zero. The reverse is the case for a non-isolated system where the entropy is a decreasing property. The effects of Prandtl number on the flow of second-grade nanofluids over a vertical heated stretchable surface when the Lorentz force, Joule heating, heat source, cubic autocatalysis kind of chemical reaction, thermophoresis, and haphazard movement of tiny particles are significant were investigated by Alsaadi et al. [26].

1. In the study carried out by Alsaadi et al. [26] on the theoretical analysis of MHD mixed convective flow of second-grade nanofluids due to a heated stretchable sheet, it was discovered that as the Prandtl number increases, there occurs a corresponding increase in the Nusselt number at the rate of 0.157361.
2. It was also established in Table 3 and in Table 3.1 that the result corroborates with that of Hayat et al. [127], Olanrewaju et al. [217], and Sithole et al. [283].

TABLE 3.1
Variation in the Nusselt Number $-\Theta'(0)$ with the Prandtl Number as Shown by Alsaadi et al. [26], Sithole et al. [283], Olanrewaju et al. [217], and Hayat et al. [127]

P_r	Nusselt Number Alsaadi et al. [26]	Nusselt Number Sithole et al. [283]	Nusselt Number Olanrewaju et al. [217]	Nusselt Number Hayat et al. [127]
0.5	0.214879	0.21441547	0.214368	0.214365
0.7	0.250986	0.24976956	0.250142	0.250132
1.0	0.289875	0.28782508	0.289161	0.288561
2.0	0.3565648	0.35519994	0.356176	0.356172

3.3.24 *Chaos, Solitons and Fractals*, 130, 109415, 2020

The significance of Prandtl number on the dynamics of two-dimensional viscoelastic fluid along a convectively heated vertical surface in the presence of the Lorentz force, thermophoresis, Brownian motion of tiny particles, thermal radiation, viscous dissipation, Joule heating, and heat source was appraised by Waqas et al. [313].

1. Figure 10 presented in the report by Waqas et al. [313] shows that the temperature distribution across the flow of two-dimensional flow of viscoelastic fluid decreases due to an increase in the Prandtl number at the rate of -0.250790068. This was estimated at $\eta = 0$.

3.3.25 *Journal of Fluid Mechanics*, 882, A10, 2020

Langham et al. [172] defined Prandtl number P_r as the ratio of viscosity to density diffusion. At a fixed value of Reynolds number ($R_e = 400$), coherent structures in a typical stratified plane Couette flow when Prandtl number $P_r \to 0$ and $P_r \to \infty$ were examined.

1. Langham et al. [172] remarked that density transport is diffusion-dominated in the case $P_r \to 0$ and is advection dominated in the case $P_r \to \infty$.
2. At higher values of Prandtl number, homogenization is inevitable (i.e., disappearance of stratification in the shear-driven flow).
3. In the case of fluid flow through a channel, as the magnitude of Prandtl number increases there exists a thorough mixture between the fluid's density at the bottom and the top.

3.3.26 *Heliyon*, 6(1), e03076, 2020

Galerkin's weighted residual method was adopted by Gbadeyan et al. [104] to examine the dynamics of non-Newtonian Casson fluids in the presence of thermal radiation, velocity slip, convectively heated surface, the Lorentz force, thermophoresis, Brownian motion of tiny particles, and viscous dissipation. The study was designed to consider the case in which plastic dynamic viscosity and thermal conductivity vary with temperature.

1. When the Brownian motion-related parameter is small ($N_b = 0.1$) and the buoyancy ratio parameter $N_r = 0$, Gbadeyan et al. [104] discovered that the Nusselt number increases with the Prandtl number at the rate of 0.005716705.
2. More so, when $N_b = 0.3$ and $N_r = 0.4$, the same property of the fluid flow increases with P_r at the rate of 0.004982615.

3.3.27 *Physica Scripta*, 95(3), 035210, 2020

When the earth's rotation becomes substantially significant, Coriolis force is as significant as inertial force and viscous forces. The gap in the literature led Koriko et al. [161] to present the combined effects of increasing Coriolis force and Prandtl number on the flow of air along a horizontal surface of a paraboloid of revolution.

1. The result of Koriko et al. [161] showed that when Coriolis force is negligible and significant ($k = 0.01$ and $k = 0.4$), an increment in the Prandtl number resulted in decrease in the vertical velocity at the rates of -0.101 and -0.085 respectively.
2. When $k = 0.01$ and $k = 0.4$, an increment in the Prandtl number leads to a decrease in horizontal velocity at the rates of -0.06 and -0.064, respectively.

3. Prandtl number has a dual effect on the shear stress between any two successive layers in the flow over a horizontal surface of parabolic revolution at the rates of -0.055 and -0.058.
4. When the ratio of momentum diffusivity to thermal diffusivity increases, the temperature decreases at the rates of -0.085 and -0.08 (when $k = 0.01$ and $k = 0.4$, respectively).

3.3.28 *Canadian Journal of Physics*, 98(1), 1–10, 2020

Owing to the realization that increasing the thermal conductivity of a material may affect the production on a large scale, Bilal et al. [61] studied the effect of Prandtl number on the temperature distribution in the flow within the thin layer adjacent to the wall of a Sutterby magneto-nanofluid experiencing drag resistance along an axially stretchable surface.

1. The temperature distribution across the flow of a Sutterby magneto-nanofluid diminishes with the Prandtl number at the rate of -0.04 estimated at $\eta = 2.2$; see Figure 8 presented by Bilal et al. [61].

3.3.29 *Physica A: Statistical Mechanics and Its Applications*, 550, 123986, 2020

The dynamics of time-dependent viscoelastic micropolar fluid conveying nanometer-sized particles due to periodically accelerating surface was investigated by Khan et al. [157]. The study was designed to consider the case in which heat flux relaxation time and mass flux relaxation time are non-zero.

1. The effective Prandtl number $Pr_{\text{eff}} = \frac{P_r}{1+R_d}$, where R_d is the thermal radiation parameter that emerged at the end of the non-dimensionalization adopted by Khan et al. [157].
2. It is worth remarking that the Nusselt number $-\theta'(0)$ and Sherwood number $-\phi'(0)$ increase with the effective Prandtl number Pr_{eff} at the rates of 0.117264286 and 0.128333516, respectively.

3.3.30 *Coatings*, 10(1), 55, 2020

In the presence of equal diffusivity kind of cubic autocatalysis chemical reaction, the Lorentz force, thermophoresis, and Brownian motion of tiny particles during the flow of second-grade and elastico-viscous liquids on a horizontal surface, Alghamdi [21] explained the observed increasing effect of the Prandtl number.

1. The outcome of the non-dimensionalization shows that the temperature distribution $\theta(\eta)$ across the fluid for viscoelastic parameter k_1^* greater than zero is higher than when it is less than zero.
2. However, at $\zeta = 2$, for $k_1^* > 0$, $\theta(\eta)$ decreases with the Prandtl number at the rate of -0.25 (Alghamdi [21]).

3.3.31 *Heat Transfer*, 49(3), 1256–1280, 2020

Nayak et al. [207] presented the importance of the Prandtl number on the motion of a Newtonian fluid along x and y due to buoyancy forces and exponential stretchable surface when thermodiffusion and diffuso-thermal are important and the Lorentz force is highly significant. The aim of the study was to account for first-order chemical reaction, heat source and sink, porosity, stretching along x- and y-directions, wall temperature T_w, and wall concentration, which are defined as

$$U_w = U_o \text{Exp}\left[\frac{x+y}{L}\right], \quad V_w = V_o \text{Exp}\left[\frac{x+y}{L}\right],$$

$$T_w = T_\infty + T_o \text{Exp}\left[\frac{A(x+y)}{2L}\right], \quad C_w = C_\infty + C_o \text{Exp}\left[\frac{B(x+y)}{2L}\right]$$

1. According to the findings of Nayak et al. [207], the local skin friction for the flow in the x- and y- directions increases at the rates of 0.001087 and 0.0015044, respectively. The Nusselt number increases negligibly at a rate of -0.000088 due to a rise in P_r, while the Sherwood number increases negligibly at a rate of $4.8E-06$.

3.3.32 *Arabian Journal for Science and Engineering*, 45(7), 5471–5490, 2020

Scientists often use the Darcy–Forchheimer model to model the nonlinear behavior of the pressure difference leading to nonlinear high-velocity flows whenever inertial effects become relevant in the case of fluid flow through a porous medium. The dynamics of the Darcy–Forchheimer flow of spinel-type ferrite (i.e., water and Casson fluids conveying $MnFe_2O_4$) due to rotating disk and buoyancy when the wall is convectively heated, thermal radiation, and entropy generation was examined by Shaw et al. [279].

1. The results illustrated as Figure 30 by Shaw et al. [279] shows that due to an increase in the Prandtl number, the temperature distribution across the flow increases near the wall (i.e., $0 \leq \eta \leq 0.35$), but decreases thereafter as $\eta \to 4$.
2. At $\eta = 0.15$, the optimal observed increase in the temperature was estimated as 0.1. Meanwhile, the observed decrease in the same property of the fluid flow at $\eta = 1$ was estimated as -0.062336449 (Shaw et al. [279]).

3.3.33 *Symmetry*, 12, 652, 2020

Mass diffusion occurs if there is a gradient in the proportions of a mixture correlated with either volume concentration or mass concentration. The dynamics of a viscous fluid on a horizontal object due to a nonlinear stretchable surface under a porous medium were investigated using the Darcy–Forchheimer model in the presence of Brownian diffusion, entropy production, thermophoresis, Joule heating, and thermal radiation by Ghulam Rasool et al. [253].

1. According to the findings of Ghulam Rasool et al. [253], the Nusselt number rises at a rate of 0.144395 with the Prandtl number, while the Sherwood number decreases at a rate of -0.052528 with the same dimensionless parameter.

3.3.34 *Applied Mathematics & Mechanics*, 41(5), 741–752, 2020

Whenever inertial effects become important in the case of fluid flow through a porous medium, scientists often use the Darcy–Forchheimer model to explore the nonlinear behavior of the pressure difference leading to nonlinear high-velocity flows. The Darcy–Forchheimer flow of nanofluid as applicable in rotor–stator spinning disk reactors and shrink fits led Hayat et al. [123] to describe the three-dimensional flow in a porous medium in the presence of thermophoresis, Brownian motion of nanoparticles, viscous dissipation, thermal slip, partial slip, and concentration slip. At the wall, the thermal jump was modeled as

$$T = T_w + L_2 \frac{\partial T}{\partial z}, \quad at \quad z = 0$$

1. Hayat et al. [123] remarked that it is not in all cases that the temperature distribution decreases across the domain due to an increase in the Prandtl number.
2. An increase in the magnitude of P_r causes an increase in the temperature distribution near the wall ($0 \leq \zeta < 2$) and a decrease near the free stream.
3. At $\eta = 0$, the function increases at the rate of 0.161333333. At $\eta = 6$, the same function decreases at the rate of -0.045.

3.3.35 *Physica Scripta*, 95(9), 095205, 2020

An analysis of the effect of increasing Prandtl number on the local skin friction coefficients and temperature distribution of the three-dimensional flow of water conveying multi-walled CNT and silicon dioxide (SiO_2) nanoparticles was presented by Nehad et al. [211]. It is very easy to extract Table 3.2 from Tables 6 to 8 presented in the study.

1. As presented by Nehad et al. [211] and as shown in Table 3.2, the rate of increase in the local skin friction coefficient $f''(0)$ due to higher P_r is $3.4402E-05$.
2. The rate of increase in the local skin friction coefficients for the flow along y-direction due to higher Prandtl number P_r is 0.000104259.
3. The temperature distribution decreases with the Prandtl number at the rate of -0.000387113.

3.3.36 *Journal of Fluid Mechanics*, 910, A37, 2021

In contrast to laminar flow, turbulent flow is a form of fluid flow in which the fluid fluctuates or mixes irregularly. A study on the significance of increasing Prandtl number on turbulent flow induced by thermal convection was carried out by Blass et al. [64].

1. Blass et al. [64] remarked that whenever the Prandtl number is large in magnitude, the momentum diffusivity is sufficiently large and therefore transforms the fluid to shear dominated but with higher heat transfer rate.

TABLE 3.2
Variations in Skin Friction Coefficients and Temperature Away from the Wall at Various Values of Prandtl Number When $c = 2.5$ as Reported by Nehad et al. [211]

P_r	$f''(0)$	$g''(0)$	$\theta(0.2)$
6.0723	−2.377269739263540	−7.683598714554352	0.089527594457403
22.9540	−2.374540457775160	−7.675184714776091	0.000328292505257
150.46	−2.37142835437 43	−7.6658377187914	0
S_{lp}	$3.4402E-05$	0.000104259	−0.000387113

2. Besides, the large P_r number enhances the shear in the boundary layer. As P_r exceeds unity, the kinematic viscosity overtakes thermal diffusivity, and the wall shear affects the flow structures in bulk more easily.

3.3.37 *Coatings*, 11(3), 353, 2021

The gradual deformation caused by shear or tensile stresses in most viscous fluid flows is determined by either the shear rate or the shear rate background. Fruitful engineering and industrial usages of the dynamics of non-Newtonian Jeffrey's nanofluid over a vertically stretchable wall of cylinders attracted Ur Rasheed et al. [301] to examine not only the increasing effect of Prandtl number, but also buoyancy forces, the Lorentz force, and Deborah number.

1. Ur Rasheed et al. [301] discovered that the temperature distribution increases with the Prandtl number. At $r = 5$, the rate of increase was estimated as 0.052523364.

3.3.38 *Alexandria Engineering Journal*, 60(3), 3073–3086, 2021

Farooq et al. [100] studied the dynamics of a non-Newtonian Carreau fluid conveying tiny particles and motile microorganisms when thermal radiation, Cattaneo–Christov heat and mass flux, heat source, and convectively heating, mass flux, and flux of motile microorganisms at the wall are significant.

1. It was discovered by Farooq et al. [100] that the temperature distribution decreases with the Prandtl number at the rate of −0.055. This was estimated at $\zeta = 0$. The increasing values of the Prandtl number result in lowering the thermal diffusivity, which consequently reduces the temperature distribution.
2. The concentration of a non-Newtonian Carreau fluid was discovered to be a decreasing property of Prandtl number at the rate of −0.0473, estimated at $\zeta = 0$.
3. The Nusselt number increases with the Prandtl number P_r at the rate of 0.0199.
4. The Sherwood number increases with the Prandtl number at the rate of 0.02915.

3.3.39 *Journal of Fluid Mechanics*, 915, A37, 2021

Either layered or uniform stratification of salt across the water, exchange of heat, oxygen, and carbon is bound to be affected because the density is temperature-dependent. In a study on two-dimensional dynamics of Kelvin–Helmholtz instability, Parker et al. [226] confirmed that the increasing magnitude of Prandtl number plays a major role in the dynamics of salt-stratified water and such transport phenomenon becomes complex as P_r grows.

1. Parker et al. [226] concluded that as the magnitude of P_r increases, there exists higher stratification near the centerline and reduction of length scales or increasing isotropy.

3.3.40 *Journal of Fluid Mechanics*, 915, A60, 2021

In the case of two dimensions, the type of convection found in a horizontal layer of fluid experiencing heating from the lower part such that a regular pattern of convection cells develops is known as Rayleigh–Benard convection. An experimental study of the increasing effects of Prandtl number on quasi-two-dimensional Rayleigh–Benard convection was communicated by Li et al. [177] using $11.7 \leq P_r \leq 650.7$.

1. Li et al. [177] maintained the assertion that due to an increase in the magnitude of Prandtl number from $P_r = 11.7$ to $P_r = 145.7$, thermal plumes pass through the central region with much less frequency.
2. More so, the self-organized large-scale motion is more confined along the periphery of the convection cell due to the associated increasing momentum diffusivity.

3.3.41 *Computers, Materials & Continua*, 68(1), 319–336, 2021

Depending on the area over which the force is applied, a given force may have a dramatically different impact. Through a study on heat transfer across a cavity when the pressure in x-direction and y-direction are significant, Wong et al. [316] updated the literature with the fact that P_r is directly proportional to specific heat. Consequently, increasing specific heat results in a reduction in the thermal boundary layer thickness.

1. The analysis presented by Wong et al. [316] indicated that a greater value of Prandtl number is capable of reducing the temperature distribution within the cavity.

3.3.42 *Scientific Reports*, 11(1), 3331, 2021

The recent acceptability of bioconvection is associated with the outcome of nonlinear pattern analysis, dispersion in shear flows, and measurements of algal cell swimming behavior. In biomedical and engineering sciences, the significance of increasing Prandtl number on the dynamics of chemically reactive couple stress conveying nanoparticles on a periodically accelerated surface was highlighted by Khan et al. [156] when thermophoresis, thermal radiation, bioconvection, and Brownian motion of particles were significant.

1. Khan *et al.* [156] discovered that with an increase in the magnitude of Prandtl number, the Nusselt number (Nu_x) increases at the rate of 0.264889286, local Sherwood number (Sh_x) increases at the rate of 0.096496429, and motile density number increases at the rate of 0.107214286.

3.3.43 *Case Studies in Thermal Engineering*, 25, 100898, 2021

The movement of electric current through a conductor such that heat is produced is known as either Joule heating or Ohmic heating. The forces convection dynamics of a stagnant Newtonian fluid were examined by Ghasemi and Hatami [107] using the differential quadrature method (DQM) when Joule heating, the Lorentz force, thermophoresis, Brownian motion of nanoparticles, nonlinear thermal radiation, and viscous dissipation are significant.

1. The effects of increasing Prandtl number on the temperature distribution in the absence of radiation ($R_d = 0$), moderate radiation ($R_d = 1$), and sufficiently large radiation $R_d = 2$ were examined by Ghasemi and Hatami [107].
2. In the aforementioned three cases, the observed decrease in the temperature distribution due to higher Prandtl number was found to be quite different at the wall. At the wall ($\eta = 0$), the temperature distribution decreases at the rate of -0.018163265 when $R_d = 0$, -0.024693878 when $R_d = 1$, and -0.029387755 when $R_d = 2$.

3.3.44 *Case Studies in Thermal Engineering*, 25, 100895, 2021

The dynamics of a visco-inelastic Williamson fluid can be classified as a non-Newtonian fluid with the shear thinning property since its viscosity decreases with the increasing shear stress rate. The analysis of Williamson fluid flow subject to an induced magnetic field when the viscosity and thermal conductivity vary with temperature was performed by Salahuddin et al. [265].

1. The temperature distribution across the fluid was found by Salahuddin et al. [265] to be a decreasing property of Prandtl number at the rate −0.2, estimated at $\eta = 1.5$. It was remarked in the study that the observed decrease is associated with the inverse relationship between Prandtl number and thermal diffusivity.
2. The Nusselt number decreases with the Prandtl number at the rate of −0.042075.
3. The Sherwood number was reported as a decreasing property of Prandtl number at the rate of −0.101525.

3.3.45 *Partial Differential Equations in Applied Mathematics*, 4, 100047, 2021

The exploration of the variation in momentum and thermal diffusivity during energy transfer in wet cooling towers and evaporation of the water surface in the industry led Megaraju et al. [194] to examine the dynamics of less viscous fluids through an exponentially accelerated isothermal vertical plate when the Lorentz and buoyancy forces, Coriolis force, and Hall effects are significant.

1. Using the finite element method, Megaraju et al. [194] were able to show that both primary and secondary flows are decreasing properties of Prandtl number.
2. In fact, at $z = 1$, the primary flow decreases with the Prandtl number at the rate of −0.275495611. However, at $z = 1.5$, the secondary flow decreases with the Prandtl number at the rate of −0.02597708. At $z = 2$, the temperature distribution decreases with the Prandtl number at the rate of −0.0961349.

3.3.46 *Ain Shams Engineering Journal*, in press, 2021

The significance of increasing Prandtl number during aerodynamic extrusion was modeled as unsteady Cattaneo–Christov double diffusion when random motion and thermo-migration of tiny particles experience convective heating and there exists a zero mass flux at the bidirectionally stretchable wall by Ahmad et al. [11].

1. The analysis by Ahmad et al. [11] testified to the fact that the Nusselt number proportional to the heat transfer rate rises at the rate of 0.031375333 due to an increment in the Prandtl number for $1 \leq P_r \leq 1.9$.

3.3.47 *Mathematical Problems in Engineering*, Article ID 6690366, 2021

In geophysics, the dynamics of a stagnant non-Newtonian Eyring–Powell fluid due to buoyancy and dual stretching when thermal radiation, magnetized pressure, haphazard motion, and thermophoresis tiny particles are significant was studied by Hussain et al. [137]. The outcome of the study shows the following:

1. The analysis carried out by Hussain et al. [137] showed that the Sherwood number decreases with the Prandtl number at the rate of −0.114821667.

2. Also, the Nusselt number increases with the Prandtl number at the rate of 0.127018333.
3. Starting from the wall to the free stream, the magnetic flux that formed the induced magnetic field parallel to the surface $h'(\gamma)$ decreases with the Prandtl number. For instance, at the wall, $h'(\gamma)$ decreases with P_r at the rate of -0.00260274.
4. The report of Hussain et al. [137] is also in agreement with the fact that the temperature distribution across the dynamics of a non-Newtonian Eyring–Powell fluid decreases with the Prandtl number. At $\gamma = 0.6$, θ decreases with P_r at the rate of -0.014520548.
5. The concentration was shown by Hussain et al. [137] to be an increasing property of P_r. At $\gamma = 0.6$, ϕ increases with P_r at the rate of 0.024657534.

3.4 Tutorial Questions

1. Use concept of exposure of hot water to air to explain (i) the major reason why heat and mass transfer occur simultaneously; (ii) heat transfer without mass transfer. At least two diagrams are required.
2. How does increasing the Prandtl number affect bioconvection?
3. How does the Sherwood number relates to mass transfer rate?
4. As the magnitude of Prandtl number increases, what happens to the viscosity of such fluid?
5. For two-dimensional flow, what is the difference between the velocity along the x-direction and y-direction?
6. What quantity does turbulent Prandtl number measure?
7. Provide insight into the changes in the induced magnetic field due to an upsurge in the magnitude of Prandtl number.

[illegible]

[illegible] parallel to [illegible]

[illegible]

[illegible]

[illegible] Tutorial Questions

[illegible]

[illegible]

[illegible]

[illegible]

4

Empirical Reviews and Meta-analysis

4.1 Background Information

The findings of slope linear regression via the data points may be used to construct scrutinization or meta-analysis in this chapter. The statistical analysis combines several scientific papers, each of which has some inaccuracy on a certain issue. The technique's goal is to derive a more important point based on examining the gathered data. Meta-analysis is an important part of a systematic review method. On the importance of meta-analysis, Berlin et al. [59], Rothman et al. [259], and O'Rourke and Greenland [220] explicitly remarked that the analysis provides relationships between the published results, an estimate of the unknown but prevailing truth, and patterns if there exist among the published results. The weighted mean is the average magnitude of the observed impact across all investigations. This article's meta-analysis/scrutinization technique is confined to detecting a difference between related published results. Such findings will assist researchers in shifting their focus from a conclusion in a single published paper to a conclusion based on numerous findings from two or more published studies.

> The statistical analysis of related findings (homogeneity) combined with the published results of slope linear regression across the data points to summarize two or more observable effects is defined in this text as meta-analysis/scrutinization.

Historically, Plackett [233] remarked that astronomists preferred the output of analysis based on the combination of related results for a reliable conclusion. After that, the potential statistician Karl Pearson was the first scholar to adopt meta-analysis to explore further several published reports on typhoid inoculation (i.e., effect of serum inoculations on the treatment of enteric fever; Nordmann et al. [215]). In 1976, the term "meta-analysis" was coined by the psychologist Gene Glass (Glass [111]). The promotion of carcinogenesis due to either radiation or radionuclide in any substance is known as a carcinogen. Kogevinas and Pearce [160] remarked on the importance of using meta-analysis to explore published articles on carcinogens. The step-by-step procedure of meta-analysis by Neyeloff et al. [212] suggests that the weighted effect size ($w \times es$), outcome size, test measures of heterogeneity, variance (Var), individual study weights (w), and standard error (SE) should be calculated to obtain a number to quantify the heterogeneity of all the selected reported studies. However, this is not applicable in the case of fluid dynamics, where a detailed explanation of the published effect of parameters on the transport phenomena is the core subject. However, the slope extraction for multiple points suggested by Pfister et al. [231] and the rates of increase and decrease quantified using the method of the slope of linear regression through the data points suggested by Shah et al. [276], Koriko et al. [164], Animasaun et al. [31,32], and Wakif et al. [308] are useful to quantify and explore the effects of a dimensionless number. After that, the weighted effect size and the scatter plot are presented.

DOI: 10.1201/9781003217374-4

4.2 Vertical and Horizontal Velocities

In two-dimensional flow, the velocity of the fluid's substance along y-direction is proportional to the vertical velocity. When the magnitude of the Prandtl number increases, it is worth observing in all the reviewed reports that the viscosity of the fluid substance seems to be an increasing property. Consequently, the velocity is expected to be a decreasing property of the Prandtl number. This fact agrees with the results of Koriko et al. [161], but contradicts the results presented by Animasaun [30]. The contradiction in the case of Animasaun [30] can be traced to the significance of melting heat transfer during the flow. The phenomenon is often modeled as

$$\kappa \frac{\partial T}{\partial y} = \rho[\lambda^* + c_s(T_m - T_o)]v(x, y = 0) \tag{4.1}$$

Modeling of melting heat transfer at the wall as shown in Eq. (4.1) implies that heat conduction to the melting surface is equivalent to the heat of melting in addition to the sensible heat needed to raise the wall's temperature T_o to its melting temperature T_m; see Yacob et al. [324], Rasekh et al. [252], and Adegbie et al. [9]. It is worthy of note that melting heat transfer is dependent on the relationship between temperature gradient and velocity along y-direction. Zhang et al. [327] observed experimentally that large latent heat is involved during the melting stage of paraffin-copper foam composite, and this causes the temperature of the copper foam to rise more than that of the paraffin. In the next subsection, it is also shown that melting heat transfer also affects the horizontal velocity. The mean (average) of the estimated rates is $\overline{X} = 1.735044246$, and the variance is $\sigma^2 = 3.342396317$.

The outcome of the scrutinization/meta-analysis shows 29 published results on the decreasing effects of the vertical velocity due to a higher Prandtl number. Hady et al. [118] observed no significant effect of Prandtl number on the velocity, while Moorthy and Senthilvadivu [196], Annasagaram et al. [38], and Ramesh and Prakash [247] reported an increase in the velocity due to a rise in the Prandtl number. The observed rate of increase in the velocity with P_r by Ramesh and Prakash [247] was estimated as 0.37143. It is worth concluding that the velocity of fluid flow is a decreasing property of the Prandtl number. Practically, an increase in Prandtl number implies an increment in the resistance to flow (i.e., enhancement in the viscosity), while the ratio of heat capacity to thermal conductivity is unity for the majority of fluids. As shown in Figure 4.1, a minimum decrease in the velocity due to the Prandtl number was reported by Ramesh and Prakash [247]. The first observed increase in the distance traveled per time was reported in the same study. The study's outcome in Figure 7 (p. 1319) shows that the velocity increases with the Prandtl number within the domain $(0 \leq y \leq 1)$. Checking through the formulation of the transport phenomenon, melting heat transfer was assumed to occur at both walls. The heat flux (heat transport per unit time and surface area from the wall to the fluid is proportional to the temperature difference between the wall and the fluid) at $Y = -H$ and $Y = H$ is defined as

$$\kappa_h \frac{\partial T}{\partial y} = -\eta_h(T_o - T), \quad \text{and} \quad \kappa_h \frac{\partial T}{\partial y} = -\eta_h(T - T_o).$$

The difference between the fluid's temperature and the reference temperature, as stated above, is the major reason behind the observed result: "increase in the velocity due to an increase in the magnitude of Prandtl number." In the report of Tanweer et al. [292] on unsteady flow over a heated square cylinder placed in a free stream, it was also remarked that the velocity is an increasing function of Prandtl number (Figures 4.2 and 4.3). According to the study, at the large magnitude of P_r, the downward deflection of incoming

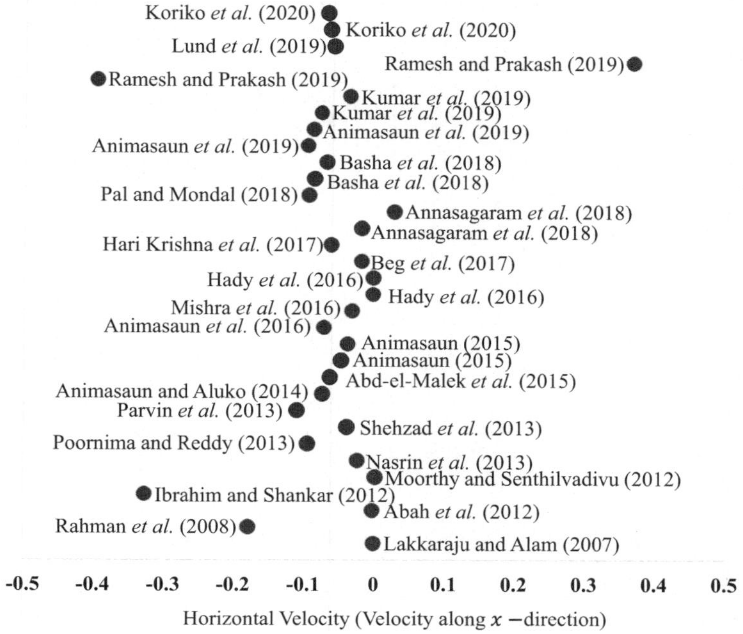

FIGURE 4.1
Forest plot for the effects of Prandtl number on the velocity along x-direction.

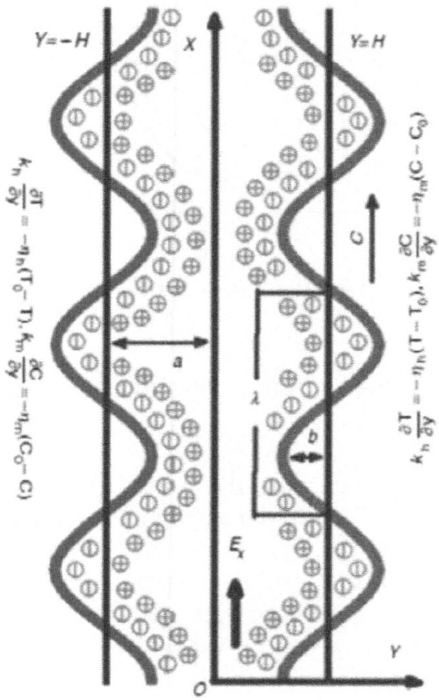

FIGURE 4.2
Illustration of transport phenomena investigated by Ramesh and Prakash [247].

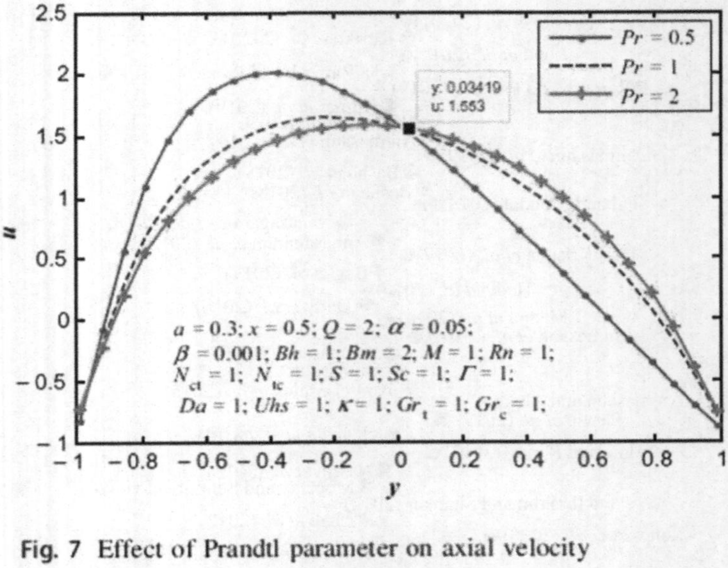

FIGURE 4.3
Observed effect of Prandtl number on the velocity by Ramesh and Prakash [247].

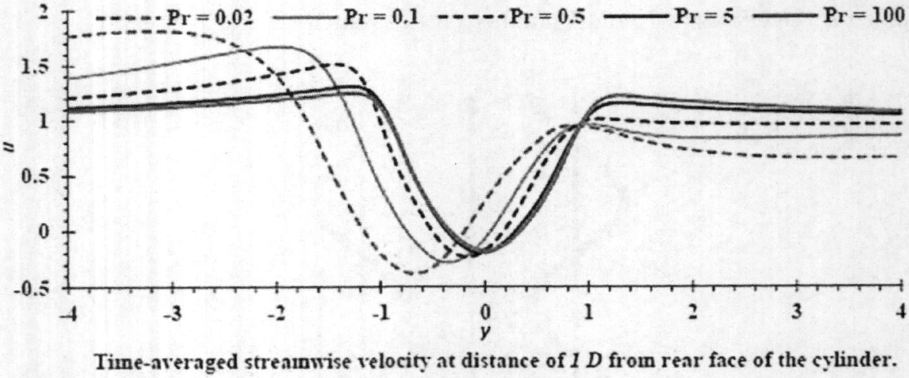

FIGURE 4.4
Observed effect of Prandtl number on flux of vorticity by Tanweer et al. [292].

streamlines decreases, and at the larger magnitude of P_r, the flow becomes nearly symmetric; Figure (4.4). At lower values of P_r, a large recirculation zone is formed near the bottom face of the cylinder, and this recirculation zone disappears at high values of P_r (Tanweer et al. [292]). More so, the flux of vorticity is also affected by the Prandtl number in the same pattern (Figure 4.5). It is important to note that when the mass flux due to the temperature gradient and the heat flux due to the concentration gradient are significant ($D_f = S_r = 3$), the maximum reduction in the velocity due to an increase in the Prandtl number is obtainable (Animasaun [29]). In the study conducted by Basha et al. [53], the velocity of the flow along the cone experienced a maximum reduction due to an increase in the magnitude of the Prandtl number. Physically, the acceleration due to gravity experienced by highly viscous fluid flow along a cone is different from the one experienced in the flow

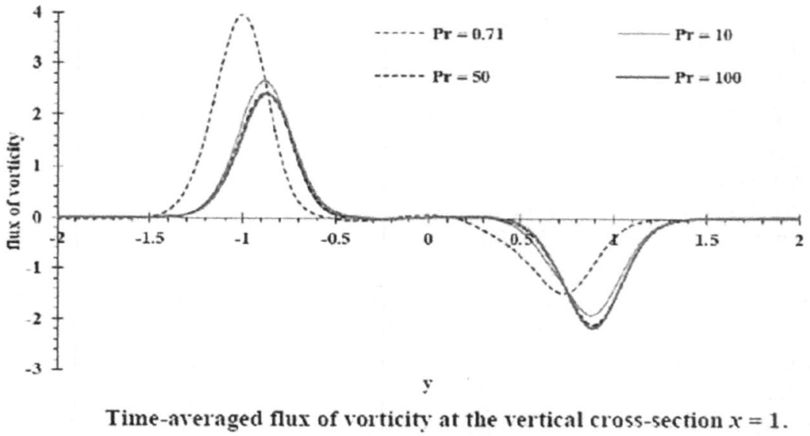

FIGURE 4.5
Observed effect of Prandtl number on the velocity by Tanweer et al. [292].

of the same fluid along a vertical flat plate. The mean (average) of the estimated rates is $\overline{X} = -0.056954361$, and the variance $\sigma^2 = 0.012840661$.

4.3 Diffusion of Microorganisms

The outcome of the scrutinization/meta-analysis shows that the diffusion of microorganisms from the wall into the fluid domain due to bioconvection increases with the Prandtl number at the rate of 0.005 as reported by Avinash et al. [42]. The macroscopic dynamics of fluid substance due to the motion of swimming of motile microorganisms associated with the density gradient that was caused during the process is called bioconvection (Makinde and Animasaun [188,191] and Saleem et al. [260]. Physically, the slow motion of fluid owing to a higher Prandtl number made the swimming of motile microorganisms more possible to cause density gradient. This is responsible for a negligible increase in the wall motile microorganisms flux observed by Avinash et al. [44] and calls for more investigation of the increasing effect of the ratio of mass diffusivity to thermal diffusivity on the bioconvection for a more insightful fact.

4.4 Dust Temperature and Temperature Distribution

The viscosity of air is approximately zero, and it is not realistic to conclude that a higher Prandtl number corresponds to an increase in the viscosity of the less viscous fluid, as earlier remarked above in this chapter. However, it is worth noting that for a less viscous fluid (i.e., gas), a higher Prandtl number $P_r = \frac{\mu c_p}{\kappa}$ indicates a decrease in the thermal conductivity κ. Based on this fact, the reduction mentioned above in the thermal conductivity κ due to a higher Prandtl number causes a reduction in the rate of heat flow from hot wall to cold domain since the wall temperature is greater than the free stream temperature (i.e., $T_w > T_\infty$). This practically leads to a reduction in the temperature distribution and justifies the outcome of the systematic review that shows 112 cases of reduction in the temperature distribution due to a higher Prandtl number with the highest decreasing rate as shown by Ramesh and Prakash [247] and followed by Christopher and Wang [79]. The lowest decreasing rate in the temperature distribution due to a higher Prandtl number was reported by Hady et al. [182], Ibrahim and Shanker [138], Olanrewaju et al. [218], and Nehad et al. [211]. We can also explain from another perspective: As viscosity increases due to higher values of Prandtl number, the rate of consumption of heat energy also increases, which causes a reduction in the temperature distribution. It is well known from elementary science that when a viscous fluid is exposed to heat energy, it would be automatically used to weaken its intermolecular forces.

The mean (average) of the estimated rates is $\overline{X} = -0.16383311$, and the variance $\sigma^2 = 23.41304943$. In the case of dusty fluids (i.e., dynamics of fluid substance such as air conveying dust particles), Kumar et al. [165] reported an increase in the dust temperature due to a higher Prandtl number at the rate of 0.051. This can be associated with the fact that the conveyor of the dust particles is a gas (highly less viscous fluid). Near the wall only did Animasaun [29] observe a very slight increase in the temperature distribution due to a higher Prandtl number when the mass flux due to the temperature gradient and the energy flux due to the concentration gradient are significant (i.e., $D_f = S_r = 3$). Convectively heating at both walls is another factor discovered to make the temperature distribution an increasing property of Prandtl number (Hayat et al. [124]). Hayat et al. [123] also confirmed that the thermal jump at the wall is also a factor in decreasing the temperature distribution due to a higher P_r. Ramesh and Prakash [247] also observed opposite results due to the significance of melting heat transfer at both channel walls. This implies dual temperature gradients capable of influencing the effect of Prandtl number on the temperature distribution. Hayat et al. [126] also observed an increase due to the melting heat transfer at the wall; see Figure 13. With an increase in the temperature within the range of $10°\text{C} \leq Temp \leq 80°\text{C}$, the Prandtl number for water decreases at the rate of -0.099 (Denbigh [90]). In other words, these results imply that the temperature decreases with an increase in the Prandtl number for water.

As the magnitude of the Prandtl number increases, there exists a change in the wall temperature T_w. Consequently, viscosity, density, and friction at the wall are affected (Young [326]), along with the formation of porous layer convection (Somerton [286]). The temperature at the wall is a tool capable of influencing the effect of the Prandtl number on the Nusselt number (Zhao et al. [328]). In the hot region, the surface temperature is a decreasing property of a higher Prandtl number (Wei et al. [314]). The observed increase in the surface temperature due to a decrease in the Prandtl number for $Pr > 1$ was traceable to a decrease in the surface speed in order to satisfy energy conservation in the thermal boundary layer whose thickness is less than that of the momentum layer (Wei et al. [314]). The observed results for $Pr < 1$ were due to a decrease in surface velocity and enhanced heat conduction through the momentum boundary layer to the pool bottom in the thick thermal boundary layer (Wei et al. [314]). Langham et al. [172] remarked that density transport is diffusion-dominated in the case $P_r \to 0$ and it is advection-dominated for the case $P_r \to \infty$. At higher values of the Prandtl number, homogenization is inevitable (i.e., disappearance of stratification in the shear-driven flow).

This does not exclude the significance of the turbulent Prandtl number. Such a scientific fact is in line with the conclusion by Heinrich [129] that reads, "When a turbulent Prandtl number is an infinity, density inversion occurs. Thus, finite value of a turbulent Prandtl number was recommended for the modeling of AGN accretion disks." Hady et al. [118] remarked that "there is instability in fluid flow when the Prandtl number is of a larger magnitude and this is associated with the generation of perturbation kinetic energy due to buoyancy force." The temperature distribution across a less viscous fluid (i.e., $P_r < 0.35$ for gases) is highly diffusive to the extent that the generation of thermal plumes is affected

(Verzicco and Camussi [306]). Consequently, fluids with small magnitudes of Prandtl number are recommended for cooling materials (Saravanan and Kandaswamy [269]). Practically, a low Prandtl number implies a larger magnitude of thermal conductivity, and such flow is diffusion-driven (Chakraborty [74]). At lower levels of the Prandtl number, as in the case of gases, the significance of convection is infinitesimal. Fluid flow characterized with a large magnitude of Prandtl number as in the case of oil (highly viscous fluid, $P_r = 70$) is the most unstable (Bera and Khalili [57]). Either for a less viscous fluid ($P_r = 0.7$) and high viscous fluid ($P_r = 70$), buoyancy force produces disturbance kinetic energy capacity to meet up with the dissipation of kinetic energy owing to surface drag (Bera and Khalili [57]).

4.5 Temperature Gradient

The direction of changes in temperature and changes in the rate of temperature are two major facts temperature gradient depicts in thermal engineering. For the classical boundary layer flow, (i) wall temperature is greater-than the free stream temperature, and temperature distribution decreases from the wall to the free stream, (ii) either thermal conductivity decreases with the Prandtl number or viscosity increases with the Prandtl number—this causes temperature distribution to decline. Based on this, the temperature gradient is expected to be an increasing property of the Prandtl number. However, the outcome of this scrutinization shows that certain phenomena may influence the significance of the Prandtl number on the temperature gradient. Koriko et al. [162] noticed in a study on the dynamics of a micropolar fluid over a horizontal melting surface that the temperature gradient increases near the wall $0 \leq \eta \leq 2$, but decreases after that till the free stream due to a higher magnitude of the Prandtl number. In other words, the melting heat transfer affects the temperature gradient far away from the wall. In Animasaun [31, 32] study, the temperature gradient was defined as $-\Theta'(\varsigma = 4)$. More so, the property's reported decreasing rates (i.e., -0.09925 and -0.08325) denote an increase. The reported decrease by Mahanti and Gaur [186] was compared with that of Crepeau and Clarksean [83]. Raza et al. [255] observed an increase in the temperature gradient due to the thermal jump at the wall and free stream. Koriko et al. [164] also observed an increase in the temperature gradient during the transport phenomenon of Eyring–Powell aluminum–water nanofluid on a thermally stratified surface. The reported increase is due to thermal stratification at the wall. It is seen that Koriko et al. [162], Animasaun [31,32], Mahanti and Gaur [186], Raza et al. [255], and Koriko et al. [164] noticed a decrease in the temperature gradient due to a rise in the ratio of momentum diffusivity to thermal diffusivity. Meanwhile, Koriko et al. [164], Christopher and Wang [79], Koriko et al. [162], and Raza et al. [255] discovered that temperature gradient increases with the Prandtl number. The mean (average) of the estimated rates is $\overline{X} = 0.049182406$, and the variance $\sigma^2 = 0.020677868$.

4.6 Stanton Number and Strouhal Number

The Stanton number quantifies the ratio of heat transfer to the heat (i.e., thermal) capacity of the liquid—the ratio of inertial forces caused by the local acceleration to the inertial forces caused by the convective acceleration. Schneider [274] noticed that the Prandtl number decreases the Stanton number at the rate of -0.960285714, while Tanweer et al. [292] reported a negligible increase in the Strouhal number at the rate of 0.000855.

4.7 Shear Stress between Two Successive Layers

The shear stress between two successive layers is proportional to the friction between layers from the wall to the free stream. It was shown by Animasaun et al. [31,32] and Koriko et al. [161] to be a decreasing property of Prandtl number. The mean (average) of the estimated rates is $\overline{X} = -0.047716667$ and the variance $\sigma^2 = 0.000888414$.

4.8 Ratio of Rayleigh Number to Critical Rayleigh Number

It is only Dan et al. [85] that observed an increase in the ratio of Rayleigh number to critical Rayleigh number due to Prandtl number at the rate of 0.377435897.

4.9 Nusselt Number Proportional to Heat Transfer

Sequel to the earlier variation in the temperature gradient $\theta'(\eta)$ due to Prandtl number, the Nusselt number is expected to occur due to Prandtl number. This fact corroborates with the outcome of 114 reports. However, Ramesh and Prakash [247], Animasaun [29], Ramesh and Gireesha [246], Animasaun [30], Ibrahim and Tulu [139], Usman et al. [303], Khan and Pop [158], Mustafa et al. [201], Adegbie et al. [9], Azhar et al. [47], Patel [228], Olanrewaju et al. [218], Makinde and Olanrewaju [189], and Nayak et al. [209] noticed that the Nusselt number is a decreasing property of Prandtl number. The Nusselt number increases negligibly with the Prandtl number (Wei and Ming-Hsiung [315]). Peripheral variation of the relative Nusselt number increases with the Prandtl number (Xin and Ebadian [320]). The wall temperature is a tool capable of influencing the effect of the Prandtl number on the Nusselt number (Zhao et al. [328]). According to Wei and Ming-Hsiung [315], the Nusselt number increases negligibly with the Prandtl number. Peripheral variation of the relative Nusselt number increases with the Prandtl number (Xin and Ebadian [320]). The wall temperature is a tool capable of influencing the effect of the Prandtl number on the Nusselt number (Zhao et al. [328]).

4.10 Mean Lift Coefficient and Magnetic Field Profile

The results presented by Atif et al. [41] and Tanweer et al. [292] indicate that magnetic field is highly increased with the Prandtl number while mean lift coefficient increases negligibly with the same parameter.

4.11 Local Skin Friction Coefficients

The outcome of scrutinization indicates 38 published instances of a decrease in the local skin friction coefficients due to a growth in the ratio of momentum diffusivity to thermal diffusivity. However, 13 cases of increased local skin friction coefficients due to a higher

Prandtl number were found. Rahman et al. [240], Avinash and Reddy [42], and Ibrahim and Tulu [139] noticed that increasing Prandtl number has no significant effect on the local skin friction coefficients.

4.12 Local Sherwood Number Proportional to Mass Transfer Rate

The movement of species in a less viscous fluid is higher than that of a high viscous fluid due to higher resistance and stronger intermolecular forces. The local Sherwood number is a parameter that quantifies the mass transfer rate where heat and mass transfer are significant. It was only the results of Usman et al. [303] that show that the Prandtl number does not affect the local Sherwood number. When the magnitude of Prandtl number was increased, the viscosity of the fluid also increased, thus reducing the movement of species from the wall with higher concentration (Ramzan et al. [249], Animasaun [30], Avinash et al. [43], Ramzan et al. [249], Avinash and Reddy [42], Sivaraj et al. [284], Buddakkagari and Kumar [71], Atif et al. [41], Shehzad et al. [280], Ghulam Rasool et al. [253], Adegbie et al. [9], Alam et al. [17], Animasaun [29], Ghadami Jadval Ghadam and Moradi [105], Bakier and Gorla [51], Ibrahim and Shankar [138], and Saleem et al. [260]). It is worth noticing that the presence of mass flux due to the temperature gradient and energy flux due to the concentration gradient and the absence of melting heat transfer are two major factors that can cause a rise in the mass transfer rate (Nayak et al. [207], Adegbie et al. [9], Avinash et al. [44], Patel [228], Mustapha et al. [201], Khan and Pop [158], and Azhar et al. [47]).

4.13 Centerline Temperature

Centerline temperature was only confirmed by Das and Mohanty [87] as a decreasing property of Prandtl number. Such an occurrence should not be expected when axial distance ξ is small as the centerline temperature remains constant as P_r increases.

4.14 Spacing Where the Heat Transfer Rate Is at Maximum

Samee AD et al. [267] pointed out that B_{opt}, which denote spacing where the heat transfer rate is at maximum, is a decreasing property of Prandtl number.

4.15 Angular Velocity, Induced Magnetic Field and Average Exit Temperature

The angular velocity is an increasing property of Prandtl number, as noted by Atif et al. [41]. The reverse is the case of induced magnetic field as it had been established by Koriko et al. [162] as a decreasing property of Prandtl number. However, Samee AD et al. [267] confirmed that the average exit temperature decreases with the Prandtl number.

4.16 Concentration and Concentration Gradient

An higher concentration implies an increase in the molecules, thus causing the viscosity to increase till the volume/domain is fully occupied with the molecules. On the contrary, the corresponding increase in the viscosity due to an increase in P_r denotes the addition of dispersed substances, consequently boosting the relative molecular mass negligible and thus capable of causing an increase in the concentration slightly. This is the exact observation of Moorthy and Senthilvadivu [196], Khan et al. [154], Animasaun [29], Lu et al. [180], Animasaun et al. [31,32], Muhammad et al. [199], Avinash and Reddy [42], Azhar et al. [47], Annasagaram et al. Ref39, Animasaun et al. [31,32], Beg et al. [55], Atif et al. [41], Nadeem et al. [203], and Ramesh and Prakash [247]. Next is to explore some reported increases in the concentration due to higher Prandtl number. In the case of homogeneous–heterogeneous chemical reaction (autocatalysis) chemical reaction, Koriko et al. [164] is the only extracted report that established the fact that the concentration of the catalyst is an increasing property of Prandtl number. Ramesh and Prakash [247] reported the highest decrease in the concentration due to a rise in Prandtl number because the flow was induced by sinusoidal peristaltic waves between $Y = -H$ and $Y = +H$ with constant speed c. Hayat et al. [126] only considered $0 \leq P_r \leq 0.3$, which denotes less viscous fluids such as gases of increasing viscosity; see Animasaun et al. [31,32]. Other cases of a decrease in the concentration due to a higher Prandtl number are associated with the existence of thermo-migration of tiny/nano-sized particles and the haphazard motion of tiny/nanosized particles; see Ibrahim and Tulu [139], Ramzan et al. [250], Nadeem et al. [203], Hayat et al. [128], Haq et al. [121], and Shah et al. [276]. Next is to report that changes in the concentration within the domain termed concentration gradient increases with the Prandtl number (Raza et al. [255]). It is realistic to relate these findings to the fact that, "if the Prandtl number of nanoparticle suspensions decreases, then there would be an increase in the concentration of the nanofluid (Wang et al. [311])".

4.17 Displacement Thickness, Drag Force, and Height of the Capillary Ridges

Schneider [274], Rahman et al. [241], and Patra and Panda [229] confirmed that a higher Prandtl number is an approach to boost the displacement thickness, drag force, and height of the capillary ridges. The formation of a viscous boundary layer is predetermined by the Prandtl number (Lam et al. [171]).

4.18 Tutorial Questions

1. In the absence and presence of melting heat transfer, to what extent does rising Prandtl number affects three-dimensional flow?
2. How does increasing the ratio of momentum diffusivity to thermal diffusivity affect the velocity of two-dimensional flow?
3. At what rate does increasing the Prandtl number affect the diffusion of motile microorganisms when Peclet number is small and large in magnitude?

4. What do you know about the impact of Prandtl number on dust temperature and temperature distribution?
5. At various levels of Prandtl number, how does the connection between temperature gradient and heat transfer rate across the domain changes?
6. When Prandtl number is negligibly small and sufficiently large, how do changes in the concentration of fluid substances affect the mass transfer rates?

[illegible]

[illegible]

[illegible]

[illegible]

5

Analysis of Self-Similar Flows I

5.1 Background Information

Any substance capable of moving from one point to another under shear stress, however small, may be gases or liquids. Such substances referred to as fluids are best categorized using the thermo-physical property called viscosity. Some of the acceptable classifications of fluids are Newtonian fluids, viscoelastic fluids, non-Newtonian fluids with time-dependent and time-independent viscosity, and nanofluids. In the first category, Newtonian fluids, the relationship between the shear stress τ and shear rate $\frac{\partial u}{\partial y}$ is linear. In fluid mechanics and rheology, it is far-fetched to present a particular fluid with no viscosity. Such a scientific fact is true because even gases possess a small magnitude of viscosity. However, this was approximated to zero by Leonard Euler and led to a certain governing equation for studying the dynamics of an inviscid fluid. After that, Claude-Louis Navier and George Gabriel Stokes independently presented the governing equation suitable to model viscous fluid flow. The conversion of Navier–Stokes equation to the governing equation for boundary layer flows using the order of magnitude is presented in Appendix I. Based on the conversion, the momentum diffusivity is inevitable in the transport phenomenon of all fluid flows. Likewise, the thermal diffusivity is significant in any fluid flow in which conduction of heat manifests.

In the literature, certain factors may either enhance or retard the dynamics of the fluids. Physically, some may be negligible due to some of the characteristics of the transport phenomenon. Sometimes, some factors are not necessary to be accounted for in the transport process. For instance, in the case of fluid flow on a horizontal surface, the effect of gravitational force is constant on any small volume of the fluid as it flows. Because of that, it is realistic to ignore the effects of buoyancy forces even though the wall temperature is higher than the free stream temperature. This chapter presents the significance of the Prandtl number on the transport phenomena of various fluids. In this chapter, Π_f denotes the rate of change in the velocity along the y-direction with η, $\Pi_{f'}$ denotes the rate of change in the motion along the x-direction with η, $\Pi_{f''}$ denotes the rate of change in the friction across the domain η, and Π_θ denotes the rate of change in the temperature distribution across the domain η.

5.2 Introduction: Stretching-Induced Flows

One of the facts that have been thoroughly described in fluid mechanics is the motion of different fluids within the thin boundary layer formed on surfaces. Such a research objective had led to a countable number of reports on the case of fluid flow due to stretching at the free stream (Blasius [63]) and the wall (Sakiadis [264]). In the case of stagnation-induced flow, stretching at the wall and the free stream is essential. An examination of Blasius and Sakiadis flows with major emphasis on the effect of uniform suction was carried out by

DOI: 10.1201/9781003217374-5

Pantokratoras [225]. It was remarked that the effectiveness of suction is based on stretching velocity that occurs either at the wall (as in the case of Sakiadis) or at the free stream (as in the case of Blasius flow). However, the asymptotic velocity state is reachable for Blasius flow and Sakiadis flow velocity profiles are independent of the Prandtl number in the case of forced convection flow. In the earlier mentioned study by Bataller [82], it was discovered that at all the levels of convective heating (denoted by the parameter a) and thermal radiation (denoted by N_R), the temperature at the wall in the case of Blasius flow is greater than that of Sakiadis flow. But, for less viscous fluids such as gases ($P_r = 0.72$ and $P_r = 1$), it is was established that $\theta_{\text{Blasius}} < \theta_{\text{Sakiadis}}$.

Ahmad et al. [12] compared the skin friction coefficients at various values of volume fraction of nanoparticles in the motion of water conveying (a) Cu, (b) TiO_2, and (c) Al_2O_3 nanoparticles forming three different nanofluids flow on a horizontal surface for Blasius flow and Sakiadis flow. With an increase in the volume fraction within the range $0 \leq \phi \leq 0.018$ for Blasius flow, it was shown that the local skin friction coefficient $Re_x^{1/2}C_f$ increases with volume fraction ϕ. For Cu–water, $Re_x^{1/2}C_f$ increases with ϕ at the rate of 1.728333333. The local skin friction coefficient $Re_x^{1/2}C_f$ in the flow of Al_2O_3–water nanofluid increases with ϕ at the rate of 0.921458333. In the case of water conveying TiO_2 nanoparticles, $Re_x^{1/2}C_f$ increases with ϕ at the rate of 0.967916667. In addition, with an increase in the volume fraction within the range $0 \leq \phi \leq 0.018$ for Sakiadis flow, it was observed that in the motion of Cu–water, Al_2O_3–water, and TiO_2–water, $Re_x^{1/2}C_f$ decreases with ϕ at the rates of -2.3, -1.234583333, and -1.295, respectively. It is worth remarking that for Blasius flow, the maximum local skin friction coefficient $Re_x^{1/2}C_f$ is ascertained at the higher levels of volume fraction of nanoparticles ϕ in the dynamics of water conveying Cu, while the most minimum local skin friction coefficients $Re_x^{1/2}C_f$ was for the case of Cu–water due to stretching at the wall (i.e., Sakiadis flow).

Pantokratoras [224] adopted the finite difference method to examine the Blasius flow and Sakiadis flow of a non-Newtonian Carreau fluid on a horizontal surface. It was concluded that the wall shear stress for shear thinning fluids decreases with longitudinal Reynolds X at the rate of -0.073421463 in the Blasius flow but increases with Reynolds X at the rate of 0.098234797 in the Sakiadis flow. The rate of decrease in wall shear stress during the Blasius flow and the increase in the same property with longitudinal Reynolds X are -0.013725528 and 0.01651187, respectively. It was also remarked that the transport phenomenon increases in the Sakiadis flow and diminishes in the Blasius flow with increasing longitudinal Reynolds X. More so, the wall shear stress is the same in Sakiadis and Blasius flows (Pantokratoras [224]). Naveed et al. [206] deliberated on the significance of thermal radiation on the two-dimensional flow of viscous fluid on the curved surface when the flow was induced by stretching at the wall and the free stream. It was shown that the Nusselt number and temperature distribution for Sakiadis flow are more enhanced than those of the Blasius flow.

Nanofluid is the name given to the colloidal mixture of a specific base fluid and nanometer-sized particles made of oxides, carbon nanotubes, metals, and carbides. Due to the intricate usefulness of nanofluids in many applications where the performance is strongly dependent on heat transfer, water, ethylene glycol, and oil are frequently used as the base fluids. The literature has proved nanofluid applications in pharmaceutical companies, hybrid-powered engines, thermal management of the engine, production of microchips, server cooling, and heat exchanger. These facts may be true because one of the outcomes of a review on nanofluids by Mohamoud Jama et al. [147] shows that with an increase in the volume fraction of the nanoparticles density, viscosity, and thermal conductivity, nanofluids are bound to enhancement. A two-step method was used by Hemmat Esfe et al. [130] to produce water-based Cu/TiO2 and ethylene glycol-based Cu/TiO2 hybrid nanofluids.

First, a mechanical mixture of the base fluid and nanoparticles was considered. Second, a magnetic stirrer of the mixture was done for 3 hours. Third, the suspension produced was poured into a 20 kHz, 400 W ultrasonic processor manufactured by Topsonic, Iran, and processed for 6 hours. Quality dispersion, minimization of agglomeration, sedimentation removal, and stable suspension production for goods are achievable after 7 days through the procedures mentioned above. According to Hemmat Esfe [130], changing the pH value, adding surface activators (surfactants), and using ultrasonic vibrations are well-known methods for establishing stable nanofluids. According to Babar and Ali [48], some of the methods of nanoparticles dispersion are single-step and two-step methods.

5.3 Fluid Flow due to Stretching

Consider the flow of an incompressible viscous fluid and nanofluid on a horizontal surface within the domain $0 \leq y < \infty$, where the effect of viscosity is highly significant; see Figure 5.1. It is assumed that the flow on the non-porous horizontal surface was induced by viscous forces, and either stretching at the free stream $U_\infty = U_o x$ (i.e., Blasius flow) or stretching at the wall $U_w = U_o x$ (i.e., Sakiadis flow) with the stretching velocity $U_\infty = U_o x$. For $q\epsilon[0, 1]$, the governing equation that models the transport phenomenon for just an ordinary Newtonian fluid $q = 0$ and for a nanofluid $q = 1$ is

$$u_x + v_y = 0, \tag{5.1}$$

$$uu_x + vu_y = (1-q)\frac{\mu}{\rho}u_{yy} + q\frac{\mu_{nf}}{\rho_{nf}}u_{yy}, \tag{5.2}$$

$$uT_x + vT_y = (1-q)\frac{\kappa}{\rho c_p}T_{yy} + q\frac{\kappa_{nf}}{(\rho c_p)_{hnf}}T_{yy}. \tag{5.3}$$

For the case of Blasius flow, Eqs. (5.1)–(5.3) are subject to the boundary conditions

$$u = 0, \quad v = 0, \quad T = T_w \quad at \quad y = 0. \tag{5.4}$$

$$u \to U_\infty, \quad T \to T_\infty, \quad as \quad y \to \infty \tag{5.5}$$

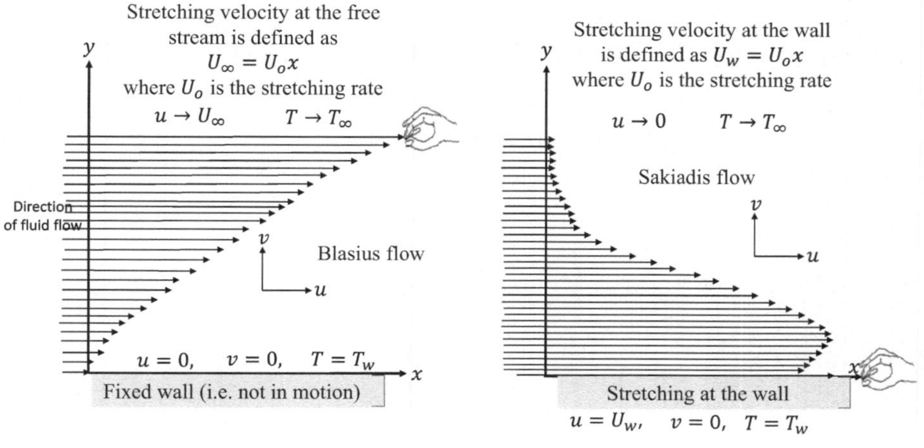

FIGURE 5.1
Illustration of fluid flow due to stretching.

For the case of Sakiadis flow, Eqs. (5.1)–(5.3) are subject to the boundary conditions

$$u = U_w, \quad v = 0, \quad T = T_w \quad at \quad y = 0. \tag{5.6}$$

$$u \to 0, \quad T \to T_\infty, \quad as \quad y \to \infty \tag{5.7}$$

For the case of $q = 1$, the model for viscosity derived by Graham [114] and refined by Gosukonda et al. [112]

$$\frac{\mu_{nf}}{\mu_{bf}} = 1 + 2.5\phi + 4.5\left[\frac{1}{\frac{h}{d_p}\left(2+\frac{h}{d_p}\right)\left(1+\frac{h}{d_p}\right)^2}\right] \tag{5.8}$$

where the radius of nanoparticles is d_p and the inter-particle spacing is h, was adopted. The thermal conductivity proposed by Maxwell [193] for nanofluid is

$$\frac{\kappa_{nf}}{\kappa_{bf}} = \frac{\kappa_{sp} + 2\kappa_{bf} - 2\phi(\kappa_{bf} - \kappa_{sp})}{\kappa_{sp} + 2\kappa_{bf} + \phi(\kappa_{bf} - \kappa_{sp})} \tag{5.9}$$

Following Pak and Cho [221] and Das et al. [88], the density of nanofluid is

$$\rho_{nf} = (1-\phi)\rho_{bf} + \phi\rho_{sp}. \tag{5.10}$$

Xuan and Roetzel [322] presented the heat capacity of nanofluid as

$$(\rho c_p)_{nf} = (1-\phi)(\rho c_p)_{bf} + \phi(\rho c_p)_{sp}. \tag{5.11}$$

The skin friction coefficient C_{fx} and Nusselt number Nu_x for the two dynamics of both fluids are defined as

$$C_{fx} = (1-q)\frac{\mu}{\rho U_o^2 x^2}\frac{\partial u}{\partial y} + q\frac{\mu_{nf}}{\rho_{bf} U_o^2 x^2}\frac{\partial u}{\partial y},$$

$$Nu_x = (1-q)\frac{-x}{\kappa(T_w - T_\infty)}\kappa\frac{\partial T}{\partial y} + q\frac{-x\kappa_{nf}}{\kappa_{bf}(T_w - T_\infty)}\frac{\partial T}{\partial y} \tag{5.12}$$

Table 5.1 presents the details of the nanoparticles and base fluids of all the nanofluids that was considered. Shell Chemicals [282] presents the viscosity of methanol at 20°C as 0.00059 Pa s. The viscosity of water at 25°C is 0.0008905 Pa s as reported by Wakif et al. [310]. The viscosity of blood is 0.0033 Pa s (Jahangiri et al. [146]). In order to obtain self-similar solution of Eqs. (5.1)–(5.3) subject to Eqs. (5.4)–(5.7), the following similarity variables were used:

$$\eta = y\sqrt{\frac{U_\infty}{\vartheta x}}, \quad \psi(x,y) = \sqrt{\vartheta x U_\infty} f(\eta), \quad \eta = y\sqrt{\frac{U_\infty}{\vartheta_{bf} x}},$$

$$\psi(x,y) = \sqrt{\vartheta_{bf} x U_\infty} f(\eta), \quad u = \frac{\partial \psi}{\partial y}, \quad R_e = \frac{U_\infty x}{\vartheta_{bf}}, \quad v = -\frac{\partial \psi}{\partial x},$$

$$\theta(\eta) = \frac{T - T_\infty}{T_w - T_\infty}, \quad P_r = \frac{\mu c_p}{\kappa} = \frac{\vartheta}{\alpha}, \quad P_r = \frac{(\mu c_p)_{bf}}{\kappa_{bf}} = \frac{\vartheta_{bf}}{\alpha_{bf}}.$$

$$A_1 = 1 + 2.5\phi + 4.5\left[\frac{1}{\frac{h}{d_p}\left(2+\frac{h}{d_p}\right)\left(1+\frac{h}{d_p}\right)^2}\right], \quad A_2 = 1 - \phi + \phi\frac{\rho_{sp}}{\rho_{bf}},$$

$$A_3 = \frac{\kappa_{sp} + 2\kappa_{bf} - 2\phi(\kappa_{bf} - \kappa_{sp})}{\kappa_{sp} + 2\kappa_{bf} + \phi(\kappa_{bf} - \kappa_{sp})}, \quad A_4 = 1 - \phi + \phi\frac{(\rho c_p)_{sp}}{(\rho c_p)_{bf}} \tag{5.13}$$

TABLE 5.1
Density, Thermal Conductivity, and Specific Heat Capacity of Some Base Fluids and Nanoparticles

Source	**Base Fluid**	ρ (kg/m^{-3})	κ (W/mK)	c_p (J/kg K)	$P_r = \frac{\mu c_p}{\kappa}$
Tlili et al. [296]	Methanol	792	0.2035	2,545	7.3786
Nayak et al. [209]	Water H_2O	997.1	0.613	4,180	6.1723
Koriko et al. [164]	Blood	1,050	0.52	3,617	22.9540
Source	**Nanoparticles**	ρ (kg/m^{-3})	κ (W/mK)	c_p (J/kgK)	
Abbas et al. [3]	MWCNT	1,600	3000	796	
Selimefendigil et al. [275]	SiO_2	2,200	1.2	703	
Mahanthesh et al. [184]	SWCNT	2,600	6600	425	
Nayak et al. [209]	Al	2,702	237	903	
Tlili et al. [296]	MgO	3,580	48.4	960	
Acharya et al. [7]	TiO_2	4,250	8.954	686.2	
Sandeep et al. [202]	Fe_3O_4	5,180	9.7	670	
Acharya et al. [7]	CuO	6,320	76.5	531.80	
Nayak et al. [209]	Zn	7,140	116	389	
Xu and Chen [321]	Cu	8,933	400	385	

The final dimensionless governing equation is

$$(1-q)\frac{d^3 f}{d\eta^3} + q\frac{A_1}{A_2}\frac{d^3 f}{d\eta^3} + \frac{1}{2}f\frac{d^2 f}{d\eta^2} = 0, \tag{5.14}$$

$$(1-q)\frac{d^2\theta}{d\eta^2} + q\frac{A_3}{A_4}\frac{d^2\theta}{d\eta^2} + \frac{1}{2}P_r f\frac{d\theta}{d\eta} = 0. \tag{5.15}$$

Dimensionless boundary conditions for Blasius flow Eqs. (5.4)–(5.5) are

$$\frac{df}{d\eta} = 0, \quad f = 0, \quad \theta = 1 \quad at \quad \eta = 0. \tag{5.16}$$

$$\frac{df}{d\eta} \to 1, \quad \theta \to 0 \quad as \quad \eta \to \infty. \tag{5.17}$$

Dimensionless boundary conditions for Sakiadis flow Eqs. (5.6)–(5.7) are

$$\frac{df}{d\eta} = 1, \quad f = 0, \quad \theta = 1 \quad at \quad \eta = 0. \tag{5.18}$$

$$\frac{df}{d\eta} \to 0, \quad \theta \to 0 \quad as \quad \eta \to \infty. \tag{5.19}$$

For $q = 0$, upon substituting the variables Eq. (5.13) into Eq. (5.12), the dimensionless physical properties are

$$C_{fx}\sqrt{Re_x} = f''(0), \qquad \frac{Nu_x}{\sqrt{Re_x}} = -\theta'(0). \tag{5.20}$$

For $q = 1$, upon substituting the variables in Eq. (5.13) into Eq. (5.12), we obtain

$$\frac{C_{fx}\sqrt{Re_x}}{A_1} = f''(0), \qquad \frac{Nu_x}{A_3\sqrt{Re_x}} = -\theta'(0). \tag{5.21}$$

5.3.1 Research Questions I

For the first case of ordinary fluid ($q = 0$), the aims and objectives of this case are to provide answers to the following research questions:

1. What is the effect of increasing the ratio of momentum diffusivity to thermal diffusivity in the Blasius and Sakiadis flows over a horizontal surface?
2. What is the pattern of decrease in the temperature distribution with the Prandtl number when stretching induced the fluid flow at the wall and the free stream?
3. Do the local skin friction coefficient $f''(0)$ and Nusselt number $-\theta'(0)$ for Blasius flow and Sakiadis flow vary with the chosen η at infinity?

   For the dynamics of nanofluids ($q = 1$), this study provides answers to the following research questions:
4. What are the variations in the skin friction coefficients and Nusselt number for Blasius and Sakiadis flows of (i) methanol, (ii) water, and (iii) blood when the radius of aluminum nanoparticles is small and large in magnitude?
5. What are the increasing effects of nanoparticles radius of (i) muti-walled CNT, (ii) silicon dioxide (SiO_2), (iii) single-walled CNT (SWCNT), (iv) aluminium (Al), (v) magnesium oxide (MgO), (vi) titanium dioxide (TiO_2), (vii) iron(III)oxide (Fe_3O_4), (viii) copper(II)oxide (CuO), (ix)zinc (Zn), and (x) copper (Cu) conveyed by methanol in Blasius and Sakiadis flows?
6. How does the distance between copper nanoparticles within blood nanofluid for Blasius and Sakiadis flows affect the velocity and local skin friction coefficients?

5.3.2 Analysis and Discussion of Results for Nanofluids ($q = 1$)

Stretching at the free stream (Blasius flow) and that at the wall (Sakiadis flow) are two possibilities of fluid flow due to stretching. The numerical solution of dimensionless governing Eqs. (5.14) and (5.15) subject to Eqs. (5.16)–(5.17) and Eq. (5.18) – Eq. (5.19) was obtained using the four-stage Lobatto IIIa formula as described in Chapter 1 using $\eta = 5$. The data presented in Table 5.2 and Table 5.3 show that the local skin friction coefficients at all the considered Prandtl number values for (i) methanol, (ii) water, and (iii) blood are higher in the case of Blasius flow due to the decreasing impact of stretching starting from the free stream and ending at the wall. The larger radius of

TABLE 5.2
Variations in Skin Friction Coefficients and Nusselt Number for Blasius and Sakiadis Flows of (i) Methanol, (ii) Water, and (iii) Blood at $h = 0.1$, $\phi = 0.1$ When the Radius of Aluminum Nanoparticles Is Small ($d_p = 0.1$)

$d_p = 0.1$	**Blasius**	**Flow**	**Sakiadis**	**Flow**
	$f''(0)$	$-\theta'(0)$	$f''(0)$	$-\theta'(0)$
Methanol $P_r = 7.3786$	0.290202104	0.575236590	−0.387814949	1.249538876
Water $P_r = 6.1723$	0.281878673	0.525535462	−0.376691913	1.099429339
Blood $P_r = 22.9540$	0.280230657	0.816258373	−0.374489200	2.215981825
S_{lp}	−0.000328959	0.016536475	0.000439629	0.064624617

aluminum nanoparticles A_l was found to be a yardstick capable of declining the local skin friction coefficients for any of the three base fluids. Physically, as the radius of nanoparticles increases, the ratio of area to volume decreases because the volume would be larger than the area. This may be the reason why $\frac{C_{fx}\sqrt{Re_x}}{A_1}$ for $d_p = 2.5$ is very low compared to the case $d_p = 0.1$ in Table 5.2. Tables 5.4–5.13 depict the variations in the skin friction coefficients $\frac{C_{fx}\sqrt{Re_x}}{A_1}$ and Nusselt number $\frac{Nu_x}{A_3\sqrt{Re_x}}$ for Blasius and Sakiadis flows of methanol conveying ten different nanoparticles of increasing densities

TABLE 5.3
Variations in Skin Friction Coefficients and Nusselt Number for Blasius and Sakiadis Flow of (i) Methanol, (ii) Water, and (iii) Blood at $h = 0.1$, $\phi = 0.1$ When the Radius of Aluminum Nanoparticles Is Large ($d_p = 2.5$)

$d_p = 2.5$	Blasius	Flow	Sakiadis	Flow
	$f''(0)$	$-\theta'(0)$	$f''(0)$	$-\theta'(0)$
Methanol $P_r = 7.3786$	0.052406210	0.326073745	−0.070598663	1.326333327
Water $P_r = 6.1723$	0.051191683	0.298629534	−0.068842418	1.174805585
Blood $P_r = 22.9540$	0.050955566	0.462860808	−0.068496839	2.286106572
S_{lp}	$-4.77325E-05$	0.009358846	$6.92779E-05$	0.064262602

TABLE 5.4
Variations in Skin Friction Coefficients and Nusselt Number for Blasius and Sakiadis Flows of Methanol Conveying Multiple-Walled (MWCNT) Nanoparticles of Different Radii When $\phi = 0.3$, $h = 0.1$, and $P_r = 7.3786$

(i) Multiple	Blasius	Flow	Sakiadis	Flow
Wall CNT	$f''(0)$	$-\theta'(0)$	$f''(0)$	$-\theta'(0)$
$d_p = 0.1$	0.260324671	0.441899218	−0.347887841	0.871375056
$d_p = 1$	0.086026707	0.306772696	−0.115212609	0.930486192
$d_p = 2$	0.059251719	0.270993552	−0.080026856	0.938585210
S_{lp}	−0.104369329	−0.088950882	0.139027552	0.034870541

TABLE 5.5
Variations in Skin Friction Coefficients and Nusselt Number for Blasius and Sakiadis Flows of Methanol Conveying Silicon Dioxide (SiO_2) Nanoparticles of Different Radii When $\phi = 0.3$, $h = 0.1$, and $P_r = 7.3786$

(ii) Silicon	Blasius	Flow	Sakiadis	Flow
dioxide SiO_2	$f''(0)$	$-\theta'(0)$	$f''(0)$	$-\theta'(0)$
$d_p = 0.1$	0.282066751	0.510271928	−0.376943249	1.047628660
$d_p = 1$	0.093210632	0.353992438	−0.124722167	1.110804815
$d_p = 2$	0.063940842	0.312270931	−0.086236243	1.119601951
S_{lp}	−0.113225004	−0.103058107	0.150890814	0.037344083

TABLE 5.6
Variations in Skin Friction Coefficients and Nusselt Number for Blasius and Sakiadis Flows of Methanol Conveying Single-Walled CNT (SWCNT) Nanoparticles of Different Radii When $\phi = 0.3$, $h = 0.1$, and $P_r = 7.3786$

(iii) Single	**Blasius**	**Flow**	**Sakiadis**	**Flow**
CNT *SWCNT*	$f''(0)$	$-\theta'(0)$	$f''(0)$	$-\theta'(0)$
$d_p = 0.1$	0.295674622	0.455998237	−0.395128323	0.844474675
$d_p = 1$	0.097707331	0.316970854	−0.130691792	0.913203777
$d_p = 2$	0.066943013	0.279535566	−0.090171624	0.922637837
S_{lp}	−0.118731538	−0.091852215	0.158289811	0.040553553

TABLE 5.7
Variations in Skin Friction Coefficients and Nusselt Number for Blasius and Sakiadis Flows of Methanol Conveying Aluminum (Al) Nanoparticles of Different Radii When $\phi = 0.3$, $h = 0.1$, and $P_r = 7.3786$

(i) Aluminium	**Blasius**	**Flow**	**Sakiadis**	**Flow**
Al	$f''(0)$	$-\theta'(0)$	$f''(0)$	$-\theta'(0)$
$d_p = 0.1$	0.299045624	0.491224013	−0.399633138	0.948307068
$d_p = 1$	0.098821267	0.341152060	−0.132172227	1.016773193
$d_p = 2$	0.067692043	0.300821377	−0.091151387	1.026251513
S_{lp}	−0.120092795	−0.099107094	0.160120108	0.04044138

TABLE 5.8
Variations in Skin Friction Coefficients and Nusselt Number for Blasius and Sakiadis Flows of Methanol Conveying Magnesium Oxide (MgO) Nanoparticles of Different Radii When $\phi = 0.3$, $h = 0.1$, and $P_r = 7.3786$

(v) Magnesium	**Blasius**	**Flow**	**Sakiadis**	**Flow**
Oxide MgO	$f''(0)$	$-\theta'(0)$	$f''(0)$	$-\theta'(0)$
$d_p = 0.1$	0.32662663	0.52970701	−0.43649157	1.01410621
$d_p = 1$	0.10793556	0.36794080	−0.14430291	1.08913123
$d_p = 2$	0.07386891	0.32434971	−0.09922759	1.09956033
S_{lp}	−0.13120447	−0.10689290	0.17506396	0.04433845

TABLE 5.9
Variations in Skin Friction Coefficients and Nusselt Number for Blasius and Sakiadis Flows of Methanol Conveying Titanium Dioxide (TiO_2) Nanoparticles of Different Radii When $\phi = 0.3$, $h = 0.1$, and $P_r = 7.3786$

(vi) Titanium	**Blasius**	**Flow**	**Sakiadis**	**Flow**
Dioxide TiO_2	$f''(0)$	$-\theta'(0)$	$f''(0)$	$-\theta'(0)$
$d_p = 0.1$	0.346198838	0.535405172	−0.462647017	0.993370677
$d_p = 1$	0.114403251	0.372151285	−0.152924871	1.073814365
$d_p = 2$	0.078280267	0.328055794	−0.105010155	1.084965561
S_{lp}	−0.139074594	−0.10793133	0.185641086	0.047524132

TABLE 5.10
Variations in Skin Friction Coefficients and Nusselt Number for Blasius and Sakiadis Flows of Methanol Conveying Iron(III)Oxide (Fe_3O_4) Nanoparticles of Different Radii When $\phi = 0.3$, $h = 0.1$, and $P_r = 7.3786$

(vii) Iron(iii)Oxide	Blasius	Flow	Sakiadis	Flow
Fe_3O_4	$f''(0)$	$-\theta'(0)$	$f''(0)$	$-\theta'(0)$
$d_p = 0.1$	0.37166156	0.56022655	−0.49667432	1.02168615
$d_p = 1$	0.12281758	0.38956222	−0.16423521	1.10858103
$d_p = 2$	0.08408836	0.34348244	−0.11311750	1.12051764
S_{lp}	−0.14927636	−0.1128213	0.19909056	0.05127709

TABLE 5.11
Variations in Skin Friction Coefficients and Nusselt Number for Blasius and Sakiadis Flows of Methanol Conveying Copper(II) Oxide (CuO) Nanoparticles of Different Radii When $\phi = 0.3$, $h = 0.1$, and $P_r = 7.3786$

(viii) Copper(II)	Blasius	Flow	Sakiadis	Flow
Oxide CuO	$f''(0)$	$-\theta'(0)$	$f''(0)$	$-\theta'(0)$
$d_p = 0.1$	0.40067268	0.56308952	−0.53544368	0.97868983
$d_p = 1$	0.13240442	0.39204582	−0.17700382	1.07406545
$d_p = 2$	0.09060634	0.34564463	−0.12164452	1.08711815
S_{lp}	−0.16095305	−0.11318926	0.21479217	0.0562554

TABLE 5.12
Variations in Skin Friction Coefficients and Nusselt Number for Blasius and Sakiadis Flows of Methanol Conveying Zinc (Zn) Nanoparticles of Different Radii When $\phi = 0.3$, $h = 0.1$, and $P_r = 7.3786$

(ix) Zinc	Blasius	Flow	Sakiadis	Flow
Zn	$f''(0)$	$-\theta'(0)$	$f''(0)$	$-\theta'(0)$
$d_p = 0.1$	0.420303886	0.556797883	−0.561678127	0.928044322
$d_p = 1$	0.138891668	0.388151546	−0.185654274	1.029649695
$d_p = 2$	0.095034508	0.342228469	−0.127454679	1.043492739
S_{lp}	−0.168845014	−0.111694957	0.22539587	0.059896655

TABLE 5.13
Variations in Skin Friction Coefficients and Nusselt Number for Blasius and Sakiadis Flows of Methanol Conveying Copper (Cu) Nanoparticles of Different Radii When $\phi = 0.3$, $h = 0.1$, and $P_r = 7.3786$

(x)Copper	Blasius	Flow	Sakiadis	Flow
Cu	$f''(0)$	$-\theta'(0)$	$f''(0)$	$-\theta'(0)$
$d_p = 0.1$	0.460321644	0.589247041	−0.615156066	0.957919750
$d_p = 1$	0.152115753	0.411108902	−0.203423046	1.070179894
$d_p = 2$	0.104158860	0.362579445	−0.140153159	1.085313600
S_{lp}	−0.184880348	−0.117993029	0.246556297	0.066091541

as presented in Table 5.1. In all the ten cases of boundary layer flow of nanofluids, due to higher radius of nanoparticles, an optimal increase in local skin friction coefficients is seen in the case of methanol conveying copper nanoparticles induced by stretching at the wall; see Table 5.13, where $S_{lp} = 0.246556297$.

For incompressible nanofluids, the change in base fluid's density is either constant or zero. However, the relationship between the nanoparticles' radius and the nanoparticles' density can be noted herein to play a significant role that affects the local skin friction coefficients. Figures 5.2 and 5.3 show that the velocity of blood conveying copper nanoparticles whose radius $d_p = 0.8$ increases with higher nanoparticle spacing for Blasius flow. Besides, the same vertical velocity is a decreasing property of h. This observation can be associated with

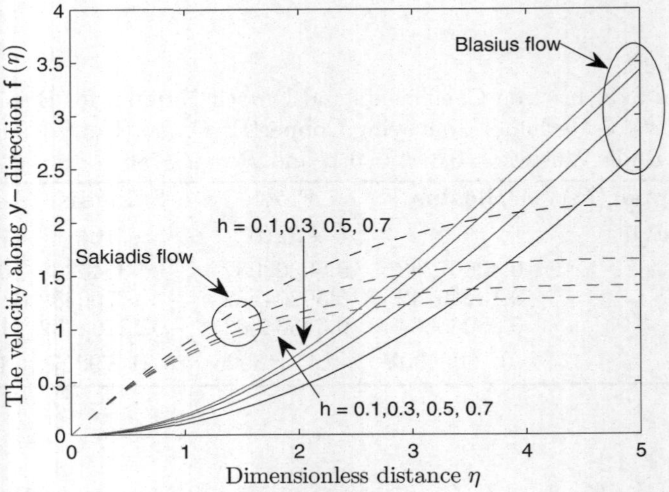

FIGURE 5.2
Variation in $f(\eta)$ due to h when $\phi = 0.3$, $d_p = 0.8$, $P_r = 22.9540$.

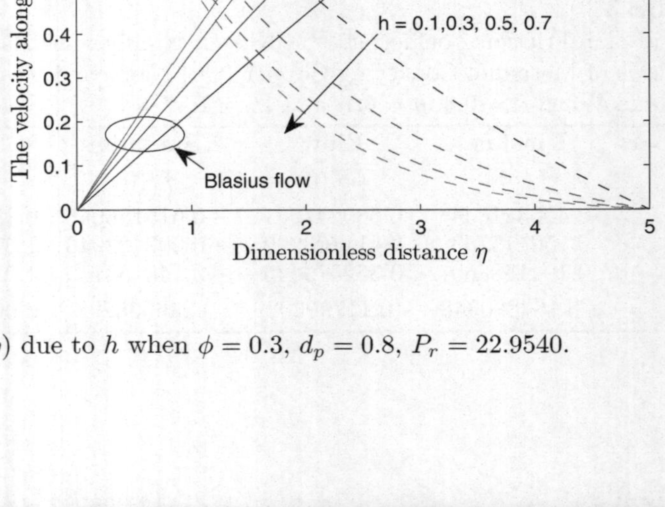

FIGURE 5.3
Variation in $f'(\eta)$ due to h when $\phi = 0.3$, $d_p = 0.8$, $P_r = 22.9540$.

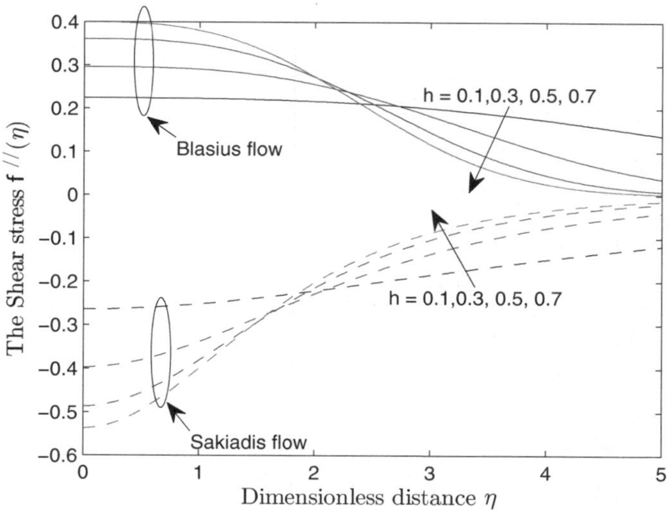

FIGURE 5.4
Variation in $f''(\eta)$ due to h when $\phi = 0.3$, $d_p = 0.8$, $P_r = 22.9540$.

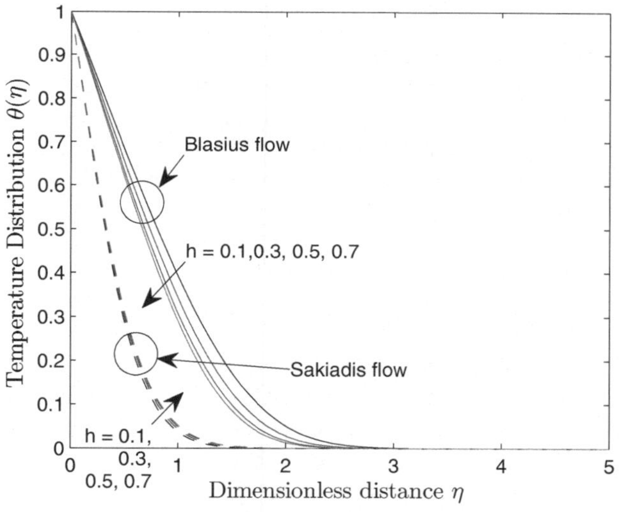

FIGURE 5.5
Variation in $\theta(\eta)$ due to h when $\phi = 0.3$, $d_p = 0.8$, $P_r = 22.9540$.

the fact that at various values of nanoparticles spacing, stretching at the free stream (Blasius flow) enhances vertical and horizontal velocities, while stretching at the wall deteriorates the same velocity along the x- and y-directions. Due to the relationship between velocity and shear stress across the layers, Figure 5.4 further reveals that the increase in the velocity at h grows for the case of Blasius flow and the shear stress also increases for the same category. When velocity is increasing, heat energy is expected to be transferred due to the conservation of energy. Such a manifestation is the major reason why the temperature distribution, as in the case of Blasius flow, decreases with h; see Figures 5.5 and 5.6. The analysis presented by Blass et al. (2021) reads, "Note that the viscous boundary layer thickness

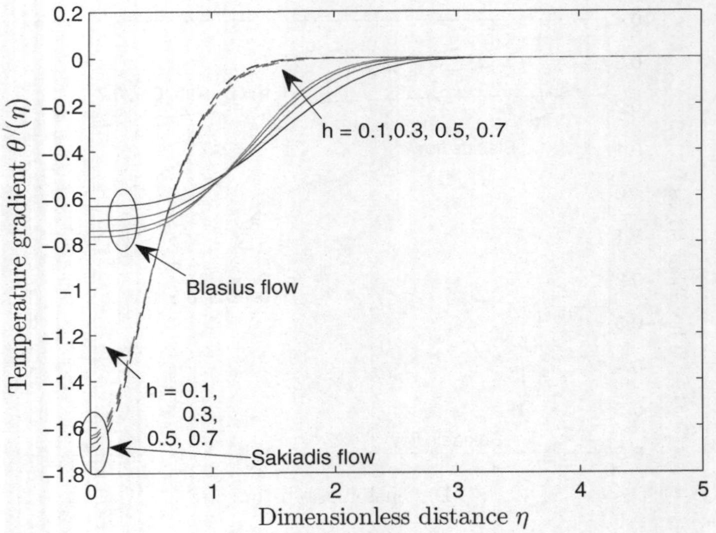

FIGURE 5.6
Variation in $\theta'(\eta)$ due to h when $\phi = 0.3$, $d_p = 0.8$, $P_r = 22.9540$.

has a stronger dependence on P_r than the thermal boundary layer thickness. Qualitatively, larger P_r reflects stronger momentum diffusivity and therefore a thicker viscous boundary layer." Based on this, it seems accurate to accept that indeed, "higher Prandtl number for $Pr \gg 1$ implies increasing momentum diffusivity when thermal diffusivity is fixed." The case of Sakiadis flow of water conveying alumina nanoparticles with an emphasis on the increasing particle radius of nanoparticles, inter-particle spacing, and volume fraction is presented in Section 5.4.

5.3.3 Analysis and Discussion of Results for Ordinary Fluids ($q = 0$)

It is shown in Table (5.14) that the local skin friction is a constant function of Prandtl number P_r in the absence of suction and injection. In most fluid flows, the variation in shear stress, mathematically denoted by $f''(\eta)$ proportional to friction between layers across the fluid is strongly determined by the nature of contact between the horizontal wall and the last layer of fluid $f''(0)$. It is seen in Table (5.14), Figures 5.13–5.16 that the temperature distribution of the fluid in both cases (Blasius flow and Sakiadis flow) is a decreasing property of Prandtl number. However, the minimum decrease in the temperature due to an increase in P_r occurs in the case of Sakiadis flow, where S_{lp} was estimated as -0.0073. This study also indicates that the Nusselt number $-\theta'(0)$ increases with the Prandtl number at the rate of 0.0125 for Blasius flow and 0.0508 for Sakiadis flow. The comparative analysis shows that the maximum increase in the Nusselt number, which is proportional to the heat transfer rate, is ascertained in the Sakiadis flow. These results corroborate with one of the findings by Olanrewaju et al. [218]. For the case of Blasius flow, the rate of increase in the Nusselt number with the Prandtl number is 0.0125, while the rate of increase in the same property with P_r for Sakiadis flow is 0.0508. This implies that an increase in $-\theta'(0)$ with P_r can be enhanced up to 306.4% from Blasius flow to Sakiadis flow. Uddin et al. [300] once concluded that the friction factor for the Sakiadis flow was found to be higher than that for the Blasius flow. More so, heat transfer rates for the Sakiadis flow are lower than that for the Blasius flow.

TABLE 5.14
Effect of Prandtl Number P_r on Some of the Properties of Blasius Flow and Sakiadis Flow of an Ordinary Fluid ($q = 0$) on a Non-Porous Horizontal Surface

	Blasius	**Flow**			**Sakiadis**	**Flow**
P_r	$f''(0)$	$\theta(0.8333)$	$-\theta'(0)$	$f''(0)$	$\theta(0.5556)$	$-\theta'(0)$
0.1	0.3362	0.8201	0.2160	−0.4540	0.8761	0.2236
0.3	0.3362	0.7947	0.2467	−0.4540	0.8493	0.2732
0.71	0.3362	0.7486	0.3026	−0.4540	0.7926	0.3797
1	0.3362	0.721	0.3362	−0.4540	0.7537	0.4540
6	0.3362	0.499	0.6158	−0.4540	0.3834	1.2725
10	0.3362	0.4143	0.7311	−0.4540	0.2518	1.6776
100	0.3362	0.0413	1.5783	−0.4540	0.0002	5.5424
S_{lp}	0	−0.0068	0.0125	0	−0.0073	0.0508

TABLE 5.15
The Rate of Decrease in $\theta(\eta)$ with η at Various Values of Prandtl Number for the Case of Blasius Flow

P_r	0.1	0.3	0.71	1	6	10	100
S_{lp}	−0.2018	−0.2042	−0.2039	−0.2039	−0.1588	−0.1476	−0.0955

TABLE 5.16
The Rate of Decrease in $\theta(\eta)$ with η at Various Values of Prandtl Number for the Case of Sakiadis Flow

P_r	0.1	0.3	0.71	1	6	10	100
S_{lp}	−0.2015	−0.2034	−0.2035	−0.1987	−0.1447	−0.1249	−0.0565

It is worth deducing from Tables 5.15 and 5.16 that the rate at which the temperature distribution decreases across the space η is very fast in less viscous fluids (as in the case of $P_r = 0.1$) compared to high viscous fluids (as in the case of $P_r = 100$). Both tables further present a significant difference in the rate of decrease in the temperature distribution across space with the Prandtl number. Since $f''(0)$ is a constant property of P_r, the answer could be sought for at any value of P_r. However, $P_r = 0.1$ was used for the analysis of Blasius and Sakiadis flows. It is evident as shown in Table 5.17 that the local skin friction coefficient $f''(0)$ and Nusselt number $-\theta'(0)$ for Blasius flow and Sakiadis flow vary with the chosen η at infinity. The answer to the third research question is yes. The outcome of the analysis for this case as illustrated in Figures 5.7–5.10 shows that the Prandtl number does not affect the motion along the y-direction and x-direction due to the stretching of fluid layers at the wall and the free stream. Exploration of the Blasius flow shows that stretching of fluid layers at the free stream $f(5)$ is estimated as 3.3171. It is seen that $f(\eta)$ increases with η at the rate of 0.689169129; see Figure 5.7. The case of Sakiadis flow was also explored, and it is seen in Figure 5.8 that $f(5)$ is estimated as 1.4874. The slope linear regression through the data points shows that the rate of increase in the vertical velocity, which is the same at all the values of Prandtl number, is 0.305023856.

However, as shown in Figure 5.9, $f'(\eta)$ increases across the domain at the rate of 0.208327041. The reverse is the case in Sakiadis flow, where the same property of the fluid flow decreases across the flow at the rate of −0.213011833; see Figure 5.10. It can be easily deduced from Figure 5.11 that the shear stress decreases from the wall till the free stream at the rate of −0.076547973. In the case of fluid flow at the wall due to stretching

TABLE 5.17
Variation in $f''(0)$ and $-\theta'(0)$ for Blasius and Sakiadis Flows as $\eta_\infty \to \infty$

η_∞	$f''(0)$ Sakiadis Flow	$-\theta'(0)$ Sakiadis Flow	$f''(0)$ Blasius Flow	$-\theta'(0)$ Blasius Flow
0.5	−2.0312037241	2.0031132370	2.0104570458	2.0010464163
1	−1.0621046366	1.0061574602	1.0211567246919	1.0021217009703
1.5	−0.7589902266	0.6757357750	0.6990130376535	0.6699229945352
2.5	−0.5489984728	0.4143116604	0.4571671242162	0.4058447904635
3.5	−0.4835568960	0.3043553913	0.3721835139903	0.2948507112820
4.5	−0.4596784759	0.2443284823	0.3417779278078	0.2356627758382
5.0	−0.4539701684	0.2235657450	0.3361523452754	0.2159962151480
6.0	−0.4480456416	0.1927195255	0.3325660058200	0.1883966200153
7.0	−0.4455675560	0.1709533384	0.3320965311445	0.1708144055777
8.0	−0.4445515738	0.1548115229	0.3320591874353	0.1593906973450
9.0	−0.4441013493	0.1424017050	0.3320572642608	0.1519581742340
12	−0.4437792355	0.1181492834	0.3320576348293	0.1423522817586
14	−0.4437542989	0.1081238640	0.3320573679630	0.1406730470317
16	−0.4437496303	0.1008420705	0.3320573887349	0.1401790161985
18	−0.4437488192	0.0953768166	0.3320573841066	0.1400583347004
25	−0.4437506385	0.0841445938	0.3320573351713	0.1400294285350
50	−0.4437514587	0.0741819338	0.3320562142211	0.1400212485695
100	−0.4437504969	0.0728852117	0.3320569625784	0.1400219872563
200	−0.4437504376	0.0728624173	0.3320569652041	0.1400219896321

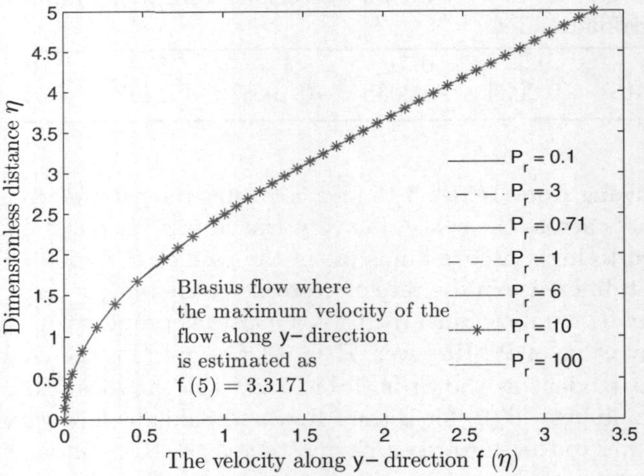

FIGURE 5.7
Velocity of Blasius flow along the y-direction at various values of Prandtl number in the absence of suction and injection.

$U_w = U_o x$ illustrated in Figure 5.12, the shear stress increases with the Prandtl number at the rate of 0.101082224. Meanwhile, it is important to remark that the temperature gradient $\theta'(\eta)$ decreases significantly near the wall and increases moderately near the free stream. For brevity, these results are not illustrated, but noteworthy of reporting. The comparative analysis between Figures 5.13 and 5.14 shows that the observed maximum temperature at the wall in the case of Sakiadis flow is very small as the viscosity of the fluid increases due to an increase in the magnitude of the Prandtl number. The reduction in the temperature

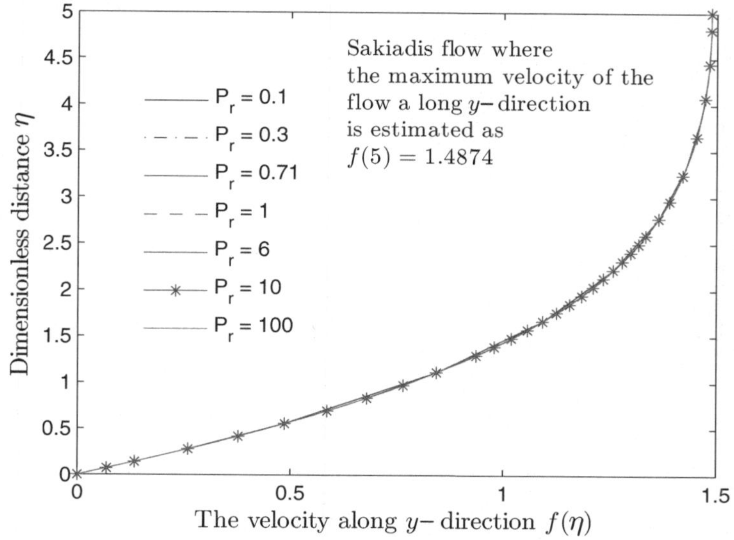

FIGURE 5.8
Velocity of Sakiadis flow along the y-direction at various values of Prandtl number in the absence of suction and injection.

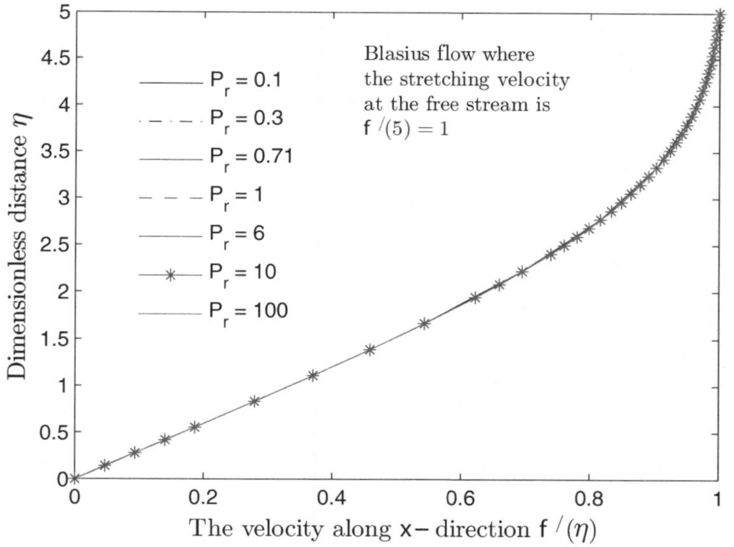

FIGURE 5.9
Velocity of Blasius flow along the x-direction at various values of Prandtl number in the absence of suction and injection.

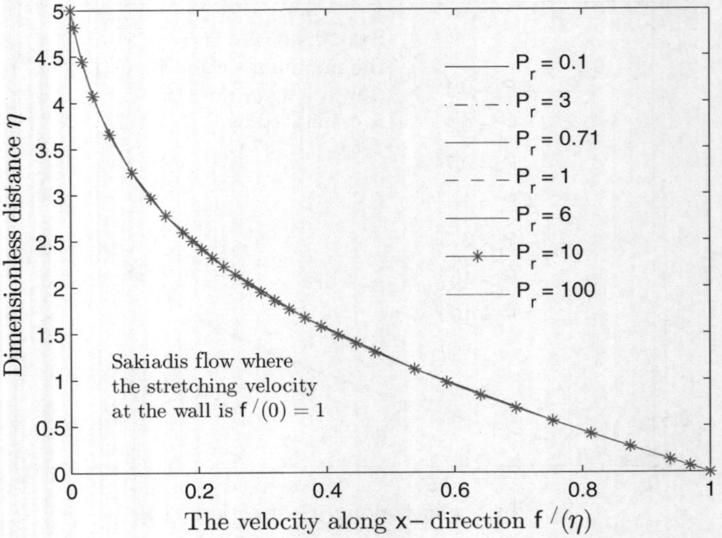

FIGURE 5.10
Velocity of Sakiadis flow along the x-direction at various values of Prandtl number in the absence of suction and injection.

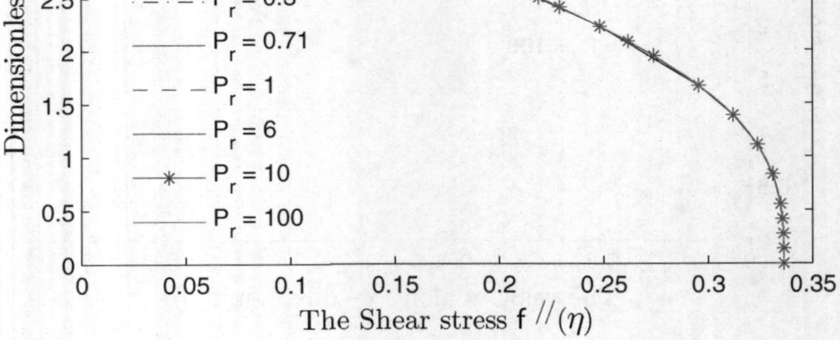

FIGURE 5.11
Shear stress profile at various values of Prandtl number: the case of Blasius flow.

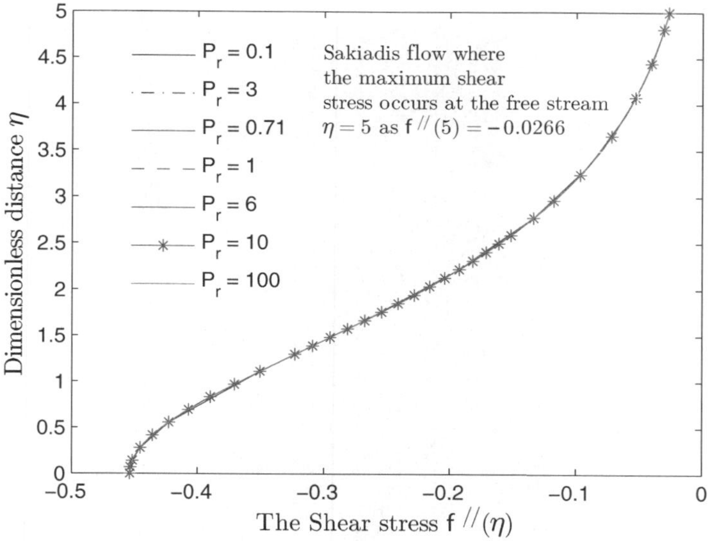

FIGURE 5.12
Shear stress profile at various values of Prandtl number: the case of Sakiadis flow.

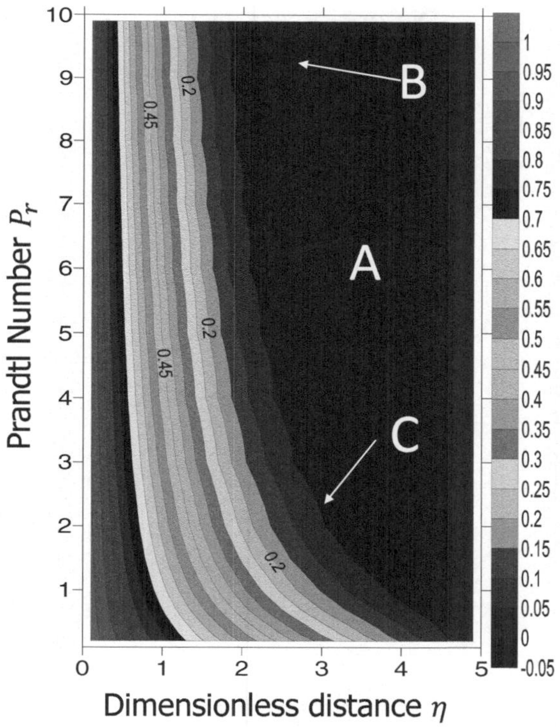

FIGURE 5.13
Variation in the effect of Prandtl number on the temperature distribution across the domain: the case of Blasius flow.

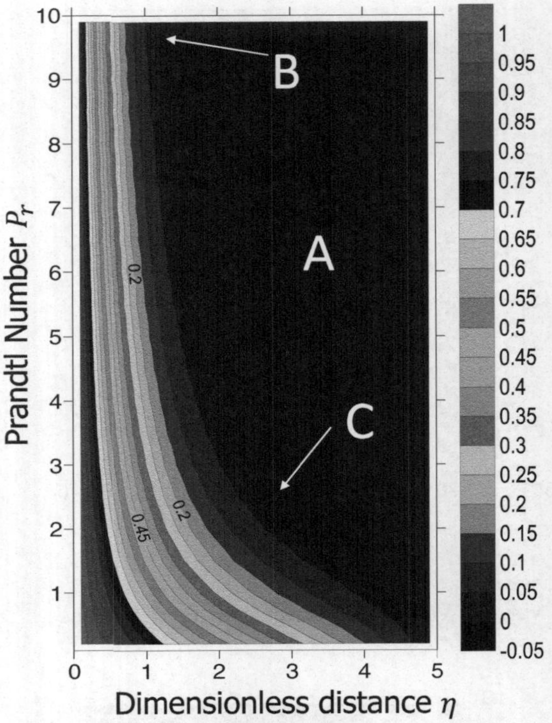

FIGURE 5.14
Variation in the effect of Prandtl number on the temperature distribution across the domain: the case of Sakiadis flow.

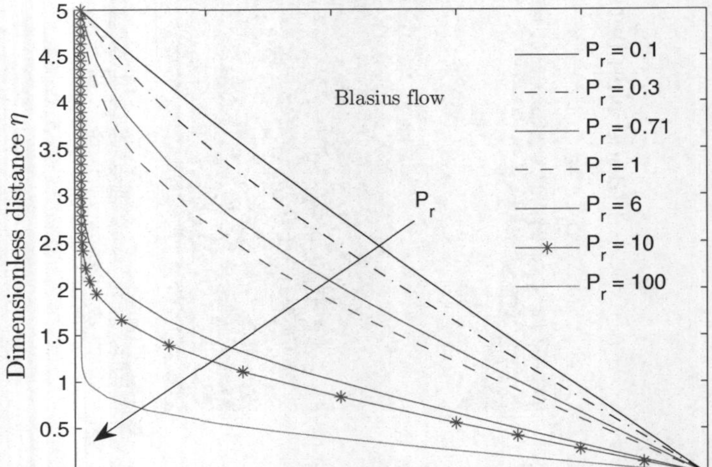

FIGURE 5.15
Temperature distribution across Blasius flow at various values of Prandtl number.

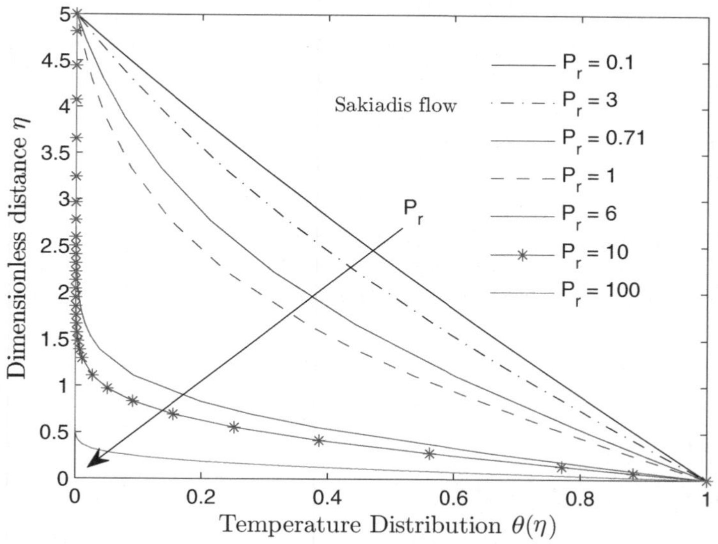

FIGURE 5.16
Temperature distribution across Sakiadis flow at various values of Prandtl number.

distribution from the wall to the free stream in the case of less viscous fluid may never be asymptotic due to its state of matter - gas. Also, a reduction in the temperature distribution across the flow of fluid with a larger magnitude of Prandtl number occurs within a wider region in the case of Blasius flow and a smaller domain in Sakiadis flow.

5.4 Introduction: Alumina Nanoparticles-Based Nanofluid

Aluminum oxide (alumina) nanoparticle is a tiny particle nanoparticles are a tiny particles well known for their high specific surface area and application as an adsorbent, desiccant, abrasive, and refractory material. It is particularly useful in capturing hydrocarbon impurities from the air and extracting fluorine from various media. The weak interaction and high degree of orientation are two properties that distinguish aluminum oxide nanoparticles. Rai et al. [242] associated the increasing usefulness of aluminum nanoparticles to their increased reactivity compared to other traditional micron-sized particles. Sahu and Hiremath [261] used the micro-EDM method to synthesize aluminum nanoparticles in deionized (DI) water and polyethylene glycol (PEG) as the base fluids, resulting in two distinct nanofluids. According to transmission electron microscopy, the size of aluminum nanoparticles suspended in DI water and PEG mixed solvent ranges from 45 to 500 nm, with a mean size of 196 nm. The latest study on aluminum (Al) remarked that the element is an abundant crystal metal due to its ability to be fully recyclable. Ali et al. [24] pointed out aluminum (Al) as one of the most abundant crystal metals on the planet, and it is regarded as a very sustainable material due to its ability to be fully recyclable.

In a study on settling characteristics of alumina nanoparticles in ethanol–water mixtures by Ilyas et al. [140], it is made known that the kinetic stability of the nanofluid gets negatively impacted due to the gravitational force that normally separates the agglomerated particles from the base fluid and causes sedimentation. Sedimentation behaviors in most unstable nanofluids can be classified as either flocculated sedimentation, dispersed sedimentation, or mixed sedimentation. Ali et al. [24] concluded at the end of the study mentioned above that a decrease in the stability of alumina-based nanofluid is bound to occur with the increase in the fabrication temperature. In a study on the dynamics of water (substance with lower density) and ethylene glycol (substance with higher density) on a convectively heated vertical surface conveying alumina/copper nanoparticles by Song et al. [287], aluminum oxide nanoparticles are used to improve the rheological and filtration properties of drilling fluids due to their high thermal conductivity, which allows them to dissipate heat from the fluid by Brownian motion effectively (Smith et al. [285]). Alumina nanoparticles have extremely high melting and boiling points, which increase the thermal inertia of the nanofluids.

5.5 Sakiadis Flow of Water-Alumina Nanofluid

The motion of a nanofluid on a horizontal surface with major emphasis on the significance of particle radius of nanoparticle, inter-particle spacing, volume fraction, and Prandtl number is presented in this section. It is assumed that the two-dimensional flow occurs along the x-direction with the velocity $u(x, y)$. The velocity $v(x, y)$ along the y-direction is perpendicular to x-direction as shown in Figure 5.17. The governing equation suitable to model the cases is of the form

$$u_x + v_y = 0, \tag{5.22}$$

$$uu_x + vu_y = \frac{\mu_{nf}}{\rho_{nf}} u_{yy}, \tag{5.23}$$

$$uT_x + vT_y = \frac{\kappa_{nf}}{(\rho Cp)_{nf}} T_{yy}. \tag{5.24}$$

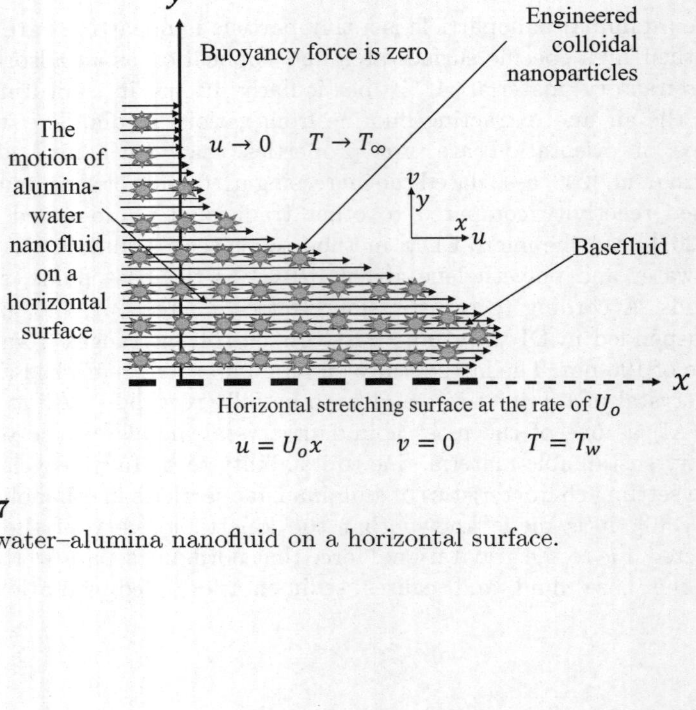

FIGURE 5.17
Illustration of water–alumina nanofluid on a horizontal surface.

For this case, Eqs. (5.22)–(5.24) are subject to the boundary conditions

$$u = U_o x, \quad v = 0, \quad T = T_w \quad at \quad y = 0. \tag{5.25}$$

$$u \to 0, \quad T \to T_\infty, \quad as \quad y \to \infty \tag{5.26}$$

The model proposed by Graham [114] for the ratio of dynamic viscosity of the water–alumina nanofluid to the dynamic viscosity of the base fluid defined as

$$\frac{\mu_{nf}}{\mu_{bf}} = 1 + 2.5\phi + 4.5\left[\frac{1}{\frac{h}{d_p}\left(2 + \frac{h}{d_p}\right)\left(1 + \frac{h}{d_p}\right)^2}\right] \tag{5.27}$$

where the particle radius of nanoparticle is h and the inter-particle spacing is d_p, was adopted. The model presented by Maxwell [193] for the ratio of thermal conductivity of the nanofluid to the based fluid was adopted:

$$\frac{\kappa_{nf}}{\kappa_{bf}} = \frac{\kappa_{sp} + 2\kappa_{bf} - 2\phi(\kappa_{bf} - \kappa_{sp})}{\kappa_{sp} + 2\kappa_{bf} + \phi(\kappa_{bf} - \kappa_{sp})} \tag{5.28}$$

Following Mohamoud Jama et al. [147], the heat capacity and density of the nanofluid are of the form

$$(\rho c_p)_{nf} = (1 - \phi)(\rho Cp)_{bf} + \phi(\rho Cp)_{sp}, \quad \rho_{nf} = (1 - \phi)\rho_{bf} + \phi\rho_{sp}. \tag{5.29}$$

Given that the base fluid is water and alumina is the nanoparticles, following Koriko et al. [164], the thermo-physical properties are $\rho_{sp} = 3970$, $\rho_{bf} = 997.1$, $\kappa_{sp} = 40$, $\kappa_{bf} = 0.613$, $Cp_{sp} = 765$, and $Cp_{bf} = 4179$. In order to obtain self-similar solution of Eqs. (5.22)–(5.26), the following similarity variables

$$\eta = y\sqrt{\frac{U_\infty}{\vartheta_{bf} x}}, \quad \psi(x, y) = \sqrt{\vartheta_{bf} x U_\infty} f(\eta), \quad u = \frac{\partial \psi}{\partial y}, \quad v = -\frac{\partial \psi}{\partial x},$$

$$\theta(\eta) = \frac{T - T_\infty}{T_w - T_\infty}, \quad P_r = \frac{\mu_{bf} Cp_{bf}}{\kappa_{bf}} = \frac{\vartheta_{bf}}{\alpha_{bf}}. \tag{5.30}$$

were used to obtain the dimensionless governing equation below:

$$\left(\frac{1 + 2.5\phi + 4.5\left[\frac{1}{\frac{h}{d_p}\left(2+\frac{h}{d_p}\right)\left(1+\frac{h}{d_p}\right)^2}\right]}{1 - \phi + \phi\frac{\rho_{sp}}{\rho_{bf}}}\right)\frac{d^3 f}{d\eta^3} + \frac{1}{2} f \frac{d^2 f}{d\eta^2} = 0, \tag{5.31}$$

$$\left(\frac{\frac{\kappa_{sp}+2\kappa_{bf}-2\phi(\kappa_{bf}-\kappa_{sp})}{\kappa_{sp}+2\kappa_{bf}+\phi(\kappa_{bf}-\kappa_{sp})}}{1 - \phi + \phi\frac{(\rho Cp)_{sp}}{(\rho Cp)_{bf}}}\right)\frac{d^2\theta}{d\eta^2} + \frac{1}{2}P_r f \frac{d\theta}{d\eta} = 0. \tag{5.32}$$

Dimensionless boundary conditions for this case are

$$\frac{df}{d\eta} = 1, \quad f = 0, \quad \theta = 1 \quad at \quad \eta = 0. \tag{5.33}$$

$$\frac{df}{d\eta} \to 0, \quad \theta \to 0 \quad as \quad \eta \to \infty. \tag{5.34}$$

5.5.1 Research Questions II

This subsection aims to provide answers to the following research questions:

1. How does the motion and temperature distribution over the domain change when the particle radius of nanoparticles, inter-particle spacing, volume fraction, and Prandtl number increase?
2. To what extent is the variation in local skin friction coefficients and heat transfer rate due to enlargement in the magnitude of particle radius of nanoparticles, inter-particle spacing, volume fraction, and Prandtl number?

5.5.2 Analysis and Discussion of Results II

Equations (5.31) and (5.32) subject to Eqs. (5.33) and (5.34) were solved using the four-stage Lobatto IIIa formula as described in Chapter 1 using $\eta = 5$. In this study of the motion of water conveying alumina nanoparticles, $6 \leq P_r \leq 7$ was considered for the simulation.

5.5.2.1 Prandtl Number and Volume Fraction of Nanoparticles

Figure 5.18 shows that at fixed values of $h = 0.1$ and $d_p = 0.5$ when $\phi = 0.1$ and $\phi = 0.8$, it is seen that as the Prandtl number increases within the interval $6 \leq P_r \leq 7$, the vertical and horizontal velocities are found to be constant functions. However, the motion along the y-direction increases from the wall to the free stream at the rate of 0.315911179 when $\phi = 0.1$ and 0.308651984 when $\phi = 0.8$. Figure 5.19 reveals that the horizontal velocity (i.e., the motion along the x-direction) decreases from the wall to the free stream at the rate of -0.184828811 when $\phi = 0.1$ and -0.195097077 when $\phi = 0.8$. As the magnitude of P_r increases, the momentum viscosity increases, but this has no effect on the transport phenomenon due to the disjoint between momentum and energy equations that model the flow. It is worth deducing from Figure 5.20 that the shear stress function $f''(\eta)$ that quantifies the friction across the flow when $\phi = 0.1$ increases with η at the rate

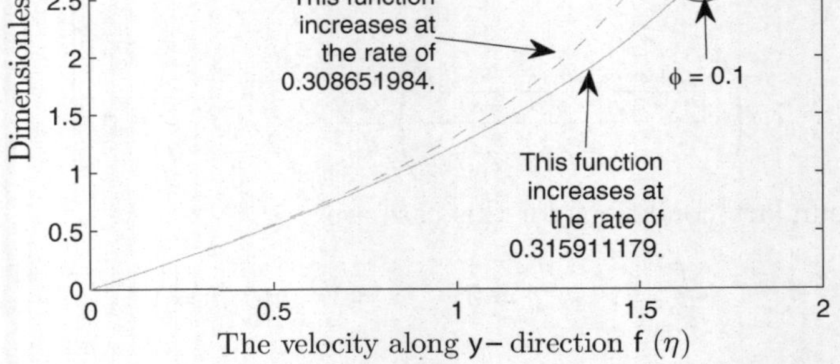

FIGURE 5.18
Variation in $f(\eta)$ due to P_r when $h = 0.1$ and $d_p = 0.5$.

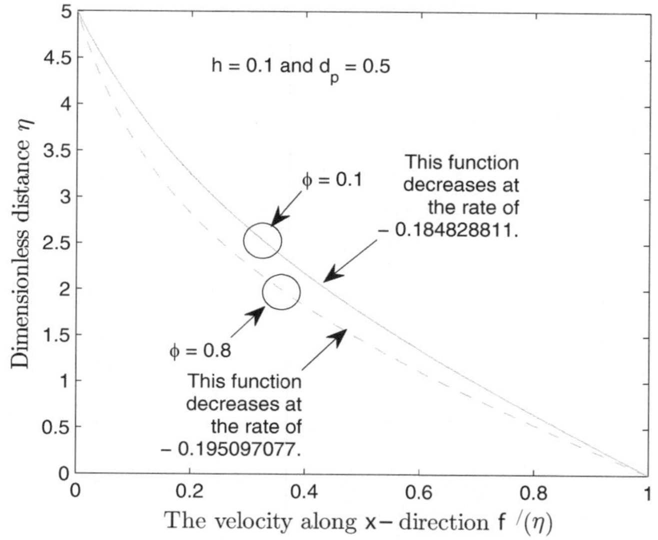

FIGURE 5.19
Variation in $f'(\eta)$ due to P_r when $h = 0.1$ and $d_p = 0.5$.

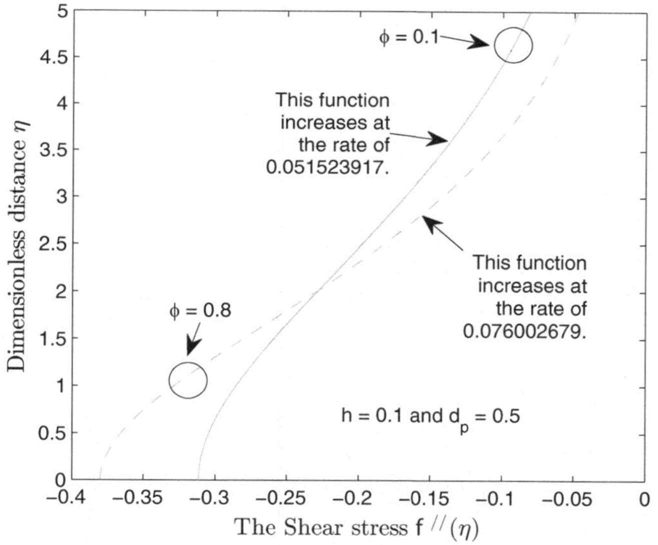

FIGURE 5.20
Variation in $f''(\eta)$ due to P_r when $h = 0.1$ and $d_p = 0.5$.

of 0.051523917. When $\phi = 0.8$, $f''(\eta)$ increases with η at the rate of 0.076002679. Changes in the shear stress across the flow independent of the temperature are minute and can be approximated to zero. This is the major reason why $f''(\eta)$ is a constant function as P_r ranges within the interval $6 \leq P_r \leq 7$. Also, the temperature function $\theta(\eta)$ across the flow decreases negligibly with the Prandtl number when $\phi = 0.1$ and $\phi = 0.8$; see Figure 5.21. However, the maximum temperature distribution is seen when $\phi = 0.8$ at various values of Prandtl number, while the observed declination in the temperature from the wall to

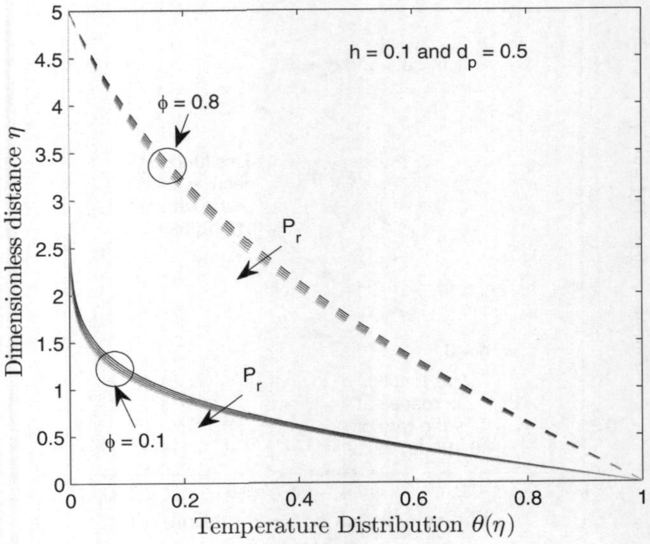

FIGURE 5.21
Variation in $\theta(\eta)$ due to P_r when $h = 0.1$ and $d_p = 0.5$.

the free stream is sharp, for instance when $\theta(\eta)$ declines with η for $P_r = 7$ at the rate of -0.077639503.

In addition, at a larger value of volume fraction ($\phi = 0.8$), $\theta(\eta)$ decreases across the domain $\eta\varepsilon[0, 5]$ for $P_r = 7$ at the rate of -0.198057266. Physically, $\phi = 0.1$ and $\phi = 0.8$ imply small amount of alumina nanoparticles and large amount of the nanoparticles in the base fluid, respectively. When the amount of alumina nanoparticles is high, the thermal conductivity is also enhanced (Mohamoud Jama et al. [147]). This caused the maximum temperature distribution to occur at a larger value of volume fraction (ϕ) because the rate at which the nanofluid itself passes heat is higher for $\phi = 0.8$. It is pertinent to remark that when the volume fraction is small in magnitude ($\phi = 0.1$), the temperature gradient $\theta'(\eta)$ increases with the Prandtl number P_r near the wall; see Figure 5.22. However, such significant effect disappears when $\phi = 0.8$. It is observed in Figure 5.23 that the maximum heat transfer rate at the wall $-\theta'(0)$ occurs at a larger value of P_r when the volume fraction ϕ is most minimum. At a larger value of volume fraction, $-\theta'(0)$ is a constant value for all the levels of Prandtl number within the interval $6 \leq P_r \leq 7$. This is true because the rate of increase in $-\theta'(0)$ with P_r when $\phi = 0.7$ is nearly zero (i.e., 0.03152) as it is evident in Table 5.18. The particle radius of nanoparticle h and the inter-particle spacing d_p have no significant relationship with the Prandtl number and volume fraction of concentration.

5.5.2.2 Prandtl Number and Inter-Particle Spacing d_p

Table 5.19 shows that with an increase in the Prandtl number, the Nusselt number $-\theta'(0)$ increases at a minimum rate when $d_p = 0.1$ and at a higher rate when $d_p = 5$. This implies that the inter-particle spacing is a factor suitable to increase heat transfer rate at each level of Prandtl number and boost the rate at which $-\theta'(0)$ increases with the Prandtl number. When the magnitude of inter-particle spacing is increased, the agglomeration of nanoparticles is minimized. However, it is important to note that the higher the agglomeration of nanoparticles in the fluid, the lower the thermal conductivity (Mohamoud Jama et al. [147]). Based on this, an increase in d_p corresponds to an increment in the

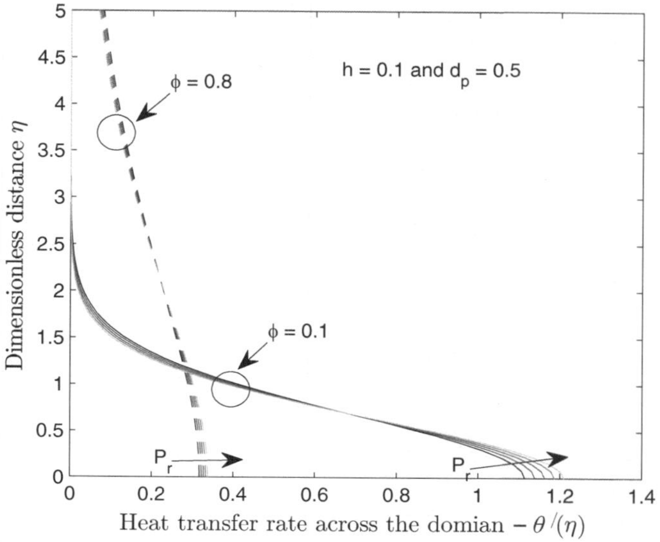

FIGURE 5.22
Variation in $-\theta'(\eta)$ due to P_r when $h = 0.1$ and $d_p = 0.5$.

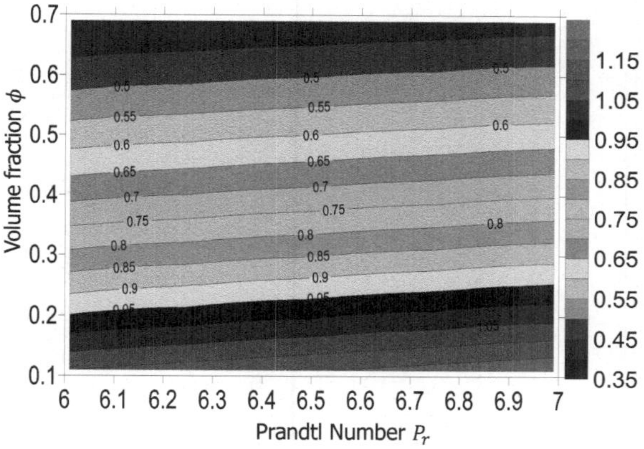

FIGURE 5.23
Variation in $-\theta'(0)$ with P_r and ϕ when $h = 0.1$ and $d_p = 0.5$.

TABLE 5.18
Variation in $-\theta'(0)$ with P_r When $\phi = 0.1$ and $\phi = 0.7$: The Case $h = 0.1$ and $d_p = 0.5$

P_r	$-\theta'(0)$ **When** $\phi = 0.1$	$-\theta'(0)$ **When** $\phi = 0.7$
6	1.1141	0.3888
6.25	1.1387	0.3967
6.5	1.1629	0.4046
6.75	1.1866	0.4125
7.0	1.2098	0.4203
S_{lp}	0.09572	0.03152

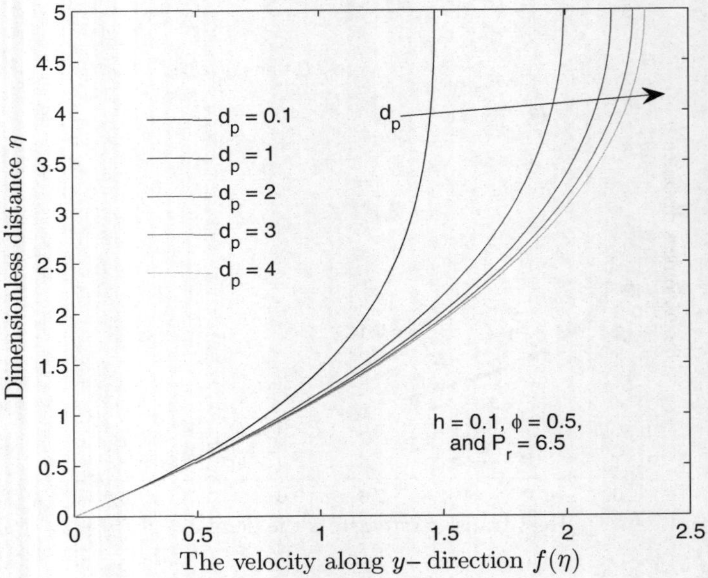

FIGURE 5.24
Variation in $f(\eta)$ with d_p when $\phi = 0.5$, $h = 0.1$ and $P_r = 6.5$.

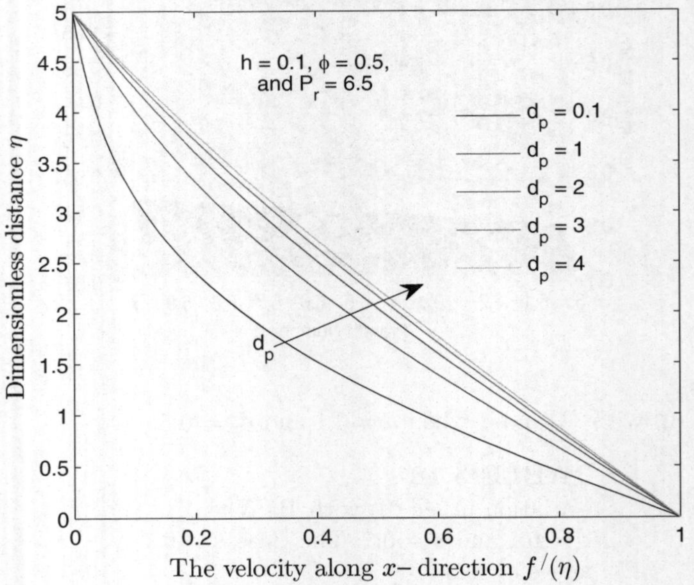

FIGURE 5.25
Variation in $f'(\eta)$ with d_p when $\phi = 0.5$, $h = 0.1$ and $P_r = 6.5$.

thermal conductivity of nanofluids, and consequently, the motion of the nanofluid along both directions (x, y) increases; see Figures 5.24 and 5.28. In such a case, the nanofluid passes heat sufficiently, and the heating reduces the intermolecular forces holding the molecules of the nanofluid together. The shear stress, which is proportional to friction, increases with d_p

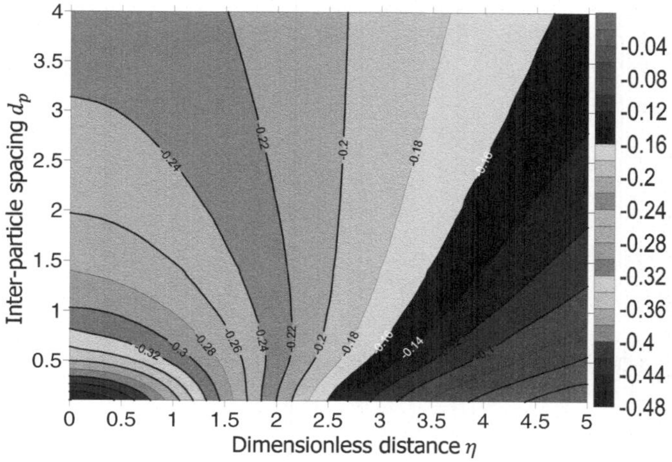

FIGURE 5.26
Variation in $f''(\eta)$ with η at various values of d_p when $\phi = 0.5$, $h = 0.1$ and $P_r = 6.5$.

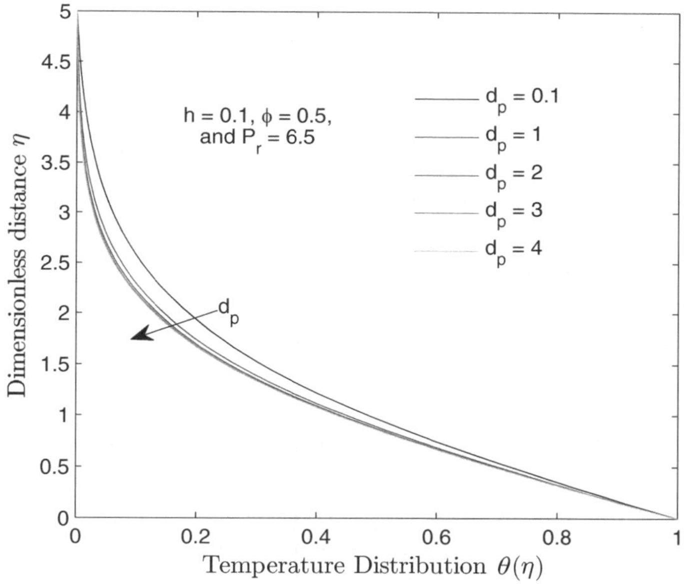

FIGURE 5.27
Variation in $\theta(\eta)$ with d_p when $\phi = 0.5$, $h = 0.1$ and $P_r = 6.5$.

near the wall, and the reverse is the case near the free stream; see Figure 5.26. In addition, it is seen that $f''(\eta)$ increases with η at the rate of 0.09235924 when $d_p = 0.1$ and at the rate of 0.017388882 when $d_p = 4$. Figure 5.27 indicates that the temperature distribution across the flow is a decreasing property of inter-particle spacing (Figure 5.25).

5.5.2.3 Prandtl Number and Particle Radius of Nanoparticle h

It is evident in Table 5.20 and Figure 5.33 that the Nusselt number proportional to heat transfer rate increases with the Prandtl number at the rate of 0.05356 when the radius of

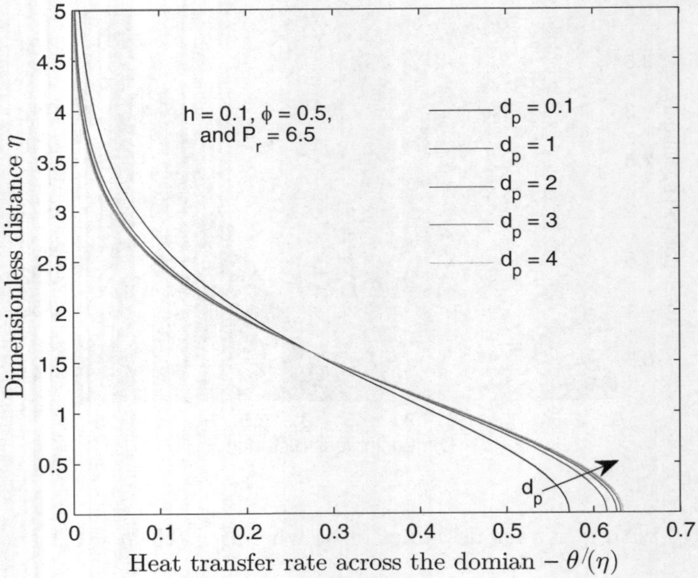

FIGURE 5.28
Variation in $-\theta'(\eta)$ with d_p when $\phi = 0.5$, $h = 0.1$ and $P_r = 6.5$.

TABLE 5.19
Variation in $-\theta'(0)$ with P_r When $\phi = 0.1$ and $\phi = 0.7$: The Case $h = 0.1$ and $\phi = 0.5$

P_r	$-\theta'(0)$ **When** $d_p = 0.1$	$-\theta'(0)$ **when** $d_p = 5$
6	0.5456	0.6081
6.25	0.5589	0.6219
6.5	0.572	0.6354
6.75	0.5849	0.6487
7.0	0.5977	0.6617
S_{lp}	0.05208	0.0536

alumina nanoparticle is small (i.e., $h = 0.1$). In the case of larger radius of nanoparticles (i.e., $h = 0.5$), the Nusselt number increases with the Prandtl number at a smaller rate. An increase in the particle radius of nanoparticles indicates enlargement in the size of alumina nanoparticles. This attributes to a decrease in the motion along the x- and y-directions shown as Figures 5.29 and (5.30). Figure 5.31 reveals that the shear stress across the flow decreases near the wall and increases after $\eta = 2.2$ due to an increase in the radius of alumina nanoparticles. The temperature distribution is seen clearly in Figure (5.32) to be an increasing property of particle radius of the nanoparticles. Mohamoud Jama et al. [147] once reported that a smaller particle possesses a larger surface area relative to its diameter, thus corresponding to the increment in the thermal conductivity of the nanofluid under consideration.

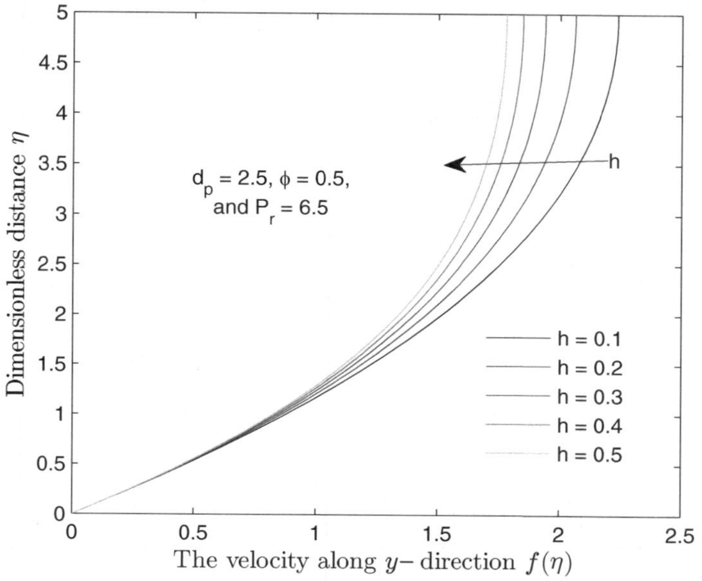

FIGURE 5.29
Variation in $f(\eta)$ with h when $\phi = 0.5$, $d_p = 2.5$ and $P_r = 6.5$.

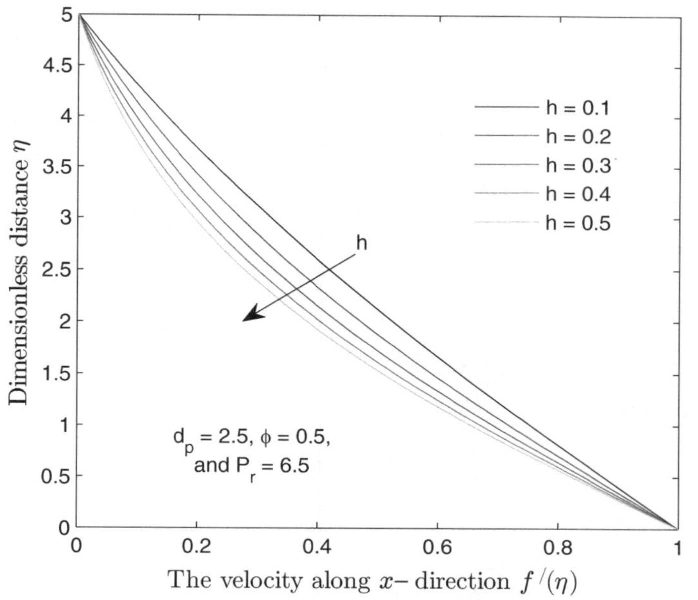

FIGURE 5.30
Variation in $f'(\eta)$ with h when $\phi = 0.5$, $d_p = 2.5$ and $P_r = 6.5$.

5.6 Introduction: Injection and Suction

In the industry, suction and injection are two different methods for introducing or reducing pressure/force on stagnant or moving fluids. Suction in flowing fluids is a force that causes

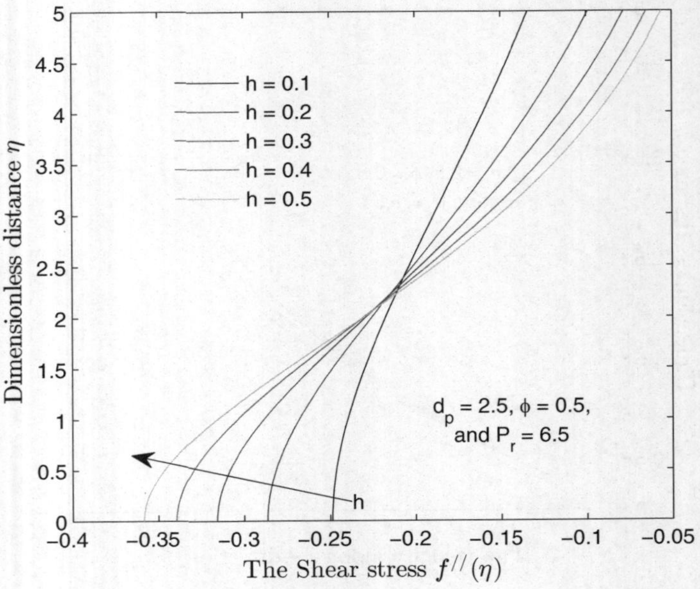

FIGURE 5.31
Variation in $f''(\eta)$ with h when $\phi = 0.5$, $d_p = 2.5$ and $P_r = 6.5$.

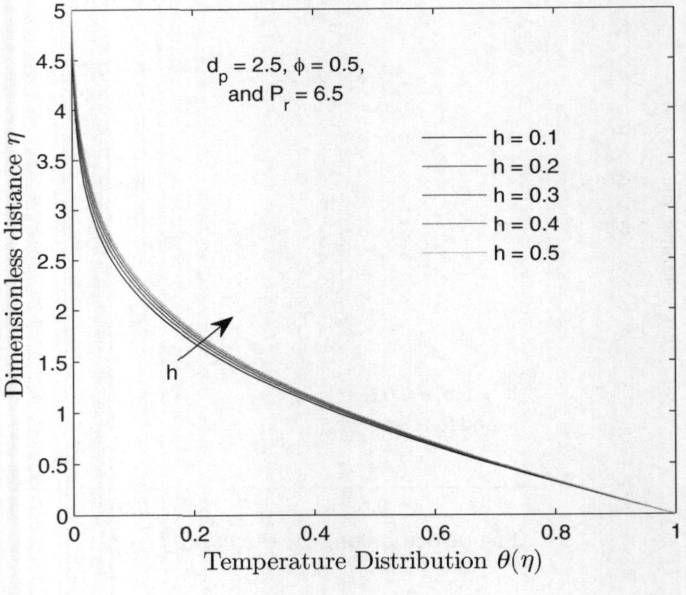

FIGURE 5.32
Variation in $\theta(\eta)$ with h when $\phi = 0.5$, $d_p = 2.5$ and $P_r = 6.5$.

the fluid to be drawn into an interior space. This method is sometimes used to remove air or impurities from a fluid produced by pressure difference. Braslow [70] explained the laminar flow control in aeronautical research by sucking a small amount of air substance through the wings and the entire surface. During the transport phenomenon, there are frictional forces

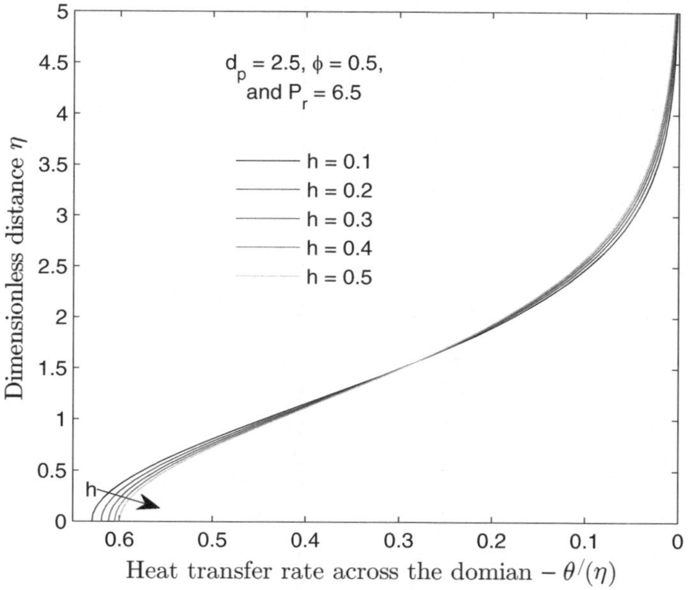

FIGURE 5.33
Variation in $-\theta'(\eta)$ with h when $\phi = 0.5$, $d_p = 2.5$ and $P_r = 6.5$.

TABLE 5.20
Variation in $-\theta'(0)$ with P_r when $h = 0.1$ and $h = 0.5$: The Case $d_p = 2.5$ and $\phi = 0.5$

P_r	$-\theta'(0)$ **When** $h = 0.1$	$-\theta'(0)$ **When** $h = 0.5$
6	0.6021	0.5728
6.25	0.6159	0.5864
6.5	0.6294	0.5998
6.75	0.6426	0.6129
7.0	0.6557	0.6259
S_{lp}	0.05356	0.05308

(viscous drag) between the surface and the last fluid layer. However, a small laminar flow can be controlled through surface cooling, removing the minute amount of air within the boundary layer, small perforations, and multiple narrow surface slots. Most importantly, suction through perforated plates is not successful due to the emergence of disturbances because of the edges at the hole. Suction or blowing slightly affects the turbulent Prandtl number (Kays [150]).

The design of the wings of an aircraft and control of the laminar transition to turbulence is based on the principle of suction (Pantokratoras [224]). An injection introduces gases or liquids under pressure into either a stagnant fluid or fluid in motion, intending to reduce its temperature and increase the pressure. As the flow develops along a vertical surface moving through the binary mixture, one way of dealing with boundary layer transition (flow separation) is to suck the thin boundary layer through the vertical porous surface (Animasaun, [29]). As this method reduces drag, heat energy escapes away from the flow regime; hence, the temperature reduces as the magnitude of suction increases. It was remarked by Shaw et al. [279] that increased suction sets the considerable amount of immeasurable quantities from the ambient to the surface of the rotating disk, thereby augmenting the viscosity, which in turn restrains the fluid motion. Moreover, axial velocity increases with suction. The effectiveness of both Reynolds number and suction is based on stretching velocity (Pantokratoras [224]).

5.7 Fluid Flow Subject to Injection or Suction

Next is to consider the two-dimensional flow of an incompressible fluid on a horizontal porous surface subject to suction and injection within the domain $0 \le y < \infty$ where the effect of viscosity is highly significant; see Figure 5.34. It is assumed that the flow is subject to either suction or injection at $y = 0$, for $V_w \neq 0$ in Eqs. (5.38) and (5.40). Viscous forces induce the transport phenomenon, either stretching at the free stream (i.e., Blasius flow for the first case) or stretching at the wall (i.e., Sakiadis flow for the second case) with the stretching velocity $U_\infty = U_o x$. The governing equation that models the transport phenomenon is

$$u_x + v_y = 0, \tag{5.35}$$

$$uu_x + vu_y = \frac{\mu}{\rho} u_{yy}, \tag{5.36}$$

$$uT_x + vT_y = \frac{\kappa}{\rho c_p} T_{yy}. \tag{5.37}$$

For the case of Blasius flow, Eqs. (5.35)–(5.37) are subject to the boundary conditions

$$u = 0, \quad v = V_w, \quad T = T_w \quad at \quad y = 0. \tag{5.38}$$

$$u \to U_\infty, \quad T \to T_\infty, \quad as \quad y \to \infty \tag{5.39}$$

For the case of Sakiadis flow, Eqs. (5.35) –(5.37) are subject to the boundary conditions

$$u = U_w, \quad v = V_w, \quad T = T_w \quad at \quad y = 0. \tag{5.40}$$

$$u \to 0, \quad T \to T_\infty, \quad as \quad y \to \infty \tag{5.41}$$

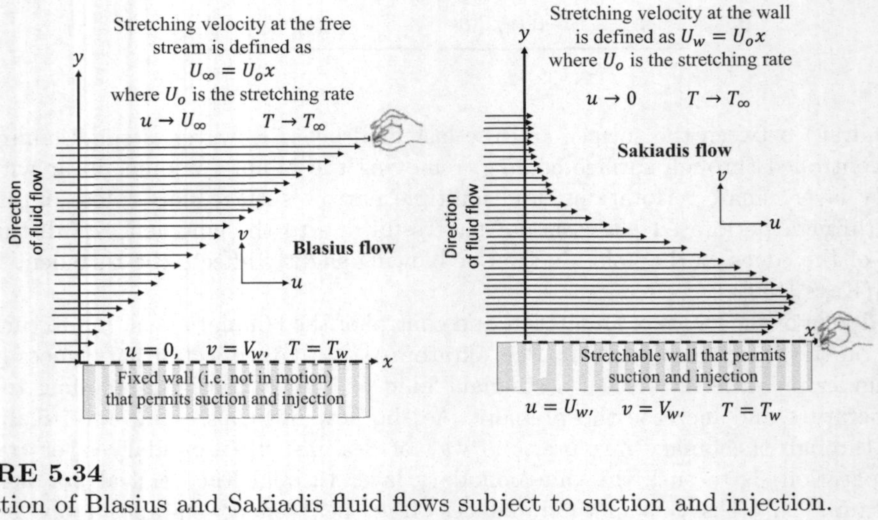

FIGURE 5.34
Illustration of Blasius and Sakiadis fluid flows subject to suction and injection.

In order to obtain self-similar solutions, the similarity variables defined below were considered:

$$\eta = y\sqrt{\frac{U_\infty}{\vartheta x}}, \quad \psi(x,y) = \sqrt{\vartheta x U_\infty} f(\eta), \quad u = \frac{\partial \psi}{\partial y}, \quad v = -\frac{\partial \psi}{\partial x},$$

$$\theta(\eta) = \frac{T - T_\infty}{T_w - T_\infty}, \quad P_r = \frac{\mu c_p}{\kappa} = \frac{\vartheta}{\alpha}, \quad f_w = -\frac{v_w}{\sqrt{\vartheta U_o}}. \tag{5.42}$$

The final dimensionless governing equation Eqs. (5.38) and (5.39) are subject to

$$\frac{df}{d\eta} = 0, \quad f = f_w, \quad \theta = 1, \quad at \quad \eta = 0. \tag{5.43}$$

$$\frac{df}{d\eta} \to 1, \quad \theta \to 0, \quad as \quad \eta \to \infty. \tag{5.44}$$

for the Blasius flow. The dimensionless boundary conditions for the Sakiadis flow Eqs. (5.40) and (5.41) are

$$\frac{df}{d\eta} = 1, \quad f = f_w, \quad \theta = 1, \quad at \quad \eta = 0. \tag{5.45}$$

$$\frac{df}{d\eta} \to 0, \quad \theta \to 0, \quad as \quad \eta \to \infty. \tag{5.46}$$

5.7.1 Research Questions III

The study in this subsection aims to provide answers to the following research questions:

1. How does injection affect the boundary layer flow due to stretching at the wall and free stream?
2. What is the influence of suction on the dynamics of Blasius flow and Sakiadis flow?
3. What is the major distinction in the effect of Prandtl number on the local skin friction coefficients and heat transfer rate during Blasius and Sakiadis flows?

TABLE 5.21
Effect of Prandtl Number P_r on Some of the Properties of Blasius and Sakiadis Flows in the Presence of Suction ($f_w = 3$)

$f_w = 3$		**Blasius Flow**			**Sakiadis Flow**	
P_r	$f''(0)$	$\theta(0.8333)$	$-\theta'(0)$	$f''(0)$	$\theta(0.88)$	$-\theta'(0)$
0.1	1.6322	0.7511	0.3183	−1.6477	0.8412	0.2985
0.3	1.6322	0.5848	0.6000	−1.6477	0.7286	0.5554
0.71	1.6322	0.3392	1.2063	−1.6477	0.5077	1.1888
1	1.6322	0.2316	1.6322	−1.6477	0.3891	1.6477
6	1.6322	0.0003	9.0499	−1.6477	0.0043	9.2695
10	1.6322	0	15.0324	−1.6477	0.0001	15.2909
100	1.6322	0	150.0036	−1.6477	0	150.3283
S_{lp}	0	−0.0039	1.4986	0	−0.0050	1.5017

TABLE 5.22
The Rate of Decrease in $\theta(\eta)$ with η at Various Values of Prandtl Number for the Case of Blasius Flow When ($f_w = 3$)

P_r	0.1	0.3	0.71	1	6	10	100
S_{lp}	−0.1960	−0.1750	−0.1357	−0.1173	−0.0460	−0.0354	−0.0125

TABLE 5.23
The Rate of Decrease in $\theta(\eta)$ with η at Various Values of Prandtl Number for the Case of Sakiadis Flow When ($f_w = 3$)

P_r	0.1	0.3	0.71	1	6	10	100
S_{lp}	−0.1963	−0.1790	−0.1368	−0.1157	−0.0452	−0.0350	−0.0144

TABLE 5.24
Effect of Prandtl Number P_r on Some of the Properties of Blasius and Sakiadis Flows in the Presence of Injection ($f_w = -3$)

$f_w = -3$		**Blasius**	**Flow**		**Sakiadis**	**Flow**
P_r	$f''(0)$	$\theta(3.3333)$	$-\theta'(0)$	$f''(0)$	$\theta(3.3333)$	$-\theta'(0)$
0.1	0.0009	0.4171	0.1348	−0.0244	0.3522	0.1648
0.3	0.0009	0.5829	0.0540	−0.0244	0.3877	0.1103
0.71	0.0009	0.8237	0.0056	−0.0244	0.4510	0.0463
1	0.0009	0.9091	0.0009	−0.0244	0.4882	0.0244
6	0.0009	1	0.0000	−0.0244	0.7439	0.0000
10	0.0009	1	0.0000	−0.0244	0.8156	0.0000
100	0.0009	1	0.0000	−0.0244	0.9966	0.0000
S_{lp}	0	0.0026	−0.0004	0	0.0051	−0.0007

TABLE 5.25
The Rate of Decrease in $\theta(\eta)$ with η at Various Values of Prandtl Number for the Case of Blasius Flow When ($f_w = -3$)

P_r	0.1	0.3	0.71	1	6	10	100
S_{lp}	−0.1988	−0.1887	−0.1559	−0.1359	−0.0438	−0.0387	−0.0168

TABLE 5.26
The Rate of Decrease in $\theta(\eta)$ with η at Various Values of Prandtl Number for the Case of Sakiadis Flow When ($f_w = -3$)

P_r	0.1	0.3	0.71	1	6	10	100
S_{lp}	−0.2024	−0.2063	−0.2119	−0.2080	−0.1919	−0.1884	−0.2580

5.7.2 Analysis and Discussion of Results III

This analysis aims to point out the relationship between the Prandtl number, suction, and injection. The objective was achieved using the four-stage Lobatto IIIa formula described in Chapter 1 using $\eta = 5$. Tables 5.21 and 5.24 corroborate the earlier observation that increasing Prandtl number does not affect the local skin friction coefficients proportional to friction at the wall for Blasius and Sakiadis flows. However, in the case of suction, increasing the Prandtl number causes a decrement in the temperature distribution. The opposite is the case of injection as temperature distribution across Blasius and Sakiadis flows increases due to higher Prandtl numbers. The rates of decrease in the temperature distribution from

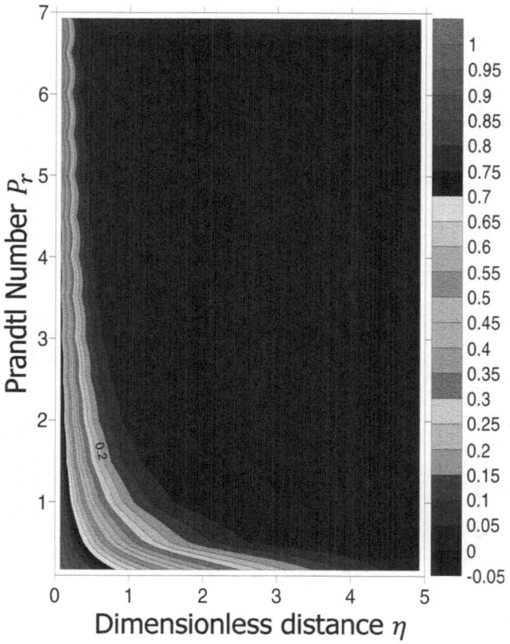

FIGURE 5.35
Variation in the temperature distribution with the Prandtl number across the domain of Blasius flow subject to suction.

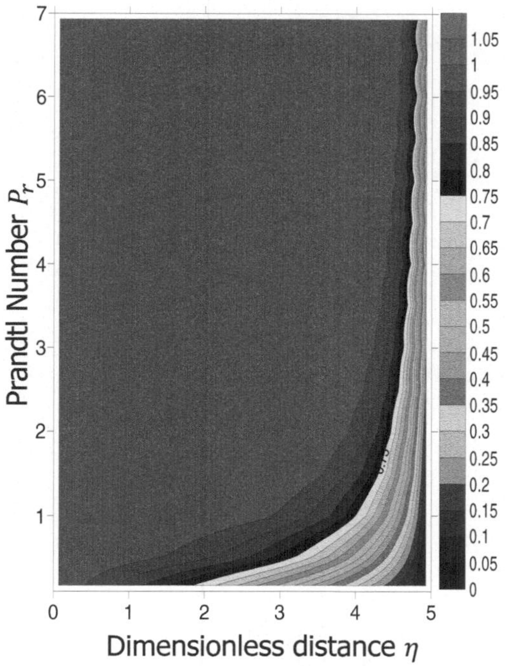

FIGURE 5.36
Variation in the temperature distribution with the Prandtl number across the domain of Blasius flow subject to injection.

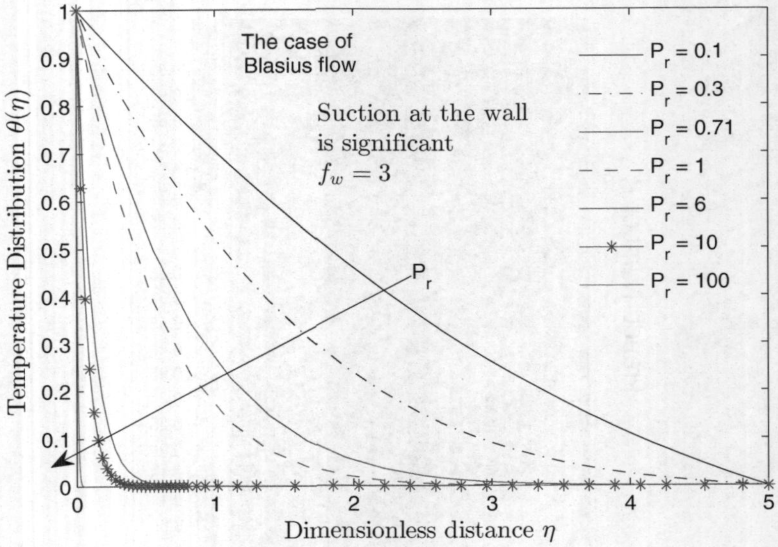

FIGURE 5.37
Temperature distribution across Blasius flow subject to suction at various Prandtl numbers.

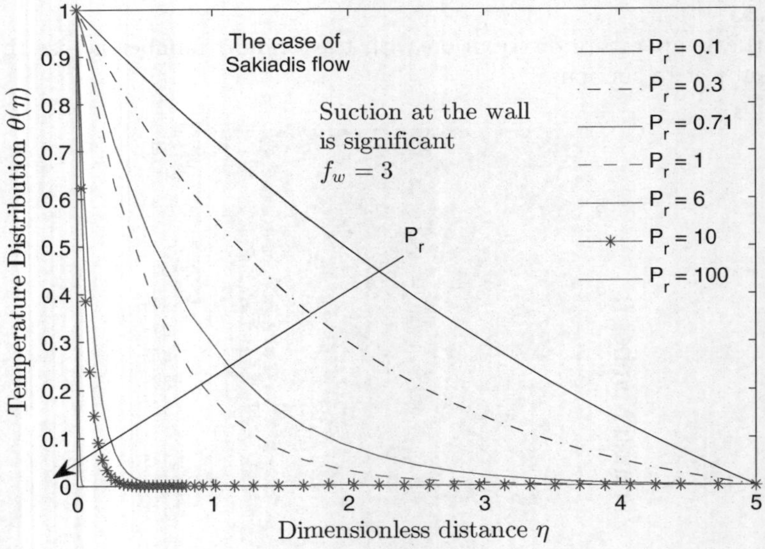

FIGURE 5.38
Temperature distribution across Sakiadis flow subject to suction at various Prandtl numbers.

the wall to the free stream at each level of Prandtl number for Blasius and Sakiadis flows subject to suction and injection are illustrated in Tables (5.22)–(5.26). The comparative analysis shows that the maximum decrease in the temperature distribution is seen in the case of Blasius flow subject to suction ($f_w = 3$) when $P_r = 100$ as -0.0125 (Table 5.22). A minimum decrease in the temperature distribution is ascertained when the stretching that induced the flow occurs at the wall subject to injection ($f_w = -3$) when $P_r = 100$

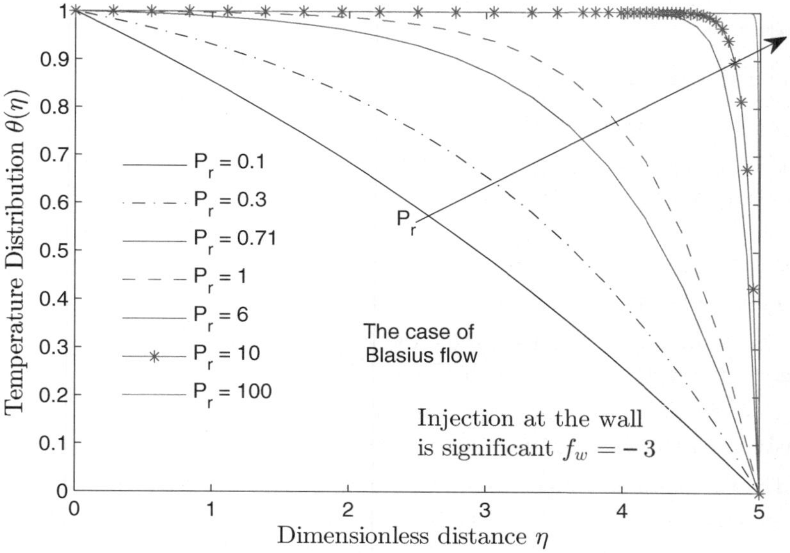

FIGURE 5.39
Temperature distribution across Blasius flow subject to injection at various Prandtl numbers.

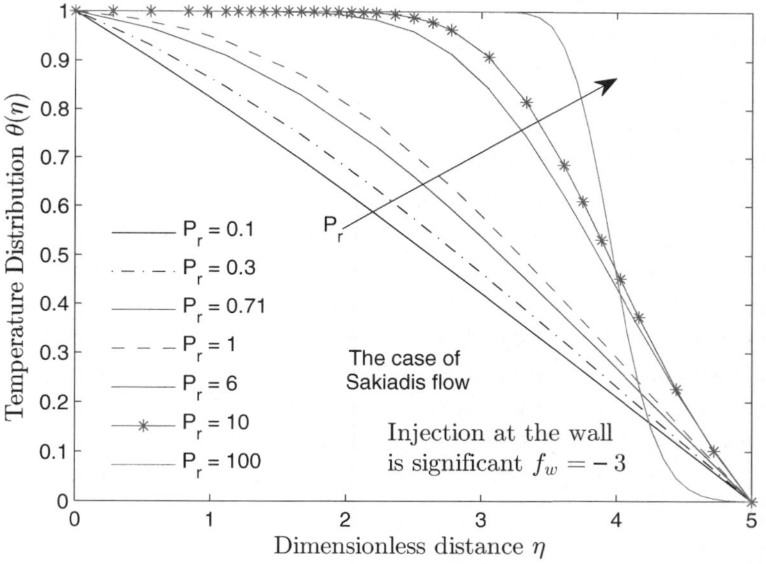

FIGURE 5.40
Temperature distribution across Sakiadis flow subject to injection at various Prandtl numbers.

as -0.2580 (Table 5.26). This is further illustrated graphically using the contour map in Figures 5.35 and 5.36 as it is seen that suction draws the temperature to zero ($\eta = 0$), while injection raises the temperature to $0.9 \leq \theta \leq 1.1$ within a larger domain near the surface; see Figures (5.37)–(5.40). Suction or blowing slightly affects the turbulent Prandtl number (Kays [150]).

5.8 Tutorial Questions

1. How does increasing Prandtl number affect the fluid flow induced by stretching?
2. Mention at least a difference in the effects of Prandtl number on pure fluids without nanoparticles and nanofluids with different nanoparticles of different diameters.
3. Explain simple nanofluid preparation and highlight an approach to break down agglomeration and prevent sedimentation.
4. Mention some methods for establishing stable nanofluid suspension and methods of dispersing nanoparticles in the base fluid.
5. What is the effect of increasing the ratio of momentum diffusivity to thermal diffusivity in the Blasius and Sakiadis flows over a horizontal surface?
6. What is the pattern of decrease in the temperature distribution with the Prandtl number when stretching induced the fluid flow at the wall and the free stream?
7. Do the local skin friction coefficient $f''(0)$ and Nusselt number $-\theta'(0)$ for Blasius flow and Sakiadis flow vary with the chosen η at infinity?
8. What are the variations in the skin friction coefficients and Nusselt number for Blasius and Sakiadis flows of (i) methanol, (ii) water, and (iii) blood when the radius of aluminum nanoparticles is small and large in magnitude?
9. What are the increasing effects of nanoparticles radius of (i) multi-walled CNT, (ii) silicon dioxide (SiO_2), (iii) single wall CNT ($SWCNT$), (iv) aluminium (Al), (v) magnesium oxide (MgO), (vi) titanium dioxide (TiO_2), (vii) iron(III)oxide (Fe_3O_4), (viii) copper(II)oxide (CuO), (ix)zinc (Zn), and (x) copper (Cu) conveyed by methanol for Blasius and Sakiadis flows?
10. How does the distance between copper nanoparticles within blood nanofluid for Blasius and Sakiadis flows affect the velocity and local skin friction coefficients?
11. What is the impact of increasing particle radius of nanoparticles, inter-particle spacing, volume fraction, and Prandtl number on the motion (i.e., velocity) and temperature distribution across the domain?
12. To what extent is the variation in local skin friction coefficients and heat transfer rate occur due to enlargement in the magnitude of particle radius of nanoparticles, inter-particle spacing, volume fraction, and Prandtl number?
13. How does injection affect the boundary layer flow due to stretching at the wall and free stream?
14. What is the influence of suction on the dynamics of Blasius flow and Sakiadis flow?
15. What is the major distinction between the effect of Prandtl number on the local skin friction coefficients and heat transfer rate during Blasius and Sakiadis flows?

6

Analysis of Self-Similar Flows II

6.1 Background Information

Whenever a volume of fluid flow in an upward direction due to the variation in density, though sufficiently small but eventually reduces the mass and consequently makes the effect of gravity to be neglected, it can be referred to as fluid flow due to buoyancy. Boussinesq [67, 68] deliberated on the mathematical model suitable to account for buoyancy-induced fluid flow along a vertical surface considering body force term (i.e., the product of acceleration due to gravity and density). There are three modes of heat transfer due to convection. Rajput [243] described buoyancy as the propensity for a body part or something submerged in a fluid to be raised as a result of weight displacement. According to Cao et al. [331], gravitational force causes natural convection, much like it causes the land breeze, a phenomena that happens around sunset. In such phenomena earlier mentioned, a less dense fluid parcel rises while a denser fluid parcel falls in the presence of gravity, Coriolis force, or centrifugal force, causing bulk fluid movement. Forced convection is a type of heat transmission that exclusively uses external heat sources, such fans or suction devices (i.e. the fluid molecules are forced to move). Mixed convection flows are induceable by the combination of buoyance and pressure forces. Nonetheless, the major parameters that govern how much each convection mode contributes to heat transmission are direction, temperature, and flow. This chapter is devoted to unraveling the significance of increasing the Prandtl number on boundary layer flows induced by forced convection, free convection, and mixed convection.

6.2 Introduction: Buoyancy-Induced Flows

According to Joseph Valentin Boussinesq's report, if ρ_∞ signifies the fluid density in the free stream at a sufficiently low temperature T_∞, the density model when there is a temperature difference between the wall (T_w) and free stream layer (T_∞) is presented as

$$\rho = \rho_\infty[1 - \beta(T - T_\infty)] \quad \text{where} \quad T_w > T_\infty. \tag{6.1}$$

The coefficient of volume expansion is denoted by β in this case. A minimal change in density occurs when temperature varies moderately; hence,

$$\begin{aligned} \delta\rho &= |\rho - \rho_\infty|, \\ \delta\rho &= |\rho_\infty[1 - \beta(T - T_\infty)] - \rho_\infty|, \\ \delta\rho &= |-\rho_\infty\beta(T - T_\infty)|, \\ \delta\rho &= \rho_\infty\beta(T - T_\infty). \end{aligned} \tag{6.2}$$

The buoyancy term

$$g(\rho - \rho_\infty) \approx g\delta\rho \tag{6.3}$$

DOI: 10.1201/9781003217374-6

is of the same order of magnitude as the inertia or viscous term; thus, it is not negligible when simulating buoyancy-induced flow over a vertical surface of uniform thickness. The buoyancy term can be simplified by equating Eq. (6.3) to Eq. (6.2) as

$$g(\rho - \rho_\infty) = g\beta\rho_\infty(T - T_\infty) \tag{6.4}$$

The body force term (buoyancy force) and pressure gradient term are algebraically integrated into the momentum equation for free convection over a vertical surface of uniform thickness or an inclined surface of uniform thickness as

$$-\frac{dp}{dx} + \rho g_x = g\beta\rho_\infty(T - T_\infty). \tag{6.5}$$

The pressure gradient term is zero in this case (i.e., $\frac{\partial p}{\partial x} = 0$). It is well understood that forced convection necessitates significant layer stretching, either in the free stream or next to the wall u_∞. The body force term (buoyancy force) and pressure gradient term are algebraically integrated into the momentum equation for forced convection over a vertical or inclined surface with uniform thickness as

$$-\frac{dp}{\partial x} + \rho g_x = u_\infty \frac{du_\infty}{dx}. \tag{6.6}$$

In this scenario, $\rho g_x = 0$. In addition, for mixed convection along a vertical surface with uniform thickness, the body force term (buoyancy force) is algebraically integrated into the momentum equation as

$$-\frac{dp}{dx} + \rho g_x = u_\infty \frac{du_\infty}{dx} + g\beta\rho_\infty(T - T_\infty). \tag{6.7}$$

Whenever buoyancy is enhanced in fluid flows, thermal dispersion is expected to be significant. Consequently, this can increase the Nusselt number proportional to the heat transfer rate (Chen [77]). This section presents the motions of a typical Newtonian fluid due to forced convection, free convection, and mixed convection.

6.3 Induced Flow due to Convection

Consider a two-dimensional flow with velocity u in the x-direction and velocity v in the y-direction. Sakiadis flow is assumed owing to stretching at the wall, as seen in Figure 6.1.

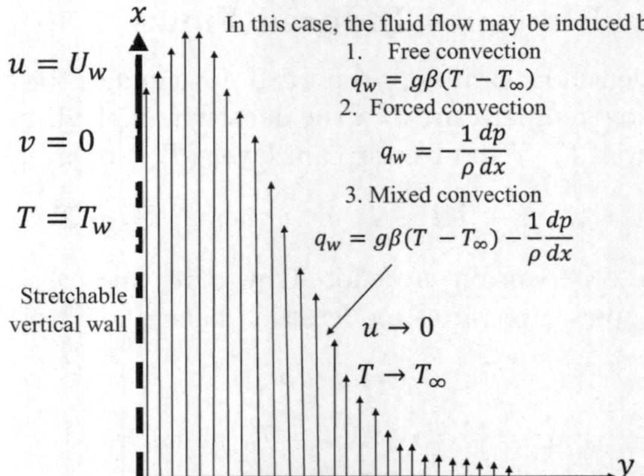

FIGURE 6.1
Graphical illustration of convectively induced transport phenomenon.

The appropriate governing equation to model the instances is of the type

$$u_x + v_y = 0, \tag{6.8}$$

$$uu_x + vu_y = \frac{\mu}{\rho} u_{yy} + q_w, \tag{6.9}$$

$$uT_x + vT_y = \frac{\kappa}{\rho c_p} T_{yy}. \tag{6.10}$$

For this case of Sakiadis flow, Eqs. (6.8)–(6.10) are subject to the boundary condition

$$u = U_\infty, \quad v = 0, \quad T = T_w \quad at \quad y = 0. \tag{6.11}$$

$$u \to T_\infty, \quad T \to \infty, \quad as \quad y \to \infty \tag{6.12}$$

In order to obtain self-similar solution of Eqs. (6.8)–(6.12), the following similarity variables were used to obtain the dimensionless governing equation:

$$\eta = y\sqrt{\frac{U_\infty}{\vartheta x}}, \quad \psi(x,y) = \sqrt{\vartheta x U_\infty} f(\eta), \quad u = \frac{\partial \psi}{\partial y}, \quad v = -\frac{\partial \psi}{\partial x},$$

$$\theta(\eta) = \frac{T - T_\infty}{T_w - T_\infty}, \quad P_r = \frac{\mu c_p}{\kappa} = \frac{\vartheta}{\alpha}, \quad Gr_x = \frac{gx\beta(T_w - T_\infty)}{U_\infty^2} \tag{6.13}$$

6.4 Forced Convective Induced Flow

For this case, q_w in Eq. (6.9) is equivalent to $-\frac{1}{\rho}\frac{dp}{dx}$ for the momentum equation for forced convective induced flow. Using Bernoulli equation, $\frac{p}{\rho} + \frac{U_\infty^2}{2}$ = constant

$$\frac{d}{dx}\left[\frac{p}{\rho} + \frac{U_\infty^2}{2}\right] = \frac{d}{dx}\text{constant}$$

$$\frac{1}{\rho}\frac{dp}{dx} + \frac{1}{2}\frac{d}{dx}(U_\infty U_\infty) = 0$$

$$\frac{1}{\rho}\frac{dp}{dx} + \frac{1}{2}\left[U_\infty \frac{du_\infty}{dx} + U_\infty \frac{dU_\infty}{dx}\right] = 0$$

Thus,

$$-\frac{1}{\rho}\frac{dp}{dx} = U_\infty \frac{dU_\infty}{dx} \tag{6.14}$$

Upon using the similarity variables listed as Eq. (6.13), the dimensionless equation of the governing equation that models forced convective induced flow is of the form

$$\frac{d^3 f}{d\eta^3} + \frac{1}{2} f \frac{d^2 f}{d\eta^2} + 1 = 0, \tag{6.15}$$

$$\frac{d^2\theta}{d\eta^2} + \frac{1}{2} P_r f \frac{d\theta}{d\eta} = 0. \tag{6.16}$$

Dimensionless boundary conditions are

$$\frac{df}{d\eta} = 1, \quad f = 0, \quad \theta = 1 \quad at \quad \eta = 0. \tag{6.17}$$

$$\frac{df}{d\eta} \to 0, \quad \theta \to 0 \quad as \quad \eta \to \infty. \tag{6.18}$$

6.4.1 Research Questions I

The examination questions intended to direct the investigation of the impact of Prandtl number on buoyancy-induced flow are outlined in this section.

1. In the case of forced convection, how do the Prandtl number influence the velocity and temperature distribution of the flow?
2. Does Prandtl number affect the motion along the x-direction and y-direction, and shear stress?

6.4.2 Analysis and Discussion of Results I

Equations (6.15) and (6.16) subject to Eqs. (6.17) and (6.18) were solved separately using the four-stage Lobatto IIIa formula as described in Chapter 1 using $\eta = 5$. The outcomes of the analysis for forced convective induced flow are outlined in this subsection. The analysis shows that the Prandtl number does not affect the velocity along the y-direction and x-direction. The slope linear regression through the data points of the velocity function f and dimensionless distance η for $\eta\varepsilon[0, 5)$ shows that the velocity along the y-direction increases at the rate of 0.966898245, while the rate of decrease in the horizontal velocity from the wall to the free stream is -0.399783516; see Figures 6.2 and 6.3. It is confirmed in Figure 6.4 that the shear stress across the forced convective induced flow decreases across the domain at the rate of -0.136399383. Figure 6.5 shows that the temperature distribution is a decreasing property of the Prandtl number. The rate of reduction in the temperature distribution from the wall to the free stream rises as the ratio of momentum diffusivity to thermal diffusivity increases, as shown in Table 6.1. A decrease in the temperature gradient near the wall due to the enhancement in the ratio of momentum diffusivity to thermal diffusivity was observed and illustrated in Figure 6.6. It can be deduced from Figure 6.6 that the heat transfer rate $-\theta'(\eta)$ increases with the Prandtl number within the fluid layers near the wall (i.e., $0 \leq \eta < 0.85$) and decreases with the same dimensionless number at the free stream.

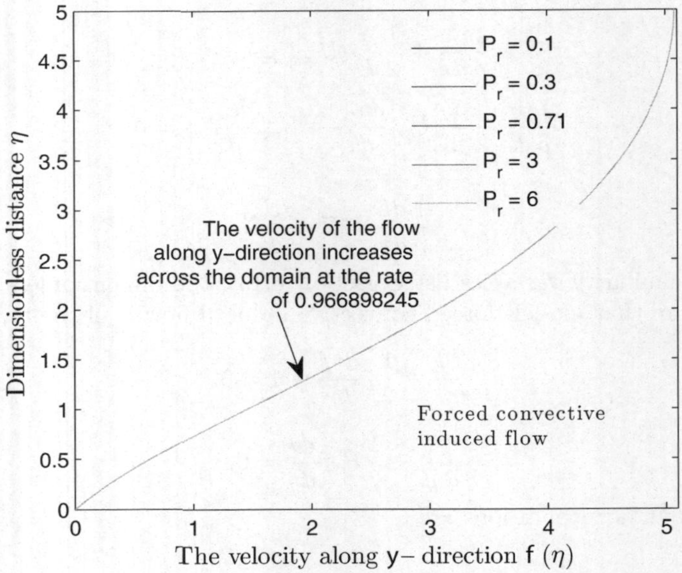

FIGURE 6.2
Variation in $f(\eta)$ due to P_r: the case of forced convection.

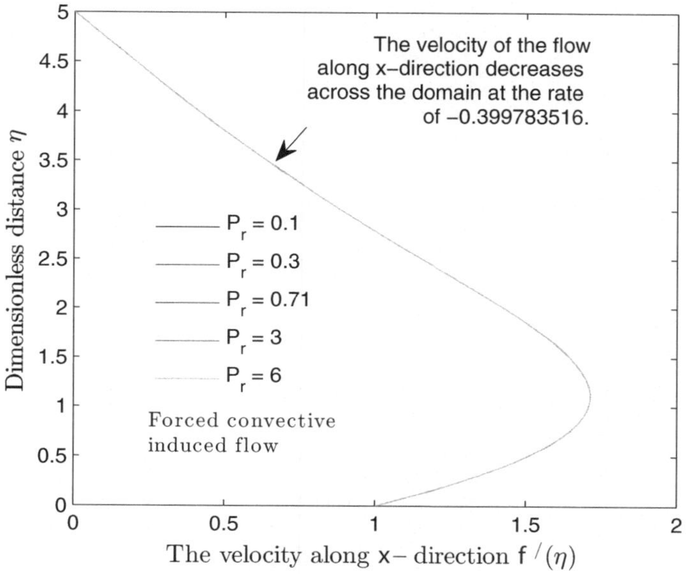

FIGURE 6.3
Variation in $f'(\eta)$ due to P_r: the case of forced convection.

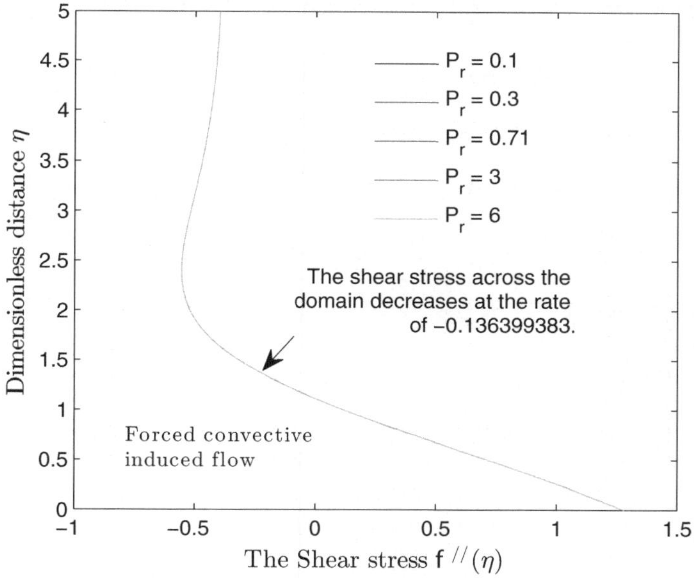

FIGURE 6.4
Variation in $f''(\eta)$ due to P_r: the case of forced convection.

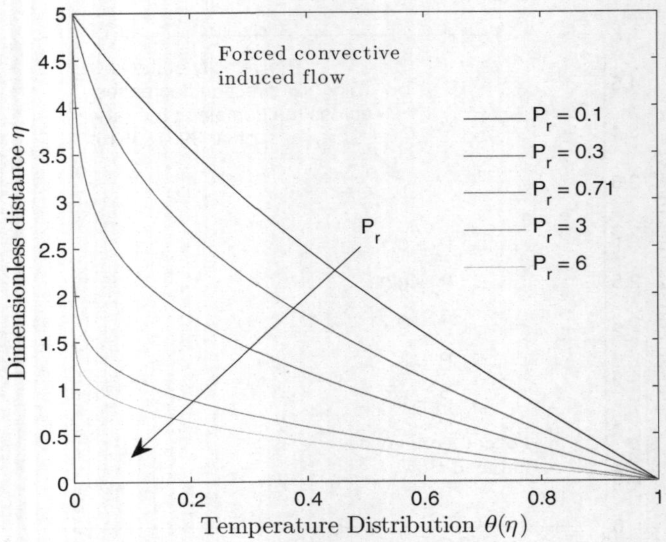

FIGURE 6.5
Variation in $\theta(\eta)$ due to P_r: the case of forced convection.

TABLE 6.1
The Rate of Decrease in $\theta(\eta)$ with η each at Various Values of Prandtl Number: The Case of Forced Convective Induced Flow

$P_r \Rightarrow$	0.1	0.3	0.71	3	6	10
S_{lp} for $\theta(\eta)$	−0.19440	−0.17750	−0.14449	−0.06527	−0.04918	−0.03573

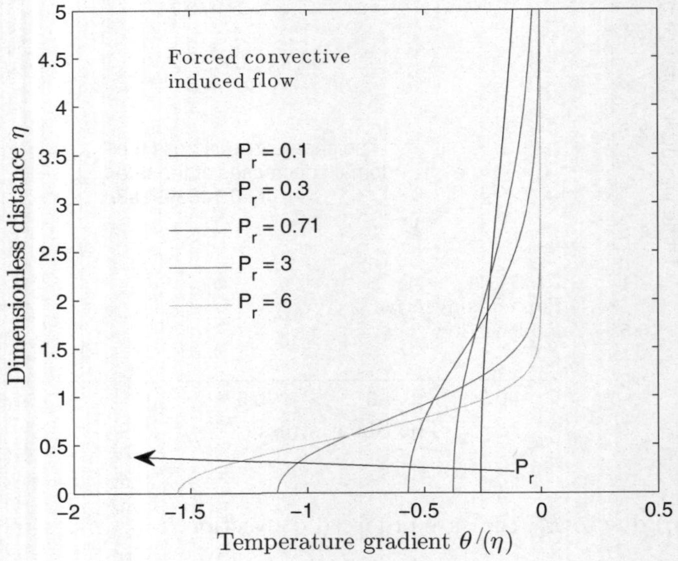

FIGURE 6.6
Variation in $\theta'(\eta)$ due to P_r: the case of forced convection.

6.5 Free Convective Induced Flow

It is important to remark that in Eq. (6.9), the momentum equation for free convective induced flow is complete when $q_w = g\beta(T - T_\infty)$. Upon using the similarity variables listed as Eq. (6.13), the dimensionless equation of the governing equation that models free convective induced flow is of the form

$$\frac{d^3 f}{d\eta^3} + \frac{1}{2} f \frac{d^2 f}{d\eta^2} + G_{rx}\theta = 0, \tag{6.19}$$

$$\frac{d^2\theta}{d\eta^2} + \frac{1}{2} P_r f \frac{d\theta}{d\eta} = 0. \tag{6.20}$$

Dimensionless boundary conditions for this case are

$$\frac{df}{d\eta} = 1, \quad f = 0, \quad \theta = 1 \quad at \quad \eta = 0. \tag{6.21}$$

$$\frac{df}{d\eta} \to 0, \quad \theta \to 0 \quad as \quad \eta \to \infty. \tag{6.22}$$

6.5.1 Research Questions II

This subsection provides answers to the accompanying examination questions:

1. In the case of free convection, how does the Prandtl number affect the temperature and velocity distribution of the flow?
2. What are the combined effects of Prandtl number and Grashof number on fluid flow induced by free convection?
3. What is the variation in velocity, temperature, local skin friction, and heat transfer rate in fluid flow induced by free convection due to increased Prandtl number?

6.5.2 Analysis and Discussion of Results II

The numerical solution of dimensionless governing Eqs. (6.19) and (6.20) subject to Eqs. (6.21) and (6.23) was obtained using the four-stage Lobatto IIIa formula as described in Chapter 1 using $\eta = 5$. In this case of the transport phenomenon, the effects of Prandtl number on free convective induced flow at different levels of buoyancy forces were simulated when $G_{rx} = 0.01$, $G_{rx} = 2$, $G_{rx} = 4$, and $G_{rx} = 6$. When $G_{rx} = 0.01$, the vertical velocity decreases negligibly near the free stream with no visible effect near the wall; see Figure 6.7. Figures 6.8–6.10, which unravel the impact of increasing Prandtl number on the motion along the y-direction when $G_{rx} = 2$, $G_{rx} = 4$, and $G_{rx} = 6$. It is seen that buoyancy forces and the increase in the ratio of momentum diffusivity to thermal diffusivity haves no impact on the vertical velocity within the interval of $0 \le y < 0.5$. In this case, $f(\eta = 5)$ was considered for further exploration. Table 6.2 shows that the velocity of the transport phenomenon decreases with the Prandtl number at the rate of -0.003070111 when buoyancy forces are small in magnitude and at the rate of -0.48374254 when $G_{rx} = 6$. Further analysis of these results is presented as Figure 6.27. As shown in Figure 6.11, when $G_{rx} = 0.01$, the horizontal velocity diminishes negligibly across the domain with no effect at the wall and at

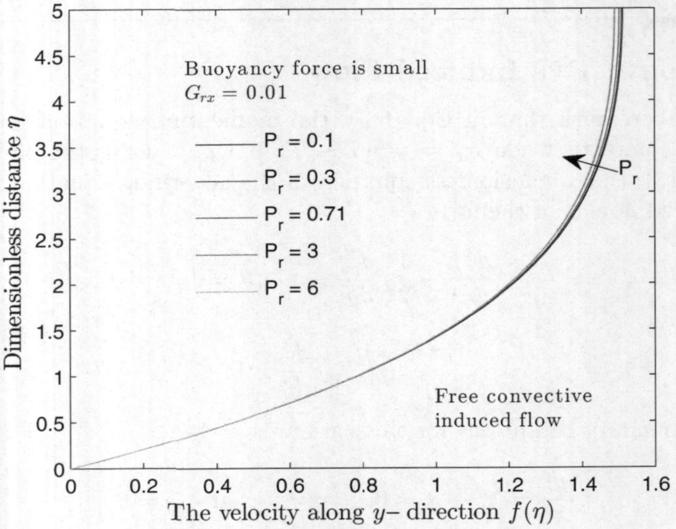

FIGURE 6.7
Variation in $f(\eta)$ due to P_r when $G_{rx} = 0.01$.

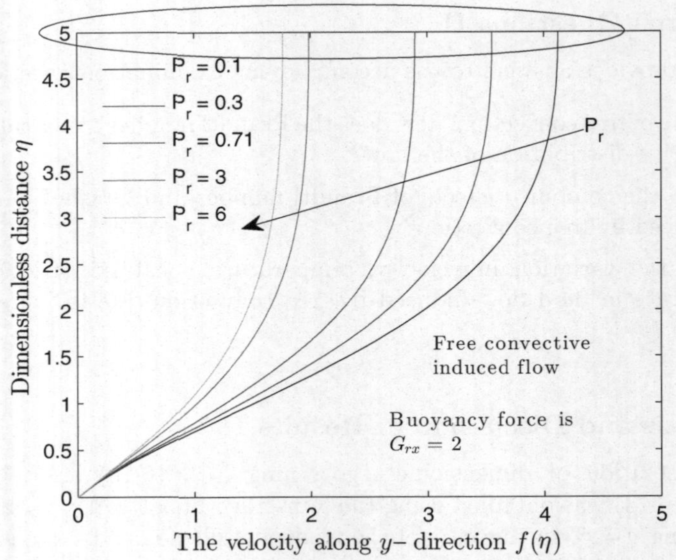

FIGURE 6.8
Variation in $f(\eta)$ due to P_r when $G_{rx} = 2$.

the free stream. Cross-examination as revealed in Table (6.3) shows that the velocity along the x-direction at $\eta = 1$ increases but negligibly with the Prandtl number at the rate of -0.003070111. This is evident as shown in Figures 6.7–6.14. For graphical illustration of these results, see Figure 6.29.

It is shown in Figure 6.15 that when $G_{rx} = 0.01$, the shear stress also decreases negligibly at the wall with no effect thereafter till the free stream. Analysis of Figures 6.15–6.18 as presented in Table 6.4 unravels the exact relationship between the two dimensionless

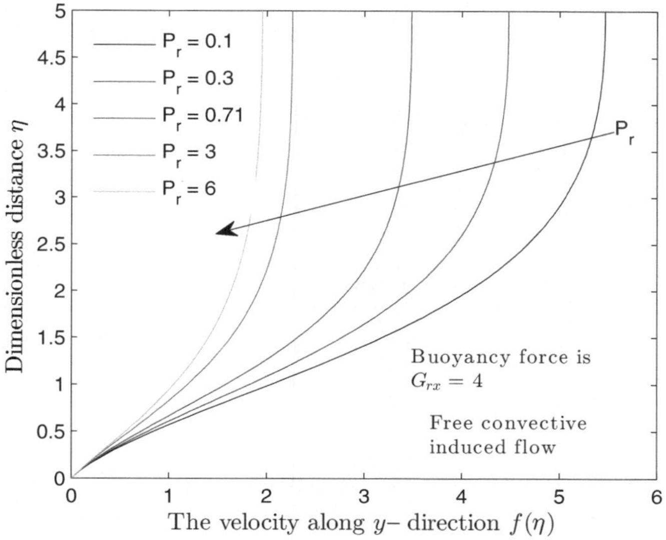

FIGURE 6.9
Variation in $f(\eta)$ due to P_r when $G_{rx} = 4$.

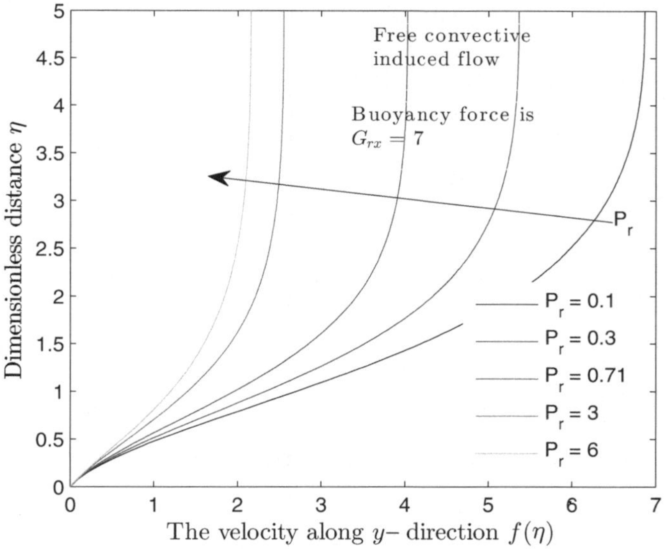

FIGURE 6.10
Variation in $f(\eta)$ due to P_r when $G_{rx} = 7$.

parameters and their effects on the local skin friction coefficients proportional to the friction between the wall and the last layer of fluid at the wall. It is seen that most minimum local skin friction coefficients at all the values of Prandtl number are seen when buoyancy forces are most minimum. The contour plot of the analysis is presented as Figure 6.31. Looking at Figures 6.19–6.22, it is worth noticing that the temperature distribution decreases with the Prandtl number at different rates. At $\eta = 2$, the temperature at various values of Prandtl number and Grashof number was extracted, analyzed, and presented as Figure 6.33. The

TABLE 6.2
Variations in the Vertical Velocity at the Free Stream with Prandtl Number P_r and Grashof Number G_{rx}: The Case of Free Convective Induced Flow

$P_r \Downarrow$	$f(\eta = 5)$ $G_{rx} = 0.01$	$f(\eta = 5)$ $G_{rx} = 2$	$f(\eta = 5)$ $G_{rx} = 4$	$f(\eta = 5)$ $G_{rx} = 6$
0.1	1.5126	4.1208	5.466	6.449
0.3	1.509	3.5556	4.4764	5.1041
0.71	1.503	2.8963	3.4807	3.87
1	1.4999	2.6307	3.1148	3.4401
2	1.4937	2.179	2.5187	2.7561
3	1.4913	1.9866	2.2632	2.4628
4	1.4902	1.8798	2.1167	2.2921
5	1.4896	1.8115	2.0198	2.1775
6	1.4892	1.7638	1.9502	2.0941
7	1.4889	1.7284	1.8974	2.03
S_{lp}	−0.003070111	−0.281579262	−0.401902445	−0.48374254

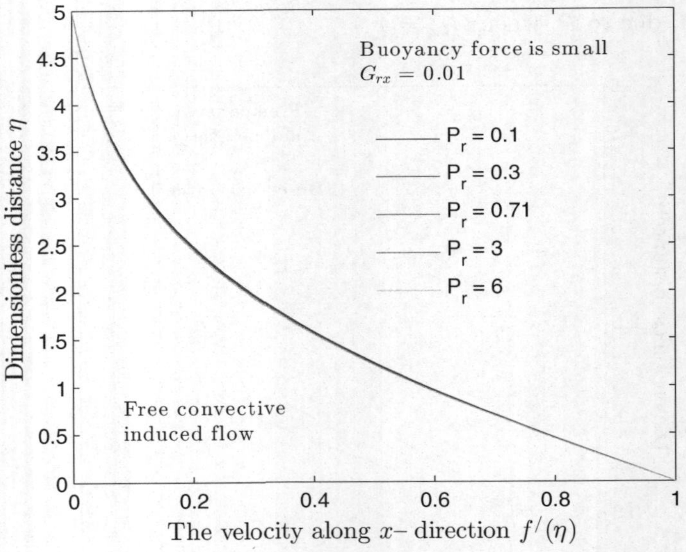

FIGURE 6.11
Variation in $f'(\eta)$ due to P_r when $G_{rx} = 0.01$.

slope linear regression through the data points shows that when $G_{rx} = 0.01$, the temperature at $\eta = 2$ decreases with the Prandtl at the rate of −0.074203341. When $G_{rx} = 2$, the same property decreases with the Prandtl number at the rate of −0.056915275. Table 6.5 further reveals that when $G_{rx} = 4$ and $G_{rx} = 6$, $\theta(\eta = 2)$ decreases with the Prandtl number at the rate of −0.049231566 and −0.044281741, respectively. Further exploration of the selected results is presented as Tables 6.6 and 6.7. It is seen that the temperature gradient decreases near the wall and increases after $\eta = 1$; see Figure 6.23–6.26. The variation in the heat transfer rate $-\theta(0)$ with the Grashof and Prandtl numbers is found to be minimum in the case of free convective induced flow as presented in Figure (6.35).

TABLE 6.3
Variations in the Horizontal Velocity at a Small Distance away from the Wall with Prandtl Number P_r and Grashof Number G_{rx}: The Case of Free Convective Induced Flow

$P_r \Downarrow$	$f'(\eta = 1)$ $G_{rx} = 0.01$	$f'(\eta = 1)$ $G_{rx} = 2$	$f'(\eta = 1)$ $G_{rx} = 4$	$f'(\eta = 1)$ $G_{rx} = 6$
0.1	0.5639	1.6903	2.4433	3.0239
0.3	0.563	1.4917	2.0702	2.4922
0.71	0.5614	1.2653	1.6607	1.9055
1	0.5605	1.1371	1.4948	1.6937
2	0.5585	0.9597	1.1906	1.3605
3	0.5576	0.8611	1.0419	1.1765
4	0.5571	0.8017	0.9512	1.0642
5	0.5568	0.7742	0.8896	0.9877
6	0.5657	0.7331	0.8449	0.9349
7	0.5655	0.7116	0.8109	0.8892
S_{lp}	0.000153477	−0.120598122	−0.196137182	−0.247501182

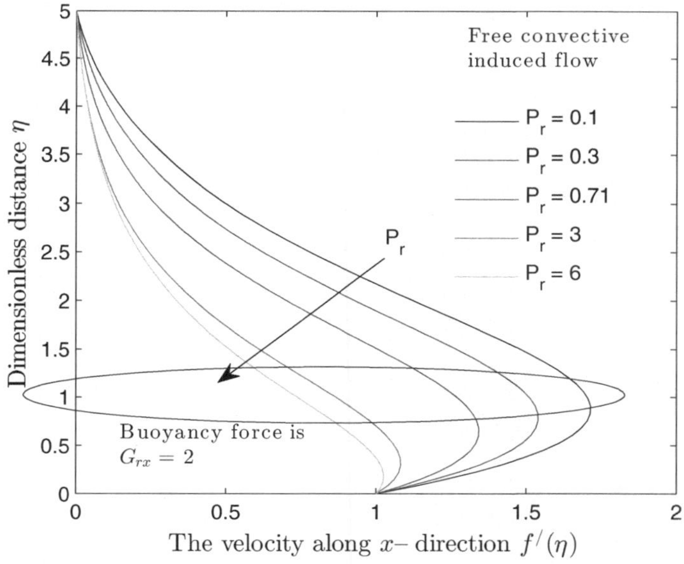

FIGURE 6.12
Variation in $f'(\eta)$ due to P_r when $G_{rx} = 2$.

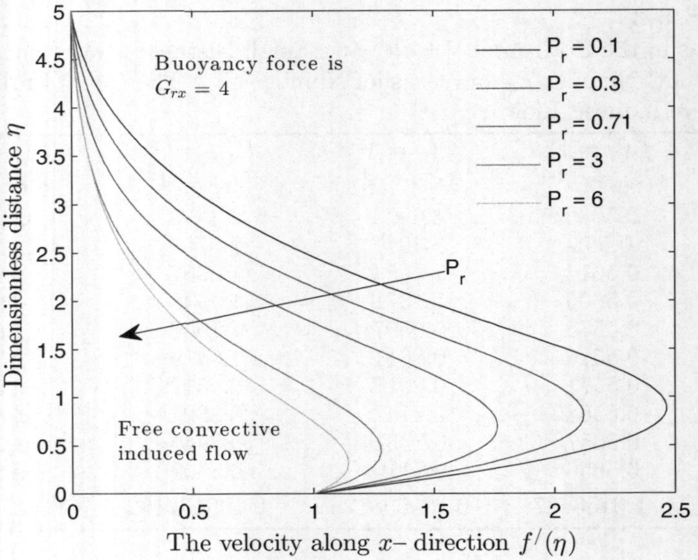

FIGURE 6.13
Variation in $f'(\eta)$ due to P_r when $G_{rx} = 4$.

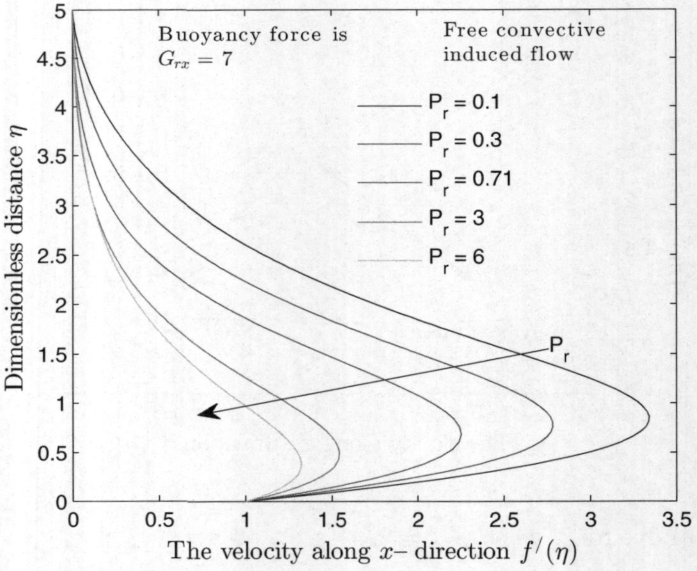

FIGURE 6.14
Variation in $f'(\eta)$ due to P_r when $G_{rx} = 7$.

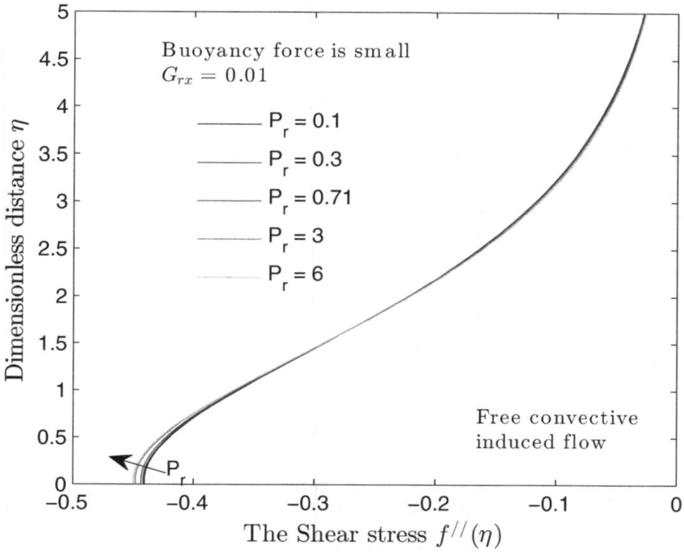

FIGURE 6.15
Variation in $f''(\eta)$ due to P_r when $G_{rx} = 0.01$.

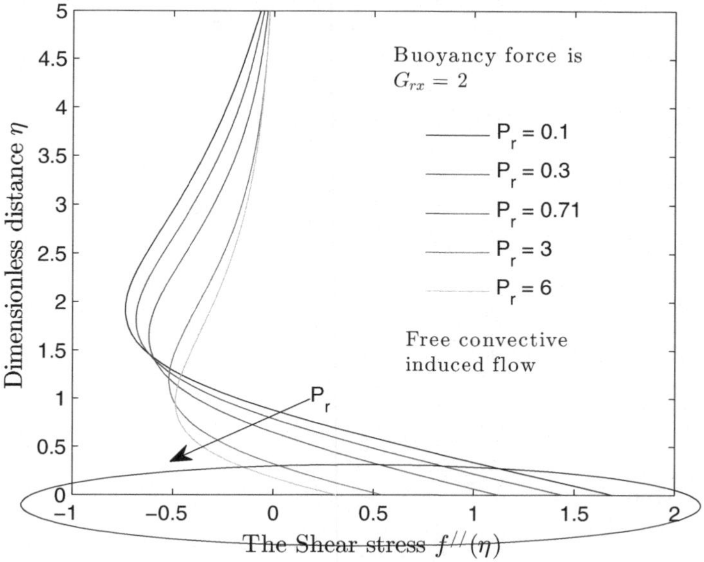

FIGURE 6.16
Variation in $f''(\eta)$ due to P_r when $G_{rx} = 2$.

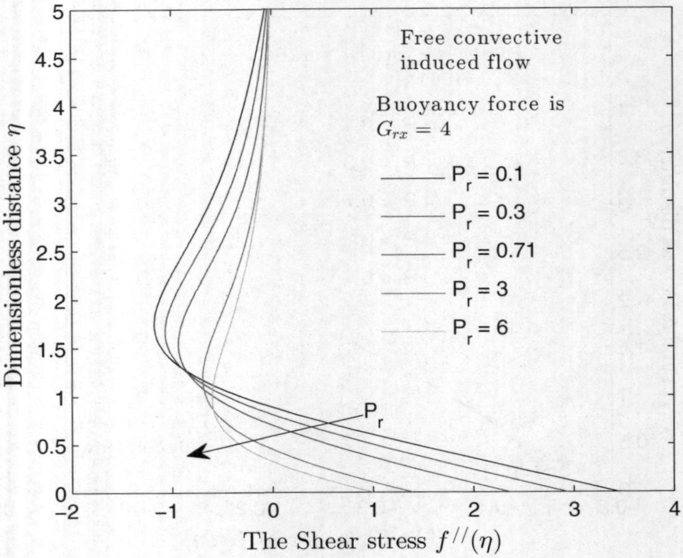

FIGURE 6.17
Variation in $f''(\eta)$ due to P_r when $G_{rx} = 4$.

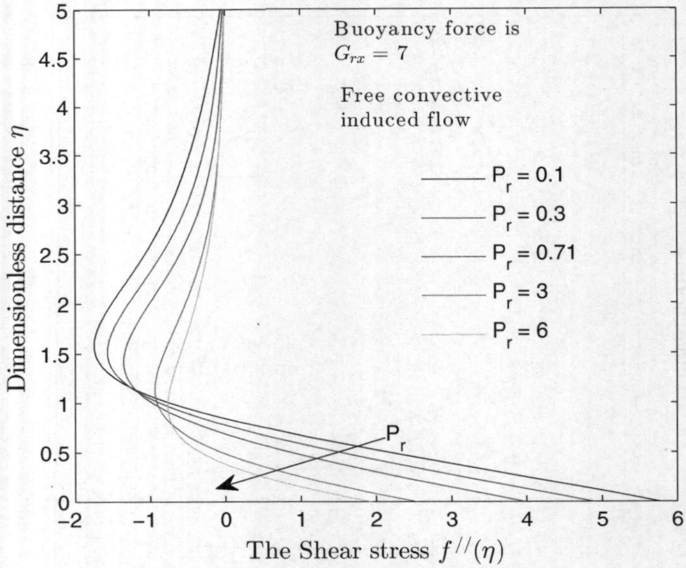

FIGURE 6.18
Variation in $f''(\eta)$ due to P_r when $G_{rx} = 7$.

TABLE 6.4
Variations in the Local Skin Friction Coefficients with Prandtl Number P_r and Grashof Number G_{rx}: The Case of Free Convective Induced Flow

$P_r \Downarrow$	$f''(\eta=0)$ $G_{rx}=0.01$	$f''(\eta=0)$ $G_{rx}=2$	$f''(\eta=0)$ $G_{rx}=4$	$f''(\eta=0)$ $G_{rx}=6$
0.1	−0.4407	1.6905	3.4426	5.0261
0.3	−0.4417	1.4424	2.9341	4.2632
0.71	−0.4435	1.127	2.3627	3.4659
1	−0.4446	0.9853	2.1224	3.1413
2	−0.4469	0.7017	1.6526	2.5135
3	−0.4482	0.5479	1.397	2.1718
4	−0.4489	0.4463	1.2258	1.9417
5	−0.4494	0.3721	1.0991	1.7705
6	−0.4498	0.3146	0.9997	1.6355
7	−0.4501	0.2682	0.9186	1.5249
S_{lp}	−0.001296757	−0.181513219	−0.313261757	−0.427586976

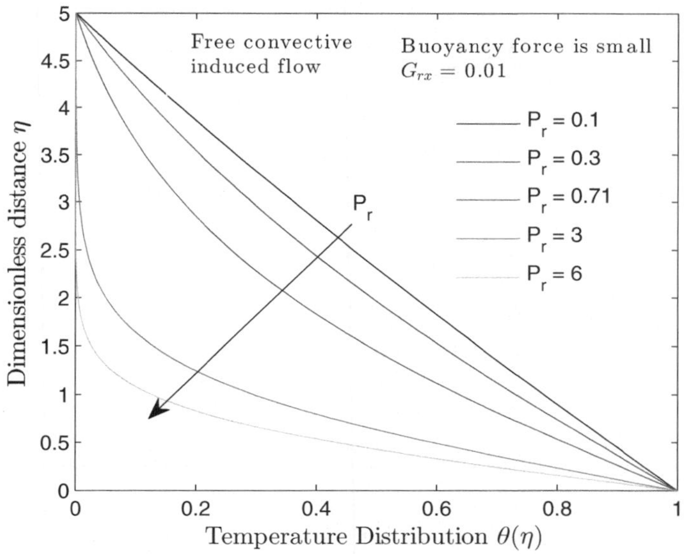

FIGURE 6.19
Variation in $\theta(\eta)$ due to P_r when $G_{rx} = 0.01$.

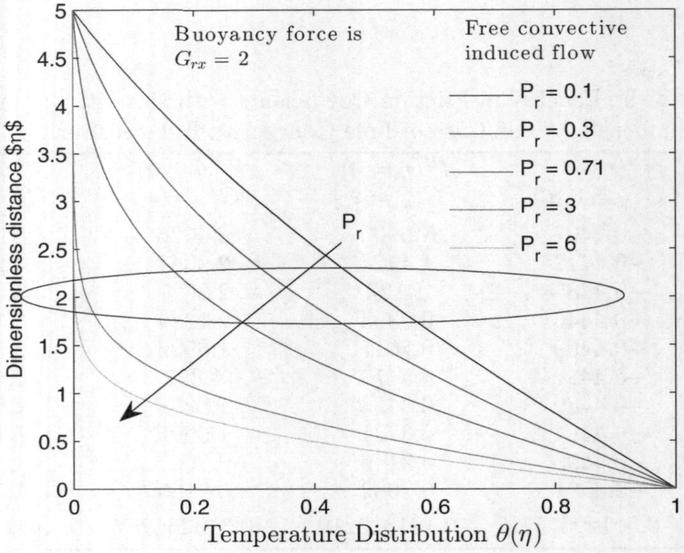

FIGURE 6.20
Variation in $\theta(\eta)$ due to P_r when $G_{rx} = 2$.

TABLE 6.5
Variations in the Temperature Distribution with Prandtl Number P_r and Grashof Number G_{rx}: The Case of Free Convective Induced Flow

$P_r \Downarrow$	$\theta(\eta = 2)$ $G_{rx} = 0.01$	$\theta(\eta = 2)$ $G_{rx} = 2$	$\theta(\eta = 2)$ $G_{rx} = 4$	$\theta(\eta = 2)$ $G_{rx} = 6$
0.1	0.5540	0.5066	0.4761	0.4535
0.3	0.4827	0.3722	0.3184	0.2830
0.71	0.3500	0.2144	0.1630	0.1288
1	0.2733	0.1518	0.1091	0.0854
2	0.1118	0.0532	0.0298	0.0236
3	0.0486	0.0208	0.0120	0.0078
4	0.0209	0.0086	0.0046	0.0028
5	0.0092	0.0037	0.0019	0.0011
6	0.0041	0.0016	0.0008	0.0004
7	0.0019	0.0007	0.0003	0.0002
S_{lp}	−0.074203341	−0.056915275	−0.049231566	−0.044281741

TABLE 6.6
The Rate of Decrease in $\theta(\eta)$ with η each at Various Values of Prandtl Number When $G_{rx} = 0.01$: The Case of Free Convective Induced Flow

$P_r \Rightarrow$	0.1	0.3	0.71	3	6	10
S_{lp} for $\theta(\eta)$	−0.19834	−0.19385	−0.18112	−0.11572	−0.08110	−0.05203

TABLE 6.7
The Rate of Decrease in $\theta(\eta)$ with η each at Various Values of Prandtl Number When $G_{rx} = 6$: The Case of Free Convective Induced Flow

$P_r \Rightarrow$	0.1	0.3	0.71	3	6	10
S_{lp} for $\theta(\eta)$	−0.19150	−0.16867	−0.12985	−0.08925	−0.05913	−0.04217

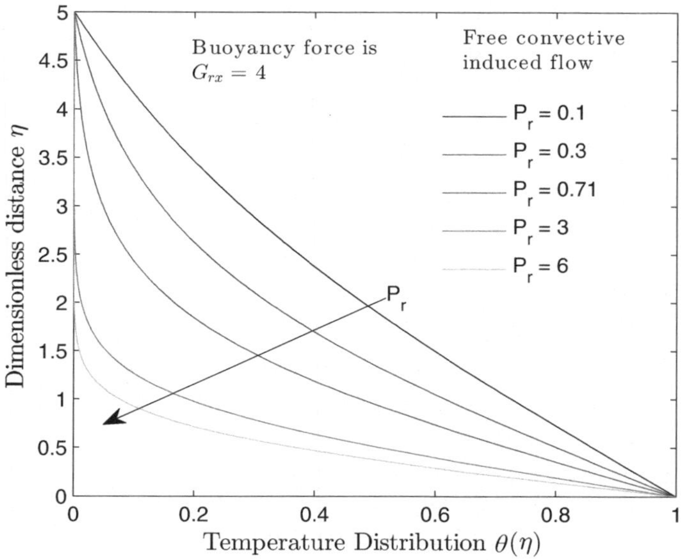

FIGURE 6.21
Variation in $\theta(\eta)$ due to P_r when $G_{rx} = 4$.

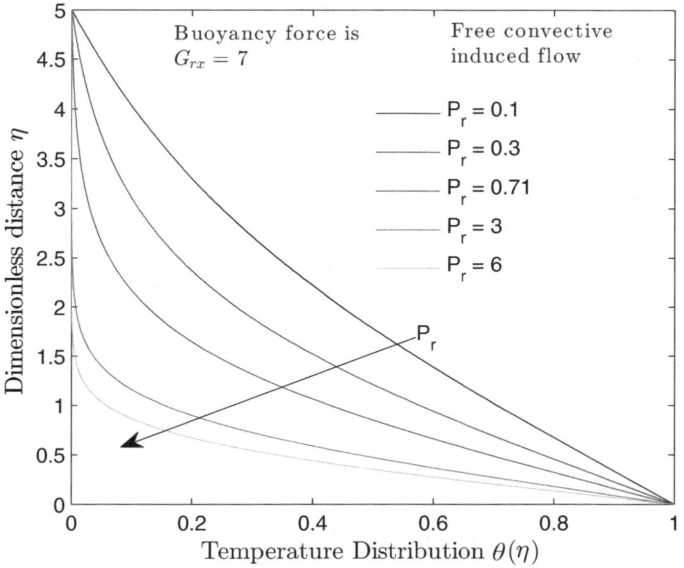

FIGURE 6.22
Variation in $\theta(\eta)$ due to P_r when $G_{rx} = 7$.

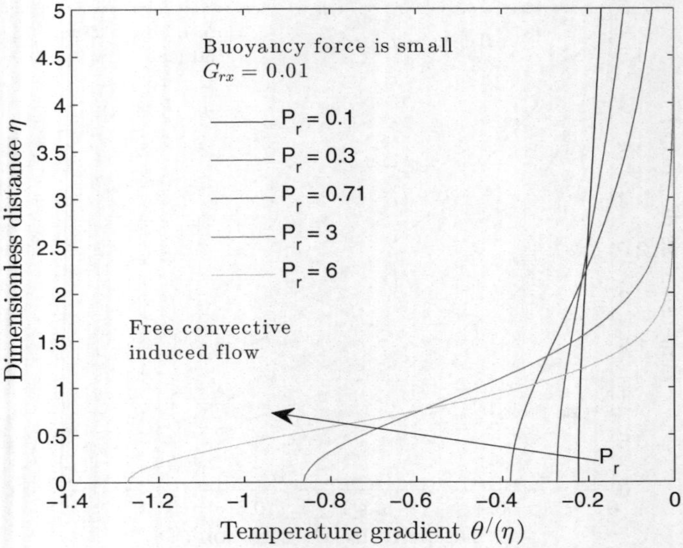

FIGURE 6.23
Variation in $\theta'(\eta)$ due to P_r when $G_{rx} = 0.01$.

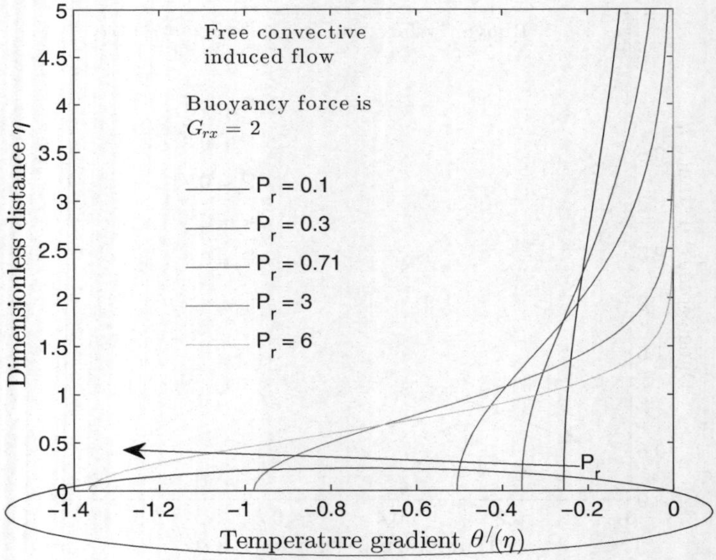

FIGURE 6.24
Variation in $\theta'(\eta)$ due to P_r when $G_{rx} = 2$.

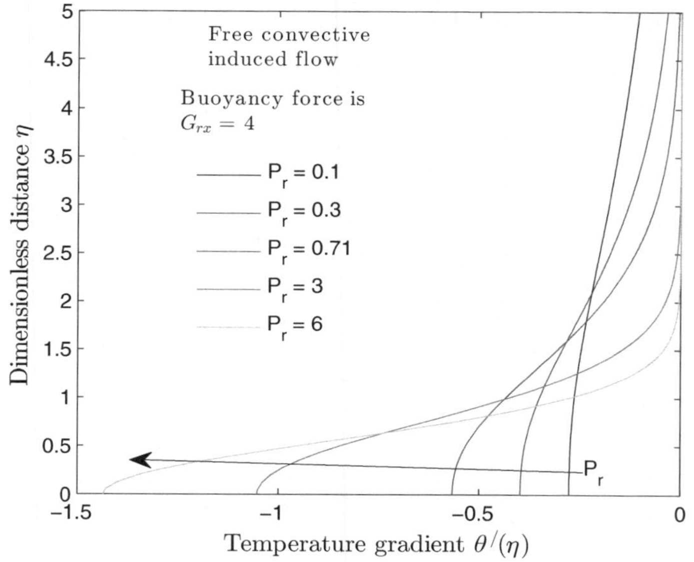

FIGURE 6.25
Variation in $\theta'(\eta)$ due to P_r when $G_{rx} = 4$.

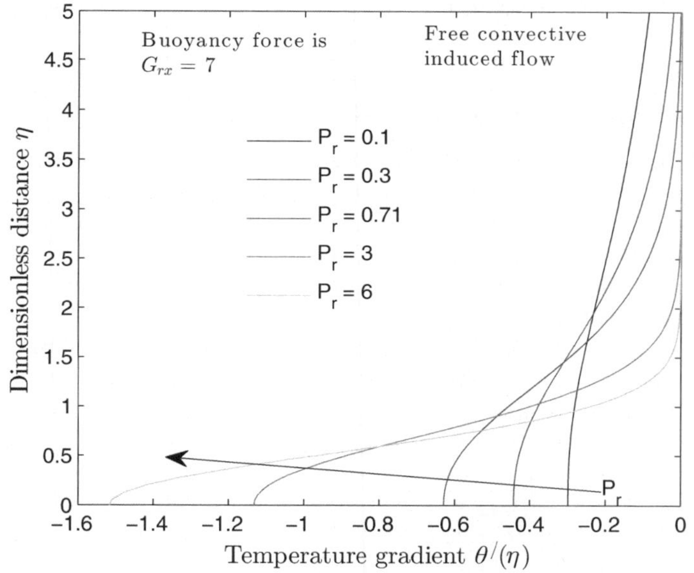

FIGURE 6.26
Variation in $\theta'(\eta)$ due to P_r when $G_{rx} = 7$.

6.6 Mixed Convective Induced Flow

Finally, $q_w = g\beta(T - T_\infty) - \frac{1}{\rho}\frac{dp}{dx}$ in Eq. (6.9) for the momentum equation for mixed convective induced flow. Upon using the similarity variables listed as Eq. (6.13), the dimensionless equation of the governing equation that models mixed convective induced flow is of the form

$$\frac{d^3 f}{d\eta^3} + \frac{1}{2} f \frac{d^2 f}{d\eta^2} + 1 + G_{rx}\theta = 0, \tag{6.23}$$

$$\frac{d^2\theta}{d\eta^2} + \frac{1}{2} P_r f \frac{d\theta}{d\eta} = 0. \tag{6.24}$$

Dimensionless boundary conditions for this case are

$$\frac{df}{d\eta} = 1, \quad f = 0, \quad \theta = 1 \quad at \quad \eta = 0. \tag{6.25}$$

$$\frac{df}{d\eta} \to 0, \quad \theta \to 0 \quad as \quad \eta \to \infty. \tag{6.26}$$

6.6.1 Research Questions III

This subsection intends to give replies to the accompanying exploration questions:

1. In the case of mixed convective induced flow, what are the variations in velocity, temperature distribution, heat transfer rate, and local skin friction coefficients due to an increase in the Prandtl number?
2. How do the Prandtl number and the Grashof number influence fluid flow induced by mixed convection?

6.6.2 Analysis and Discussion of Results III

The numerical solutions of Eqs. (6.23) and (6.24) subject to Eqs. (6.25) and (6.26) were obtained separately using four-stage Lobatto IIIa formula as described in Chapter 1 using $\eta = 5$. It is worth remarking that the observed effect of the Prandtl number on free convective flow is similar to that of mixed convective flow. The only exception is that the mixed convective flow showcases a higher magnitude of the flow's properties across the domain $\eta \varepsilon [0, \infty)$. For instance, at $\eta = 5$, Figures 6.27 and 6.28 show that the vertical velocity in mixed convective induced flow is higher than that in free convective induced flow. Likewise, the horizontal velocity, shear stress, and heat transfer rate in the case of mixed convective flow are higher than that of free convective flow; see Figures 6.29–6.36. The reverse is the case of temperature distribution as mixed convective induced flow portrays lower temperature distribution.

When the Grashof number $G_{rx} = 6$, it is evident from Tables 6.2 to 6.8 that the rate of decrease in $f(\eta = 5)$ with the Prandtl number for free convective flow is -0.48374254, while that for mixed convection is -0.26622514. Tables 6.9 and 6.10 also confirm the claim that at each value of Prandtl number, the horizontal velocity at $\eta = 1$ and $f''(0)$ in the case of mixed convection is higher than that of free convection when the Grashof number is either small or large. It is established in Table 6.11 that the temperature distribution at $\eta = 2$ decreases with the Prandtl number at the maximum rate of -0.035722007 when the Grashof number is large in magnitude. Graphically, this is presented as Figure 6.34. In this case, the

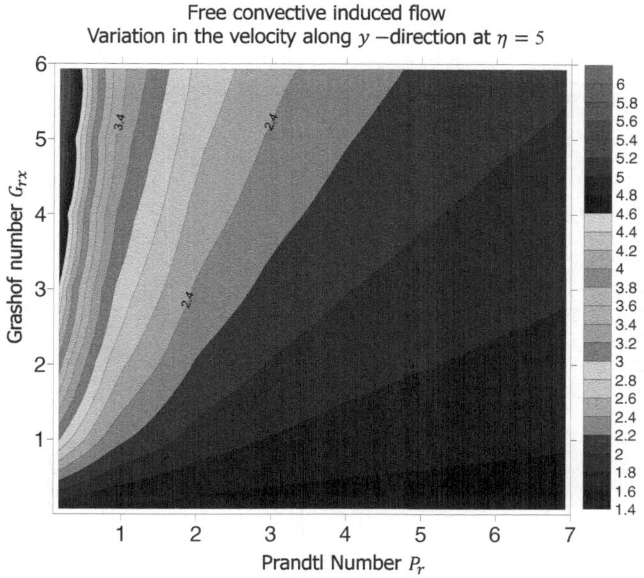

FIGURE 6.27
Variation in $f(\eta = 5)$ with P_r and G_{rx}: the case of free convection.

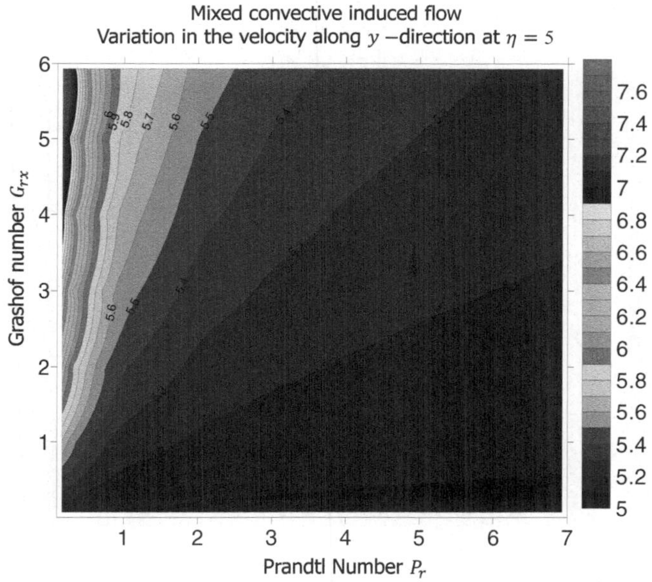

FIGURE 6.28
Variation in $f(\eta = 5)$ with P_r and G_{rx}: the case of mixed convection.

distribution of heat energy is minimum because the motion of the substance is already fast. Since the dynamics in the case of free convective induced flow is small compared to mixed convective flow, maximum heat energy across a wider domain is seen when the body force term $q_w = g\beta(T - T_\infty)$; see Figure 6.33. Tables 6.12 and 6.13 present the variation in the

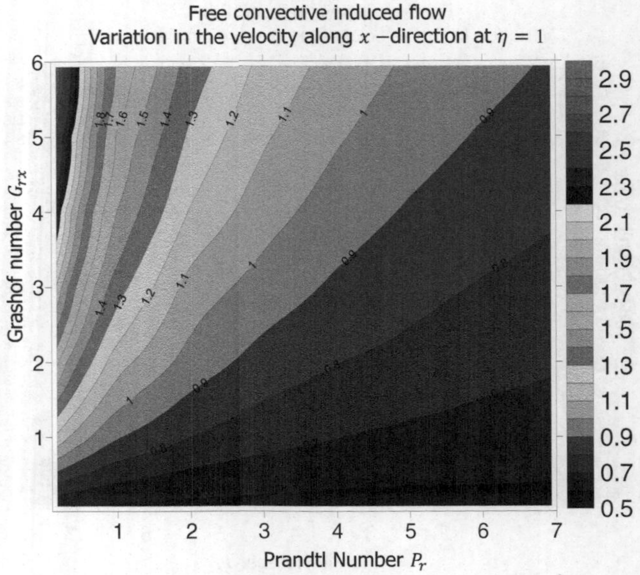

FIGURE 6.29
Variation in $f'(\eta = 1)$ with P_r and G_{rx}: the case of free convection.

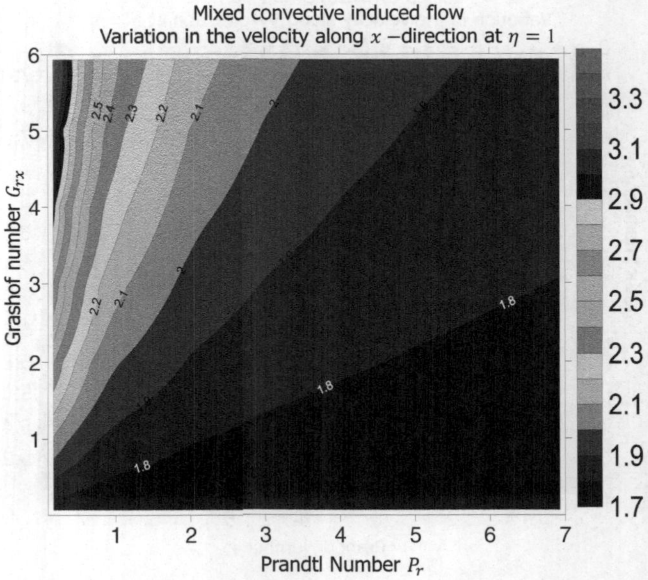

FIGURE 6.30
Variation in $f'(\eta = 1)$ with P_r and G_{rx}: the case of mixed convection.

temperature distribution across the domain when $G_{rx} = 0.01$ and $G_{rx} = 6$. Figures (6.35) and (6.36) show that the rate of heat transfer across free convective induced flow is small in magnitude while that of mixed convective induced flow is large in magnitude.

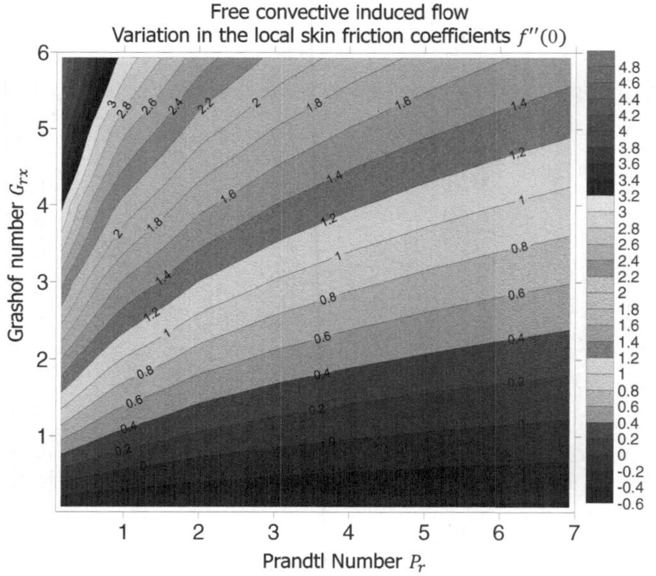

FIGURE 6.31
Variation in $f''(\eta = 0)$ with P_r and G_{rx}: the case of free convection.

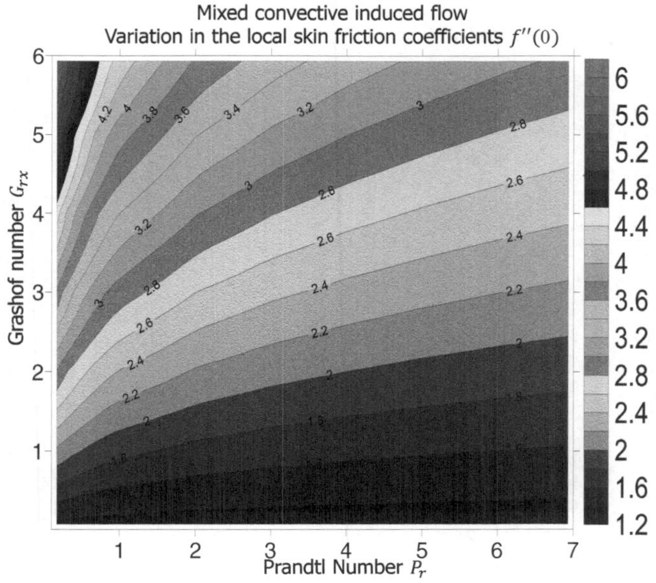

FIGURE 6.32
Variation in $f''(\eta = 0)$ with P_r and G_{rx}: the case of mixed convection.

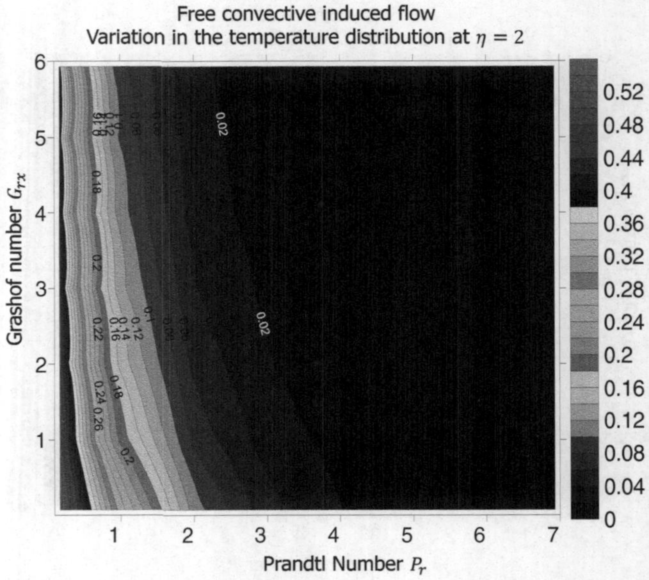

FIGURE 6.33
Variation in $\theta(\eta = 2)$ with P_r and G_{rx}: the case of free convection.

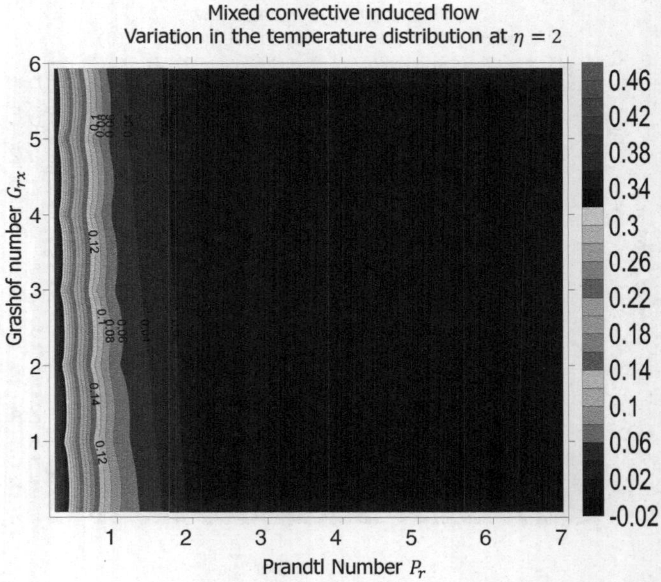

FIGURE 6.34
Variation in $\theta(\eta = 2)$ with P_r and G_{rx}: the case of mixed convection.

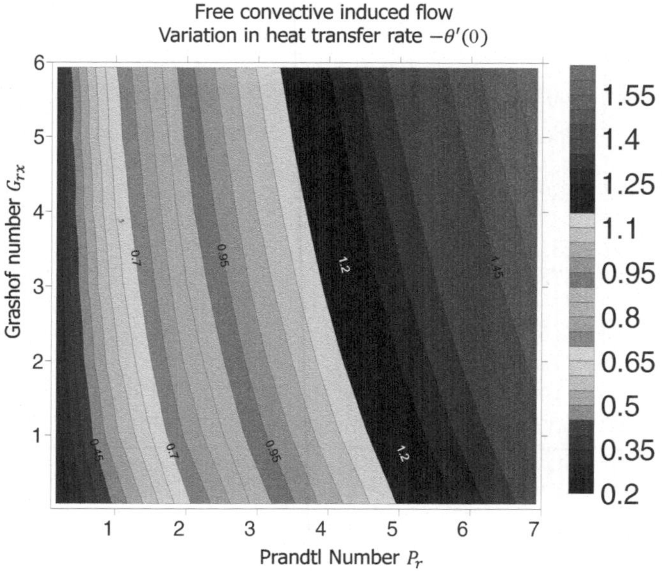

FIGURE 6.35
Variation in $-\theta'(0)$ with P_r and G_{rx}: the case of free convection.

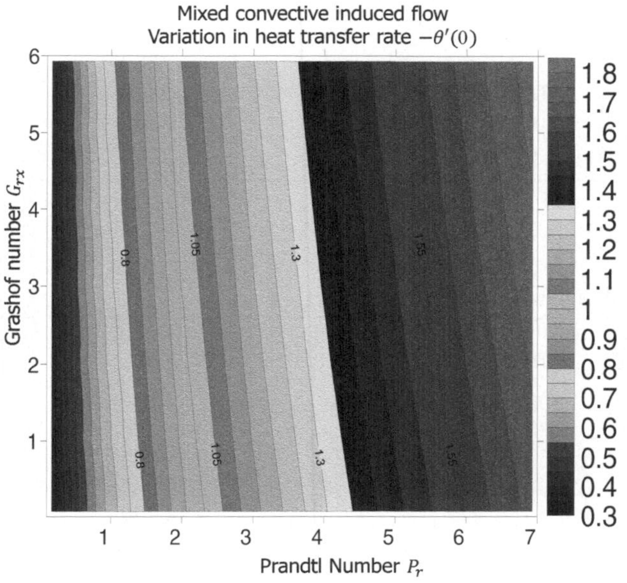

FIGURE 6.36
Variation in $-\theta'(0)$ with P_r and G_{rx}: the case of mixed convection.

TABLE 6.8
Variations in the Vertical Velocity at $\eta = 5$ with Prandtl Number P_r and Grashof Number G_{rx}: The Case of Mixed Convective Induced Flow

$P_r \Downarrow$	$f(\eta=5)$ $G_{rx}=0.01$	$f(\eta=5)$ $G_{rx}=2$	$f(\eta=5)$ $G_{rx}=4$	$f(\eta=5)$ $G_{rx}=6$
0.1	5.0848	6.3574	7.2959	8.0559
0.3	5.0821	5.8914	6.4575	6.8996
0.71	5.0796	5.5131	5.8291	6.0812
1	5.0789	5.4094	5.6591	5.8619
2	5.0781	5.2721	5.429	5.5615
3	5.0778	5.2206	5.3402	5.4435
4	5.0776	5.1924	5.2907	5.3769
5	5.0775	5.1742	5.2585	5.3332
6	5.0774	5.1614	5.2355	5.3019
7	5.0774	5.1518	5.2182	5.2781
S_{lp}	−0.000733147	−0.117969279	−0.200984549	−0.26622515

TABLE 6.9
Variations in the Horizontal Velocity at $\eta = 1$ with Prandtl Number P_r and Grashof Number G_{rx}: The Case of Mixed Convective Induced Flow

$P_r \Downarrow$	$f'(\eta=1)$ $G_{rx}=0.01$	$f'(\eta=1)$ $G_{rx}=2$	$f'(\eta=1)$ $G_{rx}=4$	$f'(\eta=1)$ $G_{rx}=6$
0.1	1.7174	2.4877	3.0886	3.5836
0.3	1.7162	2.2853	2.7133	3.0563
0.71	1.7149	2.0807	2.3596	2.583
1	1.7145	2.0086	2.2365	2.42
2	1.7138	1.8915	2.0352	2.153
3	1.7135	1.8405	1.9466	2.035
4	1.7133	1.8116	1.896	1.9673
5	1.7093	1.7929	1.8632	1.9232
6	1.7092	1.7799	1.8402	1.8922
7	1.7091	1.7704	1.8232	1.8693
S_{lp}	−0.001137462	−0.08022522	−0.13995052	−0.18786472

TABLE 6.10
Variations in the Vertical Velocity at $\eta = 5$ with Prandtl Number P_r and Grashof Number G_{rx}

$P_r \Downarrow$	$f''(\eta=0)$ $G_{rx}=0.01$	$f''(\eta=0)$ $G_{rx}=2$	$f''(\eta=0)$ $G_{rx}=4$	$f''(\eta=0)$ $G_{rx}=6$
0.1	1.2899	3.0804	4.6932	6.1902
0.3	1.2885	2.8012	4.1476	5.3894
0.71	1.2868	2.5028	3.6025	4.6249
1	1.2861	2.3884	3.3955	4.3362
2	1.2849	2.179	3.0122	3.7988
3	1.2843	2.0696	2.8094	3.5124
4	1.2839	1.9975	2.6748	3.3214
5	1.2837	1.9447	2.5756	3.1801
6	1.2834	1.9036	2.498	3.0692
7	1.2833	1.8702	2.4348	2.9787
S_{lp}	−0.000801126	−0.144661225	−0.267940107	−0.378868826

TABLE 6.11
Variations in the Vertical Velocity at $\eta = 5$ with Prandtl Number P_r and Grashof Number G_{rx}

$P_r \Downarrow$	$\theta(\eta = 2)$ $G_{rx} = 0.01$	$\theta(\eta = 2)$ $G_{rx} = 2$	$\theta(\eta = 2)$ $G_{rx} = 4$	$\theta(\eta = 2)$ $G_{rx} = 6$
0.1	0.4990	0.4686	0.4461	0.4279
0.3	0.3341	0.2868	0.2555	0.2323
0.71	0.1427	0.1128	0.0940	0.0808
1	0.0808	0.0619	0.0500	0.0418
2	0.0132	0.0100	0.0071	0.0055
3	0.0024	0.0016	0.0012	0.0009
4	0.0005	0.0003	0.0002	0.0001
5	0.0001	0.0001	0.0000	0.0000
6	0.0000	0.0000	0.0000	0.0247
7	0.0000	0.0000	0.0000	0.0000
S_{lp}	−0.049307782	−0.043695872	−0.0399121	−0.035722007

TABLE 6.12
The Rate of Decrease in $\theta(\eta)$ with η each at Various Values of Prandtl Number When $G_{rx} = 0.01$: The Case of Mixed Convective Induced Flow

$P_r \Rightarrow$	0.1	0.3	0.71	3	6	10
S_{lp} for $\theta(\eta)$	−0.19915	−0.16413	−0.12070	−0.06574	−0.04412	−0.03360

TABLE 6.13
The Rate of Decrease in $\theta(\eta)$ with η each at Various Values of Prandtl Number When $G_{rx} = 6$: The Case of Mixed Convective Induced Flow

$P_r \Rightarrow$	0.1	0.3	0.71	3	6	10
S_{lp} for $\theta(\eta)$	−0.18089	−0.13748	−0.10163	−0.05297	−0.05034	−0.04087

6.7 Tutorial Questions

1. In the case of forced convection, how and why does the Prandtl number influence the velocity and temperature distribution of the fluid as it flows?
2. How does the Prandtl number affect the motion along the x-direction and y-direction, and shear stress in the case of fluid flow induced by (i) forced convection, (ii) free convection, and (iii) mixed convection?
3. In the case of free convection, how does the Prandtl number affect the temperature and velocity distribution of the flow?
4. What is the relationship between the Prandtl number and the Grashof number on fluid flow induced by free convection?
5. What is the variation in velocity, temperature, local skin friction, and heat transfer rate in fluid flow induced by free convection due to increased Prandtl number?

6. In the case of mixed convective induced flow, what are the variations in velocity, temperature distribution, heat transfer rate, and local skin friction coefficients due to an increase in the Prandtl number?
7. What is the physical explanation behind the combined effect of Prandtl number and Grashof number on fluid flow induced by mixed convection?
8. At different levels of thermo-migration of spherical carbon nanotubes, cylindrical graphene, and platelet alumina nanoparticles, how does increasing Prandtl number affects the dynamics of such a water-based ternary-hybrid nanofluid by three modes of convection?

7

Analysis of Self-Similar Flows III

7.1 Background Information

Radiative emission is governed by the interaction between vibrational, electronic, and rotational energies of molecules and atoms that make up matter, and all things produce electromagnetic radiation. Radiative transfer is determined by the temperature of equilibrium between two objects with differing amounts of heat energy. The surface, composition, temperature, opacity, angle of interception or emission, and radiation wavelength all affect a body's radiative properties. Out-scattering and in-scattering are two types of radiation transport. When molecules and particles scatter away from their initial orientation during radiation, this is known as out-scattering. When molecules and particles are drawn at different angles to the propagation direction during radiation, this is known as in-scattering. The oscillation of waves during electromagnetic radiation is perpendicular to the propagation direction, according to wave theory. When thermal radiation, internal heating, removal of heating, thermo-effect, and thermal diffusion are relevant, this chapter elucidates the growing influence of Prandtl number on boundary layer flows.

7.2 Introduction: Thermal Radiation

One of the three ways of heat transmission that is independent of the intervening medium is radiation. This occurrence defies conduction and convection heat transfers, which require the passage of energy from one item to another through a physical medium. The transfer of heat energy essentially explains the propagation of particles from a hot to a cold environment. In the case of heat leaking and satellite temperature management, this is a regular occurrence. One of the outcomes of the experimental study by Pierre Prevost shows that all bodies do radiate heat energy, either cold or hot. Poet Josef Stefan in the year 1879 shows that the total radiant heat energy can be measured, and Boltzmann Ludwig adopted this to form the fundamental basis of Stefan–Boltzmann law. Howell et al. [136] remarked that in the absence of a medium, radiation is the only suitable mode of heat energy transfer. More so, there exists a correlation between the strength of energy that will experience emission and the internal energy state of the emitter. Due to volumetric emission from every point, the strength of radiative energy increases as the propagation proceeds in a medium. However, the transmission of heat energy through electromagnetic waves has been modeled using linearity and, recently, nonlinearity. Nonlinear transmission of heat energy through electromagnetic waves using the nonlinear Rosseland approximation was presented by Cortell [82]. The nonlinear thermal radiation model is more appropriate due to the inherent relationship between the wall temperature and thermal radiation. The concept was used by Koriko et al. [164] to examine the effect of increasing thermal

DOI: 10.1201/9781003217374-7

radiation parameter and the corresponding temperature parameter on the dynamics of the non-Newtonian Eyring–Powell fluid conveying alumina nanoparticles subject to thermal radiation. The results show that the Nusselt number increases with thermal radiation at the rate of 0.015799606, while the temperature parameter increases the Nusselt number at the rate of 0.006751067.

7.3 Fluid Flow Subject to Thermal Radiation

In this section, we present the effects of Prandtl number on the flow of viscous fluids on vertical and horizontal surfaces due to buoyancy forces, as in the case of free convection when the transmission of heat energy is nonlinear through electromagnetic waves; see Figure 7.1. Let us consider the two-dimensional flow of a Newtonian fluid where the dynamic viscosity and thermal conductivity are constant functions of temperature. The governing equation suitable to model the flow on a linearly stretchable surface is

$$u_x + v_y = 0, \tag{7.1}$$

$$uu_x + vu_y = \frac{\mu}{\rho}u_{yy} + g\beta(T - T_\infty), \tag{7.2}$$

$$uT_x + vT_y = \frac{\kappa}{\rho c_p}T_{yy} + \frac{1}{\rho c_p}\frac{\partial}{\partial y}\left(\frac{4\sigma^*}{3k^*}4T^3T_y\right). \tag{7.3}$$

In Eq. (7.3), the units of Stefan–Boltzmann constant σ^* and absorption coefficient k^* are ($Wm^{-2}K^{-4}$) and (m^{-1}), respectively. Equations (7.1)–(7.3) are subject to the boundary conditions

$$u = U_\infty = U_o x, \quad v = v_w, \quad T = T_w, \quad at \quad y = 0. \tag{7.4}$$

$$u \to 0, \quad T \to T_\infty, \quad as \quad y \to \infty. \tag{7.5}$$

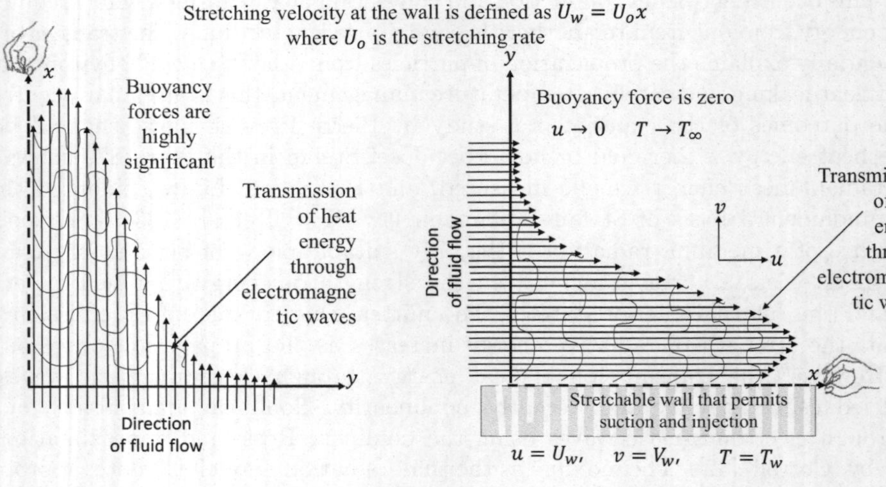

FIGURE 7.1
Illustration of fluid flow when thermal radiation and buoyancy are significant.

We can easily obtain the dimensionless governing equation for the case of self-similar solution using the following variables:

$$\eta = y\sqrt{\frac{U_\infty}{\vartheta x}}, \quad \psi(x,y) = \sqrt{\vartheta x U_\infty} f(\eta), \quad u = \frac{\partial \psi}{\partial y} \quad v = -\frac{\partial \psi}{\partial x},$$

$$\alpha = \frac{\kappa}{\rho c_p}, \quad \theta(\eta) = \frac{T - T_\infty}{T_w(x) - T_\infty}, \quad P_r = \frac{\vartheta}{\alpha}, \quad f_w = -\frac{v_w}{\sqrt{\vartheta U_o}},$$

$$Gr_x = \frac{gx\beta(T_w - T_\infty)}{U_\infty^2}, \qquad R_a = \frac{3\kappa k^*}{16\sigma^* T_\infty^3}. \tag{7.6}$$

The final dimensionless governing equation is

$$\frac{d^3 f}{d\eta^3} + \frac{1}{2} f \frac{d^2 f}{d\eta^2} + Gr_x \theta = 0, \tag{7.7}$$

$$\left[1 + \frac{(1 + \theta\theta_w - \theta)^3}{R_a}\right] \frac{d^2\theta}{d\eta^2} + \frac{1}{2} P_r f \frac{d\theta}{d\eta} + 3(\theta_w - 1)\frac{(1 + \theta\theta_w - \theta)^2}{R_a} \frac{d\theta}{d\eta}\frac{d\theta}{d\eta} = 0. \tag{7.8}$$

Dimensionless boundary conditions are

$$\frac{df}{d\eta} = 1, \quad f = f_w, \quad \theta = 1, \quad at \quad \eta = 0. \tag{7.9}$$

$$\frac{df}{d\eta} \to 0, \quad \theta \to 0, \quad as \quad \eta \to \infty. \tag{7.10}$$

7.3.1 Research Questions I

Sequel to the aims and objectives of this study, the following are the research questions:

1. What is the variation in local skin friction coefficients, temperature distribution, and Nusselt number (heat transfer rate) with thermal radiation, buoyancy, and Prandtl number?
2. When the transmission of heat energy through electromagnetic waves is nonlinear, what are the increasing effects of the ratio of momentum diffusivity to thermal diffusivity in the presence of suction and injection?
3. How does fluid transport phenomenon vary due to changes in Prandtl number when buoyancy forces and suction are significant, but the transmission of heat energy through electromagnetic waves is minimal and maximum?

7.3.2 Analysis and Discussion of Results I

Many cases were considered to capture the effects of Prandtl number on the fluid flow and its relationship to thermal radiation and buoyancy parameter (i.e., Grashof number) when suction and injection are significant. The numerical solution of dimensionless governing Eqs. (7.7) and (7.8) subject to Eqs. (7.9)–(7.10) was obtained using the four-stage Lobatto IIIa formula as described in Chapter 1 using $\eta = 5$. When buoyancy forces are significant ($G_{rx} = 1$), the nonlinear transmission of heat energy through electromagnetic waves with $\theta_w = 1.3$ and $R_a = 4$, with an increase the ratio of momentum diffusivity to thermal diffusivity, the flow along the y-direction decreases; see Figures 7.2 and 7.3. It is observed that the minimum velocity occurs in this direction in the case of injection (i.e., $f_w = -3$).

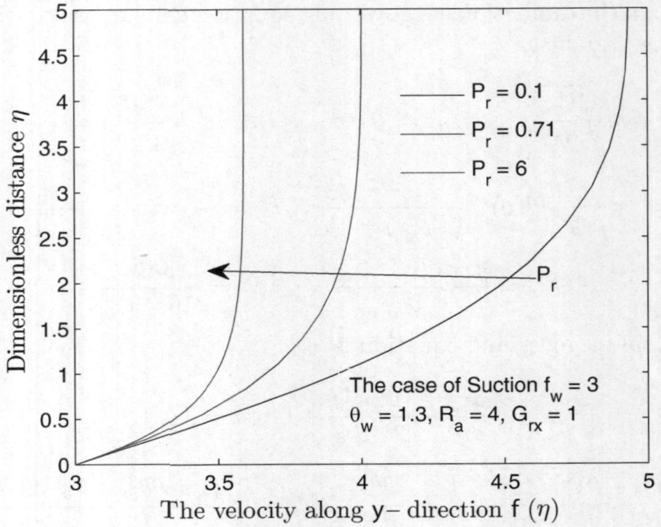

FIGURE 7.2
The flow in the y-direction when $f_w = 3$.

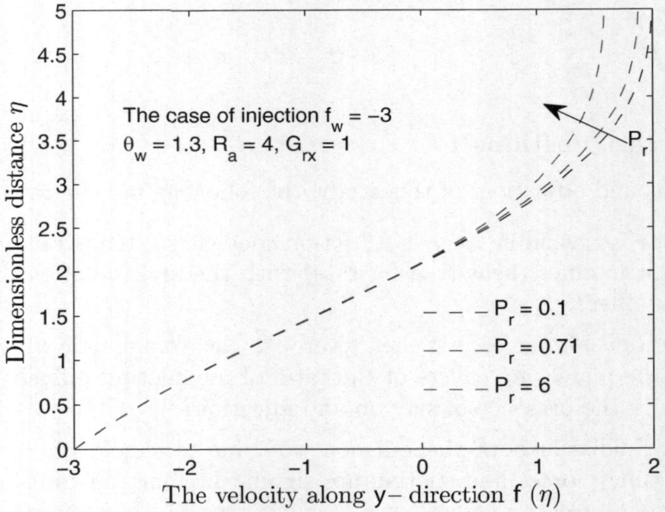

FIGURE 7.3
The flow in the y-direction when $f_w = -3$.

In the same case, the Prandtl number does not affect the vertical velocity within the fluid layers adjacent to the wall ($0 \leq \eta < 2$; Figure 7.3). At the free stream, $f(5)$ decreases with the Prandtl number at the rate of -0.168813036 for the case of suction $f_w = +3$ and -0.064911234 for the case of injection $f_w = -3$. It is important to remark that injection is in the same direction as the flow along the vertical direction. This fact justifies the constant function of the vertical flow velocity within the domain $0 \leq \eta < 2$. It is seen in Figures 7.4 and 7.5 that the horizontal velocity decreases with an increase in the Prandtl number from the wall to the free stream ($0 \leq \eta < \infty$) in the case of suction $f_w = +3$. The results

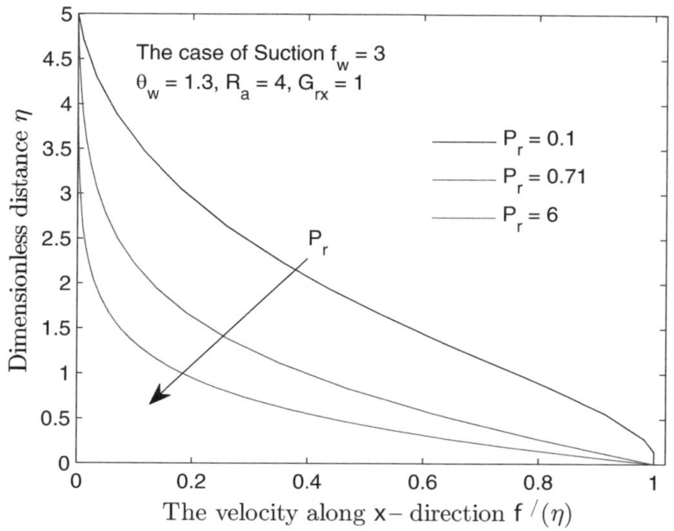

FIGURE 7.4
The flow in the x-direction when $f_w = 3$.

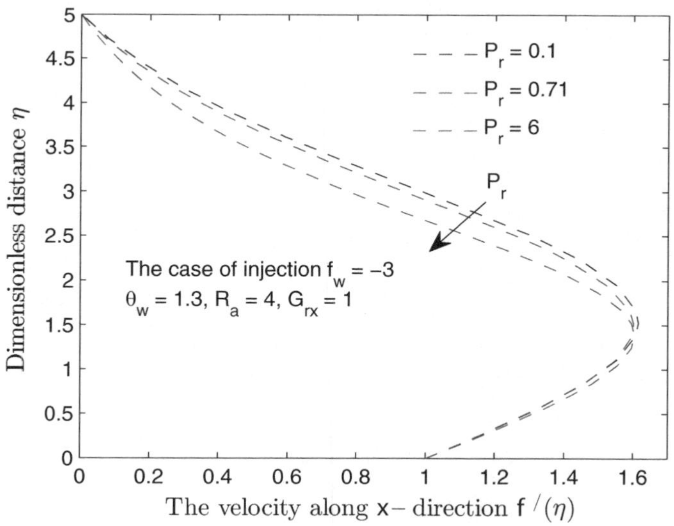

FIGURE 7.5
The flow in the x-direction when $f_w = -3$.

illustrated in Figure 7.5 reveal that injection slightly boosts the velocity of the flow within the thin layer near the wall (Figures 7.8 and 7.9).

Figures 7.6 and 7.7 indicate that there exists a significant difference in the effect of Prandtl number on the shear stress (i.e., friction across the flow in the case of suction and injection). It is worth remarking that suction is highly recommended for a significant decrease in the shear stress at the wall. Meanwhile, for higher shear stress at the wall, a property proportional to the local skin friction coefficients, injection is suggested at any value of Prandtl number. For the case of suction, it is seen that $f''(0)$ decreases with

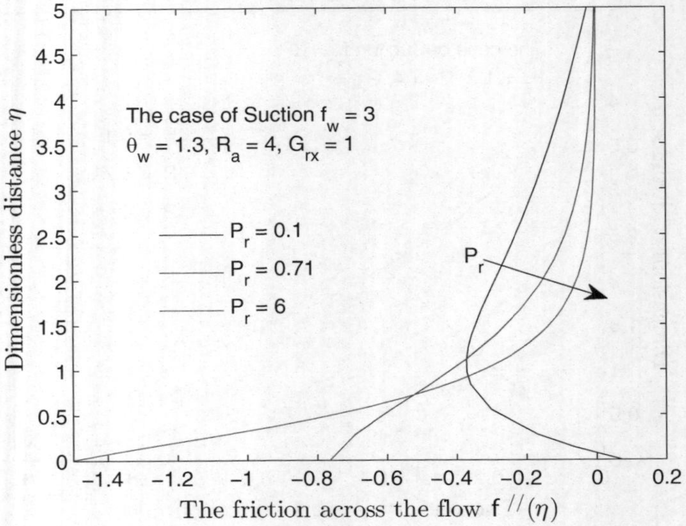

FIGURE 7.6
The friction across the flow $f_w = 3$.

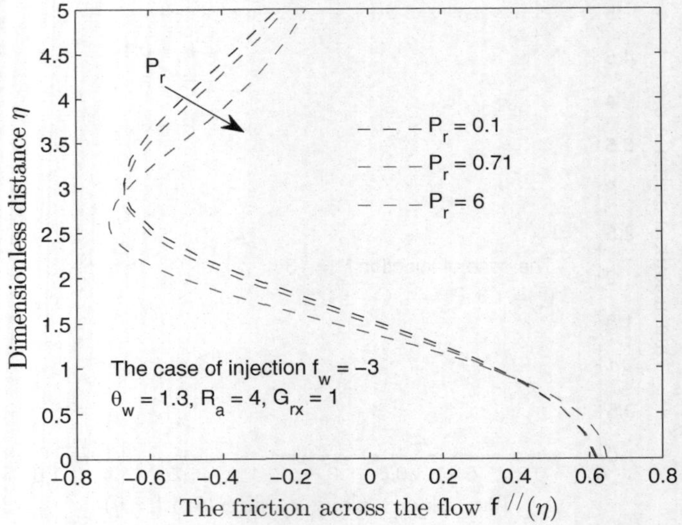

FIGURE 7.7
The friction across the flow $f_w = -3$.

the Prandtl number at the rate of −0.217953447. Meanwhile, for the case of injection, local skin friction coefficients increase with the Prandtl number at the rate of 0.004925102. Temperature distribution decreases with the Prandtl number. Table 7.1 presents the pattern of decrease in the temperature distribution across the flow for each value of Prandtl number for the case of suction and injection. It is seen that a unique pattern of decrease in $\theta(\eta)$ across the domain occurs when $f_w = -3$ (injection) and $P_r = 6$ (water). The graphical illustration of heat transfer in less viscous fluids and water for the case of injection is presented in Figure 7.10. It is seen that the heat transfer is almost linear in the flow with

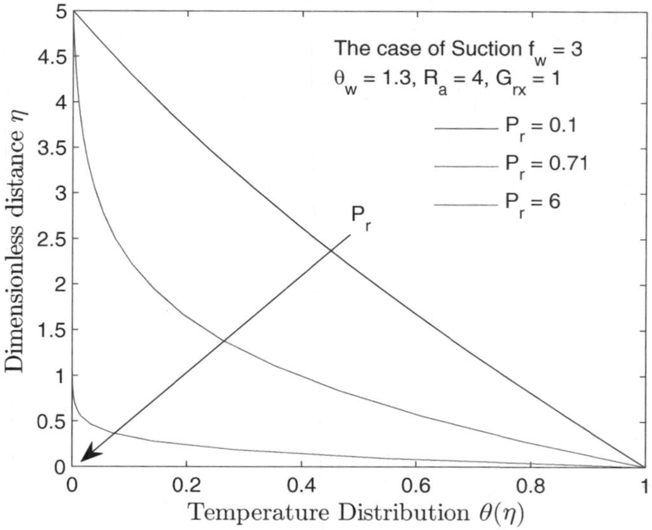

FIGURE 7.8
Temperature distribution across the flow when $f_w = 3$.

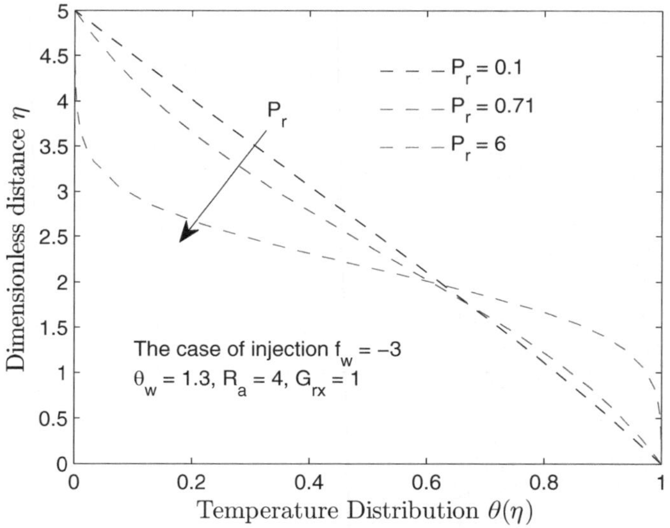

FIGURE 7.9
Temperature distribution across the flow when $f_w = -3$.

less viscous fluids (i.e., gas). Also, the heat transfer rate is parabolic across high viscous fluid. As presented in Table 7.2, it is shown that the Nusselt number $-\theta'(0)$ increases with the Prandtl number at the rate of 0.983454743 for the case of suction. Reverse is the case in injection where the Nusselt number $-\theta'(0)$ increases with the Prandtl number at the rate of -0.025365369; see Table 7.3.

To further explore the effect of Prandtl number on the flow along a horizontal surface, buoyancy forces were assumed to be zero ($G_{rx} = 0$). When $\theta_w = 1.3$ and $R_a = 4$, it is

TABLE 7.1
The Rate of Decrease in $\theta(\eta)$ with η at Various Values of Prandtl Number for Suction ($f_w = +3$) and Injection ($f_w = -3$)

	P_r	**0.1**	**0.71**	**6**
$f_w = +3$	S_{lp}	−0.20216631	−0.156261696	−0.043660321
$f_w = -3$	S_{lp}	−0.203076079	−0.219746101	−0.263711383

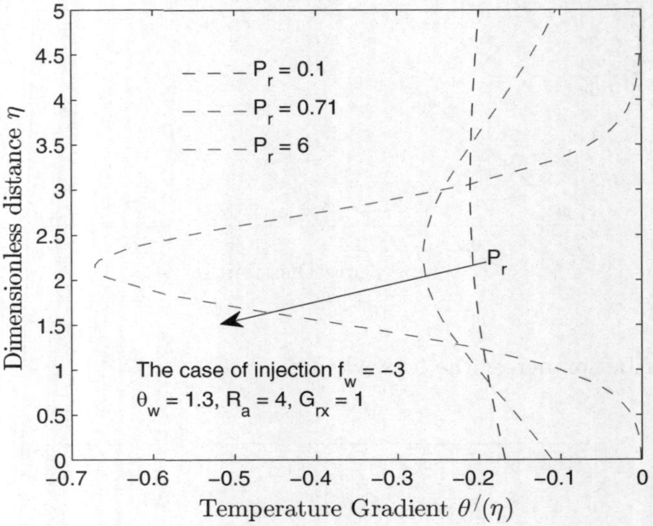

FIGURE 7.10
Temperature gradient across the flow when $f_w = -3$.

TABLE 7.2
Variation in the Vertical Velocity at the Free Stream $f(5)$, Local Skin Friction Coefficients $f''(0)$, and Nusselt Number $-\theta'(0)$ with Prandtl Number for the Case of Suction

	Suction	$f_w = 3$	
P_r	$f(5)$	$f''(0)$	$-\theta'(0)$
0.1	4.9328	0.0735	0.2514
0.71	4.003	−0.7657	0.8162
6	3.591	−1.5078	6.0391
S_{lp}	−0.168813036	−0.217953447	0.983454743

seen that the Prandtl number has no effect on the velocity along the x-direction and y-direction. More so, the temperature distribution decreases with the Prandtl number. This result implies that the significant observed effect of Prandtl number on the friction across the flow, fluid flow along the x-direction, and y-direction illustrated in Figures 7.6 and 7.7 can be traced to buoyancy forces. In the case of suction $f_w = 3$, when $\theta_w = 1.3$ and $R_a = 4$, an attempt was made to unravel the relationship between the Prandtl number and the Grashof number. It is seen that for the cooling of the surface $G_{rx} < 0$, $T - T_\infty < 0$. For this case, the local skin friction coefficients increase and decrease within certain range of Prandtl number P_r. In addition, the Nusselt number increases strictly with the Prandtl number only

TABLE 7.3
Variation in the Local Skin Friction Coefficients and Heat Transfer Rate with (i) Prandtl Number, (ii) Thermal Radiation, and (iii) Buoyancy When Suction Is Significant $f_w = +3$

$G_{rx} = 0$	$R_a = 0.1$	$\theta_w = 0.2$	$R_a = 2$	$\theta_w = 0.8$	$R_a = 4$	$\theta_w = 1.4$
P_r	$f''(0)$	$-\theta'(0)$	$f''(0)$	$-\theta'(0)$	$f''(0)$	$-\theta'(0)$
0.1	−1.6477	0.8721	−1.6477	0.2957	−1.6477	0.2253
1.1	−1.6477	2.0823	−1.6477	1.4559	−1.6477	1.0906
2.1	−1.6477	3.4066	−1.6477	2.7021	−1.6477	2.0229
3.1	−1.6477	4.7826	−1.6477	3.9282	−1.6477	2.9379
4.1	−1.6477	6.1847	−1.6477	5.1438	−1.6477	3.8444
5.1	−1.6477	7.6000	−1.6477	6.3534	−1.6477	4.7461
6.1	−1.6477	9.0196	−1.6477	7.5594	−1.6477	5.6448
S_{lp}	0	1.3663	0	1.2153	0	0.9068
$G_{rx} = 2$	$R_a = 0.1$	$\theta_w = 0.2$	$R_a = 2$	$\theta_w = 0.8$	$R_a = 4$	$\theta_w = 1.4$
P_r	$f''(0)$	$-\theta'(0)$	$f''(0)$	$-\theta'(0)$	$f''(0)$	$-\theta'(0)$
0.1	0.6375	0.8871	1.3497	0.3202	1.5506	0.2458
1.1	0.0072	2.1793	−0.4514	1.5049	−0.384	1.1324
2.1	−0.3421	3.5266	−0.9412	2.7257	−0.9012	2.0428
3.1	−0.5756	4.902	−1.1462	3.942	−1.1174	2.9495
4.1	−0.7443	6.2956	−1.2580	5.1529	−1.2355	3.852
5.1	−0.8713	7.6969	−1.3285	6.3599	−1.310	4.7515
6.1	−0.9696	9.1053	−1.3772	7.5642	−1.3615	5.6489
S_{lp}	−0.2493	1.3735	−0.3661	1.2096	−0.3901	0.9020

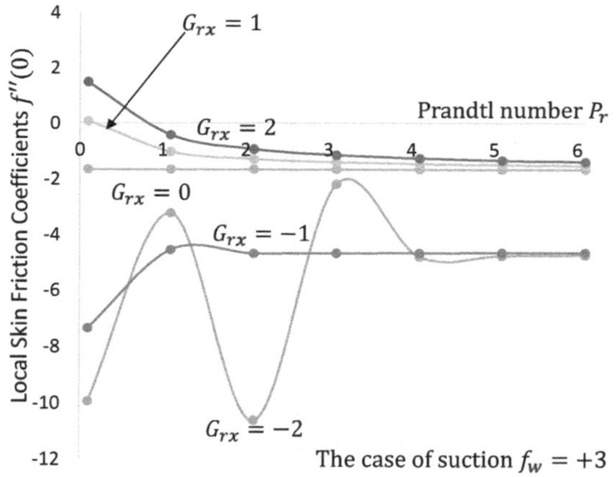

FIGURE 7.11
Variation in the local skin friction coefficients with the Prandtl number and buoyancy parameter when $f_w = +3$.

for $0 \leq G_{rx} < \infty$. Sequel to this, it is seen that the Nusselt number, proportional to heat transfer rate, increases with the Prandtl number at the rates of 0.986225, 0.983428571, and 0.981275 when $G_{rx} = 0$, $G_{rx} = 1$, and $G_{rx} = 2$, respectively; see Figures 7.11–7.14. As shown in Figure 7.13, injection is seen to be a yardstick to maintain absolute local skin friction coefficients at all the values of Prandtl number. For less viscous fluids subject to injection in the case of cooling the surface (i.e., $G_{rx} < 0$), it is worth remarking that negative heat transfer may surface when $1 \leq P_r \leq 2$; see Figure 7.14.

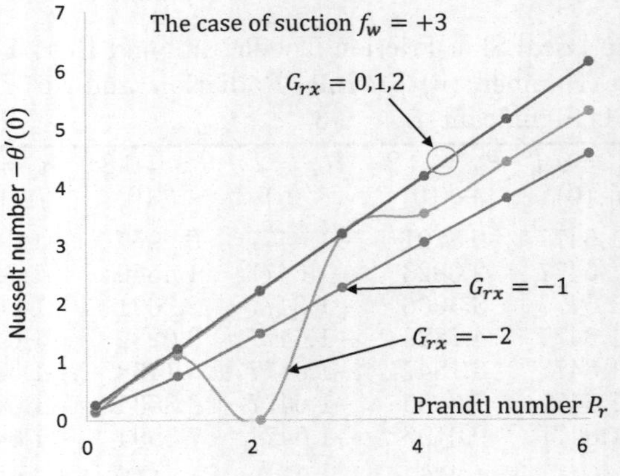

FIGURE 7.12
Variation in the Nusselt number with the Prandtl number and buoyancy parameter when $f_w = +3$.

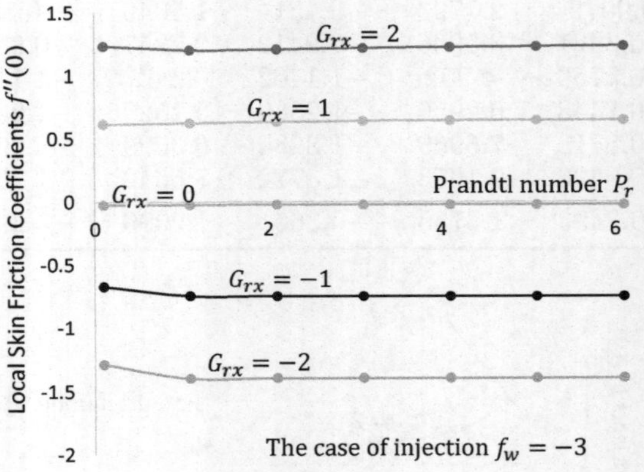

FIGURE 7.13
Variation in the local skin friction coefficients with the Prandtl number and buoyancy parameter when $f_w = -3$.

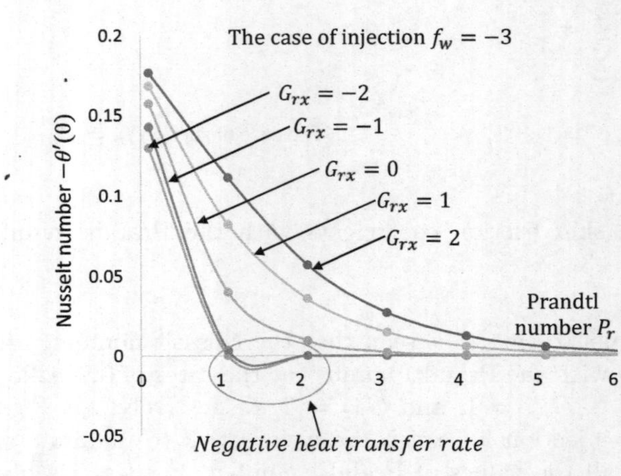

FIGURE 7.14
Variation in the Nusselt number with the Prandtl number and buoyancy parameter when $f_w = -3$.

TABLE 7.4
Variation in the Vertical Velocity at the Free Stream $f(5)$, Local Skin Friction Coefficients $f''(0)$, and Nusselt Number $-\theta'(0)$ with Prandtl Number for the Case of Injection ($f_w = -3$)

P_r	$f(5)$	$f''(0)$	$-\theta'(0)$
0.1	1.994	0.6172	0.1683
0.71	1.5784	0.6468	0.0009
6	3.591	−1.5078	6.0391
S_{lp}	−0.064911234	0.004925102	−0.025365369

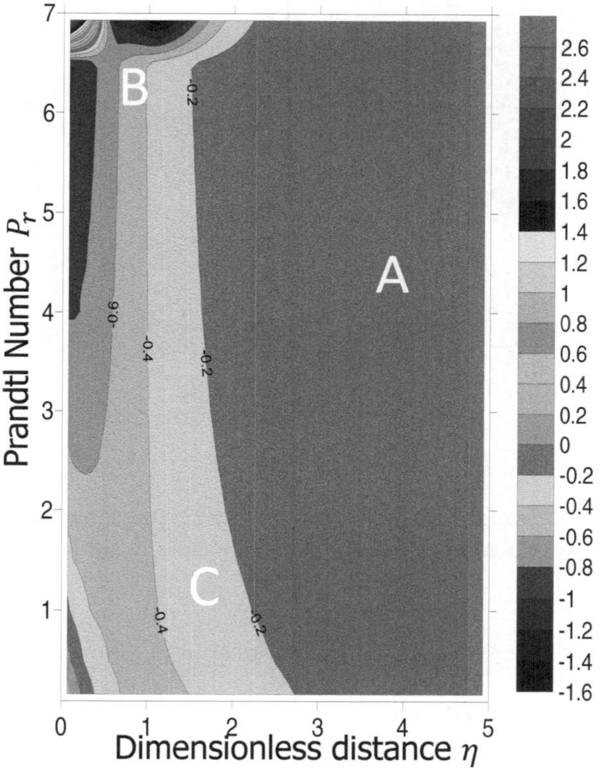

FIGURE 7.15
Variation in the effect of Prandtl number on the local skin friction coefficients across Sakiadis flow: when temperature parameter $\theta_w = 0.1$ and thermal radiation parameter $R_a = 0.1$.

In the presence of suction, little is known on the effect of increasing (i) Prandtl number, (ii) thermal radiation, and (iii) buoyancy on the local skin friction coefficients and heat transfer rate. As shown in Table 7.4, it can be deduced that the heat transfer rate increases with the Prandtl number. However, the observed increase in the heat transfer rate diminishes with thermal radiation. Meanwhile, the observed increase in the heat transfer rate was enhanced when buoyancy was enhanced. The analysis outcome shows that the minimum heat transfer rate is obtained when suction is significant, thermal radiation is enhanced, and buoyancy is highly negligible in a less viscous fluid. When the temperature parameter $\theta_w = 0.1$ and thermal radiation parameter $R_a = 0.1$, Figure 7.15 shows that the friction in the flow of less viscous fluids $0.1 \leq P_r \leq 1$ is positive but small, while the same property

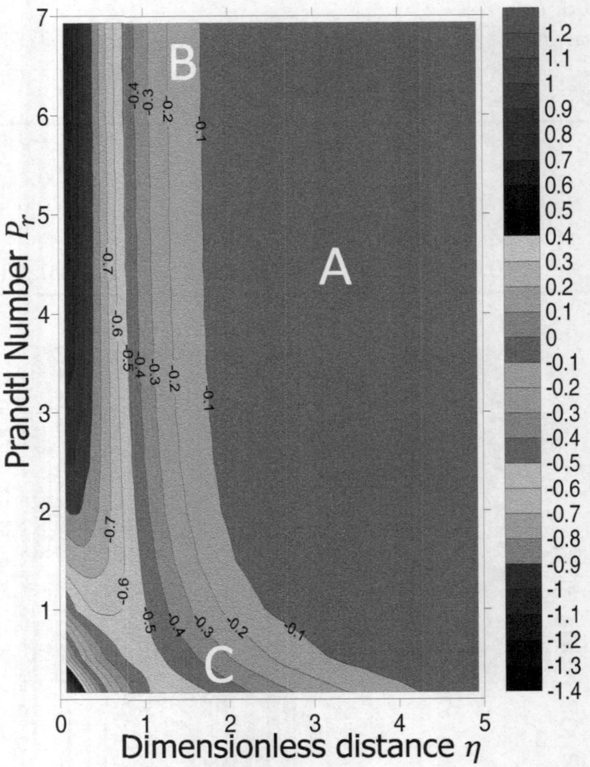

FIGURE 7.16
Variation in the effect of Prandtl number on the local skin friction coefficients across Sakiadis flow: When temperature parameter $\theta_w = 1.4$ and thermal radiation parameter $R_a = 4$.

for high viscous fluids $6.5 \leq P_r \leq 7$ is more substantial. Reverse is the case when the temperature parameter $\theta_w = 1.4$ and thermal radiation parameter $R_a = 4$, where the friction between the fluid layers near the wall for $2 \leq P_r \leq 7$ is the most minimum; see Figure 7.16.

The answers to the third research question was sought for using $Gr_x = 2$, $f_w = +3$ when $\theta_w = 0.1, R_a = 0.1$, and $\theta_w = 5, R_a = 8$ at various values of P_r. It is shown in Figures 7.17–7.23 that a rise in the magnitude of P_r causes the velocities of the motion along the x- and y-directions to decrease, the shear stress to decrease near the wall, and the temperature distribution to diminish. When the transmission of heat energy through electromagnetic waves is maximum (i.e., $\theta_w = 5, R_a = 8$), the velocity of the motion along the x-drection and y-direction was enhanced. However, fluids with lesser viscosity was greatly enhanced as shown in Figures (7.18) and (7.20). The friction near the wall and temperature distribution were also discovered to be influenced at all the chosen values of P_r as illustrated in Figures 7.22 and 7.24. One of the advantages of nonlinear model for investigating thermal radiation is the associated temperature parameter θ_w. This model has been considered to be valid and highly accurate due to the fact that the transmission of heat energy through electromagnetic waves when the wall temperature is high distinctively differ from when it is highly negligible. It is also evident from Table 7.5 that the Nusselt number $-\theta'(0)$ increases with the Prandtl number at the higher rate of 1.528329067 when the transmission of heat energy through electromagnetic waves is minimal (i.e., $\theta_w = 0.1, R_a = 0.1$).

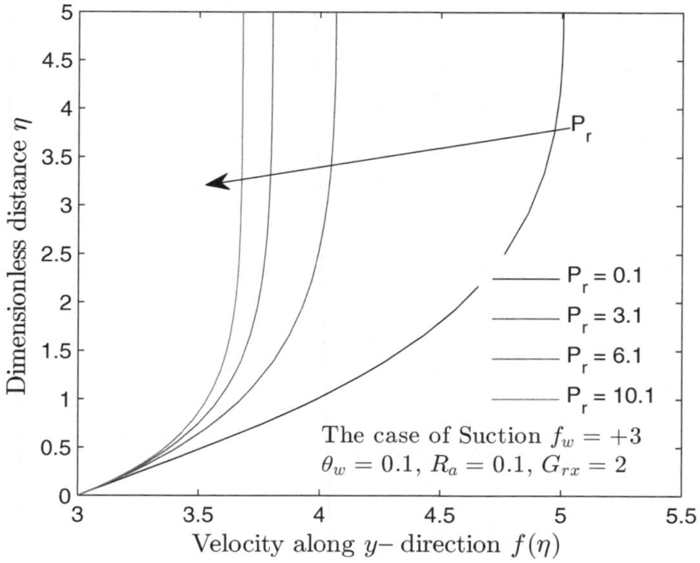

FIGURE 7.17
The flow in the y-direction when $\theta_w = 0.1$ and $R_a = 0.1$.

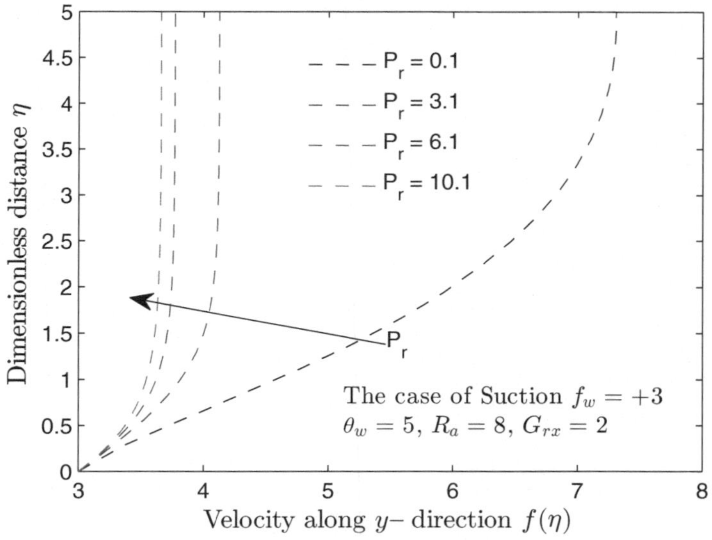

FIGURE 7.18
The flow in the y-direction when $\theta_w = 5$ and $R_a = 8$.

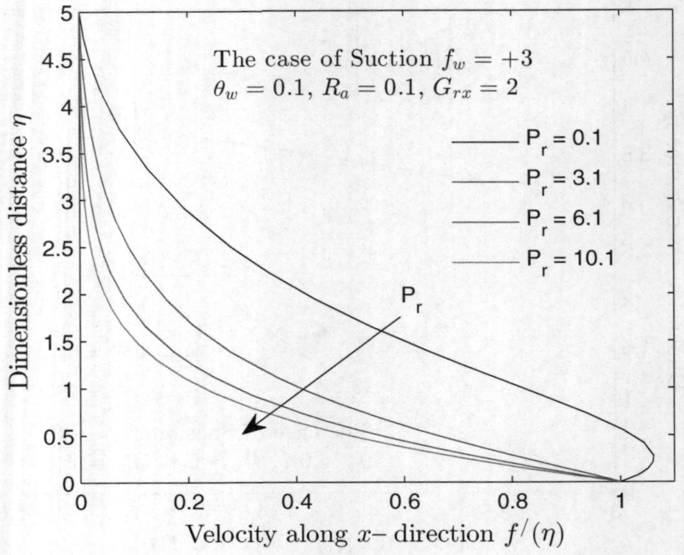

FIGURE 7.19
The flow in the x-direction when $\theta_w = 0.1$ and $R_a = 0.1$.

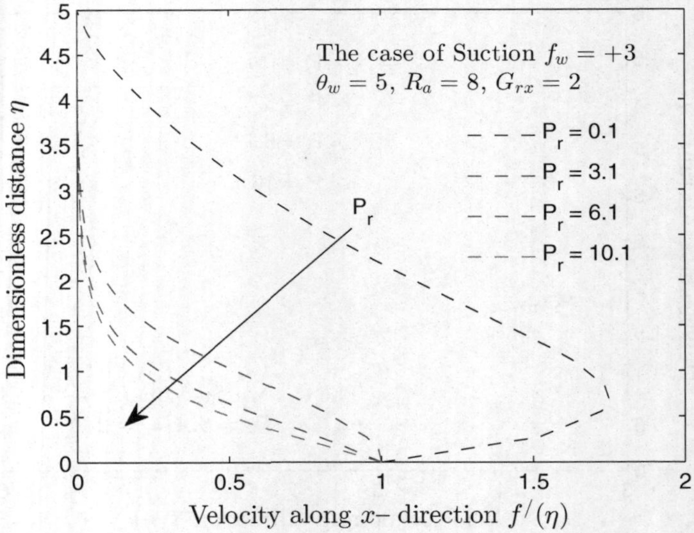

FIGURE 7.20
The flow in the x-direction when $\theta_w = 5$ and $R_a = 8$.

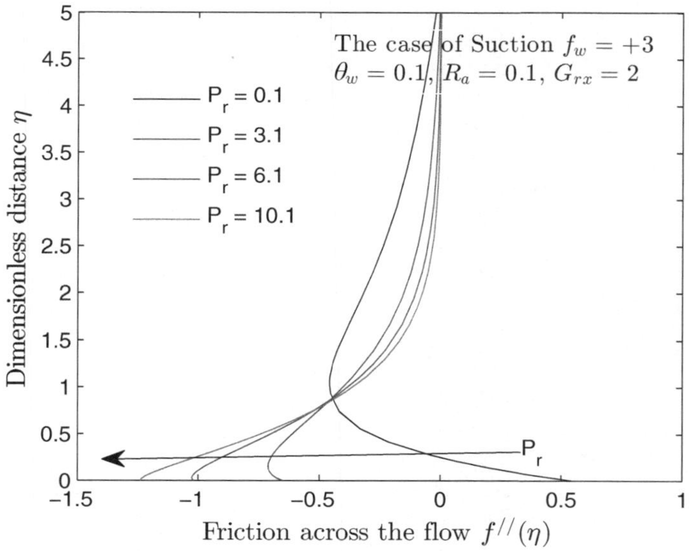

FIGURE 7.21
Friction across the flow when $\theta_w = 0.1$ and $R_a = 0.1$.

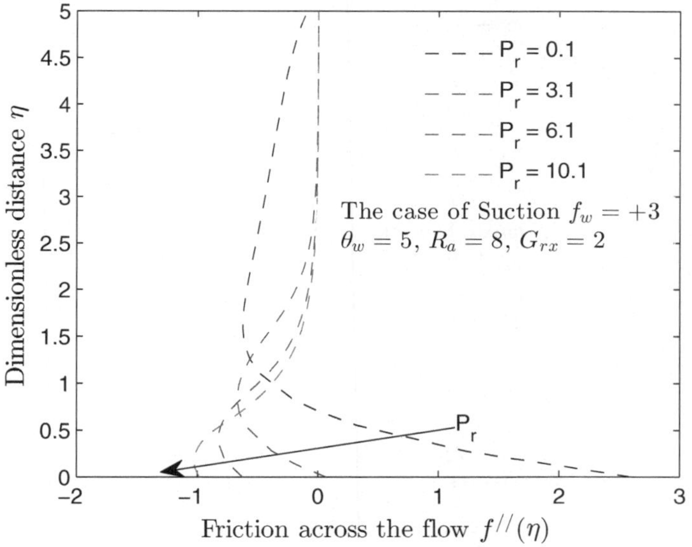

FIGURE 7.22
Friction across the flow when $\theta_w = 5$ and $R_a = 8$.

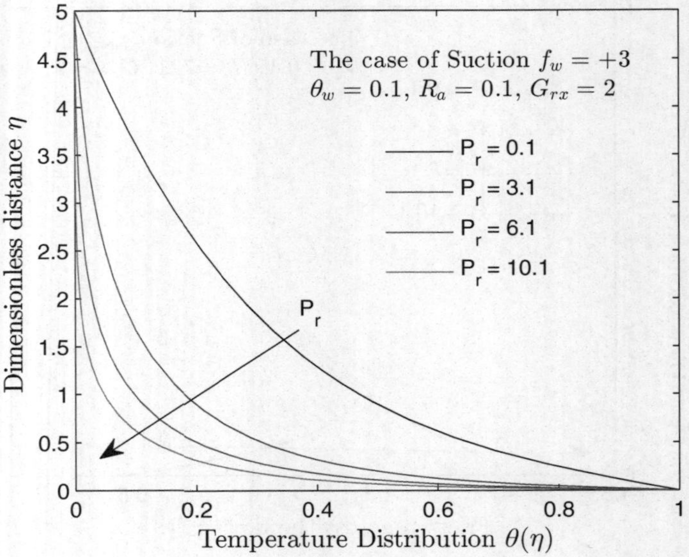

FIGURE 7.23
Temperature distribution when $\theta_w = 0.1$ and $R_a = 0.1$.

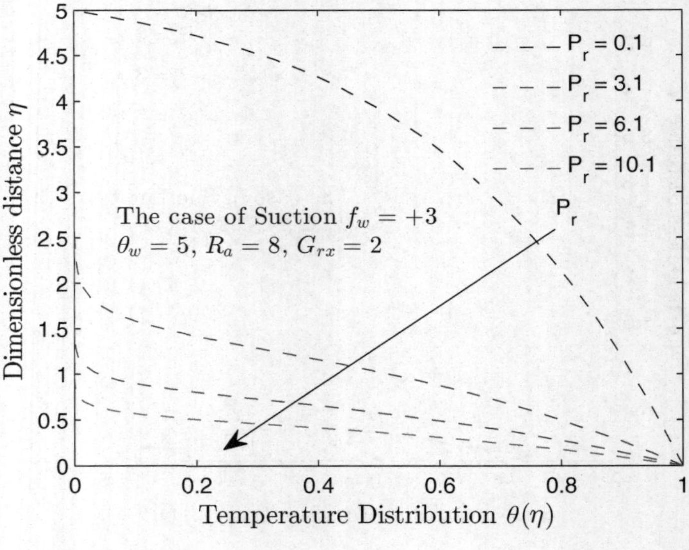

FIGURE 7.24
Temperature distribution when $\theta_w = 5$ and $R_a = 8$.

TABLE 7.5
Variation in the Nusselt Number $-\theta'(0)$ with Prandtl Number When $f_w = +3$, $Gr_x = 2$ for $\theta_w = 0.1, R_a = 0.1$ and $\theta_w = 5, R_a = 8$

P_r	$\theta_w = 0.1$ $R_a = 0.1$	$\theta_w = 5$ $R_a = 8$
0.1	0.406831502562968	0.039677993862437
3.1	5.131272278036779	0.348048633586569
6.1	9.671313951286813	0.625276026959590
10.1	15.714104781369086	0.994099703325859
S_{lp}	1.528329067	0.095033112

7.4 Introduction: Internal Heat Source and Sink

Heat source and heat sink are words widely used in thermodynamics, analysis of heat transfer, and fluid dynamics to denote the addition and removal of heat energy. The heat source is a mechanism that supplies a huge amount of energy without changing its temperature. It transfers thermal energy from a hotter source to a colder region. An example of a heat source is the sun. A heat sink can be described as a mechanism for sucking heat energy without significantly changing its temperature. An example of a heat sink is an air conditioner. Temperature-dependent and space-dependent modes are two acceptable modes of heat source and heat sink. In the case of a temperature-dependent heat source, an equal amount of heat energy (i.e., internal energy) is being introduced from the wall to the free stream. At the same time, there is variability in the case of space-dependent heat sources. Revnic et al. [257] investigated the significance of discrete heating on the dynamics of the hybrid nanofluid Cu–Al_2O_3–water within a wavy cavity. It was found by Revnic et al. [257] that the greatest temperature is accomplished in the center because of discrete warming of the fluid.

7.5 Fluid Flow Subject to Internal Heat Source or Sink

In this section, the two-dimensional flow of an incompressible fluid where both heat source and heat sink are significant is presented; see Figure (7.25). Generally, the heat source is a heat exchanger that supplies heat to the fluid flow, while a heat sink is a heat exchanger that absorbs heat during the fluid flow. Consider the flow of an incompressible viscous fluid on a horizontal surface within the domain $0 \leq y < \infty$ where the heat source is highly significant; it is assumed that viscous forces induce the flow and stretching at the wall (i.e., Sakiadis flow) with the stretching velocity $U_\infty = U_o x$. The governing equation that models the transport phenomenon is

$$u_x + v_y = 0, \tag{7.11}$$

$$uu_x + vu_y = \frac{\mu}{\rho} u_{yy}, \tag{7.12}$$

$$uT_x + vT_y = \frac{\kappa}{\rho c_p} T_{yy} + \frac{Q_o[T_w(x) - T_\infty]}{(\rho C_p)\Delta T} Exp\left(-ny\sqrt{\frac{U_\infty}{2\vartheta x}}\right). \tag{7.13}$$

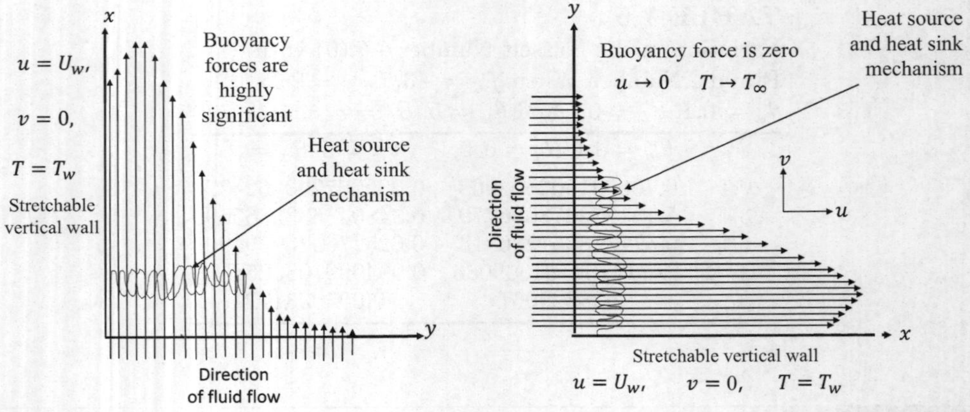

FIGURE 7.25
Illustration of fluid flow subject to heat source and sink.

For this case of Sakiadis flow, Eqs. (7.11)–(7.13) are subject to the boundary condition

$$u = U_w, \quad v = 0, \quad T = T_w \quad at \quad y = 0. \tag{7.14}$$

$$u \to 0, \quad T \to T_\infty \quad as \quad y \to \infty \tag{7.15}$$

In order to obtain self-similar solution of Eqs. (7.11)–(7.13) subject to Eqs. (7.14) and (7.15), the following similarity variables:

$$\eta = y\sqrt{\frac{U_\infty}{\vartheta x}}, \quad \psi(x,y) = \sqrt{\vartheta x U_\infty} f(\eta), \quad u = \frac{\partial \psi}{\partial y}, \quad v = -\frac{\partial \psi}{\partial x},$$

$$\theta(\eta) = \frac{T - T_\infty}{T_w - T_\infty}, \quad P_r = \frac{\mu c_p}{\kappa} = \frac{\vartheta}{\alpha} \quad \gamma = \frac{Q_o}{(\rho C_p)\Delta T U_o} \tag{7.16}$$

were used to obtain the final dimensionless governing equation

$$\frac{d^3 f}{d\eta^3} + \frac{1}{2} f \frac{d^2 f}{d\eta^2} = 0, \tag{7.17}$$

$$\frac{d^2\theta}{d\eta^2} + P_r f \frac{d\theta}{d\eta} + 2P_r \gamma Exp(-n\eta) = 0. \tag{7.18}$$

Dimensionless boundary conditions of Eqs. (7.14) and (7.15) are

$$\frac{df}{d\eta} = 1, \quad f = 0, \quad \theta = 1 \quad at \quad \eta = 0. \tag{7.19}$$

$$\frac{df}{d\eta} \to 0, \quad \theta \to 0 \quad as \quad \eta \to \infty. \tag{7.20}$$

7.5.1 Research Questions II

Provision of answers to the following research questions is the aim of this case:

1. What is the effect of Prandtl number on the temperature distribution within the fluid flow due to stretching at the wall with heat source and heat sink?

2. What is the influence of increasing the ratio of momentum diffusivity to thermal diffusivity on local skin friction coefficients and heat transfer rate during the flow due to stretching at the wall with heat source and heat sink?
3. How does the intensity of the heat source affect the boundary layer flow in the presence of the heat source and heat sink?

7.5.2 Analysis and Discussion of Results II

The numerical solutions of dimensionless governing Eqs. (7.17) and (7.18) subject to Eqs. (7.19) and (7.20) were obtained using the four-stage Lobatto IIIa formula as described in Chapter 1 using $\eta = 5$. The outcome of the analysis when the heat source is small ($\gamma = 0.01$) and large ($\gamma = 10$), but the intensity of heat is small ($n = 0.01$) is presented in Table 7.6. The temperature distribution decreases with the Prandtl number when $\gamma = 0.01$. Reverse is the case for $\gamma = 10$ where the temperature distribution increases with the Prandtl number. When the intensity of heat is large ($n = 10$), it is now observed that the temperature distribution near the wall decreases with the Prandtl number; see Table 7.7. From 0.1755 in Table 7.6 to −0.0044 in Table 7.7, it is evident that the percentage decrease in the rate of change of $\theta(0.88)$ is 102.51%. Considering heat sink, Table 7.8 reveals that the Nusselt number proportional to the heat transfer rate $-\theta'(0)$ decreases with the Prandtl number at the rates of 0.7480 and 0.1944 when the intensity is $n = 0.01$ and $n = 10$, respectively. In other words, with an increase in the intensity, the percentage decrease in the rate of increase in the Nusselt number with Prandtl number is 74.01. A comparative analysis between the effects of heat source and heat sink is presented in Table 7.9. The results show that the temperature distribution is a decreasing property of both cases. However, heat source leads to a decrease in the Nusselt number while heat sink leads to an increase in the Nusselt number (Table 7.10).

Overall, Table 7.11 shows that the temperature distribution decreases from the wall to the free stream at different rates for both the cases of heat sink and heat source. Reverse is the case for $P_r = 100$ when $\gamma = 1$ and $\gamma = -1$. For these two cases, the temperature distribution decreases with the Prandtl number for the heat source, but increases with the same dimensionless parameter for the heat sink. The combined effects of heat source, heat sink, and Prandtl number are illustrated in Figures 7.26–7.31. The temperature distribution of fluid layers near the wall increases with the Prandtl number for a heat source, but decreases with the same number for the heat sink. Near the wall, the observed effect is more enhanced when $\gamma = 10$ and $\gamma = -10$, as illustrated in Figures (7.27) and (7.31).

TABLE 7.6
Effect of Prandtl Number P_r on Some of the Properties of Sakiadis Flow When the Intensity of Heat Source Is Small ($n = 0.01$)

	$n = 0.01$	$\gamma = 0.01$		$n = 0.01$	$\gamma = 10$	
P_r	$f''(0)$	$\theta(0.88)$	$-\theta'(0)$	$f''(0)$	$\theta(0.88)$	$-\theta'(0)$
0.1	−0.6215	0.5407	0.5217	−0.6215	1.5127	−1.4699
0.3	−0.6215	0.5112	0.5660	−0.6215	3.3906	−5.4428
0.71	−0.6215	0.4525	0.6598	−0.6215	7.0433	−13.6569
1	−0.6215	0.4131	0.7274	−0.6215	9.4307	−19.4596
6	−0.6215	0.0854	1.6988	−0.6215	26.73003	−101.5491
10	−0.6215	0.0413	2.2244	−0.6215	28.0817	−149.1831
100	−0.6215	0.0272	7.1447	−0.6215	27.1616	−690.1366
S_{lp}	0	−0.0037	0.0638	0	0.1755	−6.7031

TABLE 7.7
Effect of Prandtl Number P_r on Some of the Properties of Sakiadis Flow When the Intensity of Heat Source Is Large ($n = 10$)

	$n = 10$	$\gamma = 0.01$		$n = 10$	$\gamma = 10$	
P_r	$f''(0)$	$\theta(0.88)$	$-\theta'(0)$	$f''(0)$	$\theta(0.88)$	$-\theta'(0)$
0.1	−0.6215	0.5398	0.5235	−0.6215	0.5506	0.3340
0.3	−0.6215	0.5667	0.5715	−0.6215	0.5390	0.0048
0.71	−0.6215	0.4460	0.6728	−0.6215	0.5101	−0.6585
1	−0.6215	0.4042	0.7457	−0.6215	0.4864	−1.1194
6	−0.6215	0.0589	1.7918	−0.6215	0.1389	−8.5503
10	−0.6215	0.0133	2.3594	−0.6215	0.0471	−14.2035
100	−0.6215	0.0000	7.7178	−0.6215	0.0000	−116.9950
S_{lp}	0	−0.0040	0.0694	0	−0.0044	−1.1679

TABLE 7.8
Effect of Prandtl Number P_r on Some of the Properties of Sakiadis Flow Subject to Heat Sink $\gamma = -1$ When the Intensity Is Small ($n = 0.01$) and Large ($n = 10$)

	$n = 0.01$	$\gamma = -1$		$n = 10$	$\gamma = -1$	
P_r	$f''(0)$	$\theta(0.88)$	$-\theta'(0)$	$f''(0)$	$\theta(0.88)$	$-\theta'(0)$
0.1	−0.6215	0.4424	0.7231	−0.6215	0.6519	0.5427
0.3	−0.6215	0.2201	1.1735	−0.6215	0.6223	0.6288
0.71	−0.6215	−0.2138	2.1073	−0.6215	0.5627	0.8074
1	−0.6215	−0.4986	2.7683	−0.6215	0.5220	0.9343
6	−0.6215	−2.6083	12.1373	−0.6215	0.0508	2.8375
10	−0.6215	−2.7936	17.5320	−0.6215	0.0099	4.034
100	−0.6215	−2.7162	77.6407	−0.6215	−0.0001	20.3266
S_{lp}	0	−0.0218	0.7480	0	−0.0048	0.1944

TABLE 7.9
Effect of Prandtl Number P_r on Some of the Properties of Sakiadis Flow Subject to Heat Source and Heat Sink When the Intensity of Heat Is $n = 1$

	$n = 1$	$\gamma = 1$	**Heat Source**	**Heat Sink**	$n = 1$	$\gamma = -1$
P_r	$f''(0)$	$\theta(0.88)$	$-\theta'(0)$	$f''(0)$	$\theta(0.88)$	$-\theta'(0)$
0.1	−0.6215	0.5802	0.4101	−0.6215	0.4993	0.6373
0.3	−0.6215	0.6270	0.2315	−0.6215	0.3897	0.9126
0.71	−0.6215	0.7119	−0.1285	−0.6215	0.1799	1.4768
1	−0.6215	0.7628	−0.3768	−0.6215	0.0454	1.8720
6	−0.6215	0.9411	−3.8149	−0.6215	−0.8235	7.4192
10	−0.6215	0.8706	−5.9747	−0.6215	−0.8441	10.7268
100	−0.6215	0.7139	−35.2034	−0.6215	−0.7139	50.8889
S_{lp}	0	−0.0001	−0.3495	0	−0.0077	0.4907

TABLE 7.10
Variation in the Temperature Distribution with the Prandtl Number When $n = 0.1$ but $\gamma = 1$

$\gamma = 1$	P_r	0.1	0.3	0.71	1	6	10
	S_{lp}	−0.2335	−0.3144	−0.4798	−0.5612	−1.0531	−1.1361

TABLE 7.11
Variation in the Temperature Distribution with the Prandtl Number when $n = 0.1$ but $\gamma = -1$

$\gamma = -1$	P_r	0.1	0.3	0.71	1	6	10
	S_{lp}	−0.1594	−0.0569	0.1597	0.2943	0.9774	1.0560

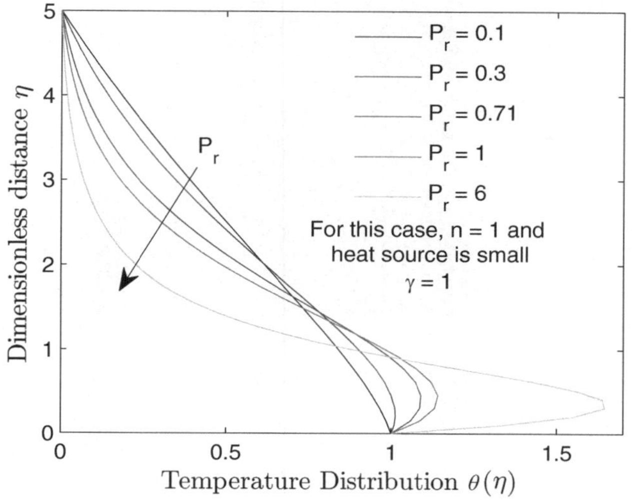

FIGURE 7.26
Variation in the temperature distribution with the Prandtl number when heat source is small $\gamma = 1$.

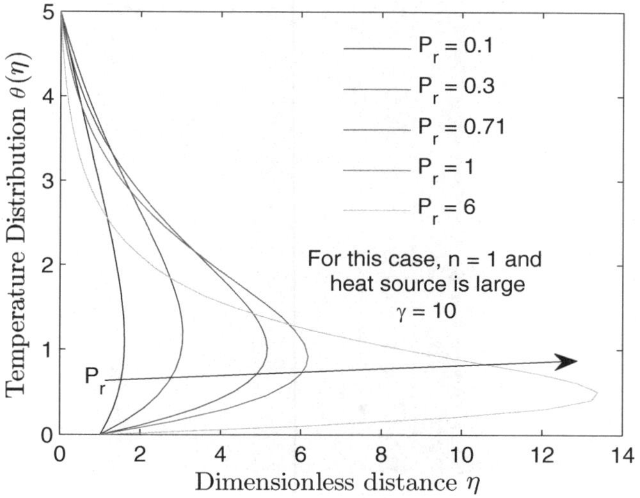

FIGURE 7.27
Variation in the temperature distribution with the Prandtl number when heat source is large $\gamma = 10$.

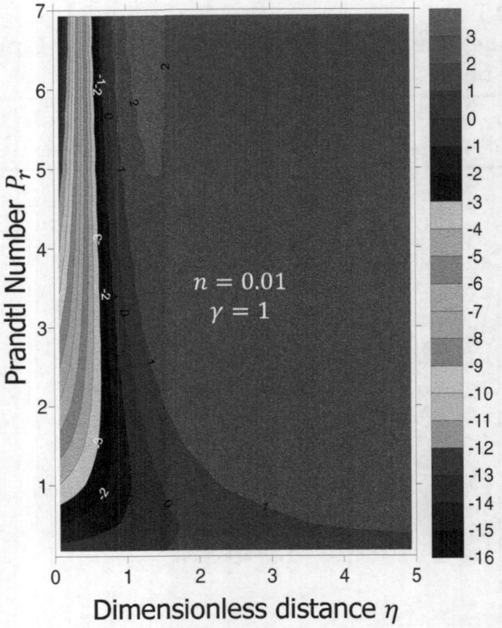

FIGURE 7.28
Variation in heat the transfer rate $-\theta'(0)$ with the Prandtl number P_r across the domain: $n = 0.01$, $\gamma = 1$.

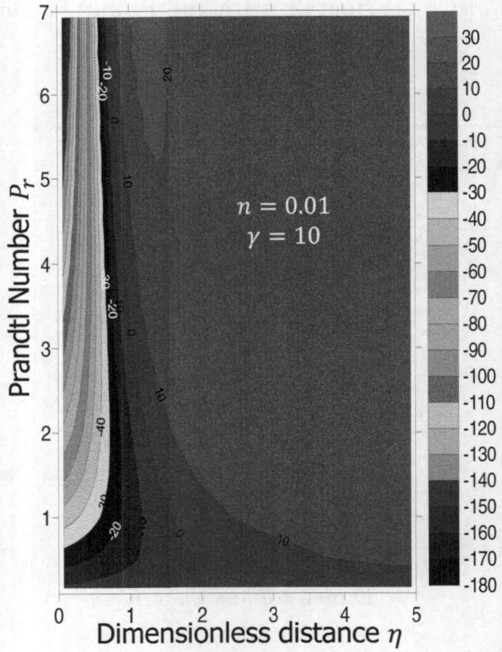

FIGURE 7.29
Variation in the heat transfer rate $-\theta'(0)$ with the Prandtl number P_r across the domain: $n = 0.01$, $\gamma = 10$.

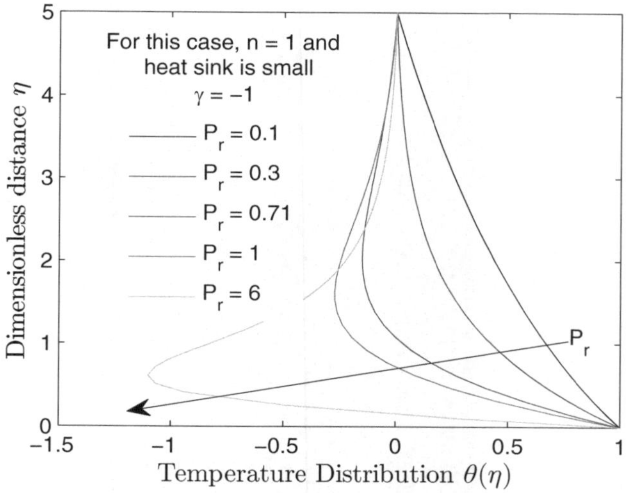

FIGURE 7.30
Variation in the temperature distribution with the Prandtl number when $\gamma = -1$.

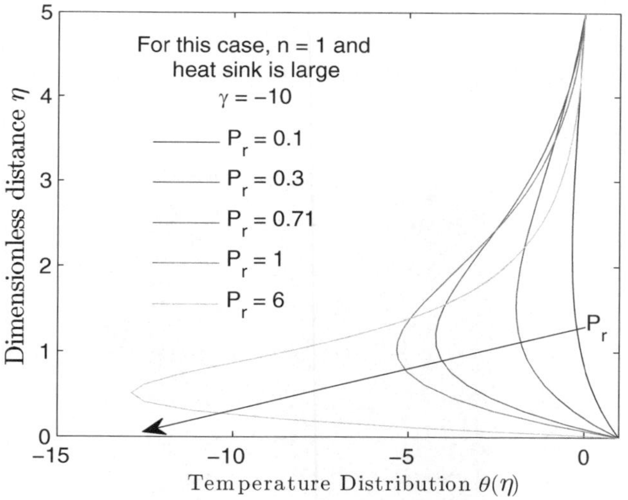

FIGURE 7.31
Variation in the temperature distribution with the Prandtl number when $\gamma = -10$.

The comparative analysis between Figures (7.28) and (7.29) show that a higher heat transfer rate is achievable when the heat source is higher. Also, the heat transfer across the domain is widespread in the dynamics of less viscous fluids, whether the heat source is small or large. For the dynamics of moderate viscous forces such as that of water, the results presented in Figures 7.28 and 7.29 reveal that there is a unique pattern of heat transfer at some distance away from the wall. It is worth remarking that when the intensity is large, there exists a significant difference between the heat transfer for the case of heat sink and that for the case of heat source (Figures 7.32 and 7.33).

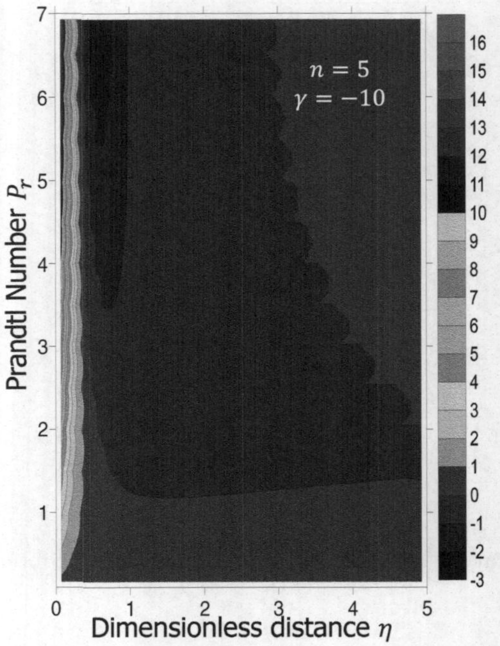

FIGURE 7.32
Variation in the heat transfer rate $-\theta'(0)$ with the Prandtl number P_r across the domain: $n = 5$, $\gamma = -10$.

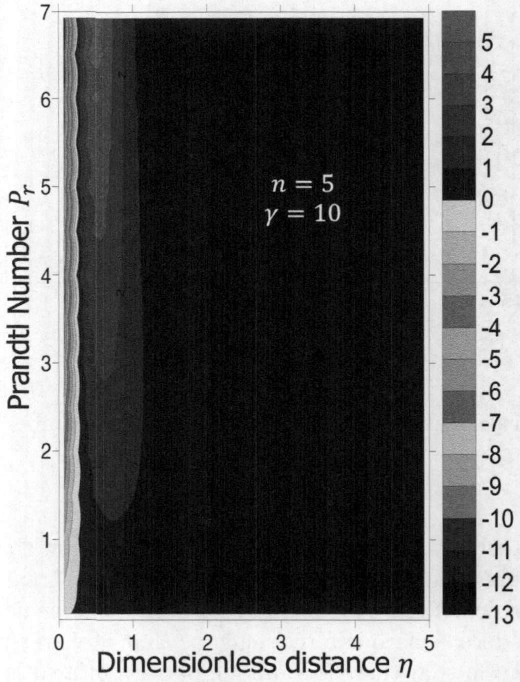

FIGURE 7.33
Variation in the heat transfer rate $-\theta'(0)$ with the Prandtl number P_r across the domain: $n = 5$, $\gamma = 10$.

7.6 Fluid Flow Subject to Internal Heating or Sinking and Buoyancy

Next is to explore the effect of heat source and sink on the same fluid dynamics along a vertical surface. For this case, the momentum equation Eq. (7.12) is replaced with Eq. (7.21).

$$uu_x + vu_y = \frac{\mu}{\rho}u_{yy} + g\beta(T - T_\infty), \tag{7.21}$$

Following the same procedure as stated in the previous section, the final dimensionless governing equation is

$$\frac{d^3 f}{d\eta^3} + \frac{1}{2}f\frac{d^2 f}{d\eta^2} + G_{rx}\theta = 0, \tag{7.22}$$

$$\frac{d^2\theta}{d\eta^2} + P_r f\frac{d\theta}{d\eta} + 2P_r\gamma e^{(-n\eta)} = 0. \tag{7.23}$$

Dimensionless boundary conditions are

$$\frac{df}{d\eta} = 1, \quad f = 0, \quad \theta = 1 \quad at \quad \eta = 0. \tag{7.24}$$

$$\frac{df}{d\eta} \to 0, \quad \theta \to 0 \quad as \quad \eta \to \infty. \tag{7.25}$$

The solutions of Eqs. (7.22)–(7.25) were obtained using the method described in the previous chapter.

7.6.1 Research Questions III

Provision of answers to the following research questions is the aim of this case:

1. In the presence of buoyancy forces to induce the flow along a vertical wall, how does the Prandtl number affect the temperature distribution within the fluid flow due to stretching at the wall with heat source and heat sink?
2. In the case of free convective flow, what is the influence of increasing the ratio of momentum diffusivity to thermal diffusivity on local skin friction coefficients and heat transfer rate during the flow due to stretching at the wall with heat source and heat sink?
3. How does the intensity of the heat source and buoyancy forces affect the dynamics in the presence of heat source and heat sink?

7.6.2 Analysis and Discussion of Results III

As shown in Tables 7.12 and 7.13, when the space-dependent heat source is small ($\gamma = 0.01$), the heat transfer rate increases with the Prandtl number at the rate of 0.1930 when the intensity is small and at the rate of 0.2121 when the intensity is large. It is seen clearly that the temperature distribution decreases at $\eta = 2$ when ($n = 0.01$ and $\gamma = 0.01$), ($n = 10$ and $\gamma = 0.01$), and ($n = 10$ and $\gamma = 5$). Reverse is the case when ($n = 0.01$ and $\gamma = 5$) as the function increases with the Prandtl number at the rate of $\eta = 2$. In the case of the same transport phenomenon along a vertical surface, the augmentation of Prandtl

TABLE 7.12
Effect of P_r on Some of the Properties of Sakiadis Flow When $n = 0.01$, Grashof Number $G_{rx} = 1$, Heat Source Parameter $\gamma = 0.01$, and $\gamma = 5$

	$n = 0.01$	$\gamma = 0.01$		$n = 0.01$	$\gamma = 5$	
P_r	$f''(0)$	$\theta(\eta = 2)$	$-\theta'(0)$	$f''(0)$	$\theta(\eta = 2)$	$-\theta'(0)$
0.1	0.6457	0.4679	0.2794	2.6971	2.7971	−2.0039
0.3	0.4676	0.3131	0.4130	3.9492	4.0139	−5.0725
0.71	0.2758	0.1435	0.6134	4.7773	4.0436	−9.2445
1	0.2009	0.0952	0.7233	5.0837	3.9789	−11.5754
6	−0.0905	0.0358	1.7400	6.5224	3.8417	−35.7382
10	−0.1443	0.0358	2.2396	6.8814	3.8844	−49.0029
S_{lp}	−0.0660	−0.0287	0.1930	0.3327	0.0258	−4.6339

TABLE 7.13
Effect of P_r on Some of the Properties of Sakiadis Flow When $n = 10$, Grashof Number $G_{rx} = 1$, Heat Source Parameter $\gamma = 0.01$, and $\gamma = 5$

	$n = 10$	$\gamma = 0.01$		$n = 10$	$\gamma = +5$	
P_r	$f''(0)$	$\theta(2)$	$-\theta'(0)$	$f''(0)$	$\theta(2)$	$-\theta'(0)$
0.1	0.6401	0.4628	0.2839	0.6490	0.4670	0.1871
0.3	0.4536	0.3011	0.4253	0.4747	0.3206	0.1394
0.71	0.2490	0.1228	0.6392	0.2874	0.1201	−0.0241
1	0.1673	0.0704	0.7577	0.2151	0.0684	−0.1670
6	−0.1595	0.0000	1.8768	−0.0319	0.0000	−3.2127
10	−0.2205	0.0000	2.4334	−0.0598	0.0000	−5.7279
S_{lp}	−0.0727	−0.0313	0.2121	−0.0578	−0.0320	−0.6041

number decreases the motion along the y-direction negligibly when the heat source is small ($\gamma = 0.01$) and significantly when the heat source is sufficiently large ($\gamma = 5$); see Figures 7.34 and 7.35. It is also observed that an increase in the heat source γ strongly influences the vertical velocity near the free stream compared to near the wall. Figures 7.36 and 7.37 show that the velocity of the transport phenomenon parallel to the x-axis decreases with the Prandtl number from the wall till the free stream when $\gamma = 0.01$. In this case of a small heat source, the maximum horizontal velocity is found when P_r is small in magnitude near the wall; see Figure 7.36. When the heat source is sufficiently large ($\gamma = 5$), the horizontal velocity increases near the wall and decreases near the free stream due to increased Prandtl number. The results presented in Figure 7.37 show that the turning point for each value of Prandtl number occurs immediately after $\eta = 1$.

As shown in Figures 7.38 and 7.39, the graphical illustrations of the shear stress function proportional to the friction across the flow show that $f''(\eta)$ is negative within the domain. However, positive increasing values of the shear stress are ascertained when $\gamma = 5$ near the wall; see Figure 7.39. It is worth noticing from Figures 7.40 and 7.41 that the temperature distribution is dependent on the heat source parameter γ. As shown in Figure 7.40, when $\gamma = 0.01$, the temperature function is widespread across the domain of fluid flow with less viscous fluids $P_r \ll 1$. When $\gamma = 0.01$, the temperature distribution function $\theta(\eta)$ decreases with P_r across the domain. Figure 7.41 indicates that the temperature distribution is an increasing function of Prandtl number near the wall ($0 \leq \eta \leq 1$). Due to the relationship between the temperature distribution and heat transfer rates across the domain,

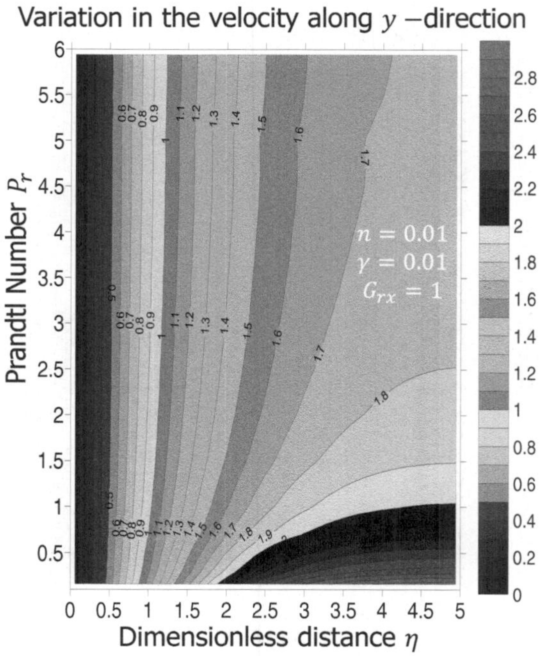

FIGURE 7.34
Variation in the velocity along the y-direction with Prandtl number P_r across the domain: $n = \gamma = 0.01$, $G_{rx} = 1$.

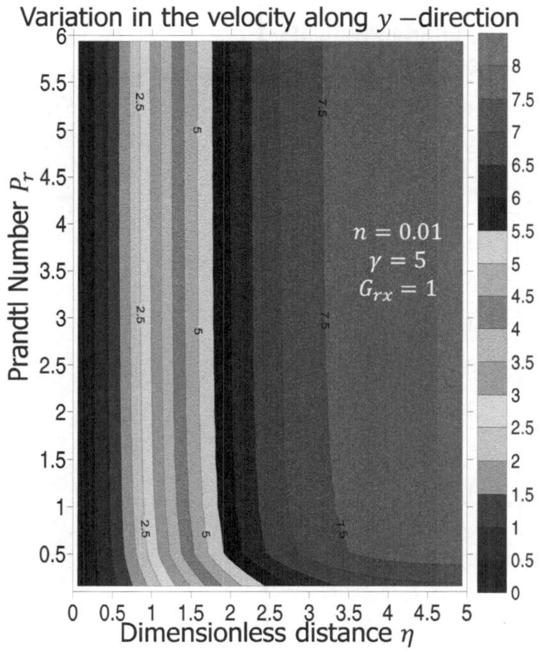

FIGURE 7.35
Variation in the velocity along the y-direction with Prandtl number P_r across the domain: $n = 0.01$, $\gamma = 5$, $G_{rx} = 1$.

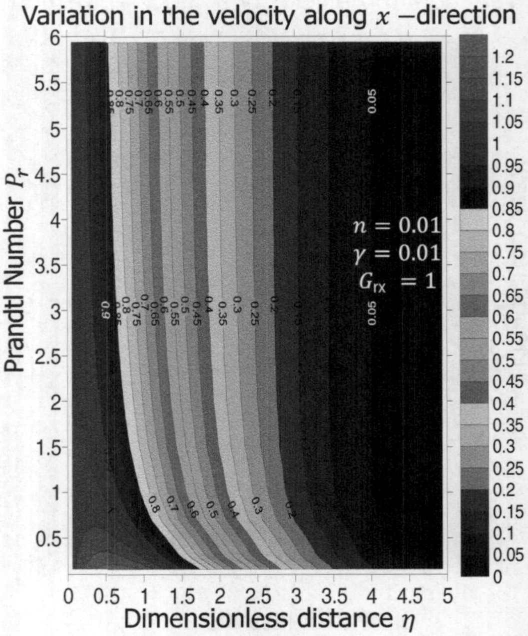

FIGURE 7.36
Variation in the velocity along the x-direction with Prandtl number P_r across the domain: $n = \gamma = 0.01$, $G_{rx} = 1$.

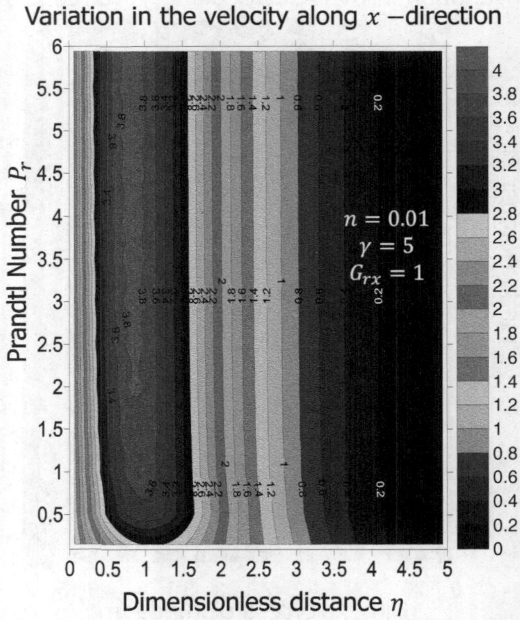

FIGURE 7.37
Variation in the velocity along the x-direction with Prandtl number P_r across the domain: $n = 0.01$, $\gamma = 5$, $G_{rx} = 1$.

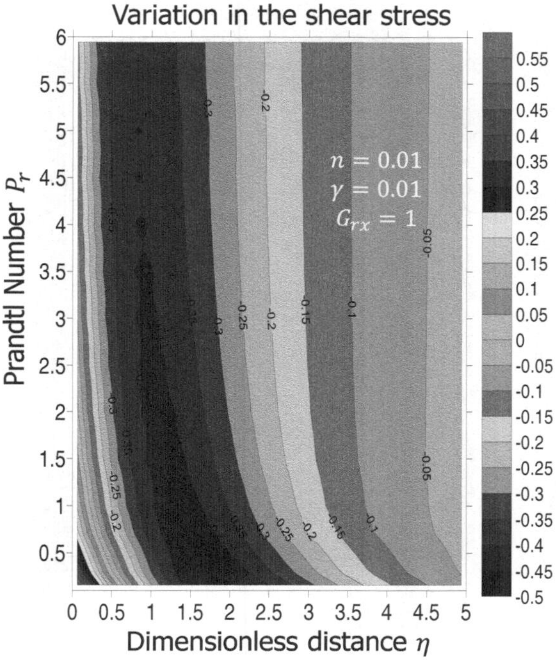

FIGURE 7.38
Variation in the shear stress with Prandtl number P_r across the domain: $n = \gamma = 0.01$, $G_{rx} = 1$.

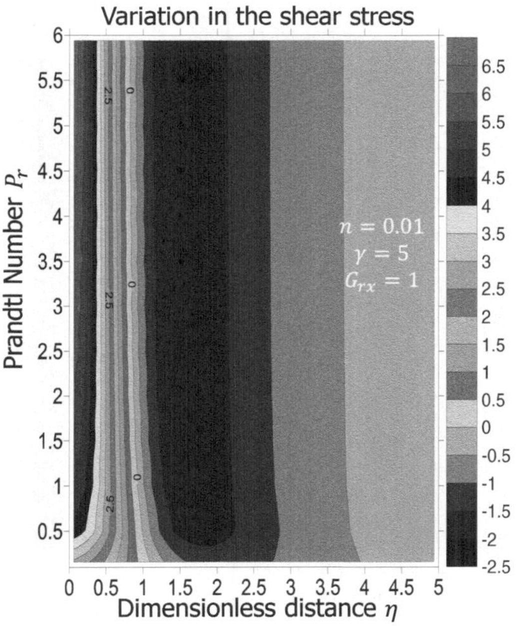

FIGURE 7.39
Variation in the shear stress with Prandtl number P_r across the domain: $n = 0.01$, $\gamma = 5$, $G_{rx} = 1$.

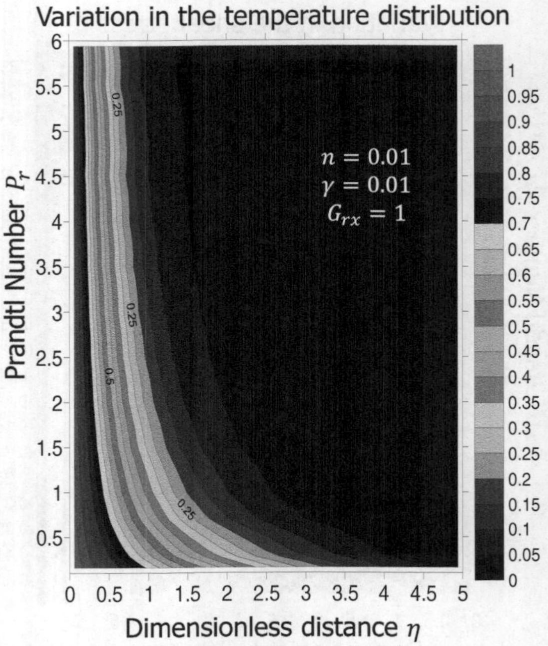

FIGURE 7.40
Variation in the temperature distribution with Prandtl number P_r across the domain: $n = \gamma = 0.01$, $G_{rx} = 1$.

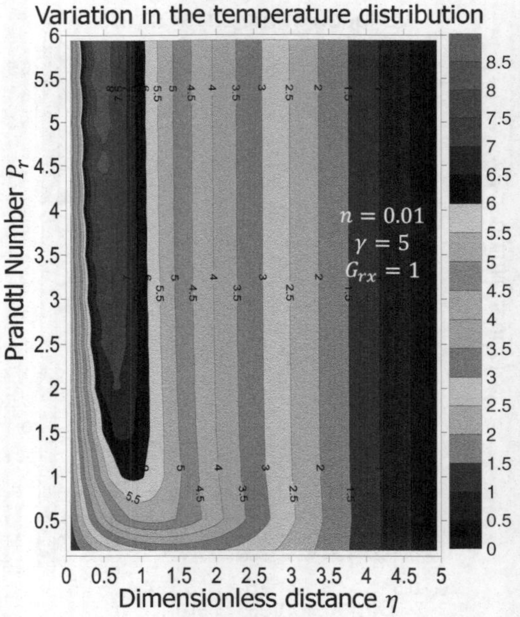

FIGURE 7.41
Variation in the temperature distribution with Prandtl number P_r across the domain: $n = 0.01$, $\gamma = 5$, $G_{rx} = 1$.

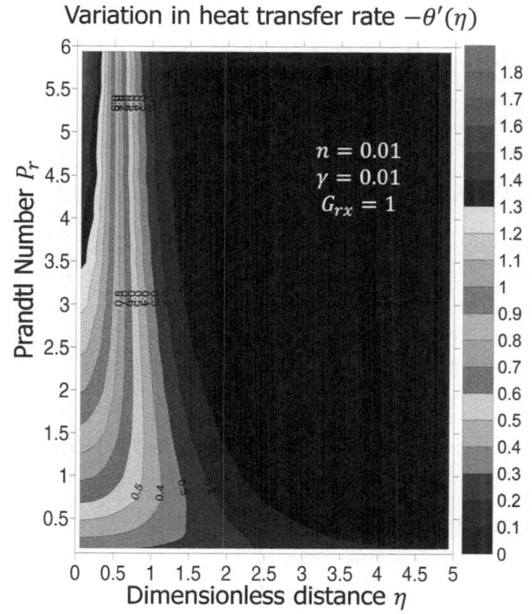

FIGURE 7.42
Variation in the heat transfer rate $-\theta'(\eta)$ with Prandtl number P_r across the domain: $n = \gamma = 0.01$, $G_{rx} = 1$.

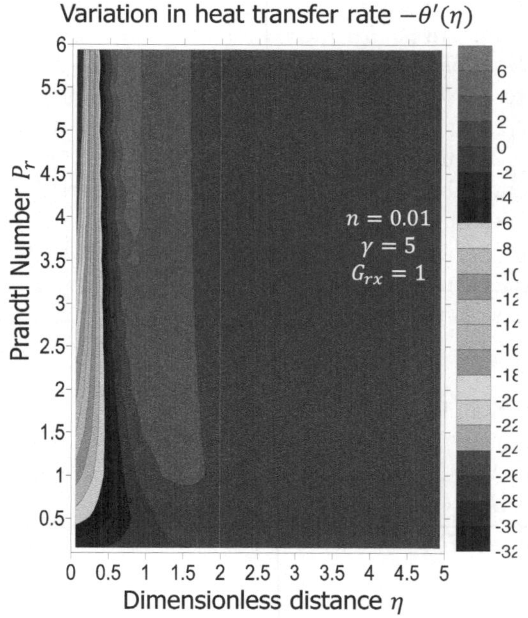

FIGURE 7.43
Variation in the heat transfer rate $-\theta'(\eta)$ with Prandtl number P_r across the domain: $n = 0.01$, $\gamma = 5$, $G_{rx} = 1$.

Figures 7.42 and 7.43 reveal that there exists a significant difference in the heat transfer rate across the domain when $\gamma = 0.01$ and $\gamma = 5$. Such phenomenon can be traced to the observed maximum and minimum of $-\theta'(0)$.

7.7 Introduction: Thermo Effect and Thermal Diffusion

Heat and mass transfer is a common occurrence in the study of fluid flow over several planes such as vertical, horizontal, cone, paraboloid of revolution, cylinder, and wedge. Both heat and mass transfer occurs simultaneously in fluid flow and thus leads to two types of fluxes (i.e., energy flux and mass flux). Energy flux is the rate of energy transfer per unit area, while mass flux is the mass transfer rate per unit area (Table 7.14).

7.7.1 Energy Flux Due to Concentration Gradient

The term "concentration gradient" implies a given area (i.e., a parcel or a domain) that contains many constituents/spices compared to another. In such a case, the constituents migrate from the higher concentration region to the lower concentration region. Also, this frequently occurs in heat and mass transfer as a coupled effect of an irreversible process. Fourier's law of heat conduction describes the relation between energy flux and temperature gradient; see Hetnarski [132] and Liu [179]. G. Henri Dufour presented a comprehensive report on energy flux due to concentration gradient in the year 1872. However, this was not properly understood until Clusius and Waldmann [80] further clarified the emergence of transient temperature gradient due to an initial concentration gradient (Boushehri and Afrashtehfar [66]). Hort et al. [135] noted that the generation of temperature fluctuations by concentration fluxes is known as the Dufour effect, and this directly influences the stability behavior of binary fluid mixtures. An increase in Dufour number but with a decrease in Soret number is a threshold of enhancing the temperature distribution and diminishing the concentration in the case of Newtonian fluid flows at the middle of a porous medium when energy flux due to concentration gradient and mass flux due to temperature gradient are non-zero (Moorthy and Senthilvadivu [196]).

Based on the fact that whenever the variations in temperature and concentration are significant, the occurrences of energy flux due to concentration gradient and mass flux due to temperature gradient are ascertained, the results before Ramzan et al. [250] led to the conclusion that an increase in Dufour number corresponds to an increase in temperature distribution, but a decrease in the concentration of viscoelastic nanofluid due to stretching at

TABLE 7.14
Variation in $f''(0)$, $-\theta'(0)$, and $-\phi'(0)$ with the Lewis Number When $G_t = G_c = 3$, $D_f = 1$, $S_r = 0.1$ at Prandtl Number $P_r = 0.3$

L_e **when** $P_r = 0.3$	$f''(0)$	$-\theta'(0)$	$-\phi'(0)$
1.62	4.5475	−0.2019	1.0909
3.62	4.2176	−1.7595	2.6639
5.62	4.0874	−4.6546	5.5677
7.62	4.0149	−12.4025	13.3199
9.62	3.9676	−101.9320	102.8511
S_{lp}	−0.068125	−10.70516	10.70882

the wall. However, it is worth deducing that the governing equation presents the combination of Brownian motion, thermophoresis, Soret effect, and Dufour effect. In this case, repetition is seen because no significant difference exists between thermophoresis and the Soret effect. Ramzan et al. [250] also concluded that with an increase in the Dufour number, the Nusselt number decreases, while the Sherwood number increases.

7.7.2 Mass Flux due to Temperature Gradient

Fick's law was determined by the correlation of mass flux and concentration gradient; see Poirier and Geiger [234]. Whenever the temperature gradient in a medium is significant, a relative motion of the components of a gaseous mixture—thermal diffusion—is bound to occur. Swiss scientist Charles Soret discovered this in 1879–1881. There is a relationship between mass fluxes and thermal energy, which conceivably produces a significant temperature inversion fluid flow (Delancey and Chiang [89]). Also, mass flux due to temperature gradient is inevitable when small molecules separate from heavy molecules. Chapman Sydney and Enskog David, in separate reports, not only predicted thermal diffusion in gases, but also confirmed the fact pointed out by Charles Soret (Pidduck [232]; Chapman & Cowling [75]; Kaljasin [148]).

Experimentally, Chaston and Hulton [76] and Ayres [46] confirmed thermal diffusion using the kinetic theory of gases. In the absence of external force but constant pressure, the total mass flux in gases is non-zero. The homogeneity of mixture composition is something that can be affected by thermal diffusion. Massive movement of constituents in a given mixture or fluid due to temperature gradient action normally leads to the separation of these constituents. Either optical/electrical or direct concentration is a good technique for measuring the separation between constituents placed in a vertical domain subject to a temperature gradient due to the Soret effect (Sani and Koster [268]). The concentration of viscoelastic nanofluids is an increasing property of Soret number (Ramzan et al. [250]), and it was also discovered that with an increase in the magnitude of Soret number, the Nusselt number increases, but the Sherwood number decrease (Figure 7.44).

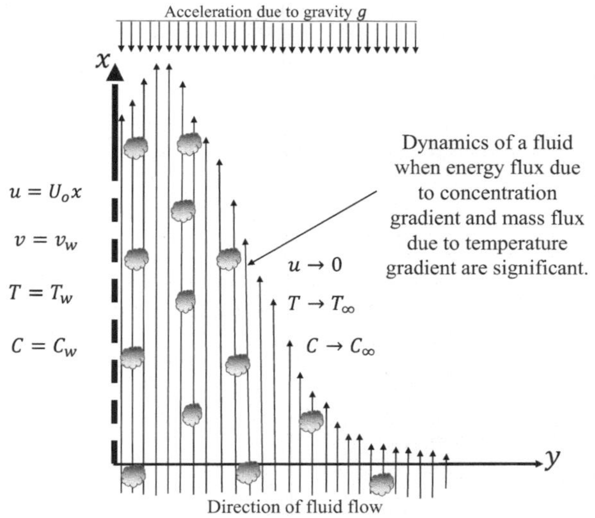

FIGURE 7.44
Graphical illustration of fluids when thermo-effect and thermal diffusion are significant.

7.8 Fluid Flow Subject to Thermo-Effect and Thermal Diffusion

As shown in Figure 7.26, the mobility of fluids on a horizontal surface susceptible to energy flux due to concentration gradient and mass flux owing to temperature gradient is relevant in this section. The two-dimensional flow occurs on a vertical surface with $u(x, y)$ velocity along the x-direction and $v(x, y)$ velocity along the y-direction. The governing equation appropriate for modeling the situations is stated as

$$u_x + v_y = 0, \tag{7.26}$$

$$uu_x + vu_y = \frac{\mu}{\rho} u_{yy} + g\beta(T - T_\infty) + g\beta^*(C - C_\infty), \tag{7.27}$$

$$uT_x + vT_y = \frac{\kappa}{\rho Cp} T_{yy} + \frac{D_m k_t}{c_s c_p} C_{yy}, \tag{7.28}$$

$$uC_x + vC_y = D_m C_{yy} + \frac{D_m k_t}{t_m} T_{yy}. \tag{7.29}$$

For this case, Eqs. (7.26)–(7.29) are subject to the boundary conditions

$$u = U_o x, \quad v = v_w, \quad T = T_w(x) \quad C = C_w(x) \quad at \quad y = 0. \tag{7.30}$$

$$u \to 0, \quad T \to T_\infty, \quad C \to C_\infty, \quad as \quad y \to \infty \tag{7.31}$$

In order to obtain the self-similar solution of Eq. (7.26) - Eq. (7.31), the following similarity variables:

$$\eta = y\sqrt{\frac{U_\infty}{\vartheta x}}, \quad \psi(x, y) = \sqrt{\vartheta x U_\infty} f(\eta), \quad u = \frac{\partial \psi}{\partial y}, \quad v = -\frac{\partial \psi}{\partial x},$$

$$D_f = \frac{D_m k_t}{c_s c_p \vartheta} \frac{(C_w - C_\infty)}{(T_w - T_\infty)}, \quad \theta(\eta) = \frac{T - T_\infty}{T_w - T_\infty},$$

$$G_t = \frac{gx\beta(T_w - T_\infty)}{U_\infty^2}, \quad G_c = \frac{gx\beta^*(C_w - C_\infty)}{U_\infty^2}, \quad \phi(\eta) = \frac{C - C_\infty}{C_w - C_\infty},$$

$$L_e = \frac{\alpha}{D_m}, \quad P_r = \frac{\mu Cp}{\kappa} = \frac{\vartheta}{\alpha}, \quad f_w = -\frac{v_w}{\sqrt{\vartheta U_o}}, \quad S_r = \frac{D_m k_t}{t_m \alpha} \frac{(T_w - T_\infty)}{(C_w - C_\infty)}. \tag{7.32}$$

were used to obtain the dimensionless governing equation below:

$$\frac{d^3 f}{d\eta^3} + \frac{1}{2} f \frac{d^2 f}{d\eta^2} + G_t \theta + G_c \phi = 0, \tag{7.33}$$

$$\frac{d^2 \theta}{d\eta^2} + \frac{1}{2} P_r f \frac{d\theta}{d\eta} + D_f \frac{d^2 \phi}{d\eta^2} = 0, \tag{7.34}$$

$$\frac{d^2 \phi}{d\eta^2} + \frac{1}{2} P_r L_e f \frac{d\phi}{d\eta} + L_e S_r \frac{d^2 \theta}{d\eta^2} = 0. \tag{7.35}$$

Dimensionless boundary conditions are

$$\frac{df}{d\eta} = 1, \quad f = f_w, \quad \theta = 1, \quad \phi = 1 \quad at \quad \eta = 0. \tag{7.36}$$

$$\frac{df}{d\eta} \to 0, \quad \theta \to 0, \quad \phi \to 0 \quad as \quad \eta \to \infty. \tag{7.37}$$

7.8.1 Research Questions IV

This section was designed to provide the answer to the following research questions:

1. In the instance of suction, how well does increasing the Lewis number impact the transport phenomena when the energy flux due to concentration gradient is significantly greater than the mass flux owing to temperature gradient?
2. What is the relevance of increasing mass flux as a result of a temperature gradient?

7.8.2 Analysis and Discussion of Results IV

The numerical solution of dimensionless governing Eqs. (7.33)–(7.35) subject to Eqs. (7.36) and (7.37) was obtained using the four-stage Lobatto IIIa formula as described in Chapter 1 using $\eta = 5$. When $G_t = G_c = 3$, $D_f = 1$, $S_r = 0.1$, it is seen that as the magnitude of Lewis number increases, the temperature distribution within the fluid layers near the wall increases. This observed effect is more significant in the case of gases; see Figure 7.47. Physically, this leads to higher randomness of species and lower concentration of the fluid as it flows. However, the obstruction of transport phenomenon occurs. This is the major reason why the velocity of the motion along both directions (x, y) and concentration decrease as shown in Figures 7.45–7.48. Such a result is more observable in the case of less viscous fluids (i.e., $P_r < 0.72$). For a higher viscous fluid, it appears as if the Lewis number has no significant effect on the fluid dynamics. Based on the observed results illustrated in Table 7.15 for $P_r = 0.3$ and $P_r = 0.71$, the local skin friction coefficients decrease. The results illustrated in Figure 7.48 is in good agreement with Figure 6 presented by Farooq et al. [99], where it was remarked that the major reason why the concentration of fluid decreases with the Lewis number L_e is that higher values of the dimensionless number imply lower

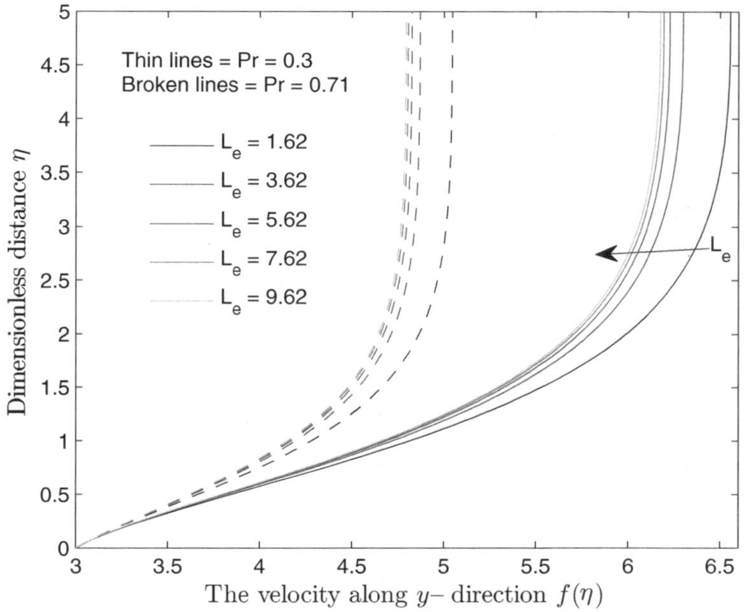

FIGURE 7.45
Variation in $f(\eta)$ due to L_e when $f_w = +3$.

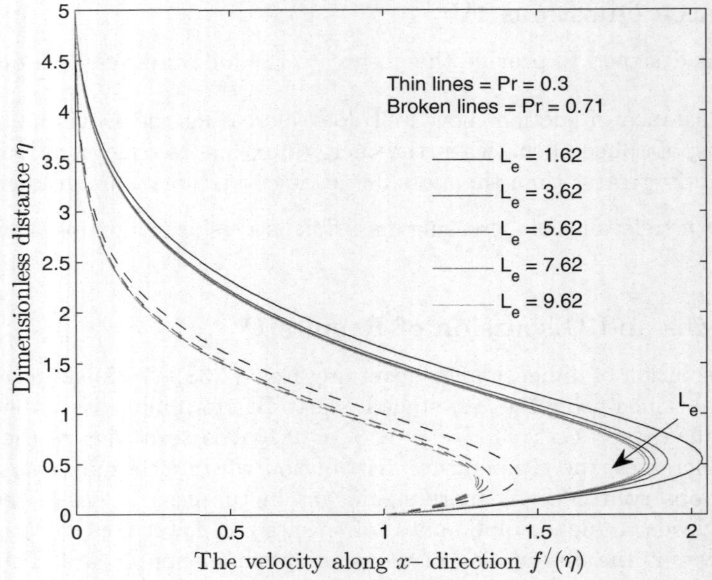

FIGURE 7.46
Variation in $f'(\eta)$ due to L_e when $f_w = +3$.

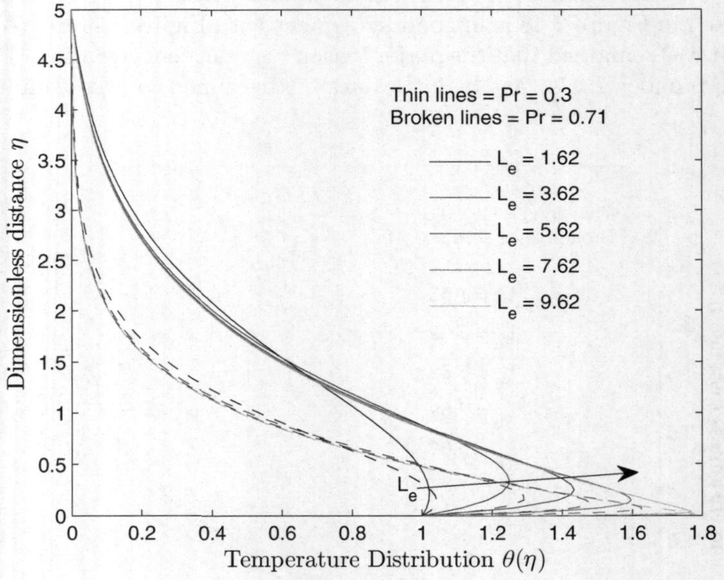

FIGURE 7.47
Variation in $\theta(\eta)$ due to L_e when $f_w = +3$.

mass diffusivity. It is noteworthy that this result is also in good agreement with Figure 16 presented by Sarojamma and Vendabai [272]. The result obtained by Acharya et al. [7] also confirms that the temperature function increases with the Lewis number. The nanoparticle concentration was noticed by Acharya et al. [7] to be an increasing function of Lewis number,

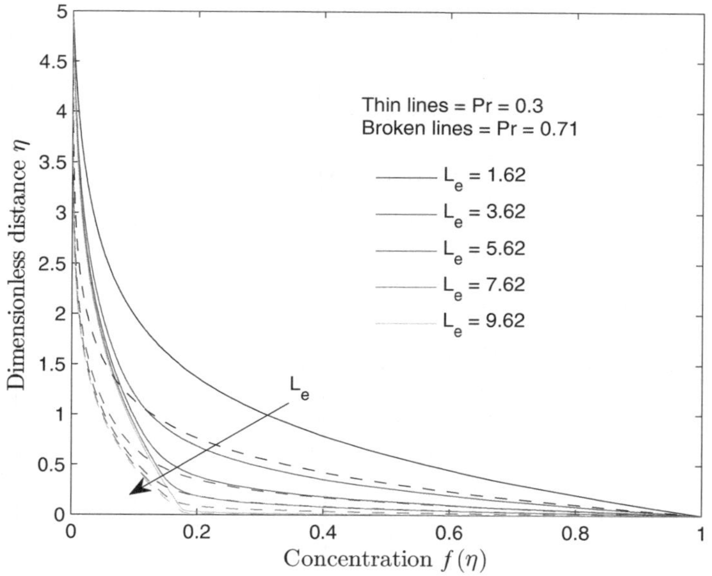

FIGURE 7.48
Variation in $\phi(\eta)$ due to L_e when $f_w = +3$.

TABLE 7.15
Variation in $f''(0)$, $-\theta'(0)$, and $-\phi'(0)$ with the Lewis Number When $G_t = G_c = 3$, $D_f = 1$, $S_r = 0.1$ at Prandtl Number $P_r = 0.71$

L_e **when** $P_r = 0.71$	$f''(0)$	$-\theta'(0)$	$-\phi'(0)$
1.62	2.6360	−0.6236	2.1582
3.62	2.3075	−4.2273	5.7690
5.62	2.1930	−11.0838	12.6285
7.62	2.1336	−29.4664	31.0123
9.62	2.0970	−241.4232	242.9695
S_{lp}	−0.062595	−25.341915	25.343295

which contradicts the observed related result (Figure 7.48). This contradiction may be traced to the presence of bioconvection (Figure 7.16).

The results presented in Figures 7.49 and 7.50 when $G_t = G_c = 3$, $P_r = 0.71$, $D_f = 1$, and $L_e = 1.62$ unravel the functional dependence of increasing mass flux due to temperature gradient (Soret number) on the velocity of the dynamics. The motion along the x- and y-directions is found to accelerate with the increasing Soret number S_r. The temperature distribution increases within the tiny layers of the fluid under consideration near the wall with S_r; see Figure 7.51. As shown in Figure 7.52, an increment in the concentration of the fluid at some distance away from the wall is seen due to a constant growth in the Soret number. Ramzan et al. [250] noticed that the concentration of the fluid decreases with an increase in the Lewis number. It was also noticed that the Nusselt number $-\theta'(0)$ increases with the Prandtl number at all the chosen values of Lewis number. The rate of increase in the Nusselt number $-\theta'(0)$ is highly significant when the Lewis number is small. When the heat flux due to concentration gradient is negligible ($D_f = 0.1$) and sufficiently larger than the mass flux due to temperature gradient ($D_f = 1$), the relationship between the Soret

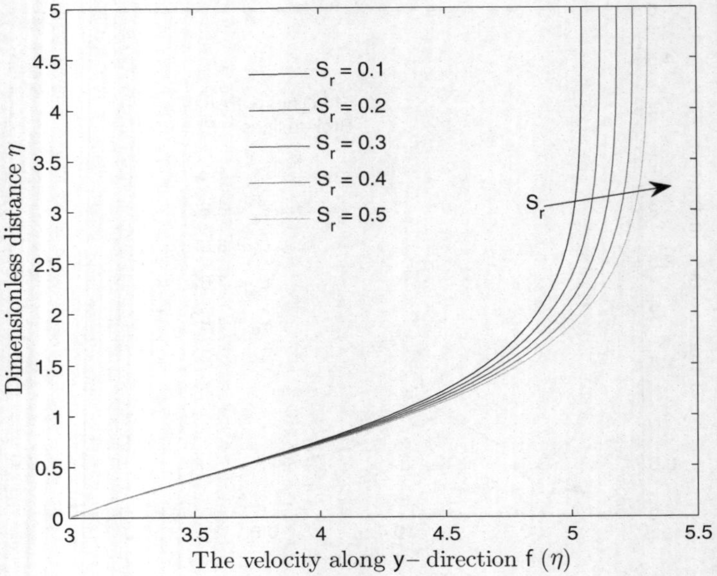

FIGURE 7.49
Variation in $f(\eta)$ due to S_r when $f_w = +3$.

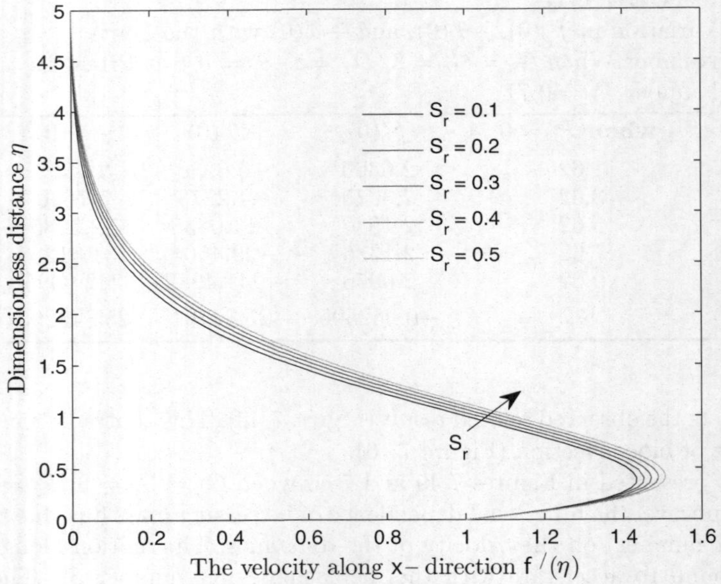

FIGURE 7.50
Variation in $f'(\eta)$ due to S_r when $f_w = +3$.

number and Prandtl number is presented in Tables 7.17 and 7.19. It is worth observing that the local skin friction coefficients $Cf_x\sqrt{Re_x}$ increases with Soret number when $D_f = 0.1$, $D_f > S_r$, for either $P_r0.1$ or $P_r0.71$. The outcome of the analysis indicates that the higher increasing rate of $Cf_x\sqrt{Re_x}$ is due to the higher level of mass flux due to temperature

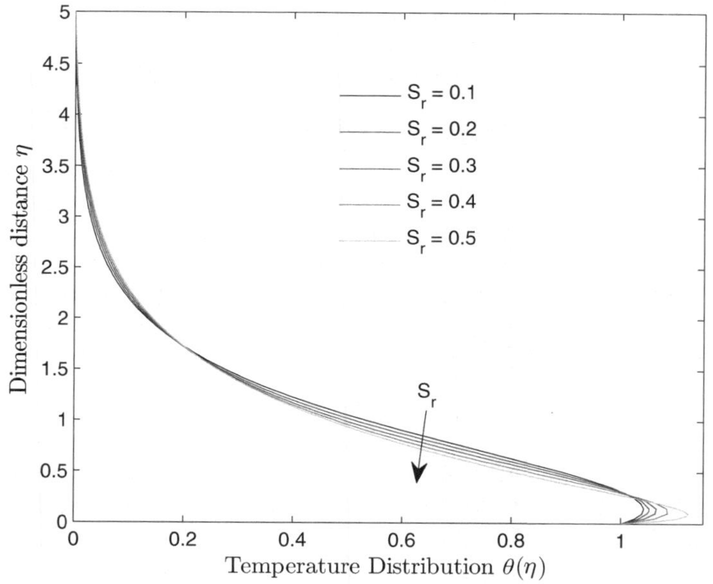

FIGURE 7.51
Variation in $\theta(\eta)$ due to S_r when $f_w = +3$.

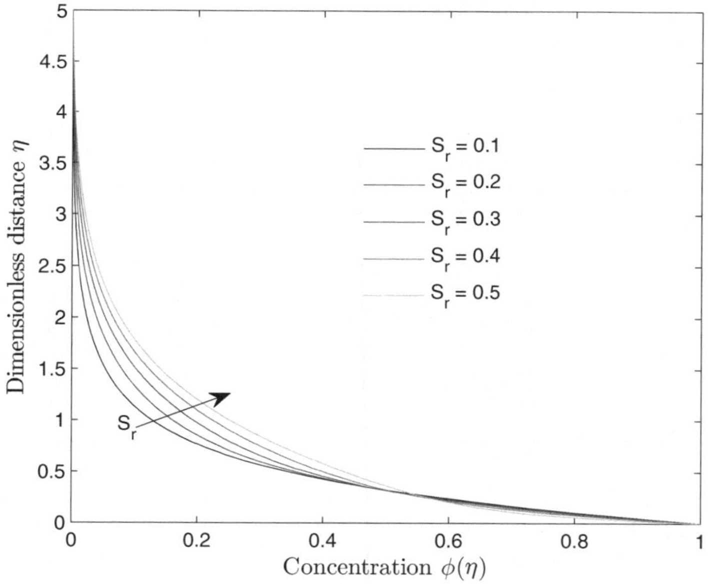

FIGURE 7.52
Variation in $\phi(\eta)$ due to S_r when $f_w = +3$.

gradient. When $D_f = 0.1$, it is glaring that a higher Soret number leads to a higher heat transfer rate $Nu_x Re_x^{-1/2}$. Reverse is the case when $D_f > S_r$, as $Nu_x Re_x^{-1/2}$ diminishes for a higher Soret number. The opposite effect of increasing S_r on $Sh_x Re_x^{-1/2}$ when $D_f = 0.1$ and $D_f > S_r$ is worth deducing from Tables 7.17 to 7.19.

TABLE 7.16
Variation in $f''(0)$, $-\theta'(0)$, and $-\phi'(0)$ with the Lewis Number When $\eta_\infty = 10$, $G_t = G_c = 3$, $D_f = 0.1$, $L_e = 1.62$ at Prandtl Number $P_r = 0.3$

S_r **when** $P_r = 0.3$	$f''(0)$	$-\theta'(0)$	$-\phi'(0)$
0.1	3.627450701987	0.594743652148	0.931926131202
0.2	3.757559966178	0.606574611965	0.859547516379
0.3	3.885113602279	0.618361088163	0.784451839249
0.4	4.010446109408	0.630152076518	0.706555643702
0.5	4.133842471726	0.641989570244	0.625762753360
S_{lp}	1.265669683	0.118069301	−0.765318628

TABLE 7.17
Variation in $f''(0)$, $-\theta'(0)$, and $-\phi'(0)$ with the Lewis Number When $\eta_\infty = 10$, $G_t = G_c = 3$, $D_f = 0.1$, $L_e = 1.62$ at Prandtl Number $P_r = 0.71$

S_r **when** $P_r = 0.71$	$f''(0)$	$-\theta'(0)$	$-\phi'(0)$
0.1	1.668131101179	1.127802438451	1.841233935680
0.2	1.784137974978	1.148536493239	1.683837441854
0.3	1.897862069371	1.169531271581	1.521032643856
0.4	2.009562707515	1.190841738506	1.352566026896
0.5	2.119458748541	1.212519843675	1.178164184765
S_{lp}	1.128080027	0.211740056	−1.657410917

TABLE 7.18
Variation in $f''(0)$, $-\theta'(0)$, and $-\phi'(0)$ with Lewis Number When $\eta_\infty = 10$, $G_t = G_c = 3$, $D_f = 1$, $L_e = 1.62$ at Prandtl Number $P_r = 0.3$

S_r **when** $P_r = 0.3$	$f''(0)$	$-\theta'(0)$	$-\phi'(0)$
0.1	4.728920467599	−0.217257443287	1.094442945395
0.2	4.790532556326	−0.289632001192	1.170094496458
0.3	4.850442904337	−0.411599882622	1.294684366179
0.4	4.908928174343	−0.651623561330	1.536696068281
0.5	4.966027899574	−1.313440377179	2.199760851567
S_{lp}	0.592610482	−2.554357428	2.577237384

TABLE 7.19
Variation in $f''(0)$, $-\theta'(0)$, and $-\phi'(0)$ with Lewis Number When $\eta_\infty = 10$, $G_t = G_c = 3$, $D_f = 1$, $L_e = 1.62$ at Prandtl Number $P_r = 0.71$

S_r **when** $P_r = 0.71$	$f''(0)$	$-\theta'(0)$	$-\phi'(0)$
0.1	2.647619223080	−0.624319185857	2.158448796391
0.2	2.715636209149	−0.804301583845	2.341990410104
0.3	2.781761624311	−1.103071790733	2.643540826158
0.4	2.846353997660	−1.684631891405	3.227118848074
0.5	2.909702961932	−3.273289051539	4.816984441260
S_{lp}	0.654885266	−6.178270039	6.202199728

7.9 Tutorial Questions

1. What is the variation in local skin friction coefficients, temperature distribution, and Nusselt number (heat transfer rate) with thermal radiation, buoyancy, and Prandtl number?
2. When the transmission of heat energy through electromagnetic waves is nonlinear, what are the increasing effects of the ratio of momentum diffusivity to thermal diffusivity in the presence of suction and injection?
3. How does fluid transport phenomenon vary due to changes in Prandtl number when buoyancy forces and suction are significant, but the transmission of heat energy through electromagnetic waves is minimal and maximum?
4. What is the effect of Prandtl number on the temperature distribution within the fluid flow due to stretching at the wall with heat source and heat sink?
5. What is the influence of increasing the ratio of momentum diffusivity to thermal diffusivity on local skin friction coefficients and heat transfer rate during the flow due to stretching at the wall with heat source and heat sink?
6. What effect does the heat source's intensity have on the dynamics in the presence of a heat source and a heat sink?
7. How does the Prandtl number influence the temperature distribution inside fluid flow owing to stretching at the wall with heat source and heat sink in the presence of buoyant forces to induce flow along a vertical wall?
8. What impact does stretching at the wall with heat source and heat sink have on local skin friction coefficients and heat transfer rate during free convective flow when the ratio of momentum diffusivity to thermal diffusivity is increased?
9. What effect do the intensity of the heat source and buoyant forces have on the dynamics in the presence of a heat source and a heat sink?
10. In the event of suction, how does increasing the Lewis number impact the transport phenomena when the energy flux due to concentration gradient is significantly greater than the mass flux owing to temperature gradient?

Critical Questions

1. [illegible]

2. [illegible]

3. [illegible]

4. [illegible]

5. [illegible]

6. [illegible]

7. [illegible]

8. [illegible]

9. [illegible]

10. [illegible]

8

Analysis of Self-Similar Flows IV

8.1 Background Information

This chapter discusses the increasing influence of Prandtl number on boundary layer flows caused by Marangoni convection. It is known that variations in surface tension brought about by temperature changes over the surface of a thin liquid layer are what trigger the Marangoni flow, also known as thermocapillary convection. In this case, spontaneous convection that was driven by liquid gravity gradually disappeared in a microgravity environment, but surface tension took over at the liquid-liquid interface, producing a surface tension gradient; see Tsukada (2015). Zheng and Zhang (2017) stated that the Marangoni convection may be observed in several situations, including the production of vapor bubbles and their expansion as a result of variations in surface tension brought on by changes in temperature and/or concentration along the bubble's surface. The temperature distribution and/or concentration of a species influences electron beam melting of metals, crystal development, and surface tension during welding.

8.2 Introduction: Thermo-Capillary Convection Flow

In 1855, the tears of wine scenario were scrutinized and led to the first report on the Marangoni effect by physicist James Thomson (Lord Kelvin's brother). Italian physicist Carlo Marangoni later deliberated upon the concept in his doctoral dissertation published in the year 1865. Wuest and Chun [318] remarked that Marangoni convection occurs during the formation of bubbles and films in boiling, condensation or vaporization of drops and films, and migration of drops or bubbles in an inhomogeneous temperature field. The surface tension within a fluid flow is primarily affected by temperature and chemicals in the fluid. The molecules within the fluid become more agitated when its temperature increases, causing more collisions and reducing the fluid's surface tension. The chemical reactions that occur when chemicals are added to the fluid also affect the surface tension. Impurity is also a factor that affects the surface tension of the fluid. For instance, when an impurity highly soluble in the fluid is added, surface tension increases, but decreases with a sparingly soluble impurity.

The total amount of energy needed to create a specific surface area or degree of imbalance of the intermolecular forces of a surface depends on the nature of the material at the surface of liquids is known as free surface energy (Rapp [251]). In the study of liquids, surface tension is synonymous to free surface energy. Most insects can walk on water because the surface tension force is equivalent to the gravitational weight of the insect. Temperature, surfactants, and tears of wine are major factors suitable to cause changes in the free surface energy. Inhomogeneities in free surface energy of either a solid or liquid surface often lead to the Marangoni effect's special force. Physically, when the Marangoni effect is significant, slip velocity in the tangential direction is inevitable due to gradients in

DOI: 10.1201/9781003217374-8

the surface tension coefficient. Residual convection flow is present in the absence of gravity, and the gradient of surface tension is responsible for Marangoni convection (Wuest and Chun [318]). The surface tension decreases when the temperature is increased. Because of this, a lower degree of surface tension at the liquid–gas interface is ascertained at higher temperatures (Gambaryan-Roisman [103]).

Can the addition of nanoparticles affect surface tension? At least one empirical review is needed for justification. The answer is yes. Addition of nanoparticles to the base fluid influences surface tension and surface wettability. This fact is one of the observed phenomena discovered by Vafaei et al. [304]. Bhuiyan et al. [60] examined the changes in the surface tension of nanofluids (i.e., distilled water conveying Al_2O_3, TiO_2, and SiO_2 nanoparticles) at various values of concentration of nanoparticles and discovered that as the concentration of nanoparticles increases, the surface tension also increases. The nanofluid with higher surface tension is 21 nm TiO_2, followed by 50 nm Al_2O_3, and 13 nm Al_2O_3, 10–20 nm SiO_2. The nanofluid with the most minimum surface tension is 5–10 nm SiO_2. Based on these two results, it can be concluded that surface tension is bound to increase due to the addition of nanoparticles laterally. Such an occurrence is associated with the fact that as the volume of nanoparticles increases in the base fluid, a higher volume of nanoparticles is driven to the liquid–solid surface, which is bound to cause agglomeration. Exertion of cohesive force between the molecules of nanoparticles leads to a higher magnitude of surface tension of nanofluids (Tanvir and Qiao [291]). Besides, the surface tension of nanofluids increases linearly due to an enlargement in particle size and an increase in concentration.

8.3 Fluid Flow on Horizontal Walls due to Surface Tension

Consider the incompressible viscous fluid flow with surface tension on a horizontal surface within the domain $0 \leq y < \infty$, where the effect of viscosity is highly significant. As shown in Figure 8.1, it is assumed that the flow on the porous horizontal surface was induced by viscous forces when suction and injection are non-zero. The governing equation that models the transport phenomenon is

$$u_x + v_y = 0, \tag{8.1}$$

$$uu_x + vu_y = \frac{\mu}{\rho} u_{yy}, \tag{8.2}$$

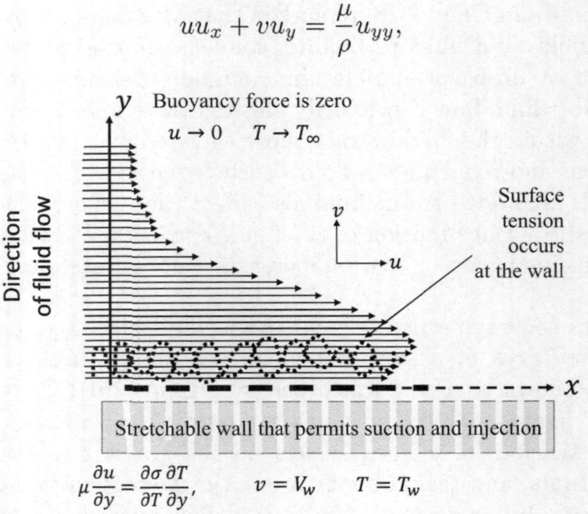

FIGURE 8.1
Illustration of fluid flow along a horizontal wall when surface tension is significant.

$$uT_x + vT_y = \frac{\kappa}{\rho c_p} T_{yy}. \tag{8.3}$$

subject to the boundary condition

$$\mu u_y = \sigma_T T_x, \quad v = v_w, \quad T = T_w + ax^2 \quad at \quad y = 0. \tag{8.4}$$

$$u \to 0, \quad T \to T_\infty \quad as \quad y \to \infty \tag{8.5}$$

In order to obtain self-similar solution of Eqs. (8.1)–(8.3) subject to Eq. (8.4) and (8.5), the following similarity variables were used

$$\eta = c_2 y, \quad \psi(x,y) = c_1 x f(\eta), \quad u = \frac{\partial \psi}{\partial y}, \quad v = -\frac{\partial \psi}{\partial x}, \quad c_1 = \left(\frac{\sigma_0 \gamma a \mu}{\rho^2}\right)^{\frac{1}{3}},$$

$$c_2 = \left(\frac{\sigma_0 \gamma a \rho}{\mu}\right)^{\frac{1}{3}}, \quad \theta(\eta) = \frac{T - T_\infty}{ax^2}, \quad P_r = \frac{\mu c_p}{\kappa} = \frac{\vartheta}{\alpha}, \quad f_w = \frac{-v_w}{c_1}. \tag{8.6}$$

The final dimensionless governing equation is

$$\frac{d^3 f}{d\eta^3} - \frac{df}{d\eta}\frac{df}{d\eta} + f\frac{d^2 f}{d\eta^2} = 0, \tag{8.7}$$

$$\frac{d^2\theta}{d\eta^2} - 2P_r\theta\frac{df}{d\eta} - P_r f\frac{d\theta}{d\eta} = 0. \tag{8.8}$$

Dimensionless boundary conditions Eqs. (8.9) and (8.10) are

$$\frac{d^2 f}{d\eta^2} = -2, \quad f = f_w, \quad \theta = 1 \quad at \quad \eta = 0. \tag{8.9}$$

$$\frac{df}{d\eta} \to 0, \quad \theta \to 0 \quad as \quad \eta \to \infty. \tag{8.10}$$

8.3.1 Research Questions I

Sequel to the aims and objectives of this study, the following are the research questions:

1. During the flow of a Newtonian fluid on a horizontal object due to surface tension, what is the effect of increasing the ratio of momentum diffusivity to thermal diffusivity?
2. What is the combined effect of suction, injection, and Prandtl number on fluid dynamics due to surface tension?

8.3.2 Analysis and Discussion of Results I

When the dynamics are along a horizontal wall, a possible effect of surface tension on the fluid flow in the presence of suction and injection was determined through the numerical solution of dimensionless governing Eqs. (8.1) and (8.3) subject to Eqs. (8.4) and (8.5) was obtained using the four-stage Lobatto IIIa formula as described in Chapter 1 using $\eta = 5$. In the case of fluid flow due to Marangoni convection along a horizontal surface, $f'(0) = 0.6258$ when $f_w = +3$ and $f'(0) = 2.7320$ when $f_w = -3$. In the case of suction and injection, it was discovered that the velocity and local skin friction are constant functions of Prandtl number. According to Gambaryan-Roisman [103], the surface tension gradient in Marangoni convection is the source of sufficiently large shear stress. As shown in Figures (8.2) and (8.3),

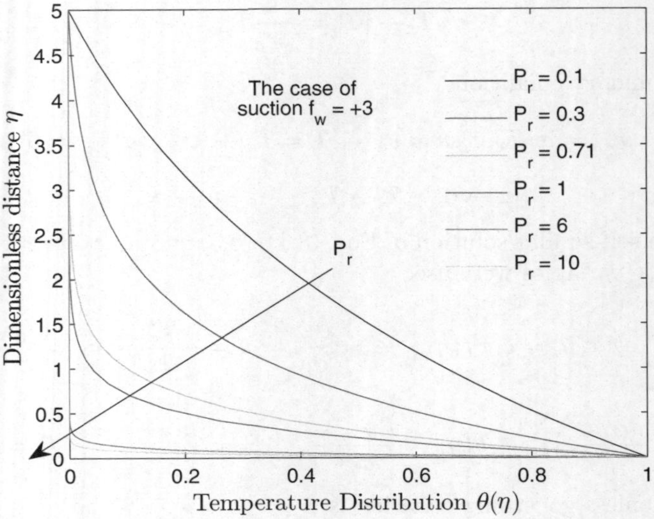

FIGURE 8.2
Variation in the temperature distribution with the Prandtl number in the presence suction $f_w = +3$.

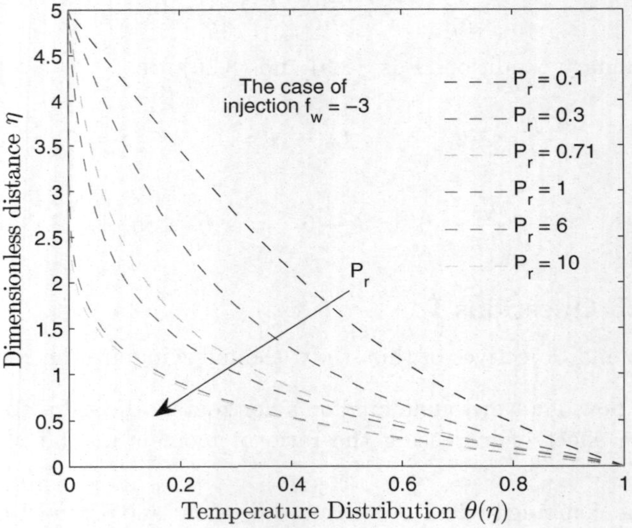

FIGURE 8.3
Variation in the temperature distribution with the Prandtl number in the presence suction $f_w = -3$.

suction greatly reduced the temperature distribution while injection boosted the heating in the flow of high viscous fluids; see Figure (8.2). This true because when $P_r = 10$, the temperature distribution is approximately zero ($\theta \approx 0$) over the interval $0.3 \leq \eta \leq 5$ when $f_w = +3$.

In other words, the changes in the heat energy are highly restricted within a thin domain near the wall due to suction. But, in the case of suction $f_w = -3$, when $P_r = 10$, $\theta \approx 0$ over the interval of $2.1 \leq \eta \leq 5$. Figure 8.2 also reveals that suction ($f_w = +3$) greatly affects the temperature distribution across the dynamics. It is seen in Tables 8.1 and 8.2 that the

TABLE 8.1
Rate of Decrease in $\theta(\eta)$ with η at Various Values of Prandtl Number for the Case of Suction $f_w = +3$

P_r	**0.1**	**0.3**	**0.71**	**1**	**6**	**10**
S_{lp}	−0.1984	−0.1677	−0.1075	−0.0824	−0.0343	−0.0263

TABLE 8.2
Rate of Decrease in $\theta(\eta)$ with η at Various Values of Prandtl Number for the Case of Injection $f_w = -3$

P_r	**0.1**	**0.3**	**0.71**	**1**	**6**	**10**
S_{lp}	−0.1812	−0.16196	−0.14446	−0.13793	−0.11606	−0.10921

TABLE 8.3
Effect of Prandtl Number P_r on Some of the Properties of the Flow in the Presence of Suction ($f_w = +3$)

		Suction $f_w = +3$		
P_r	$\theta(\eta = 1)$	$\theta(\eta = 2)$	$-\theta'(0)$	$-\theta'(0.3)$
0.1	0.6590	0.3963	0.4321	0.3759
0.3	0.3800	0.1315	1.0417	0.7540
0.71	0.1068	0.0095	2.3711	1.2078
1	0.0435	0.0014	3.2908	1.3086
6	0.0000	0.0000	18.5168	0.0580
10	0.0000	0.0000	30.5543	11.9263
S_{lp}	−0.0397	−0.0186	3.0422	0.9026

TABLE 8.4
Effect of Prandtl Number P_r on Some of the Properties of the Flow in the Presence of Injection ($f_w = -3$)

		Injection $f_w = -3$		
P_r	$\theta(\eta = 1)$	$\theta(\eta = 2)$	$-\theta'(\eta = 0)$	$-\theta'(\eta = 0.3)$
0.1	0.6812	0.4321	0.5076	0.4127
0.3	0.5323	0.2677	0.8337	0.6240
0.71	0.4082	0.1464	1.1405	0.8045
1	0.3652	0.1098	1.2578	0.8684
6	0.1927	0.0171	1.6748	0.9282
10	0.2135	0.0106	1.7291	1.1030
S_{lp}	−0.0368	−0.0292	0.0982	0.0477

temperature distribution decreases with P_r at different rates due to suction and injection. The results of this case indicate that the Nusselt number $-\theta'(0)$ increases with the Prandtl number at the rate of 0.0982 due to injection and 3.0422 due to suction (Tables 8.3 and 8.4). When $P_r = 10$ in Tables 8.3 and 8.4, the comparative analysis shows that $-\theta'(0)$ which is proportional to heat transfer rate, is 30.5543 in the case of suction and 1.7291 in the case of injection. Meanwhile, at some distance away from the wall ($\eta = 0.3$), it is important to remark that the temperature gradient $-\theta'$ increases significantly in the case of suction, but moderately in the case of injection (Tables 8.3 and 8.4). Next, consider the dynamics along a vertical surface when surface tension is apparent.

8.4 Fluid Flow on Vertical Walls due to Surface Tension

Consider the same flow on a vertical surface as shown in Figure 8.4. The governing equation that models the transport phenomenon becomes

$$uu_x + vu_y = \frac{\mu}{\rho}u_{yy} + g\beta(T - T_\infty), \tag{8.11}$$

In order to obtain the self-similar solution of Eq. (8.11) under the same conditions and with the same similarity variables, the final dimensionless governing equation is

$$\frac{d^3 f}{d\eta^3} - \frac{df}{d\eta}\frac{df}{d\eta} + f\frac{d^2 f}{d\eta^2} + G_{rx}\theta = 0. \tag{8.12}$$

Here, the buoyancy parameter G_{rx} is defined as

$$G_{rx} = \frac{g\rho\beta ax}{c_1 c_2 (c_2)^2}. \tag{8.13}$$

8.4.1 Research Questions II

Sequel to the aims and objectives of this study, the following are the research questions:

1. During the flow of a Newtonian fluid on a vertical object due to surface tension and buoyancy, what is the effect of increasing Grashof number and the ratio of momentum diffusivity to thermal diffusivity?
2. What is the combined effect of suction, injection, and Prandtl number on fluid dynamics due to surface tension?

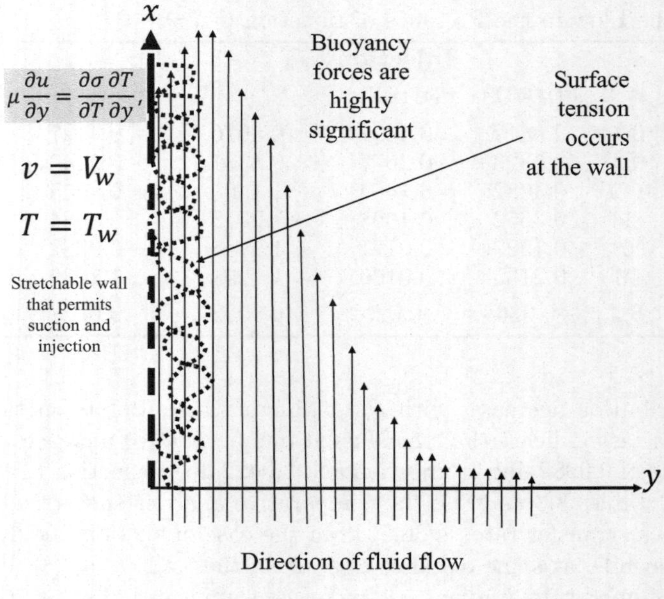

FIGURE 8.4
Illustration of fluid flow along a vertical wall when surface tension is significant.

8.4.2 Analysis and Discussion of Results II

Next is to examine the same flow along a vertical wall to capture the combined effects of Prandtl number, surface tension, and Grashof number in the presence of suction and injection. The numerical solutions of dimensionless governing equations Eqs. (8.12) and (8.8) subject to Eqs. (8.9) and (8.10) were obtained using the four-stage Lobatto IIIa formula as described in Chapter 1 using $\eta = 5$. The velocity of the dynamics along the y-direction decreases with higher Prandtl number when buoyancy forces are small ($G_{rx} = 0.1$) and suction is apparent; see Figure 8.5. When the magnitude of the buoyancy forces was increased ($G_{rx} = 5$) and suction is still apparent, the vertical velocity decreases more significantly with P_r as shown in Figure 8.6. It is also evident, as illustrated in Figures 8.7 and 8.8, that a rise in the Prandtl number causes the horizontal velocity to diminish. However, the rate of decrement of the dynamics along the x-direction is more significant; see Figure 8.8. It is worth noticing that the Prandtl number affects the velocity of highly viscous fluids due to buoyancy forces.

Table 8.5 shows that the Nusselt number increases with the Prandtl number. The observed Nusselt number at each value of Prandtl number when the dynamics are subjected to suction is higher than that of injection. Also, as buoyancy forces increase at a fixed ratio of momentum diffusivity to thermal diffusivity, the Nusselt number increases in the cases of suction and injection. A higher Nusselt number was observed when $P_r = 10$, $f_w = +3$, and $G_{rx} = 5$. In the case of Marangoni convection along a vertical surface, the Boussinesq approximation was incorporated into the momentum equation. The comparative analysis shows that the maximum increase in the Nusselt number proportional to heat transfer rate is ascertained due to suction at reduced Grashof number in the flow. These results corroborate one of the findings by Olanrewaju et al. [218]. The next question worth asking is, "Why is heat transfer rate very large at larger magnitude of Prandtl number?" The large Prandtl number as shown in Table 8.5 implies fluid with higher viscosity if the thermal diffusivity is fixed. Laterally, higher values of the Nusselt number imply a minimum heat

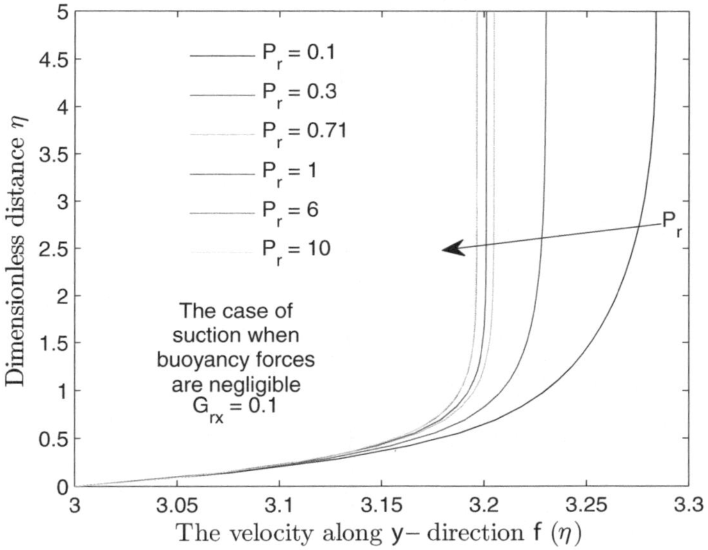

FIGURE 8.5
Velocity along the y-direction at various values of Prandtl number in the presence of suction $f_w = +3$ when $G_{rx} = 0.1$.

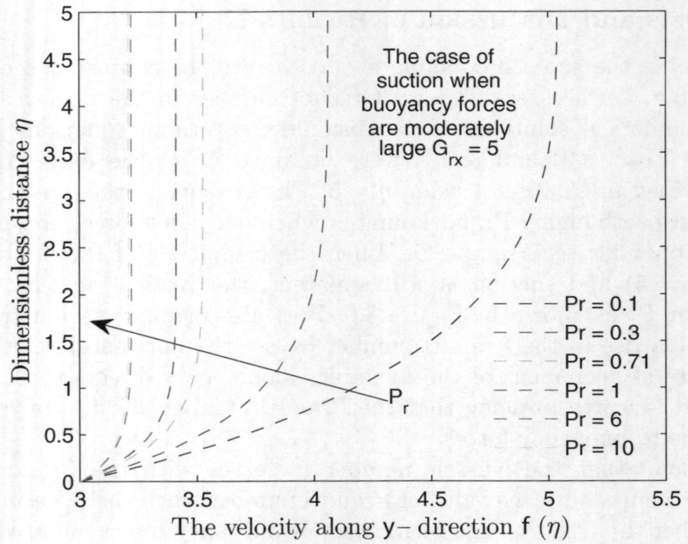

FIGURE 8.6
Velocity along the y-direction at various values of Prandtl number in the presence of suction $f_w = +3$ when $G_{rx} = 5$.

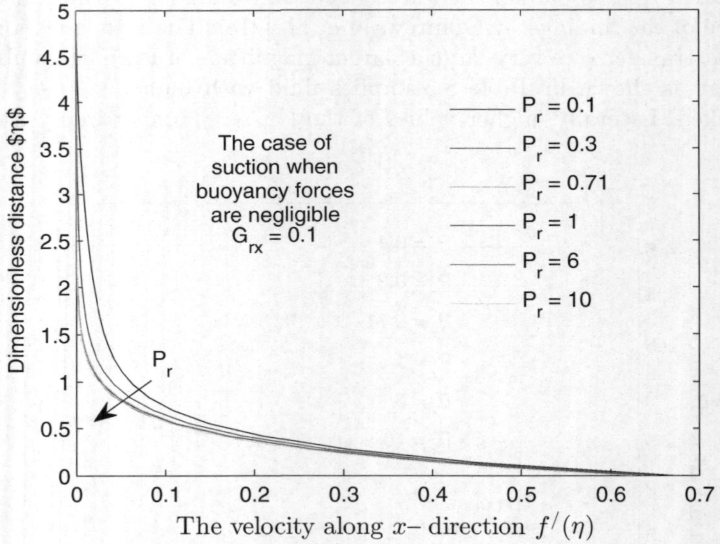

FIGURE 8.7
Velocity along the x-direction at various values of Prandtl number in the presence of suction $f_w = +3$ when $G_{rx} = 0.1$.

transfer rate. Non-uniformity of thermal conductivity that induced Marangoni convection suggested by Gambaryan-Roisman [103] is responsible. When buoyancy forces are in the range $0.1 \leq G_{rx} \leq 5$ for $f_w = +3$, $f_w = -3$, and Prandtl number was increased, it is seen that the temperature decreases; see Tables 8.6 and 8.7.

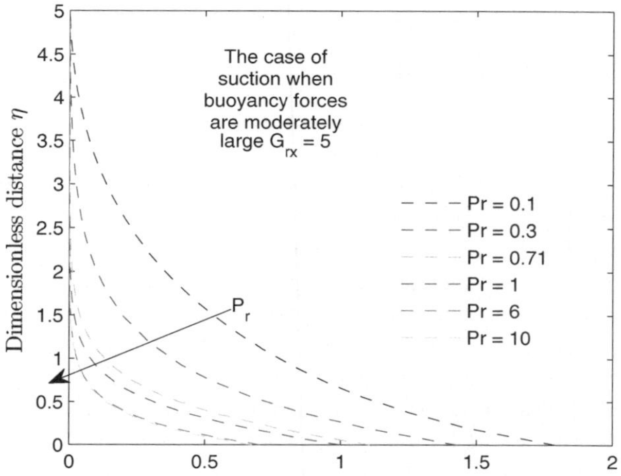

FIGURE 8.8
Velocity along the x-direction at various values of Prandtl number in the presence of suction $f_w = +3$ when $G_{rx} = 5$.

TABLE 8.5
Effect of Prandtl Number P_r on Nusselt Number $-\theta'(0)$

	$G_{rx} = 0.1$	$G_{rx} = 2$	$G_{rx} = 5$	$G_{rx} = 0.1$	$G_{rx} = 2$	$G_{rx} = 5$
	$f_w = +3$	$f_w = +3$	$f_w = +3$	$f_w = -3$	$f_w = -3$	$f_w = -3$
P_r	$-\theta'(0)$	$-\theta'(0)$	$-\theta'(0)$	$-\theta'(0)$	$-\theta'(0)$	$-\theta'(0)$
0.1	0.4451	0.5818	0.6876	0.5115	0.5722	0.6396
0.3	1.0556	1.2287	1.3811	0.8389	0.9227	1.0223
0.71	2.3785	2.4996	2.6433	1.1465	1.2484	1.3761
1	3.2965	3.3945	3.5221	1.2642	1.3739	1.5133
6	18.5182	18.5438	18.5835	1.6828	1.8225	2.0075
10	30.5552	30.5716	30.5974	1.7373	1.8806	2.0712
S_{lp}	3.041026	3.027089	3.014866	0.098569	0.105266	0.115379

TABLE 8.6
Rate of Decrease in $\theta(\eta)$ with η at Various Values of Prandtl Number for the Case of Suction $f_w = +3$

When ⇓	P_r	**0.1**	**0.3**	**0.71**	**1**	**6**	**10**
$Gr = 0.1$	S_{lp}	−0.1983	−0.1664	−0.10823	−0.0849	−0.03314	−0.0265
$Gr = 1$	S_{lp}	−0.1897	−0.1526	−0.1048	−0.0671	−0.0342	−0.0239
$Gr = 5$	S_{lp}	−0.1750	−0.1275	−0.1080	−0.0803	−0.0241	−0.0229

TABLE 8.7
Rate of Decrease in $\theta(\eta)$ with η at Various Values of Prandtl Number for the Case of Injection $f_w = -3$

When ⇓	P_r	**0.1**	**0.3**	**0.71**	**1**	**6**	**10**
$Gr = 0.1$	S_{lp}	−0.1811	−0.1617	−0.1441	−0.1375	−0.1043	−0.1017
$Gr = 1$	S_{lp}	−0.1823	−0.1592	−0.1397	−0.1328	−0.1011	−0.0875
$Gr = 5$	S_{lp}	−0.1719	−0.1476	−0.1239	−0.1164	−0.0850	−0.0661

8.5 Introduction: Magnetohydrodynamics

The dynamics of fluids that are electrical conductor within a bounded magnetic field is known as magnetohydrodynamics. Early reports on the subject matter were reported by Alfven [18], Rossow [258], and Liron and Wilhelm [178] and have recently been embedded in the introduction of Makinde and Animasaun [188]. Alfven [18] concluded that for every dynamics of conducting fluids subject to a magnetic field, a special force called electromotive force is reliable for producing electric currents and the Lorentz force. The generation of spots at the central part of the sun and its transportation to the surface is attributed to magnetohydrodynamic waves (Alfven [18,19,20]). In the presence of a magnetic field, rotation is less effective, as is evident in the case of baroclinic instability (Acheson and Hide [8]). In the case of fluid flows along an object that looks like a cone, Makinde et al. [190] and Makinde and Animasaun [188,191] noted that there exists a significant difference between the observed effects of the Lorentz force on the transport phenomenon at the leading edge and as x grows large since the magnetic field strength varies along the same direction.

Moreover, the sun generates a magnetic field, which flips roughly every 11 years (Zita [329]). On the surface of the earth, the magnetic field is not always horizontal. The magnetic inclination's angle may be up or down. For both the dark and bright components within the inner part of the penumbra, smaller inclination angles are inevitable (Langhans et al. [173]). Elnaby and El-Shamy [95] examined the effect of an inclined magnetic field on the transport phenomenon of a Newtonian fluid in a horizontal cylindrical tube, and it was concluded that the pressure on the motion is an increasing property of the magnetic field. In a study of electrically conducting Newtonian fluids over an inclined surface with primary and secondary flows, Ghosh et al. [109] once noted that a higher inclination of the applied magnetic field enhances the secondary flow, but it retards the primary flow. In another study on the significance of increasing Lorentz force on the motion of a non-Newtonian Casson fluid at an inclination angle γ, it was shown that deceleration of the motion is ascertained due to growth in the inclination angle γ and magnetic parameter (Abdul Hakeem et al. [5]).

8.6 Dynamics of Alumina-Water Nanofluid Subject to Joule Heating

Consider a steady two-dimensional electrically conducting fluid flow across a permeable elongating/shrinking sheet due to viscous forces and stretching or shrinking when Joule heating is taken into account. The surface is along the x-axis, and the y-axis is perpendicular to the wall. An inclined magnetic field of strength B_o with an acute angle γ is applied along the y-axis. At $\gamma = 90°$, this magnetic field acts like a transverse magnetic field. The fluid flow is restricted to $y > 0$, which is caused by the linear stretching of the sheet with velocity $u = u_w(x) = ax$ keeping the origin fixed. The magnetic Reynolds number of the fluid is very small. The induced magnetic field effects are neglected compared to the applied magnetic field, and the buoyancy forces are neglected. Based on the above assumptions, the equations of continuity, momentum, energy, and species concentrations of the transport phenomenon are given by

$$u_x + v_y = 0, \tag{8.14}$$

$$uu_x + vu_y = \frac{\mu_{nf}}{\rho_{nf}} u_{yy} - \frac{\sigma_{nf} B_o^2 u}{\rho_{nf}} sin^2(\gamma), \tag{8.15}$$

$$uT_x + vT_y = \frac{\kappa_{nf}}{(\rho c_p)_{nf}}T_{yy} + \frac{\sigma_{nf}B_o^2}{(\rho c_p)_{nf}}u^2 sin^2(\gamma), \tag{8.16}$$

$$uC_x + vC_y = D_m C_{yy}. \tag{8.17}$$

subject to the boundary conditions

$$u = \lambda u_w(x), \quad v = v_w, \quad T = T_w, \quad C = C_w \quad at \quad y = 0 \tag{8.18}$$

$$u \to 0, \qquad T \to T_\infty, \qquad C \to C_\infty \quad as \quad y \to \infty \tag{8.19}$$

The model proposed by Graham [114] and Gosukonda et al. [112] for the ratio of dynamic viscosity of the nanofluid to the dynamic viscosity of base fluid defined as

$$\frac{\mu_{nf}}{\mu_{bf}} = 1 + 2.5\phi + 4.5\left[\frac{1}{\frac{h}{d_p}\left(2+\frac{h}{d_p}\right)\left(2+\frac{h}{d_p}\right)^2}\right] \tag{8.20}$$

where the radius of nanoparticle is d_p and the inter-particle spacing is h, was adopted. The effective nanofluid properties are given by

$$(\rho c_p)_{nf} = (1-\phi)(\rho Cp)_{bf} + \phi(\rho Cp)_s, \quad \rho_{nf} = (1-\phi)\rho_{bf} + \phi\rho_s,$$

$$\frac{\sigma_{nf}}{\sigma_{bf}} = \left[1 + \frac{3\left(\frac{\sigma_s}{\sigma_{bf}}-1\right)\phi}{\left(\frac{\sigma_s}{\sigma_{bf}}+2\right) - \left(\frac{\sigma_s}{\sigma_{bf}}-1\right)\phi}\right], \quad \frac{\kappa_{nf}}{\kappa_{bf}} = \frac{\kappa_s + 2\kappa_{bf} - 2\phi(\kappa_{bf}-\kappa_s)}{\kappa_s + 2\kappa_{bf} + \phi(\kappa_{bf}-\kappa_s)} \tag{8.21}$$

where ϕ is the solid volume fraction, μ_{bf} is the dynamic viscosity of the base fluid, ρ_{bf} and ρ_s are the densities, $(\rho c_p)_{bf}$ and $(\rho c_p)_s$ are the heat capacitances, κ_{bf} and κ_s are the thermal conductivities, and σ_f and σ_s are the electrical conductivities of the base fluid and nanoparticles, respectively. In order to obtain self-similar solution of Eqs. (8.14)– (8.19), the following similarity variables were used:

$$\eta = y\sqrt{\frac{U_\infty}{\vartheta_{bf}x}}, \quad \psi(x,y) = \sqrt{\vartheta_{bf}xU_\infty}f(\eta), \quad \theta(\eta) = \frac{T-T_\infty}{T_w-T_\infty}, \quad \phi(\eta) = \frac{C-C_\infty}{C_w-C_\infty},$$

$$P_r = \frac{\mu_{bf}Cp}{\kappa_{bf}}, \quad E_c = \frac{U_\infty^2}{Cp_{bf}(T_w-T_\infty)}, \quad f_w = \frac{v_w}{-(a\vartheta_{bf})^{1/2}}, \quad L_e = \frac{\alpha_{bf}}{D_m},$$

$$A_1 = 1 + 2.5\phi + 4.5\left[\frac{1}{\frac{h}{d_p}\left(2+\frac{h}{d_p}\right)\left(2+\frac{h}{d_p}\right)^2}\right], \quad A_2 = 1 - \phi + \phi\frac{\rho_s}{\rho_{bf}}, \quad M = \frac{B_o^2\sigma_{bf}}{U_o\rho_{bf}},$$

$$A_3 = \left[1 + \frac{3\left(\frac{\sigma_s}{\sigma_{bf}}-1\right)\phi}{\left(\frac{\sigma_s}{\sigma_{bf}}+2\right) - \left(\frac{\sigma_s}{\sigma_{bf}}-1\right)\phi}\right], \quad A_5 = 1 - \phi + \phi\frac{(\rho Cp)_s}{(\rho Cp)_{bf}},$$

$$A_4 = \frac{\kappa_s + 2\kappa_{bf} - 2\phi(\kappa_{bf}-\kappa_s)}{\kappa_s + 2\kappa_{bf} + \phi(\kappa_{bf}-\kappa_s)}. \tag{8.22}$$

The dimensionless governing equation is now of the form

$$\frac{A_1}{A_2}\frac{d^3f}{d\eta^3} - \frac{df}{d\eta}\frac{df}{d\eta} + f\frac{d^2f}{d\eta^2} - \frac{A_3}{A_2}M\frac{df}{d\eta}Sin^2(\gamma) = 0, \tag{8.23}$$

$$\frac{A_4}{A_5}\frac{d^2\theta}{d\eta^2}+P_r f\frac{d\theta}{d\eta}+P_r\frac{A_3}{A_5}ME_c\frac{df}{d\eta}\frac{df}{d\eta}sin^2(\gamma)=0, \tag{8.24}$$

$$\frac{d^2\phi}{d\eta^2}+P_r L_e f\frac{d\phi}{d\eta}=0. \tag{8.25}$$

subject to

$$f=f_w,\quad \frac{df}{d\eta}=\lambda,\quad \theta=1,\quad \phi=1 \quad at \quad \eta=0 \tag{8.26}$$

$$\frac{df}{d\eta}\to 0,\quad \theta\to 0,\quad \phi\to 0 \quad as \quad \eta\to\infty. \tag{8.27}$$

The skin friction coefficient Cf_x, Nusselt number Nu_x, and Sherwood number Sh_x are defined as

$$Cf_x=\frac{\tau_w}{\rho_{bf}u_w^2}=\frac{\mu_{bf}}{\rho_{bf}a^2x^2}\frac{\partial u}{\partial y},\quad Nu_x=\frac{xq_w}{\kappa(T_w-T_\infty)}=\frac{-x\kappa_{bf}}{\kappa_{bf}(T_w-T_\infty)}\frac{\partial T}{\partial y},$$

$$Sh_x=\frac{xq_m}{D_B(C_w-C_\infty)}=\frac{-xD_B}{D_B(C_w-C_\infty)}\frac{\partial C}{\partial y}. \tag{8.28}$$

where the Reynolds number $\sqrt{Re_x}=\frac{U_o^{1/2}x}{\vartheta_{bf}^{1/2}}$, the shear stress is τ_w, the heat flux is q_w, the mass flux is q_m, $\rho_{bf}=997.1$ kg m^{-3}, $(Cp)_{bf}=4179$ Jkg^{-1} K^{-1}, $\kappa_{bf}=0.613$ Wm^{-1} K^{-1}, $\sigma_{bf}=5.5\times10^{-6}$ sm^{-1}, $\rho_s=3,970$ kg m^{-3}, $(Cp)_s=765$ kg^{-1} K^{-1}, $\kappa_s=40$ Wm^{-1} K^{-1}, and $\sigma_s=35\times10^6$ sm^{-1}. The dimensionless local skin friction, heat transfer rate, and mass transfer rate are

$$Cf_x\sqrt{Re_x}=f''(0),\quad Nu_xRe_x^{-0.5}=-\theta(0),\quad Sh_xRe_x^{-0.5}=-\phi(0). \tag{8.29}$$

8.6.1 Research Questions III

This section is designed to provide answers to the following research questions:

1. What are the impacts of magnetic strength on the dynamics of alumina–water owing to mild stretching in the presence of Joule heating and minimal suction at two distinct levels of Prandtl number?
2. When Joule heating and suction are moderate, what is the significance of increasing stretching at the wall on the dynamics of water–alumina nanofluid?

8.6.2 Analysis and Discussion of Results III

Equations (8.23) and (8.25) subject to Eqs. (8.26) and (8.27) were solved using the four-stage Lobatto IIIa formula as described in Chapter 1 using $\eta=5$. The entire magnetic field is neutralized in the absence of inclination (i.e., $\gamma=0$). At an inclination angle of 30°, inter-particle spacing $h=1$, radius of alumina nanoparticles $d_p=2.5$, volume fraction $\phi=0.02$, inclination of the magnetic field $\gamma=30°$, Lewis number $L_e=3$, $P_r=6$, viscous dissipation term—Eckert number $E_c=0.9$, stretching-related parameter $\lambda=0.5$, and suction $S=0.5$, the significance of higher magnetic strength M on the transport phenomenon was investigated and is illustrated i Figures 8.9–8.12. As indicated in Figures 8.9 and 8.10, the motions of the fluid along the x- and y-directions decrease with higher strength of the magnetism within the field M. Sequel to the observed decrease in the velocity due to the higher magnitude of the Lorentz force (M), the local skin friction coefficients

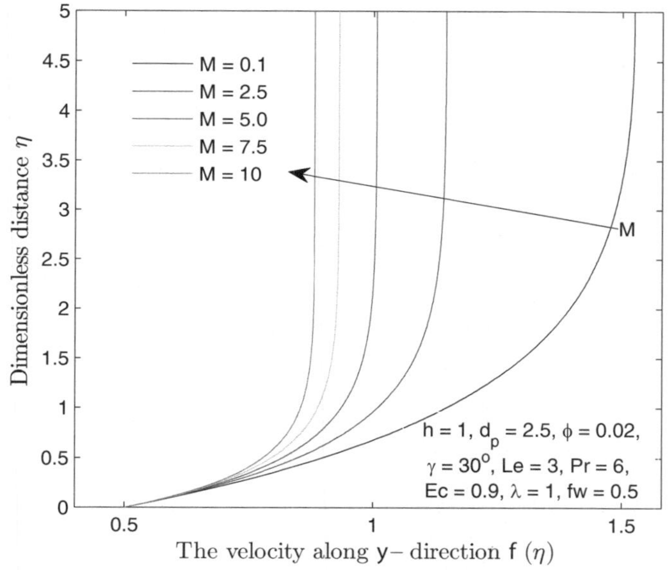

FIGURE 8.9
Variation in $f(\eta)$ due to M.

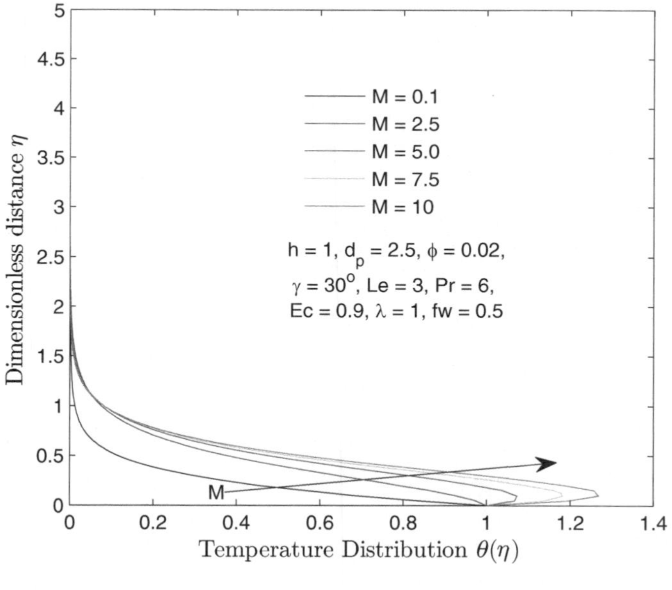

FIGURE 8.10
Variation in $f'(\eta)$ due to M.

$Cf_x\sqrt{Re_x}$ decrease with M within the domain near the wall and increase after $\eta = 0.85$; see Figure 8.11 and Table 8.8.

As shown in Figure 8.12, a higher estimation of magnetic strength M yields a higher distribution of temperature and concentration of the fluid. The observed parabolic nature near the wall is attributed to the occurrence of viscous dissipation $E_c = 0.9$. It is worth remarking that increasing M has no effect on the overall concentration of the water–alumina

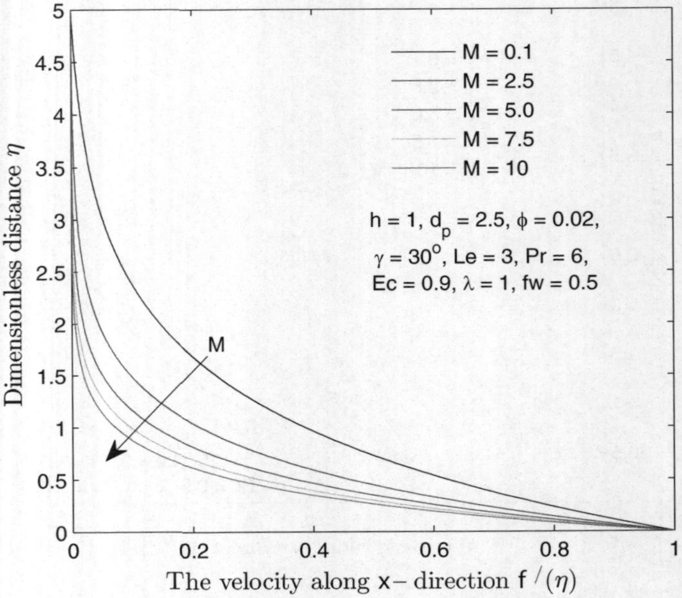

FIGURE 8.11
Variation in $f''(\eta)$ due to M.

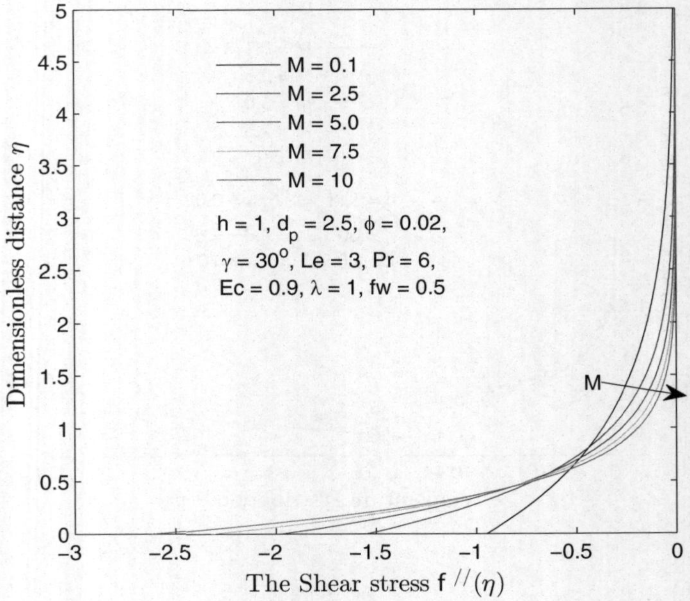

FIGURE 8.12
Variation in $\theta(\eta)$ due to M.

nanofluid. As shown in Table 8.9, a higher Prandtl number has no significant effect on the local skin friction coefficients $Cf_x\sqrt{Re_x}$ when $P_r = 6$ and $P_r = 9$. It is worthy of note that for smaller magnetic strength, maximum heat transfer rate $Nu_x Re_x^{-0.5}$ and mass transfer

TABLE 8.8
Variation in $Cf_x\sqrt{Re_x}$, $Nu_xRe_x^{-0.5}$, and $Sh_xRe_x^{-0.5}$ with M When $\eta_\infty = 100$, $\gamma = 30°$, $L_e = 3$, $E_c = 3$, $f_w = 0.5$, $h = 1$, $d_p = 2.5$, $\phi = 0.02$, and $\lambda = 1$ when $P_r = 6$

M	$Cf_x\sqrt{Re_x}$	$Nu_xRe_x^{-0.5}$	$Sh_xRe_x^{-0.5}$
0.1	−0.944826934985821	3.164934299191911	10.443739612490699
2.5	−1.548664752208164	−8.042786776538872	10.393513594236154
5.0	−1.977324955788753	−15.594890887209379	10.359072425993505
7.5	−2.323308706716123	−21.576072768691951	10.332037028479860
10	−2.621472713525406	−26.676255903828793	10.309292981709721
S_{lp}	−0.166254951	−2.948367223	−0.013304021

TABLE 8.9
Variation in $Cf_x\sqrt{Re_x}$, $Nu_xRe_x^{-0.5}$, and $Sh_xRe_x^{-0.5}$ with M When $\eta_\infty = 100$, $\gamma = 30°$, $L_e = 3$, $E_c = 3$, $f_w = 0.5$, $h = 1$, $d_p = 2.5$, $\phi = 0.02$, and $\lambda = 1$ when $P_r = 9$

M	$Cf_x\sqrt{Re_x}$	$Nu_xRe_x^{-0.5}$	$Sh_xRe_x^{-0.5}$
0.1	−0.944826934985821	4.404180322561667	15.064572549971908
2.5	−1.548664752208163	−11.979527443834689	15.023001200375125
5.0	−1.977324955788754	−23.098644922169271	14.994333097711733
7.5	−2.323308706716124	−31.935723712710761	14.971705233116186
10	−2.621472713525404	−39.488379595288336	14.952569858847079
S_{lp}	−0.166254951	−4.338799999	−0.011086684

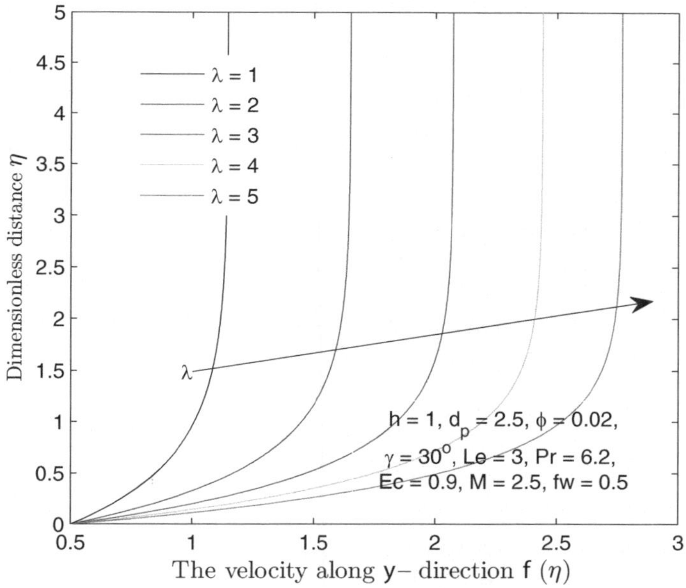

FIGURE 8.13
Variation in $f(\eta)$ due to M.

rate $Sh_xRe_x^{-0.5}$ are achievable in high viscous fluids. A higher rate of decrease in $Nu_xRe_x^{-0.5}$ with M is ascertained when Prandtl number is small in magnitude ($P_r = 6$). Figures (8.13) and (8.14) reveal that a larger magnitude of λ implies an enhancement in stretching at the wall and this leads to a higher velocity along the x-direction and y-direction with maximum

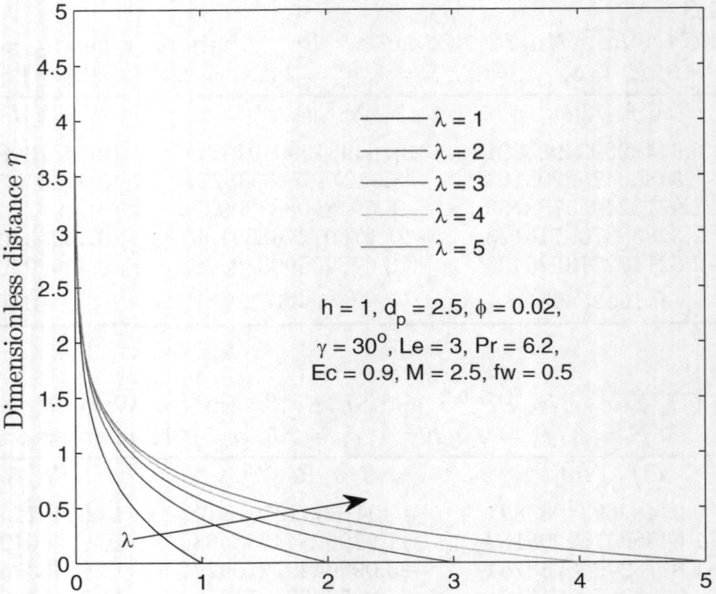

FIGURE 8.14
Variation in $f'(\eta)$ due to M.

TABLE 8.10
Variation in $Cf_x\sqrt{Re_x}$, $Nu_xRe_x^{-0.5}$, and $Sh_xRe_x^{-0.5}$ with λ when $\eta_\infty = 100$, $\gamma = 30°$, $L_e = 3$, $E_c = 0.9$, $f_w = 0.5$, $h = 1$, $d_p = 2.5$, $\phi = 0.02$, and $M = 2.5$ when $P_r = 6$

λ	$Cf_x\sqrt{Re_x}$	$Nu_xRe_x^{-0.5}$	$Sh_xRe_x^{-0.5}$
1	−1.548664752208163	0.237179003601730	10.393503222130651
2	−3.476045099279190	−7.292724546034097	11.433091285050175
3	−5.721543763226718	−18.056631071313284	12.302639416713802
4	−8.245955762609160	−31.350107131292255	13.065983567984611
5	−11.021410746609524	−46.774891368651829	13.754858684354913
S_{lp}	−2.371540265	−11.80815233	0.835560321

velocity at $f(\eta = 5)$ and $f'(\eta = 0)$ (Table 8.10). The shear stress is found to be a decreasing property of λ as shown in Figure (8.15). Figure 8.16 indicates that higher stretching at the wall is reliable to boost the temperature distribution near the wall. It is obvious in Table 8.11 that the mass transfer rate $Sh_xRe_x^{-0.5}$ increases due to growth in stretching at the wall. In fact, the observed increase in $Sh_xRe_x^{-0.5}$ with λ has a percentage increase of 16.87% due to the increase in P_r from 6 to 9. The porosity parameter defined as λ by Ghulam Rasool et al. [253] was found to decrease the Nusselt number and Sherwood number at the rates of −0.190935 and −0.2662475 respectively.

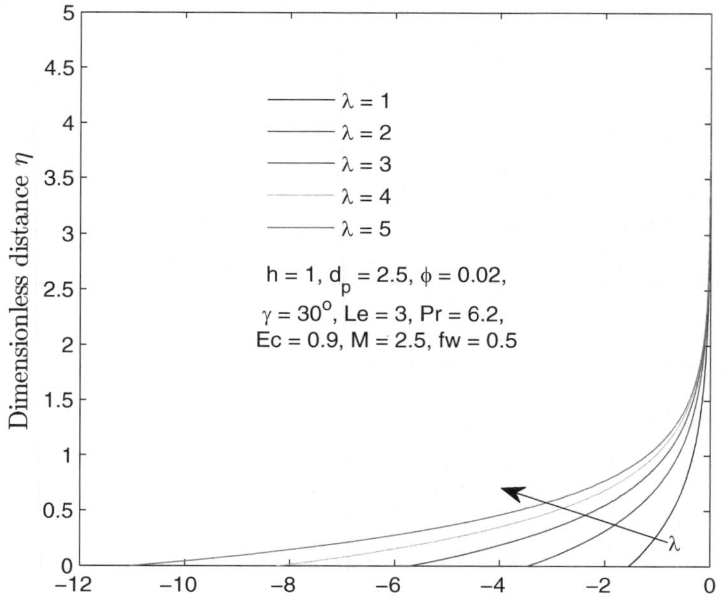

FIGURE 8.15
Variation in $f''(\eta)$ due to M.

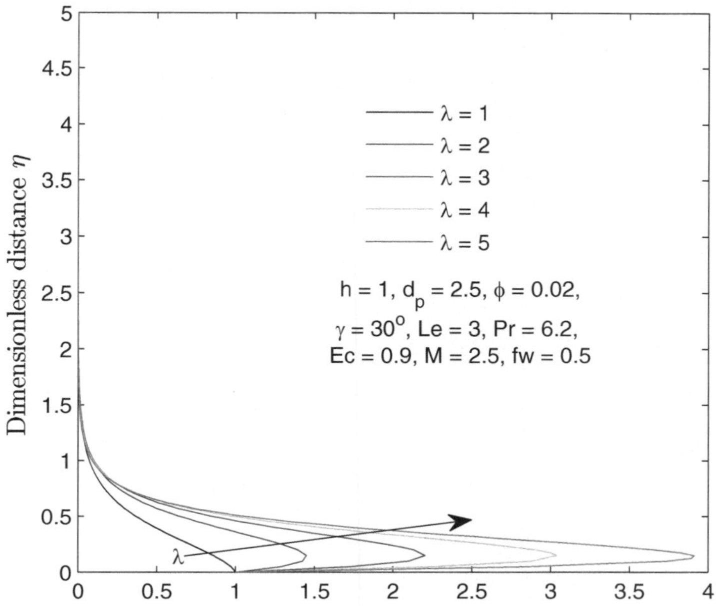

FIGURE 8.16
Variation in $\theta(\eta)$ due to M.

TABLE 8.11
Variation in $Cf_x\sqrt{Re_x}$, $Nu_xRe_x^{-0.5}$, and $Sh_xRe_x^{-0.5}$ with λ when $\eta_\infty = 100$, $\gamma = 30°$, $L_e = 3$, $E_c = 0.9$, $f_w = 0.5$, $h = 1$, $d_p = 2.5$, $\phi = 0.02$, and $M = 2.5$ when $P_r = 9$

M	$Cf_x\sqrt{Re_x}$	$Nu_xRe_x^{-0.5}$	$Sh_xRe_x^{-0.5}$
1	−1.548664752208164	0.150267465854456	15.023001200453940
2	−3.476045099279191	−10.871200417685449	16.212267784353852
3	−5.721543763226716	−26.379357752867840	17.225941892459495
4	−8.245955762609155	−45.382876687557079	18.128788124896381
5	−11.021410746609520	−67.324817323555678	18.947601066904998
S_{lp}	−2.371540265	−16.94618458	0.976572007

8.7 Tutorial Questions

1. What is the impact of increasing the ratio of momentum diffusivity to heat diffusivity during the flow of a Newtonian fluid on a horizontal object owing to surface tension?
2. What is the combined effect of suction, injection, and Prandtl number on fluid dynamics due to surface tension?
3. During the flow of a Newtonian fluid on a vertical object due to surface tension and buoyancy, what is the effect of increasing Grashof number and the ratio of momentum diffusivity to thermal diffusivity?
4. At two different levels of Prandtl number, what are the effects of magnetic strength on the dynamics of alumina–water due to moderate stretching in the presence of Joule heating and little suction?
5. When Joule heating and suction are moderate, what is the significance of increasing stretching at the wall on the dynamics of water–alumina nanofluid?

9

Analysis of Self-Similar Flow V

9.1 Background Information

Fluid dynamics and thermodynamics are related by a quantity known as viscous energy dissipation, which is the energy generated by the deformation of fluids (sometimes known as just dissipation). When a fluid is flowing viscously, its viscosity absorbs kinetic energy from the fluid's motion and converts it to internal fluid energy. That entails warming the fluid. Scientifically, the rate at which mechanical energy is transformed into heat in a viscous fluid per unit volume is represented by the quadratic function of spatial derivatives of fluid velocity components. Chan et al. (2015) claim that viscous dissipation is a progression of heat produced as a result of the work done by viscous friction working inside a fluid as it is capable of significantly raising fluid temperature led by the existence of substantial velocity gradients. In operations including metal forming, glass fiber drawing, heat exchangers, and extrusion, Tso et al. (2015) verified the occurrence of viscous dissipation due to induction of heat in micro-channel fluids and a slight oscillating motion of the lower plate. All motions and activities are propelled by energy. According to energy conservation, energy cannot be generated or destroyed, but it may be changed from one form to another. Among the different forms of energy are mechanical energy, electrical energy, and magnetic energy.

9.2 Introduction: Viscous Dissipation

The process of converting energy from one form to another is known as energy conversion. For example, an electric generator converts kinetic/mechanical energy to electrical energy. Similarly, the annihilation of a changing velocity gradient owing to viscous stress (kinetic energy to internal energy), also known as viscous dissipation, is another notion based on energy conversion from one form to another. The sticky, thick, and semi-fluid state in thick flowing due to internal energy is called viscosity (i.e., it measures fluid resistance to flow). The temperature field is dependent on the velocity field when viscous dissipation is considerable, as seen by the nonlinear component in the energy equation. The process of eliminating the unsteadiness of a velocity gradient caused by viscous stress is known as viscous dissipation. The Eckert number, the relationship between a flow's kinetic energy and boundary layer enthalpy difference, is inevitable in studying viscous dissipation. The dimensionless number is frequently used to characterize heat transfer dissipation. Sultan et al. [289] remarked that the Eckert number increases temperature distribution while the Nusselt number $NuRe^{-1/2}$ decreases with the Eckert number at the rate of -0.084255. Another related study on the dynamics of a viscoelastic liquid with significant elastic deformation subject to the Lorentz force and heat source on a horizontal convectively heated surface by Olanrewaju et al. [217] showed that the temperature distribution is an increasing function of viscous dissipation. Such a scientific fact also agrees with the result illustrated

DOI: 10.1201/9781003217374-9

in Figure 11 by Sithole et al. [283]. The results illustrated as Figure 23 by Shaw et al. [279] reveal that the temperature distribution increases with the Eckert number.

9.3 Fluid Flow Subject to Viscous Dissipation

In this section, the two-dimensional boundary layer fluid flow on a horizontal porous surface (i.e., in the presence of suction and injection) is considered. Viscous dissipation in fluid flow is a process whereby the fluid's viscosity takes energy from the motion of the fluid (i.e., kinetic energy) and converts it to internal energy (i.e., heat energy by heating the fluid). The graphical illustration of the transport phenomenon in the presence of the partially irreversible process is illustrated in Figure 9.1. Consider the flow of an incompressible viscous fluid on a horizontal surface within the domain $0 \leq y < \infty$ where the effect of viscosity is highly significant; see Figure 9.1. It is assumed that the flow on the horizontal surface was induced by viscous forces with stretching at the wall $U_w = U_o x$. The governing equation that model the transport phenomenon is

$$u_x + v_y = 0, \tag{9.1}$$

$$uu_x + vu_y = \frac{\mu}{\rho} u_{yy}, \tag{9.2}$$

$$uT_x + vT_y = \frac{\kappa}{\rho c_p} T_{yy} + \frac{\mu}{\rho c_p} u_y u_y. \tag{9.3}$$

For this case, Eqs. (9.1)–(9.3) are subject to the boundary condition

$$u = U_w, \quad v = v_w, \quad T = T_w \quad at \quad y = 0. \tag{9.4}$$

$$u \to 0, \quad T \to T_\infty, \quad as \quad y \to \infty \tag{9.5}$$

In order to obtain self-similar solution of Eqs. (9.1)–(9.5), we used the following similarity variables:

$$\eta = y\sqrt{\frac{U_\infty}{\vartheta x}}, \quad \psi(x,y) = \sqrt{\vartheta x U_\infty} f(\eta), \quad u = \frac{\partial \psi}{\partial y}, \quad v = -\frac{\partial \psi}{\partial x},$$

$$\theta(\eta) = \frac{T - T_\infty}{T_w - T_\infty}, \quad P_r = \frac{\mu c_p}{\kappa} = \frac{\vartheta}{\alpha} \quad E_c = \frac{U_0^2 x^2}{C_p(T_w - T_\infty)}, \quad f_w = -\frac{v_w}{\sqrt{\vartheta U_o}}. \tag{9.6}$$

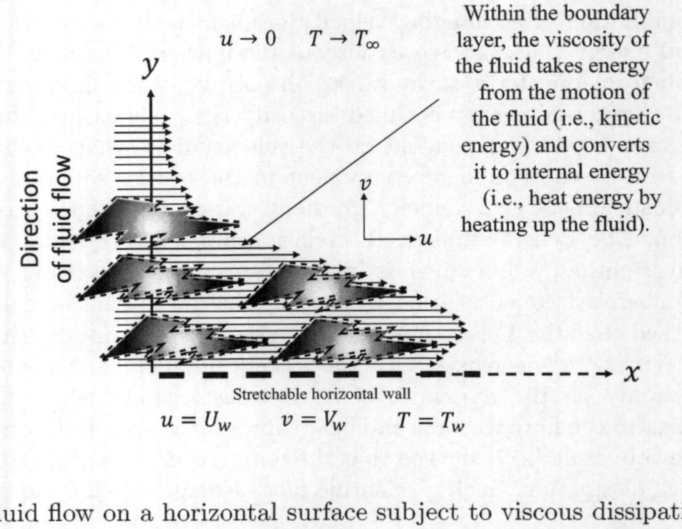

FIGURE 9.1
Illustration of fluid flow on a horizontal surface subject to viscous dissipation.

The final dimensionless governing equation is

$$\frac{d^3f}{d\eta^3} + \frac{1}{2}f\frac{d^2f}{d\eta^2} = 0, \tag{9.7}$$

$$\frac{d^2\theta}{d\eta^2} + \frac{1}{2}P_r f\frac{d\theta}{d\eta} + P_r E_c\frac{d^2f}{d\eta^2}\frac{d^2f}{d\eta^2} = 0. \tag{9.8}$$

Dimensionless boundary conditions for this case are

$$\frac{df}{d\eta} = 1, \quad f = f_w, \quad \theta = 1 \quad at \quad \eta = 0. \tag{9.9}$$

$$\frac{df}{d\eta} \to 0, \quad \theta \to 0 \quad as \quad \eta \to \infty. \tag{9.10}$$

9.3.1 Research Questions I

The purpose of this case is to address the following research questions:

1. When viscous dissipation is minimal, how does the Prandtl number influence the velocity, friction throughout the domain, and temperature distribution of a flow caused by stretching at the wall?
2. What is the effect of Prandtl number on Sakiadis flow of a Newtonian fluid when viscous dissipation is significant in the presence of suction?

9.3.2 Analysis and Discussion of Results I

The numerical solution of dimensionless governing Eqs. (9.7) and (9.8) subject to Eqs. (9.9) and (9.10) was obtained using the four-stage Lobatto IIIa formula as described in Chapter 1 using $\eta = 5$. First, the analysis of the flow along a horizontal surface in the presence of injection ($f_w = -3$) and when viscous dissipation is negligible ($E_c = 0.1$) is presented. It is worth remarking that with an increase in Prandtl number P_r within the range of $0.1 \le P_r \le 10$, $\amalg_f$ increases at the rate of 0.64652, $\amalg_{f'}$ decreases at the rate of -0.21771, and $\amalg_{f''}$ decreases at the rate of -0.06165. When viscous dissipation is significant ($E_c = 5$), $\amalg_f$ increases at the rate of 0.687538, $\amalg_{f'}$ decreases at the rate of -0.20229, and $\amalg_{f''}$ decreases at the rate of -0.06638. Second, $f_w = 0$ implies neither suction nor injection, when viscous dissipation is negligible ($E_c = 0.1$). It is discovered that with an increase in Prandtl number P_r within the same range mentioned above, $\amalg_f$ increases at the rate of 0.27375, $\amalg_{f'}$ decreases at the rate of -0.19467, and $\amalg_{f''}$ increases at the rate of 0.09847. Considering the sane case when viscous dissipation is significant ($E_c = 5$), $\amalg_f$ increases at the rate of 0.285498, $\amalg_{f'}$ decreases at the rate of -0.20079, and $\amalg_{f''}$ increases at the rate of 0.095903.

Third, when viscous dissipation is also negligible ($E_c = 0.1$) and suction is significant ($f_w = +3$), it is seen that $\amalg_f$ increases at the rate of 0.075648, $\amalg_{f'}$ decreases at the rate of -0.13177, and $\amalg_{f''}$ increases at the rate of 0.222362. When viscous dissipation is more enhanced ($E_c = 5$), $\amalg_f$ increases at the rate of 0.07414, $\amalg_{f'}$ decreases at the rate of -0.12941, and $\amalg_{f''}$ increases at the rate of 0.219304. When viscous dissipation is negligible ($E_c = 0.1$) and significant ($E_c = 5$), Table 9.1 reveals that the local skin friction coefficients $f''(0)$ is a constant function of Prandtl number during the flow along a horizontal surface. In the case of injection, when however, maximal local skin friction coefficients are found at any Prandtl number. In the case of injection $f_w = -3$, it is seen in Figure 9.2 that the temperature distribution increases with the Prandtl number. Figures 9.6 and 9.10 show that the temperature distribution diminishes with the Prandtl number. Starting from the wall where the maximum temperature is inevitable since $T_w > T_\infty$, the rate of decrease in the

TABLE 9.1
Effect of Prandtl Number P_r on Some of the Properties of the Flow When Viscous Dissipation Is Negligible $E_c = 0.1$ and Significant $E_c = 5$

$f_w = -3$		$E_c = 0.1$			$E_c = 5$	
P_r	$f''(0)$	$\theta(1.6667)$	$-\theta'(0)$	$f''(0)$	$\theta(1.6667)$	$-\theta'(0)$
0.1	−0.0244	0.6965	0.1641	−0.0244	0.7516	0.1329
0.3	−0.0244	0.7499	0.1090	−0.0244	0.8918	0.0432
0.71	−0.0244	0.8344	0.0448	−0.0244	1.0785	0.0282
1	−0.0244	0.8770	0.0232	−0.0244	1.1516	0.0366
10	−0.0244	1.0019	0.0001	−0.0244	1.0968	0.0026
S_{lp}	0	0.0236	−0.0099	0	0.0164	−0.0066
$f_w = 0$		$E_c = 0.1$			$E_c = 5$	
P_r	$f''(0)$	$\theta(1.6667)$	$-\theta'(0)$	$f''(0)$	$\theta(1.6667)$	$-\theta'(0)$
0.1	−0.4540	0.7553	0.2212	−0.4540	0.7140	0.1028
0.3	−0.4540	0.7088	0.2660	−0.4540	0.7950	−0.0862
0.71	−0.4540	0.6132	0.3631	−0.4540	0.9052	−0.4471
1	−0.4540	0.5501	0.4313	−0.4540	0.9462	−0.6810
10	−0.4540	0.0177	1.5325	−0.4540	0.5916	−5.5745
S_{lp}	0	−0.0685	0.1286	0	−0.0243	−0.5612
$f_w = +3$		$E_c = 0.1$			$E_c = 5$	
P_r	$f''(0)$	$\theta(1.6667)$	$-\theta'(0)$	$f''(0)$	$\theta(1.6667)$	$-\theta'(0)$
0.1	−1.6477	0.5666	0.2904	−1.6477	0.6404	−0.1070
0.3	−1.6477	0.3727	0.5306	−1.6477	0.5336	−0.6830
0.71	−1.6477	0.1276	1.1300	−1.6477	0.2923	−1.7480
1	−1.6477	0.0571	1.5653	−1.6477	0.1834	−2.4715
10	−1.6477	0.0002	14.4887	−1.6477	0.0110	−24.8233
S_{lp}	0	−0.0333	1.4366	0	−0.0456	−2.4892

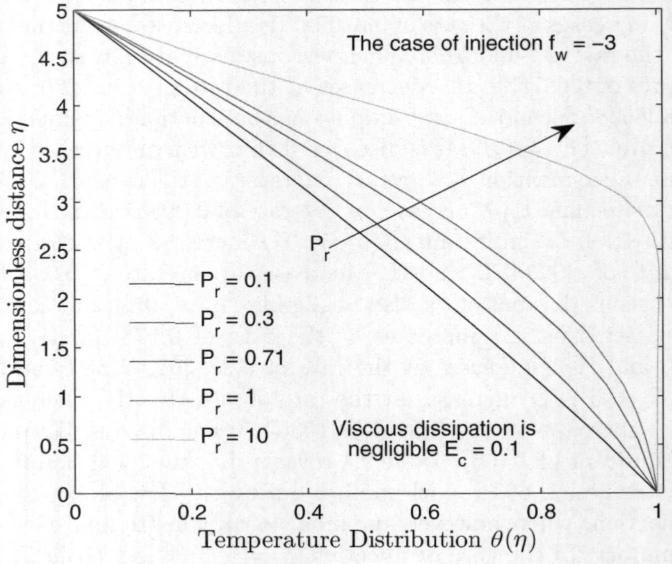

FIGURE 9.2
Variation in $\theta(\eta)$ due to P_r when $E_c = 0.1$ and $f_w = -3$.

TABLE 9.2
Rate of Decrease in $\theta(\eta)$ with η at Various Values of Prandtl Number When $E_c = 0.1$ and $f_w = [-3, 0, 3]$

	P_r	**0.1**	**0.3**	**0.71**	**1**
$f_w = -3$	S_{lp}	−0.2023053	−0.2060263	−0.2112890	−0.2055620
$f_w = 0$	S_{lp}	−0.201554499	−0.203640397	−0.204182902	−0.196085261
$f_w = 3$	S_{lp}	−0.1991863	−0.1798100	−0.1383437	−0.117627

TABLE 9.3
Rate of Decrease in $\theta(\eta)$ with η at Various Values of Prandtl Number When $E_c = 5$ and $f_w = [-3, 0, +3]$

	P_r	**0.1**	**0.3**	**0.71**	**1**
$f_w = -3$	S_{lp}	−0.1967369	−0.1884858	−0.1458080	−0.1178448
$f_w = 0$	S_{lp}	−0.205980553	−0.220157882	−0.249068269	−0.268437157
$f_w = +3$	S_{lp}	−0.2165938	−0.2447985	−0.2467729	−0.2567906

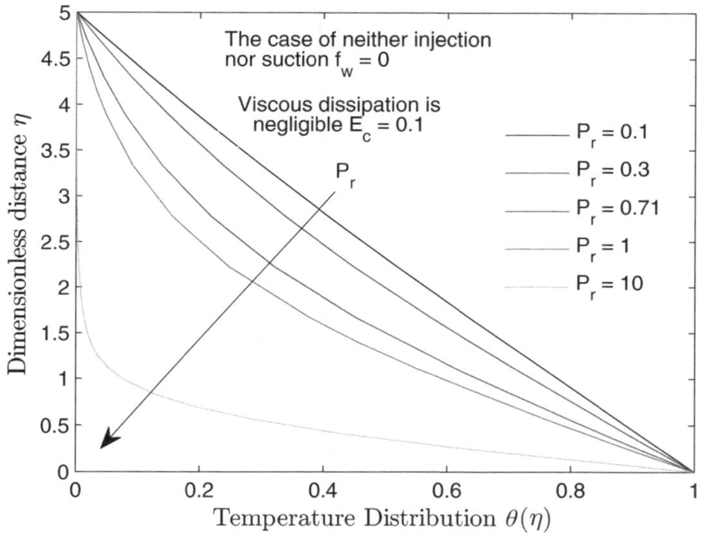

FIGURE 9.3
Variation in $\theta'(\eta)$ due to P_r when $E_c = 0.1$ and $f_w = -3$.

temperature as $\eta \to \infty$ was estimated for each value of the Prandtl number and is presented in Tables 9.2 and 9.3. It is seen that when viscous dissipation is negligible ($E_c = 0.1$), the maximum decrease in the temperature distribution is found in the case of a less viscous fluid $P_r = 0.1$ when injection is significant as 0.2023053.

It is seen in Figure 9.3 that heat transfer rate in the case of injection is unique compared to the heat transfer rate when $f_w = 0$ as shown in Figure 9.7 and when suction is significant ($f_w = +3$) as shown in Figure 9.11. Next is to examine the transport phenomenon when viscous dissipation is significant ($E_c = 5$). It is seen that the temperature distribution in a highly viscous fluid boosts near the free stream in the case of injection Figure 9.4 and near the wall in the case of suction Figure 9.8. For the case of suction, Figure 9.20 shows that the heat transfer at the wall $-\theta'(0)$ is the most minimum when the Prandtl number is small at all the values of viscous dissipation. However, the same property of the fluid flow

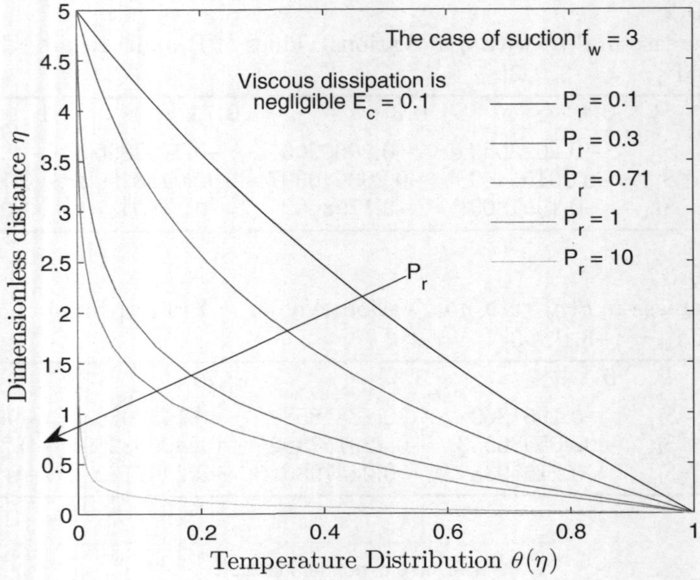

FIGURE 9.4
Variation in $\theta(\eta)$ due to P_r when $E_c = 5$ and $f_w = -3$.

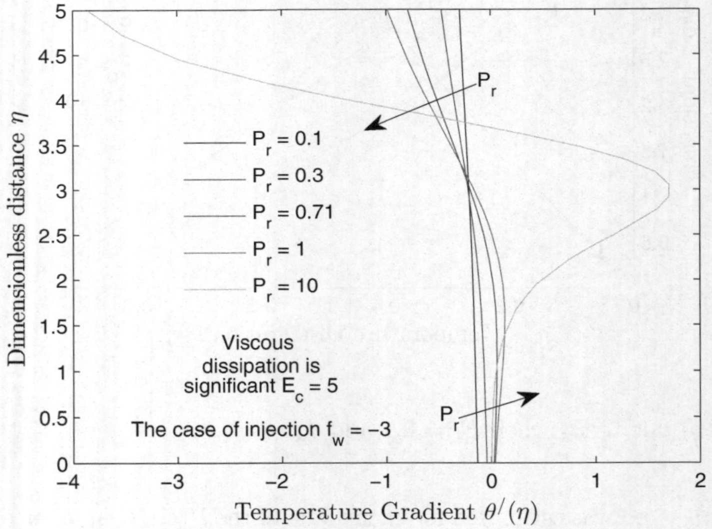

FIGURE 9.5
Variation in $\theta'(\eta)$ due to P_r when $E_c = 5$ and $f_w = -3$.

is maximum only at larger values of Prandtl number when viscous dissipation is small. The nature of heat transfer through the wall at larger values of viscous dissipation is negative for $2 \leq P_r < 7.5$. Figure 9.20 indicates that injection only makes the heat transfer at the wall to be maximum for less viscous fluids. The increasing impacts of Prandtl number, viscous dissipation, suction, and injection on the transport phenomenon described in subsection 9.3 are further shown in Figures 9.9, 9.10, 9.12–9.19, and Figure 9.21. For instance, when $f_w =$

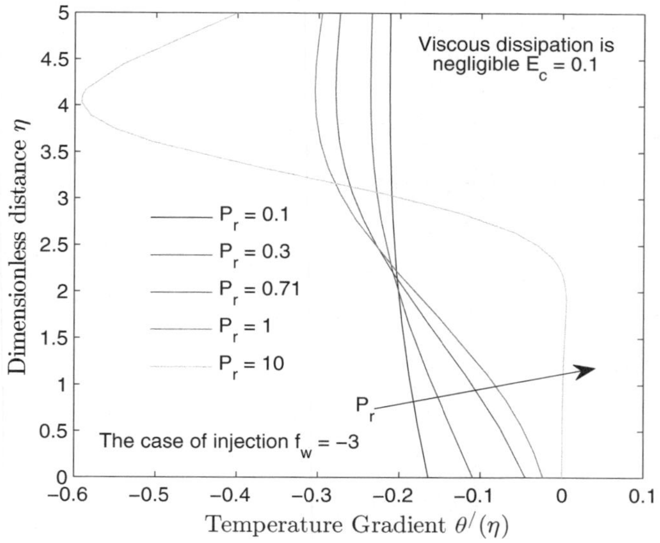

FIGURE 9.6
Variation in $\theta(\eta)$ due to P_r when $E_c = 0.1$ and $f_w = 0$.

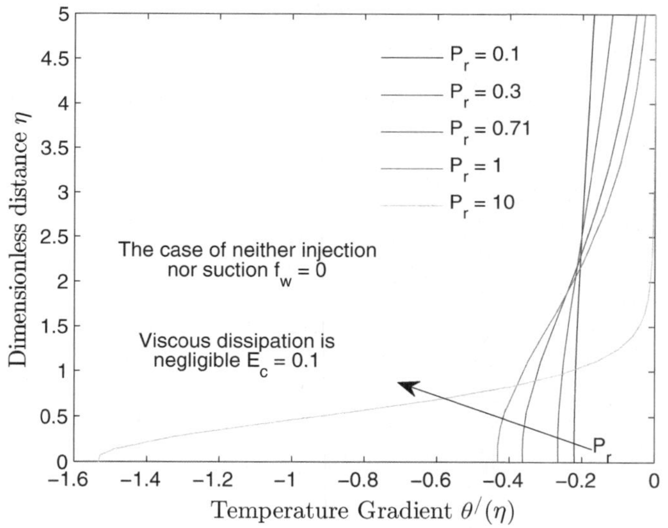

FIGURE 9.7
Variation in $\theta'(\eta)$ due to P_r when $E_c = 0.1$ and $f_w = 0$.

+3 and viscous dissipation is minimal, Figure 9.10 and Figure 9.12 show that temperature distribution decreases with increasing Prandtl number. However, temperature distribution and heat transfer rate close to the wall are significantly impacted when viscous dissipation is sufficiently high. It is noteworthy to conclude that Prandtl number, viscous dissipation, suction, and injection influence the velocity, temperature distribution, and heat transfer rates of the transport phenomenon based on the graphical representation of the observed results shown in Figures 9.9, 9.10, 9.12–9.19, and Figure 9.21.

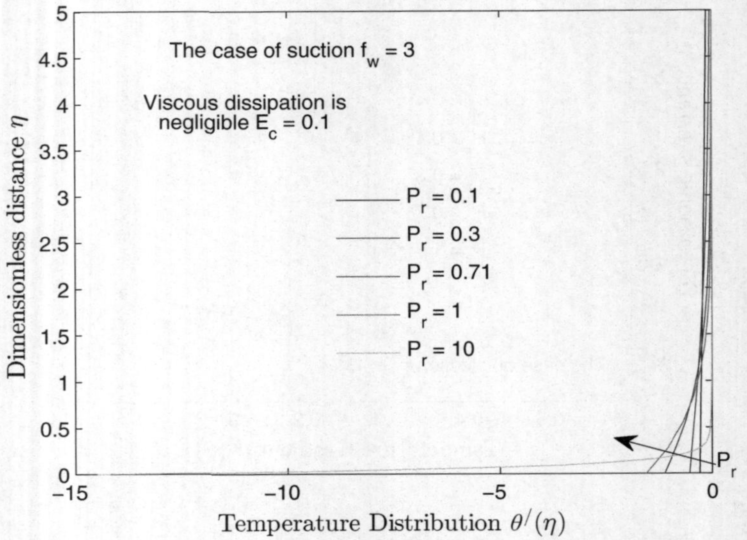

FIGURE 9.8
Variation in $\theta(\eta)$ due to P_r when $E_c = 5$ and $f_w = 0$.

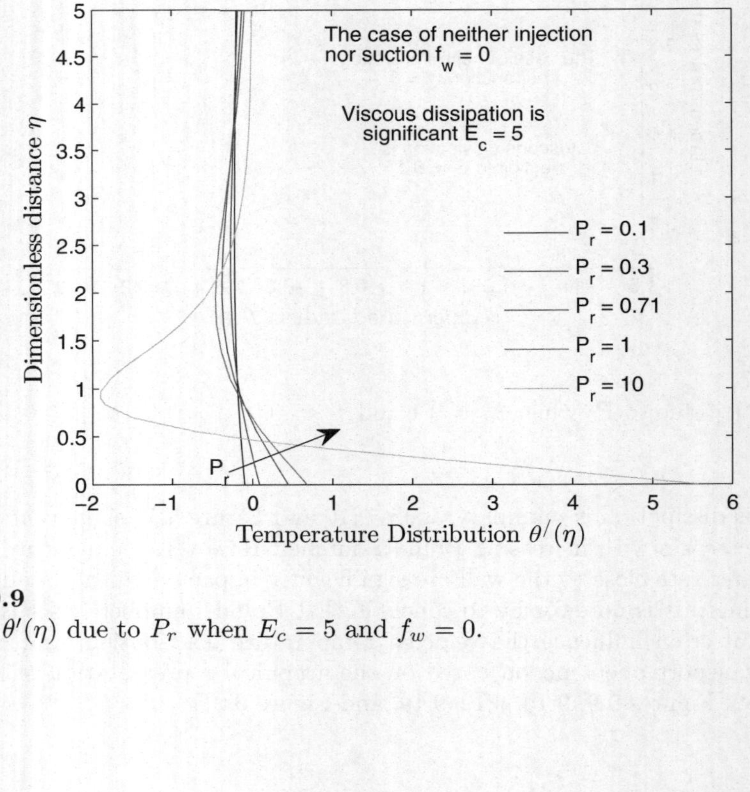

FIGURE 9.9
Variation in $\theta'(\eta)$ due to P_r when $E_c = 5$ and $f_w = 0$.

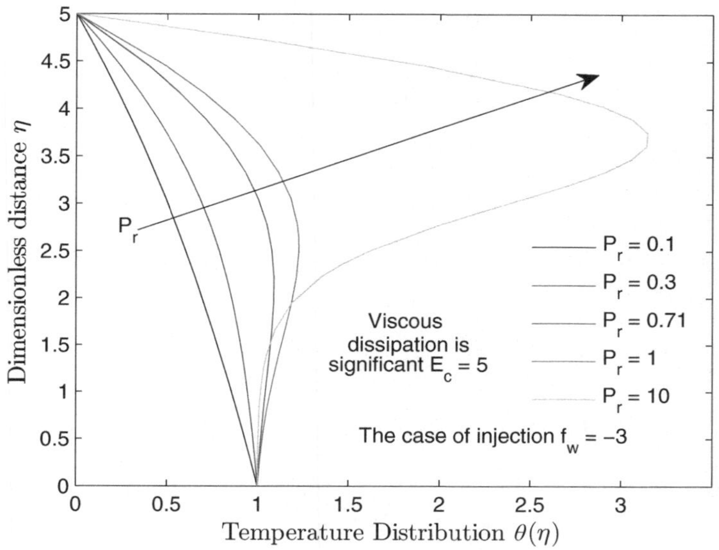

FIGURE 9.10
Variation in $\theta(\eta)$ due to P_r when $E_c = 0.1$ and $f_w = +3$.

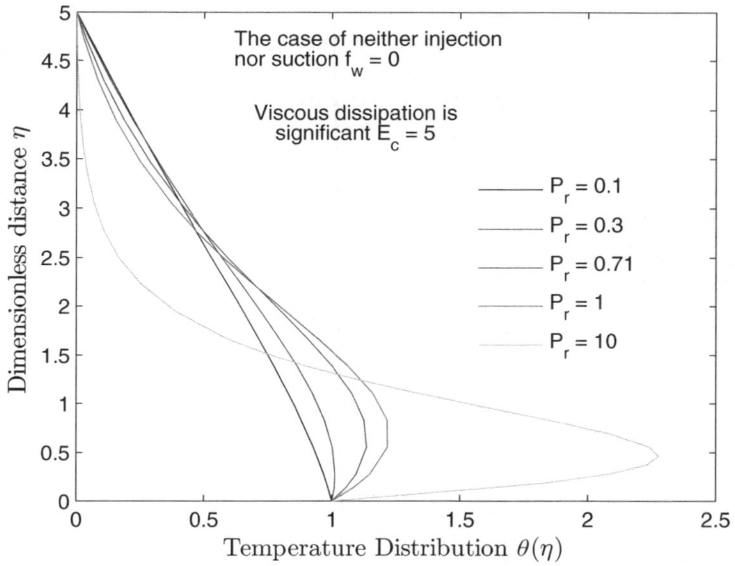

FIGURE 9.11
Variation in $\theta'(\eta)$ due to P_r when $E_c = 0.1$ and $f_w = +3$.

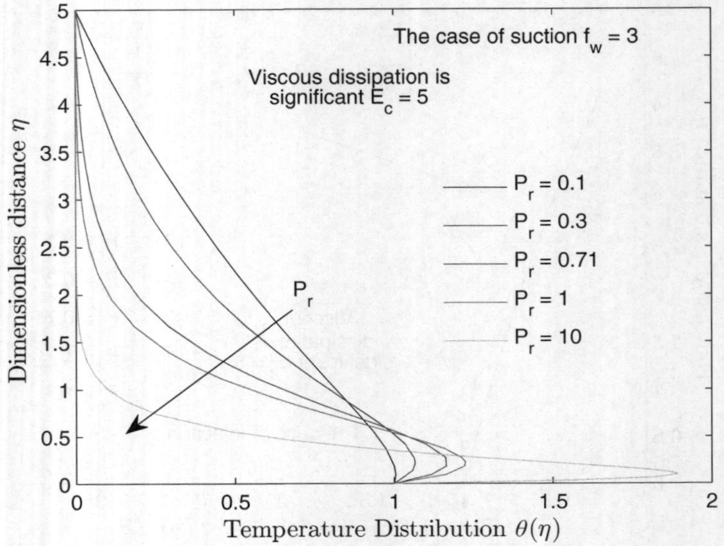

FIGURE 9.12
Variation in $\theta(\eta)$ due to P_r when $E_c = 5$ and $f_w = +3$.

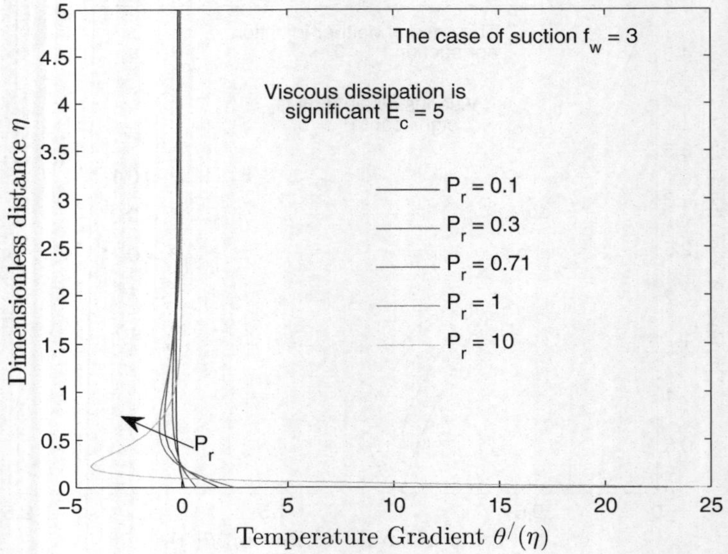

FIGURE 9.13
Variation in $\theta'(\eta)$ due to P_r when $E_c = 5$ and $f_w = +3$.

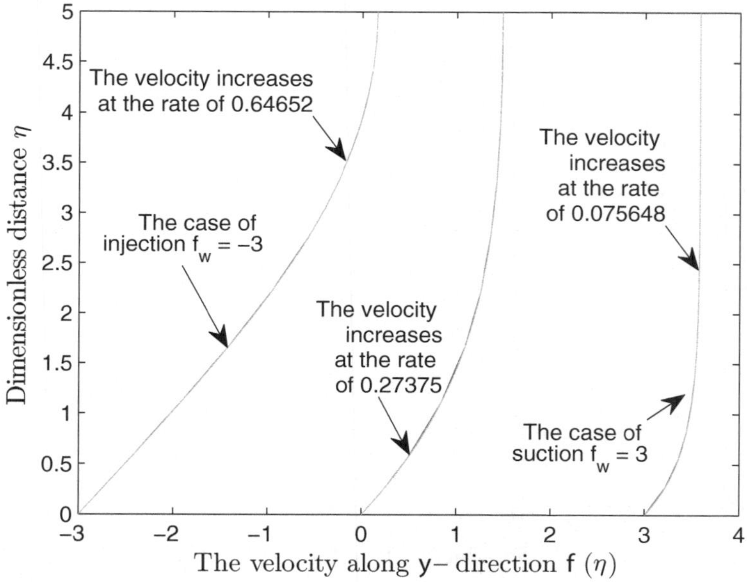

FIGURE 9.14
Variation in $f(\eta)$ due to P_r when $E_c = 0.1$.

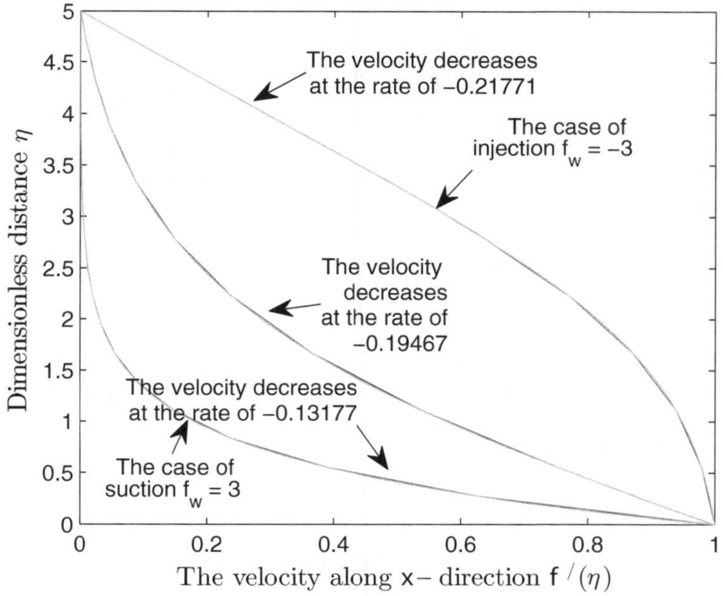

FIGURE 9.15
Variation in $f'(\eta)$ due to P_r when $E_c = 0.1$.

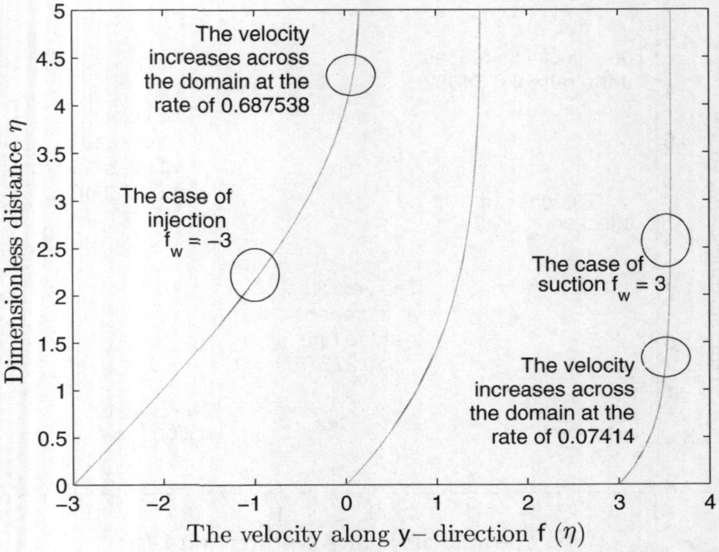

FIGURE 9.16
Variation in $f(\eta)$ due to P_r when $E_c = 5$.

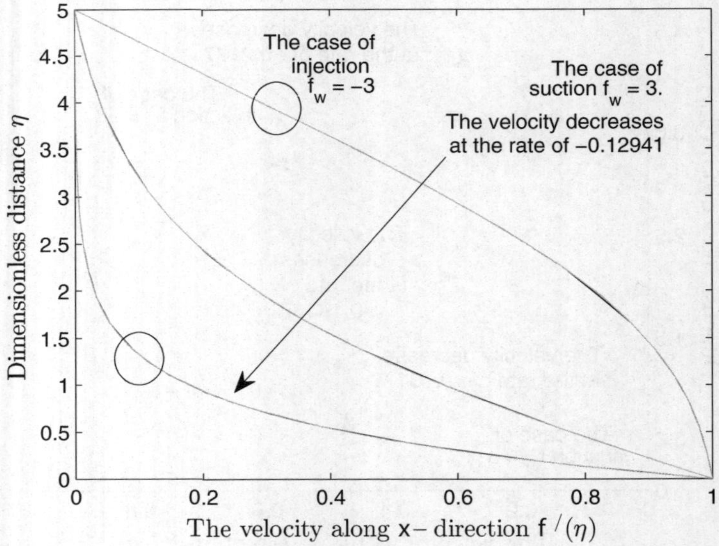

FIGURE 9.17
Variation in $f'(\eta)$ due to P_r when $E_c = 5$.

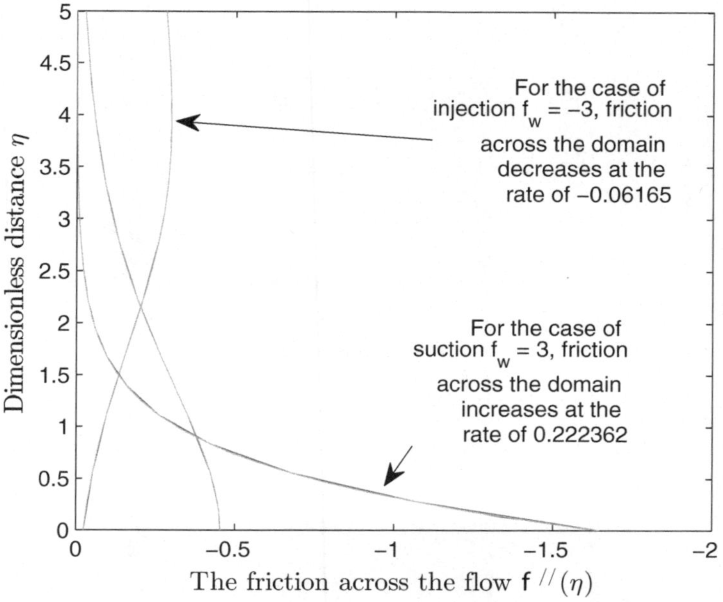

FIGURE 9.18
Variation in $f''(\eta)$ due to P_r when $E_c = 0.1$.

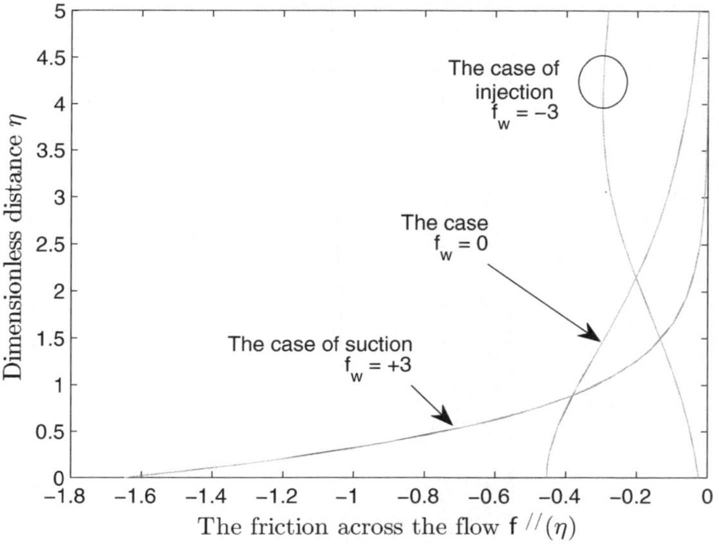

FIGURE 9.19
Variation in $f''(\eta)$ due to P_r when $E_c = 5$.

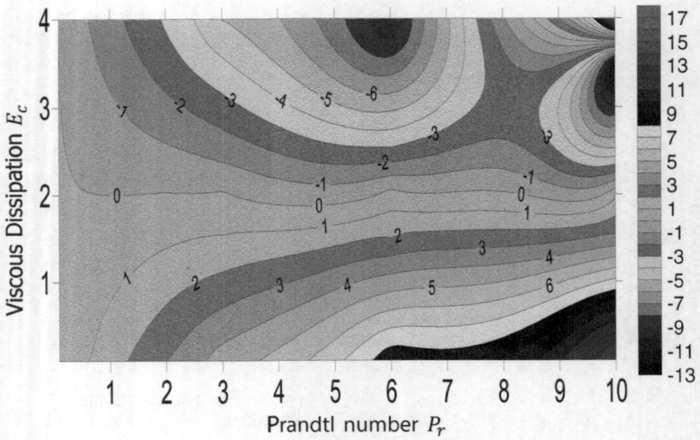

FIGURE 9.20
Variation in heat transfer rate $-\theta'(0)$ with Prandtl number P_r and viscous dissipation E_c: The case of suction $f_w = +3$.

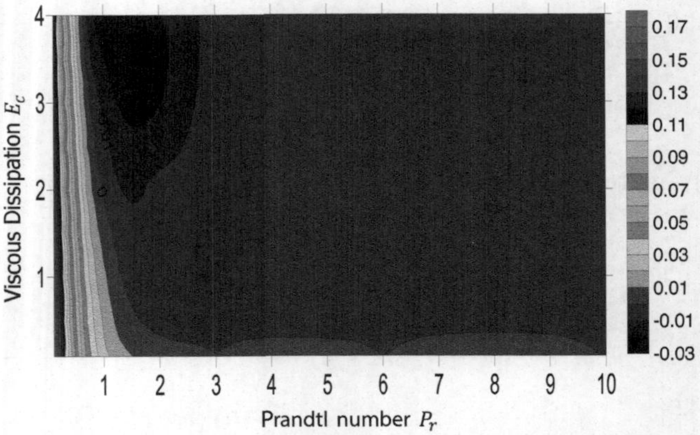

FIGURE 9.21
Variation in heat transfer rate $-\theta'(0)$ with Prandtl number P_r and viscous dissipation E_c: The case of injection $f_w = -3$.

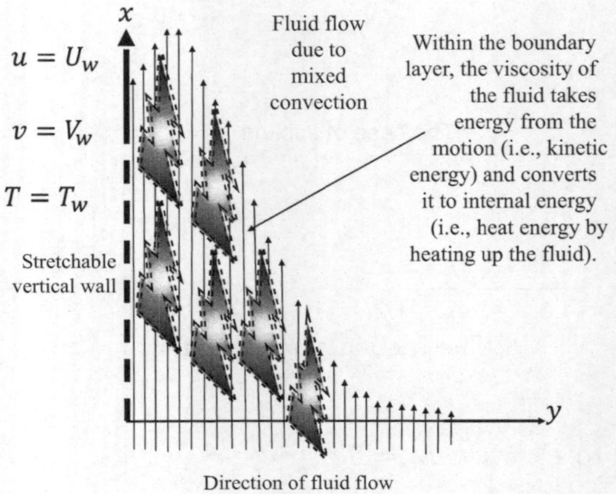

FIGURE 9.22
Illustration of fluid flow on a vertical surface subject to viscous dissipation

9.4 Mixed Convective Induced Flow Subject to Viscous Dissipation

When viscous dissipation in the fluid flow owing to mixed convection is considerable, the two-dimensional flow over a horizontal porous surface (i.e., in the presence of suction and injection) is significant, as shown in Figure 9.22. Consider the dynamics of an incompressible viscous fluid on a horizontal surface due to mixed convection within the domain $0 \leq y < \infty$, where the effect of viscosity is highly significant and the free stream velocity $U_\infty = U_o x$ as shown in Figure 9.22. The governing equation is

$$u_x + v_y = 0, \tag{9.11}$$

$$uu_x + vu_y = U_\infty \frac{dU_\infty}{dx} + \frac{\mu}{\rho} u_{yy} + g\beta(T - T_\infty), \tag{9.12}$$

$$uT_x + vT_y = \frac{\kappa}{\rho c_p} T_{yy} + \frac{\mu}{\rho c_p} u_y u_y. \tag{9.13}$$

For this case, Eqs. (9.11)–(9.13) are subject to the boundary conditions

$$u = U_w, \quad v = v_w, \quad T = T_w \quad at \quad y = 0. \tag{9.14}$$

$$u \to 0, \quad T \to T_\infty, \quad as \quad y \to \infty \tag{9.15}$$

The self-similar solution of Eqs. (9.11)–(9.15), was obtained using the following similarity variables:

$$\eta = y\sqrt{\frac{U_\infty}{\vartheta x}}, \quad \psi(x, y) = \sqrt{\vartheta x U_\infty} f(\eta), \quad u = \frac{\partial \psi}{\partial y}, \quad v = -\frac{\partial \psi}{\partial x}, \quad Gr_x = \frac{gx\beta(T_w - T_\infty)}{U_\infty^2}$$

$$\theta(\eta) = \frac{T - T_\infty}{T_w - T_\infty}, \quad P_r = \frac{\mu c_p}{\kappa} = \frac{\vartheta}{\alpha} \quad E_c = \frac{U_0^2 x^2}{C_p(T_w - T_\infty)}, \quad f_w = -\frac{v_w}{\sqrt{\vartheta U_o}}. \tag{9.16}$$

The final dimensionless governing equation is

$$\frac{d^3 f}{d\eta^3} + \frac{1}{2} f \frac{d^2 f}{d\eta^2} + 1 + G_{rx}\theta = 0, \tag{9.17}$$

$$\frac{d^2\theta}{d\eta^2} + \frac{1}{2} P_r f \frac{d\theta}{d\eta} + P_r E_c \frac{d^2 f}{d\eta^2} \frac{d^2 f}{d\eta^2} = 0. \tag{9.18}$$

Dimensionless boundary conditions are

$$\frac{df}{d\eta} = 1, \quad f = 0, \quad \theta = 1 \quad at \quad \eta = 0. \tag{9.19}$$

$$\frac{df}{d\eta} \to 0, \quad \theta \to 0 \quad as \quad \eta \to \infty. \tag{9.20}$$

9.4.1 Research Questions II

The purpose of this case study is to address the following research questions:

1. What is the impact of raising the Prandtl number on the transport phenomena in the situation of fluid flow owing to mixed convection when viscous dissipation is minimal and significant?
2. What is the change in the local skin friction coefficients and heat transfer rate owing to the Prandtl number when viscous dissipation is minimal and significant?

9.4.2 Analysis and Discussion of Results II

The numerical solution of dimensionless governing Eqs. (9.17) and (9.18) subject to Eqs. (9.19) and (9.20) was obtained using the four-stage Lobatto IIIa formula as described in Chapter 1 using $\eta = 5$. The wall temperature distribution (T_w) is known to be maximum. Meanwhile, as $y \to \infty$, the wall's temperature decreases differently. When the conversion of kinetic energy to internal energy is small and large in magnitude, Tables 9.4–9.6 show that the wall temperature decreases at different rates for different viscous fluids. It is worth deducing from Table 9.4 that the optimal rate of decrease ($S_{lp} = 0.610353$) occurs in the case of injection ($f_w = -3$) when viscous dissipation is sufficiently large ($E_c = 0.9$) and more viscous fluids ($P_r = 2$). It was also observed that as the Prandtl number increases, the rate of decrease in the temperature distribution from the wall to the free stream enhances when injection is significant ($f_w = -3$) and diminishes when suction is significant ($f_w = +3$). Table 9.7 further unravels the effect of increasing the magnitude of Prandtl number on the transport phenomenon subject to (i) injection ($f_w = -3$), (ii) absence of suction and injection ($f_w = 0$), and (iii) suction ($f_w = +3$) when viscous dissipation is minimal and maximal in magnitude. In the case of injection, it is seen that the temperature distribution near the wall increases with the Prandtl number moderately when viscous dissipation is small ($E_c = 0.1$); see Figure 9.33. When the viscous dissipation is more enhanced ($E_c = 0.9$), a significant difference in the nature of temperature distribution near the wall is seen as illustrated in Figure 9.34. Hayat et al. [123] concluded that the temperature distribution increases with an increase in the Eckert number.

Based on these results, it can be remarked that in the case of fluid flow due to mixed convection when the injection is significant, an increase in the magnitude of Prandtl

TABLE 9.4
Rate of Decrease in $\theta(\eta)$ with η at Various Values of Prandtl Number When $G_{rx} = 1$, $f_w = -3$, $E_c = 0.1$, and $E_c = 0.9$

$E_c = 0.1$, $P_r \Rightarrow$	0.1	0.3	0.71	1	2
S_{lp}	−0.2044	−0.2115	−0.2224	−0.2260	−0.2254
$E_c = 0.9$, $P_r \Rightarrow$	0.1	0.3	0.71	1	2
S_{lp}	−0.2211	−0.2681	−0.4158	−0.4810	−0.6103

TABLE 9.5
Rate of Decrease in $\theta(\eta)$ with η at Various values of Prandtl Number When $G_{rx} = 1$, $f_w = 0$, $E_c = 0.1$, and $E_c = 0.9$

$E_c = 0.1$, $P_r \Rightarrow$	0.1	0.3	0.71	1	2
S_{lp}	−0.1996	−0.1821	−0.1545	−0.1379	−0.1009
$E_c = 0.9$, $P_r \Rightarrow$	0.1	0.3	0.71	1	2
S_{lp}	−0.2095	−0.2218	−0.2307	−0.2337	−0.2255

TABLE 9.6
Rate of Decrease in $\theta(\eta)$ with η at Various Values of Prandtl Number When $G_{rx} = 1$, $f_w = +3$, $E_c = 0.1$, and $E_c = 0.9$

$E_c = 0.1$, $P_r \Rightarrow$	0.1	0.3	0.71	1	2
S_{lp}	−0.1957	−0.1612	−0.1119	−0.0893	−0.0407
$E_c = 0.9$, $P_r \Rightarrow$	0.1	0.3	0.71	1	2
S_{lp}	−0.2020	−0.1757	−0.1362	−0.1210	−0.08924

TABLE 9.7
Effect of Prandtl Number P_r on Some of the Properties of the Flow When $G_{rx} = 1$, Viscous Dissipation Is Negligible ($E_c = 0.1$) and Significant ($E_c = 0.9$)

$f_w = -3$		$E_c = 0.1$			$E_c = 0.9$	
P_r	$f''(0)$	$\theta(0.5)$	$-\theta'(0)$	$f''(0)$	$\theta(1.6667)$	$-\theta'(0)$
0.1	1.3651	0.9029	0.1804	1.4278	0.9782	−0.0102
0.3	1.3604	0.9131	0.1406	1.5447	1.1557	−0.4395
0.71	1.3595	0.9389	0.0631	1.7427	1.4836	−1.2522
1	1.3628	0.9395	0.0167	1.8542	1.7009	−1.7490
2	1.3785	1.0048	−0.0819	2.1768	2.4738	−3.1816
S_{lp}	0.00829	0.05266	−0.13668	0.38726	0.78232	−1.65442
$f_w = 0$		$E_c = 0.1$			$E_c = 0.9$	
P_r	$f''(0)$	$\theta(3)$	$-\theta'(0)$	$f''(0)$	$\theta(3)$	$-\theta'(0)$
0.1	2.2218	0.2909	0.2495	2.3116	0.3760	0.0608
0.3	2.0925	0.1484	0.3469	2.3073	0.3125	−0.1909
0.71	1.9440	0.0447	0.4995	2.2824	0.2171	−0.6299
1	1.8866	0.0268	0.5781	2.2737	0.1857	−0.9200
2	1.7833	0.0165	0.7668	2.2744	0.1465	−1.8960
S_{lp}	−0.21459	−0.12055	0.26465	−0.02024	−0.11379	−1.02033
$f_w = +3$		$E_c = 0.1$			$E_c = 0.9$	
P_r	$f''(0)$	$\theta(3)$	$-\theta'(0)$	$f''(0)$	$\theta(3)$	$-\theta'(0)$
0.1	2.4586	0.2354	0.3350	2.5324	0.2740	0.2133
0.3	2.1062	0.0687	0.6540	2.2469	0.1253	0.3622
0.71	1.7953	0.0094	1.3002	1.9729	0.0616	0.7673
1	1.6954	0.0064	1.7388	1.8841	0.0565	1.0520
2	1.5380	0.0059	3.2141	1.7463	0.0537	2.0106
S_{lp}	−0.43226	−0.08931	1.51272	−0.37279	−0.08843	0.95554

number is a yardstick to diminish the flow along the y-direction and x-direction when viscous dissipation is infinitesimal (Figures 9.23 and 9.25). When viscous dissipation is more enhanced, the velocities of the flow along both directions are increased significantly (Figures 9.24 and 9.26). In fact, with an increase in the ratio of momentum diffusivity to thermal diffusivity, Figures 9.29 and 9.30 indicate that the viscous dissipation is a suitable yardstick to control the friction across the flow induced by mixed convection. For instance, when $E_c = 0.9$, the shear stress increases significantly near the wall $0 \leq \eta < 0.9$ with the Prandtl number; see Figure 9.30. This can be easily deduced from the first row in Table 9.7, where the local skin friction coefficient $f''(0)$ increases with P_r at the rate of 0.008292 when $E_c = 0.1$ and at the rate of 0.387266 when $E_c = 0.9$. Figures 9.35 and 9.36 present the effect of Prandtl number on the temperature gradient profiles $\theta'(\eta)$, which is inversely proportional to the heat transfer rate across the fluid domain from the wall till the free stream. With an increase in the Prandtl number, it is worth remarking that viscous dissipation is capable to influence heat transfer rate within the domain $1 \leq \eta \leq 2$. Figures 9.45, 9.46, 9.55 and 9.56 show that the Prandtl number has a distinct effect on the temperature gradient $\theta'(\eta)$ when $f_w = 0$ and $f_w = +3$. However, a significant difference in the observed effect of the Prandtl number on the heat transfer rate at the wall and the free stream is achievable when $f_w = 0$ and $E_c = 0.1$.

It is worth remarking that when (i) $f_w = 0$ and $E_c = 0.1$ (Figure 9.43), (ii) $f_w = +3$ and $E_c = 0.1$ Figure 9.53), and (iii) $f_w = +3$ and $E_c = 0.9$ (Figure 9.54), the temperature distribution decreases with the Prandtl number at different rates as presented in Tables 9.5 and 9.6. It is worth noticing that when $f_w = 0$ and $E_c = 0.9$, an increase in the magnitude

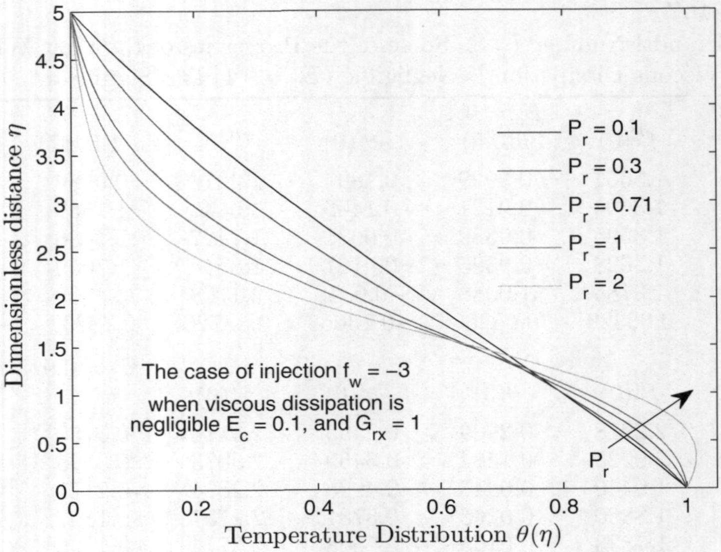

FIGURE 9.23
Variation in $f(\eta)$ with P_r when $f_w = -3$ and $E_c = 0.1$.

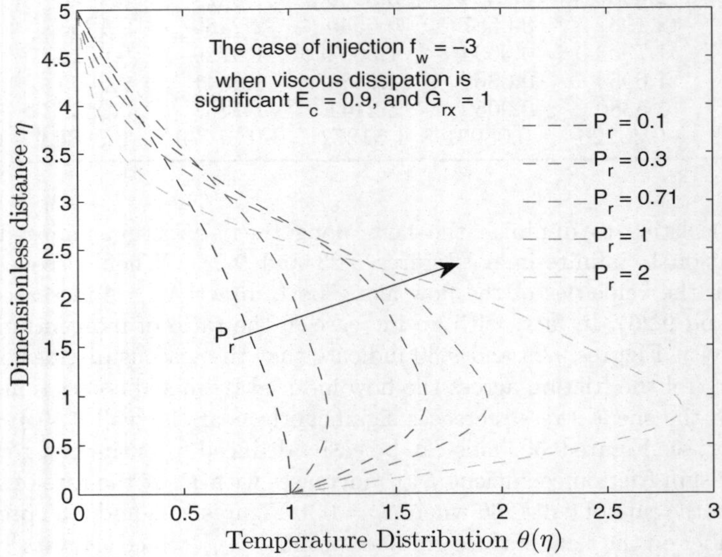

FIGURE 9.24
Variation in $f(\eta)$ with P_r when $f_w = -3$ and $E_c = 0.9$.

of Prandtl number corresponds to an enhancement in the temperature distribution within the fluid layer near the wall; see Figure (9.44). When $f_w = 0$ and $f_w = +3$, the results of this study illustrated as Figure 9.37–9.40 and 9.47–9.50 show that the motion of the fluid along the x-direction and y-direction decreases with the Prandtl number. Figures 9.27 and 9.28 further unravel the significance of viscous dissipation on the less viscous fluid within

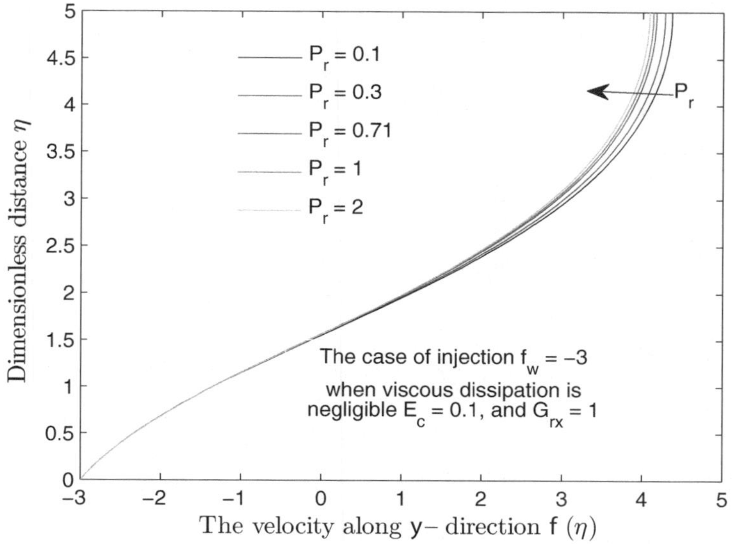

FIGURE 9.25
Variation in $f'(\eta)$ with P_r when $f_w = -3$ and $E_c = 0.1$.

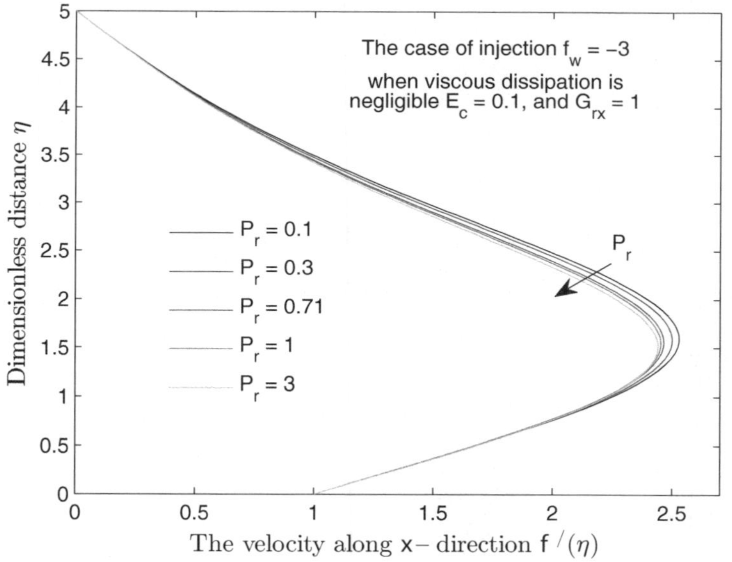

FIGURE 9.26
Variation in $f'(\eta)$ with P_r when $f_w = -3$ and $E_c = 0.9$.

the boundary layer. One thing that is paramount is that maximum velocity is achievable for $0 \leq P_r < 1$ within the domain $0.4 \leq \eta \leq 1.2$. The same trend is seen in the variation of friction across the domain as shown in Figures 9.31 and 9.32. When viscous dissipation is low, the presence of a countable number of tiny hills inside the fluid domain shows the area where considerable fluctuation occurs (Figure 9.31). However, an enhancement in viscous dissipation is seen to maintain the effect for all the values of Prandtl number $0 \leq P_r \leq 7$. See Figures 9.41, 9.42, 9.51, and 9.52 for more outcomes of simulation.

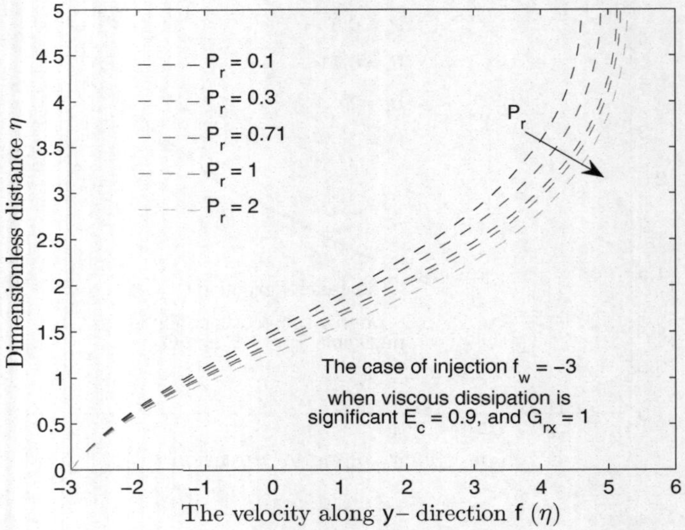

FIGURE 9.27
Variation in velocity of various less viscous fluids when viscous dissipation is small ($E_c = 0.1$) and $f_w = +3$, $G_{rx} = 3$.

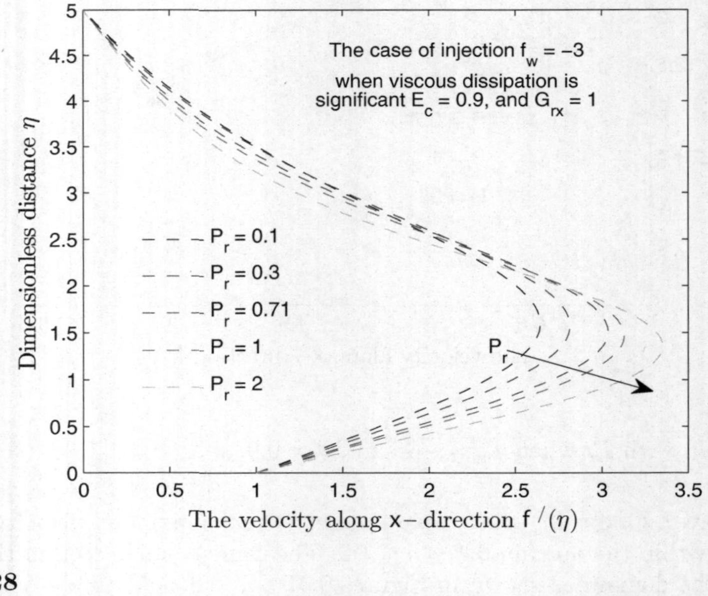

FIGURE 9.28
Variation in velocity of various less viscous fluids when viscous dissipation is large ($E_c = 1.5$) and $f_w = +3$, $G_{rx} = 3$.

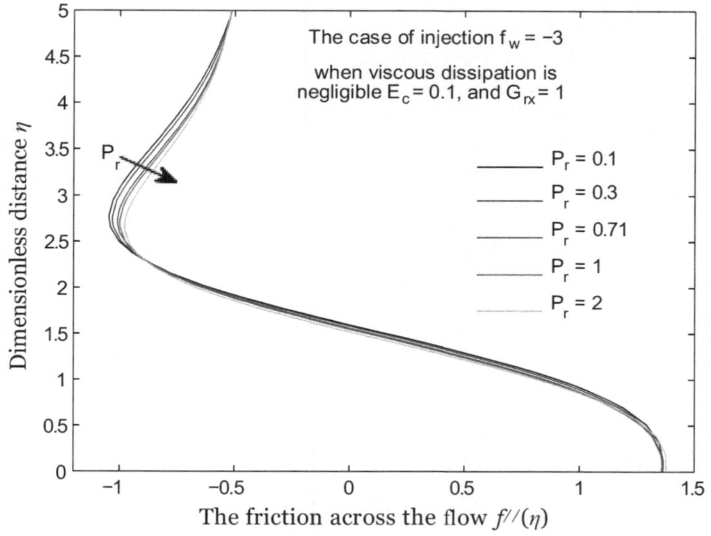

FIGURE 9.29
Variation in $f''(\eta)$ with P_r when $f_w = -3$ and $E_c = 0.1$.

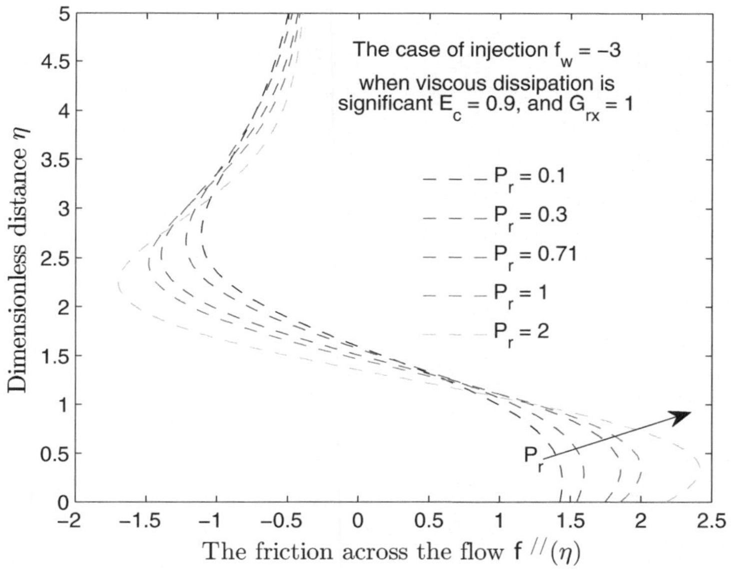

FIGURE 9.30
Variation in $f''(\eta)$ with P_r when $f_w = -3$ and $E_c = 0.9$.

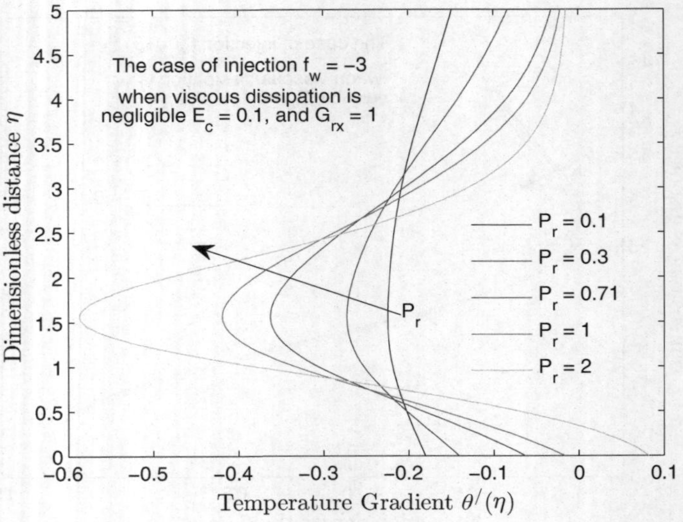

FIGURE 9.31
Variation in friction across the domain of various less viscous fluids when viscous dissipation is small ($E_c = 0.1$) and $f_w = +3$, $G_{rx} = 3$.

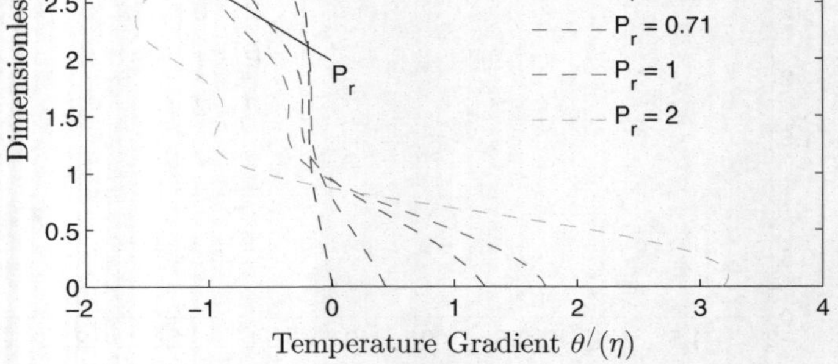

FIGURE 9.32
Variation in friction across the domain of various less viscous fluids when viscous dissipation is large ($E_c = 1.5$) and $f_w = +3$, $G_{rx} = 3$.

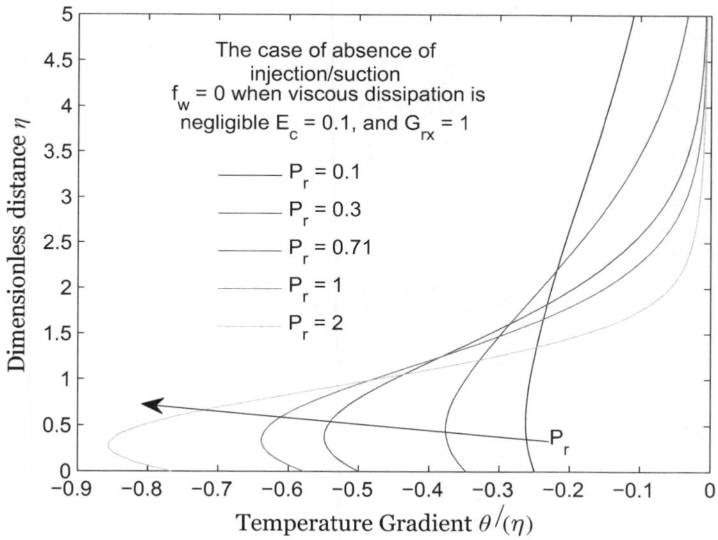

FIGURE 9.33
Variation in $\theta(\eta)$ with P_r when $f_w = -3$ and $E_c = 0.1$.

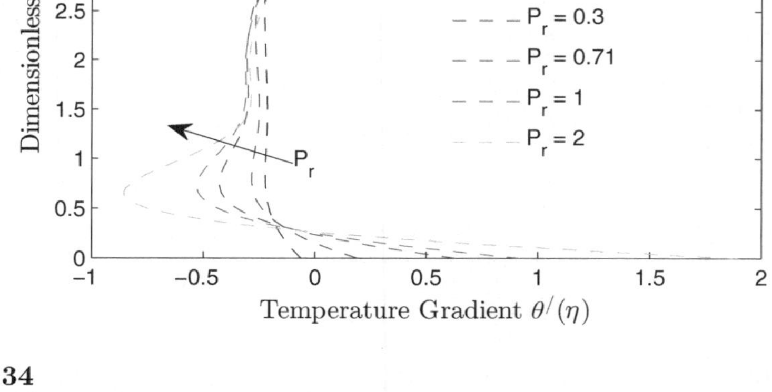

FIGURE 9.34
Variation in $\theta(\eta)$ with P_r when $f_w = -3$ and $E_c = 0.9$.

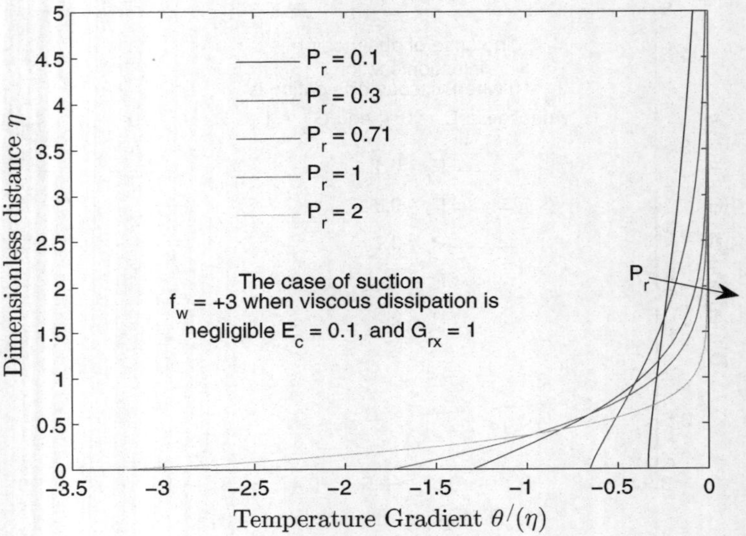

FIGURE 9.35
Variation in $\theta'(\eta)$ with P_r when $f_w = -3$ and $E_c = 0.1$.

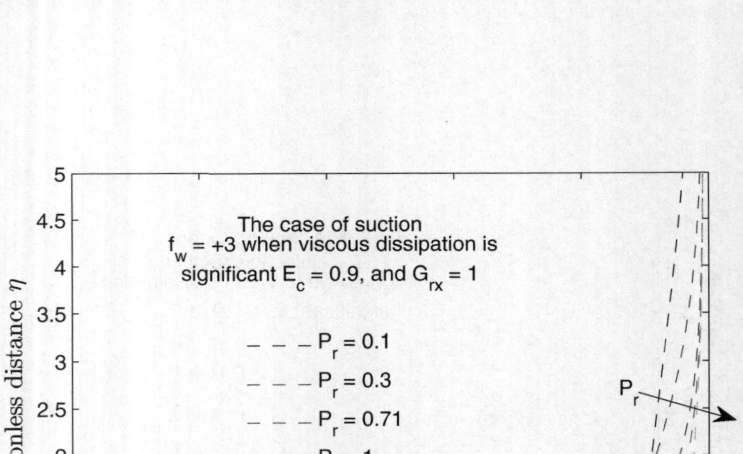

FIGURE 9.36
Variation in $\theta'(\eta)$ with P_r when $f_w = -3$ and $E_c = 0.9$.

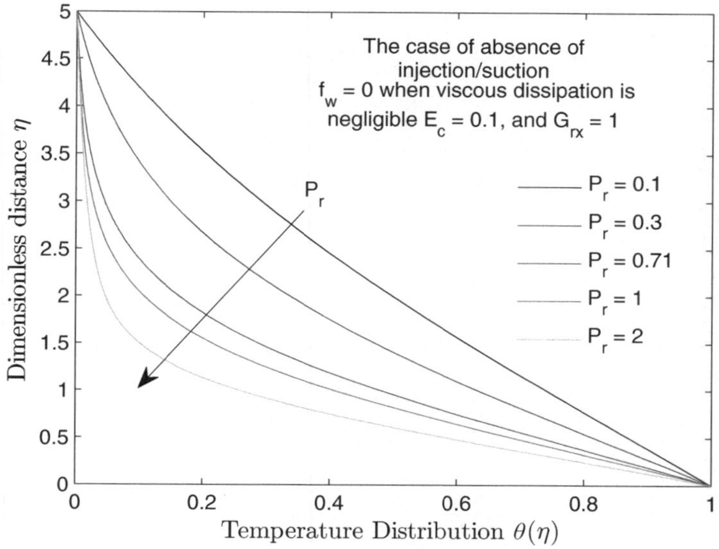

FIGURE 9.37
Variation in $f(\eta)$ with P_r when $f_w = 0$ and $E_c = 0.1$.

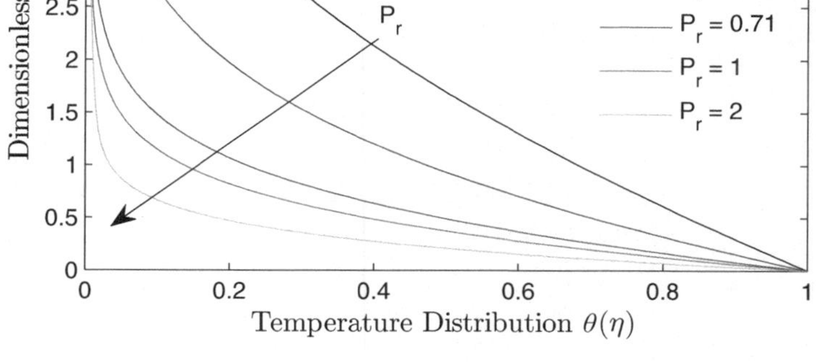

FIGURE 9.38
Variation in $f(\eta)$ with P_r when $f_w = 0$ and $E_c = 0.9$.

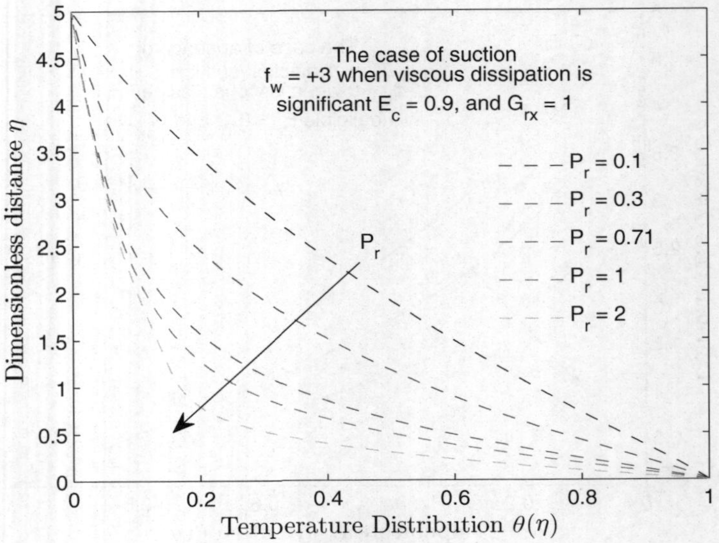

FIGURE 9.39
Variation in $f'(\eta)$ with P_r when $f_w = 0$ and $E_c = 0.1$.

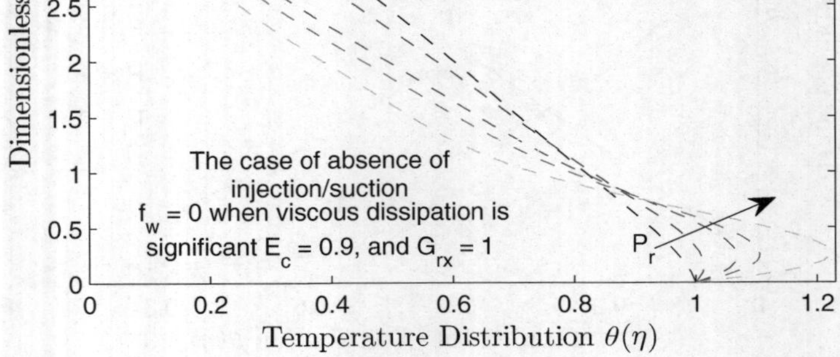

FIGURE 9.40
Variation in $f'(\eta)$ with P_r when $f_w = 0$ and $E_c = 0.9$.

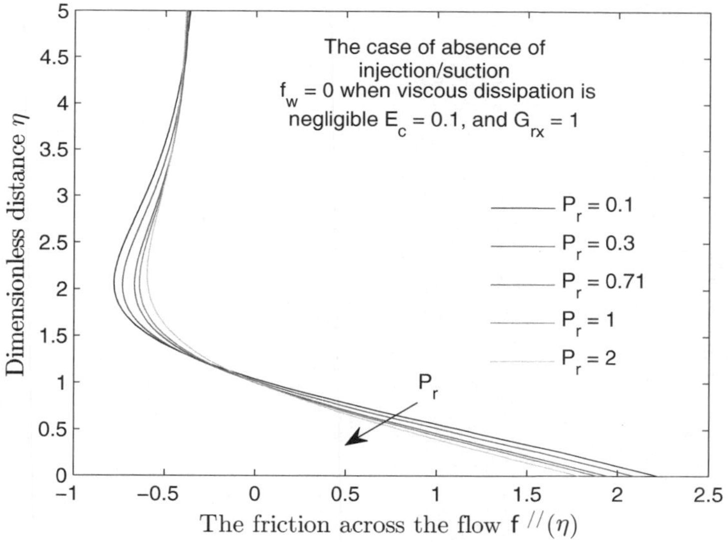

FIGURE 9.41
Variation in $f''(\eta)$ with P_r when $f_w = 0$ and $E_c = 0.1$.

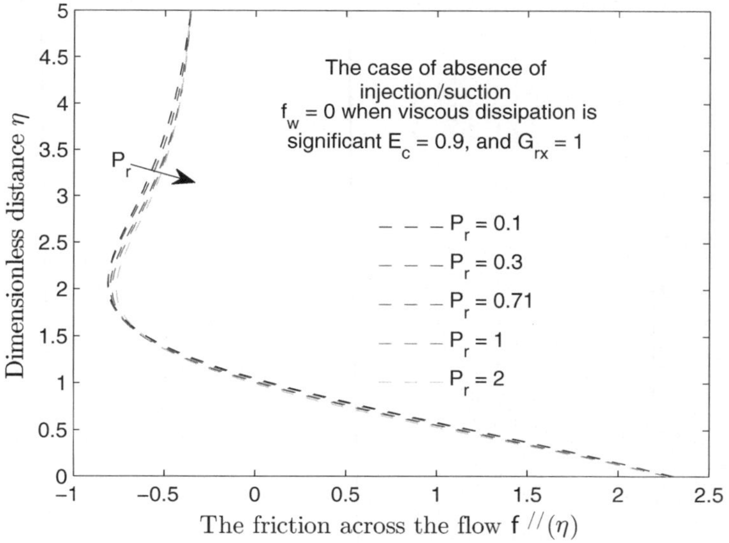

FIGURE 9.42
Variation in $f''(\eta)$ with P_r when $f_w = 0$ and $E_c = 0.9$.

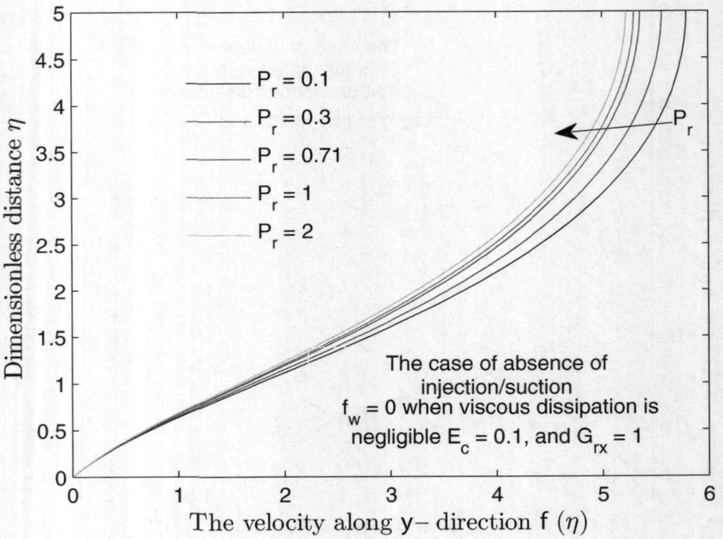

FIGURE 9.43
Variation in $\theta(\eta)$ with P_r when $f_w = 0$ and $E_c = 0.1$.

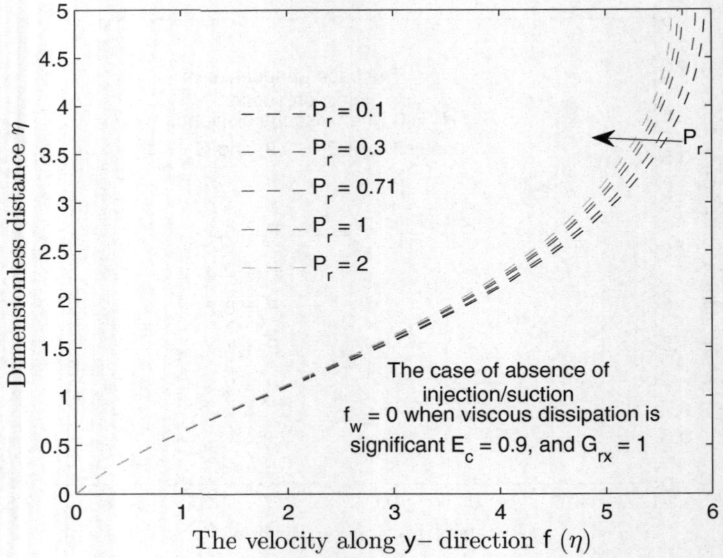

FIGURE 9.44
Variation in $\theta(\eta)$ with P_r when $f_w = 0$ and $E_c = 0.9$.

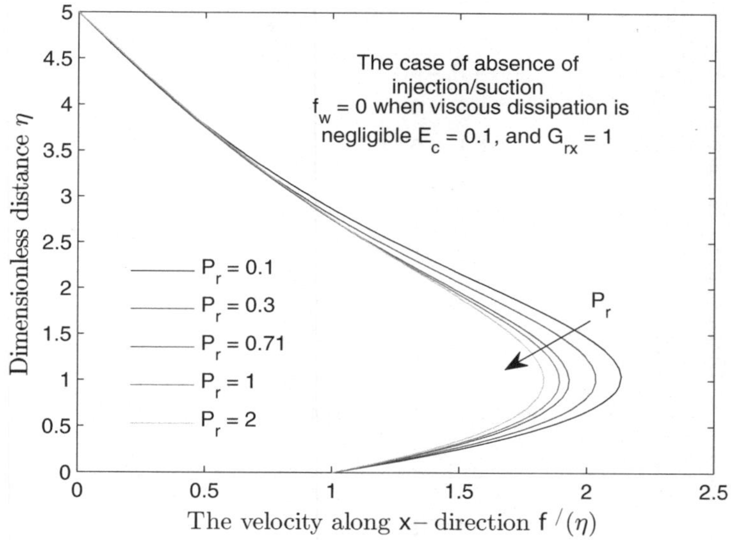

FIGURE 9.45
Variation in $\theta'(\eta)$ with P_r when $f_w = 0$ and $E_c = 0.1$.

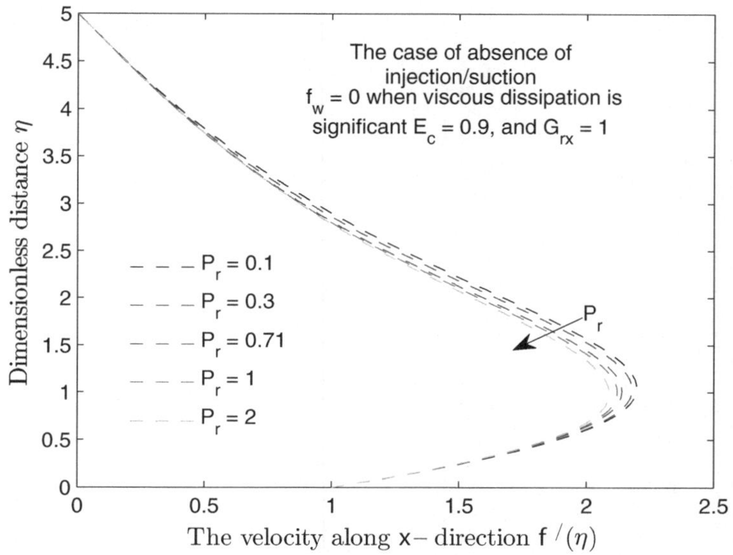

FIGURE 9.46
Variation in $\theta'(\eta)$ with P_r when $f_w = 0$ and $E_c = 0.9$.

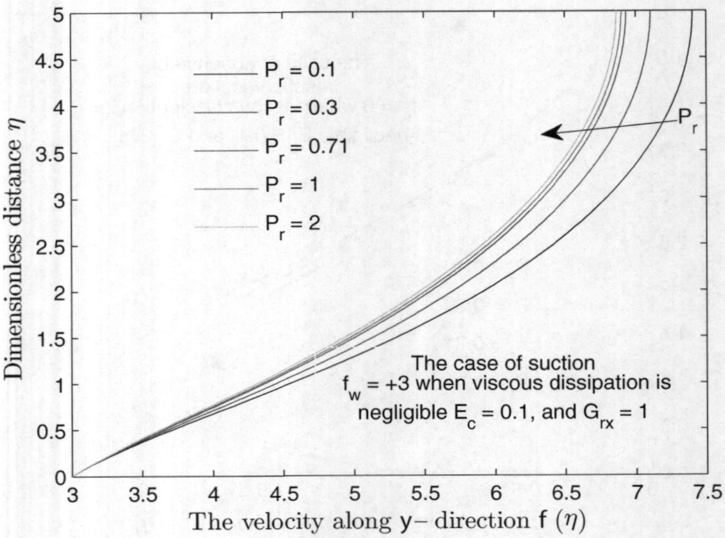

FIGURE 9.47
Variation in $f(\eta)$ with P_r when $f_w = +3$ and $E_c = 0.1$.

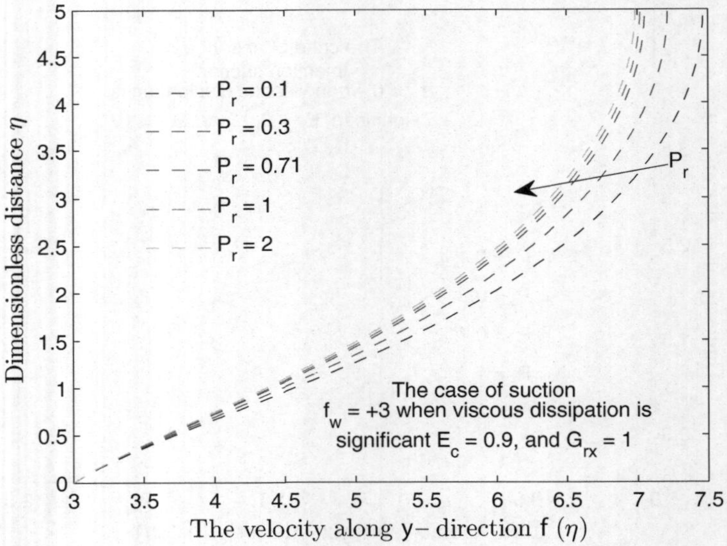

FIGURE 9.48
Variation in $f(\eta)$ with P_r when $f_w = +3$ and $E_c = 0.9$.

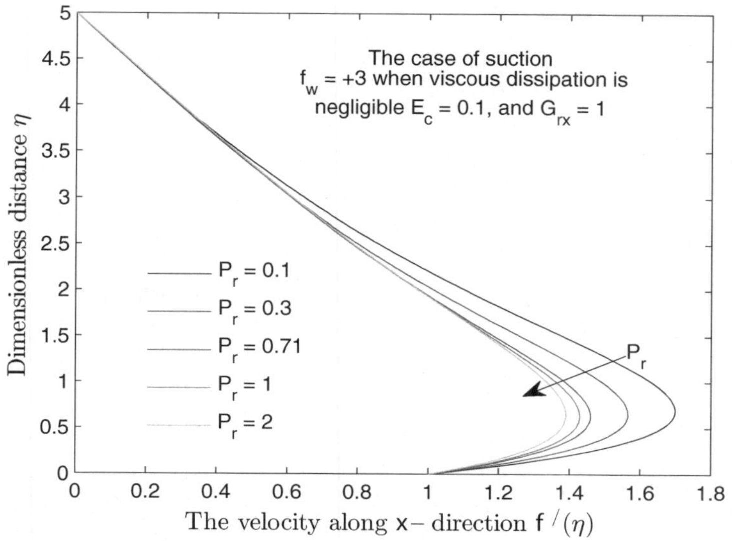

FIGURE 9.49
Variation in $f'(\eta)$ with P_r when $f_w = +3$ and $E_c = 0.1$.

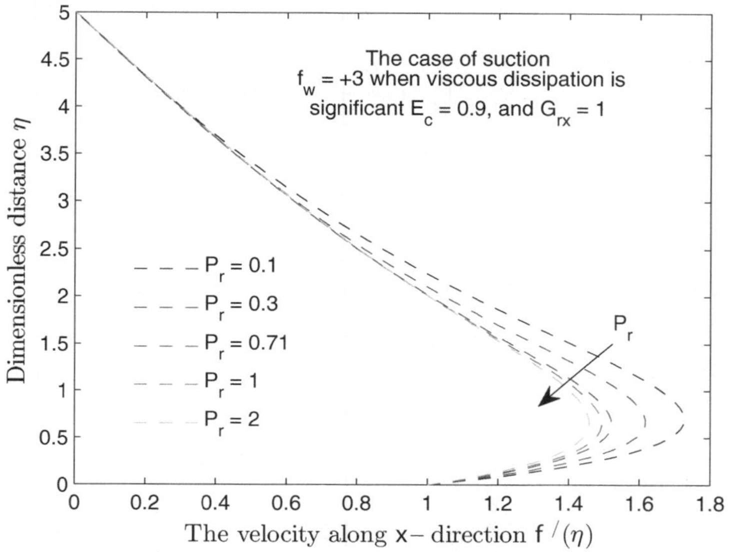

FIGURE 9.50
Variation in $f'(\eta)$ with P_r when $f_w = +3$ and $E_c = 0.9$.

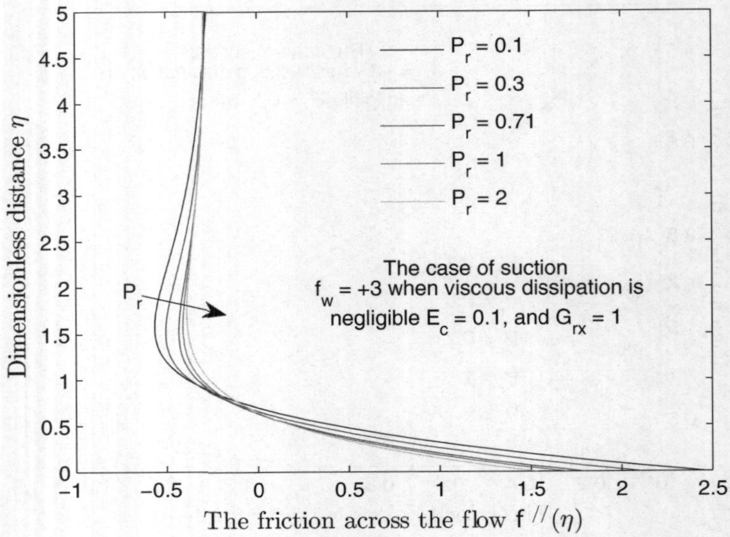

FIGURE 9.51
Variation in $f''(\eta)$ with P_r when $f_w = +3$ and $E_c = 0.1$.

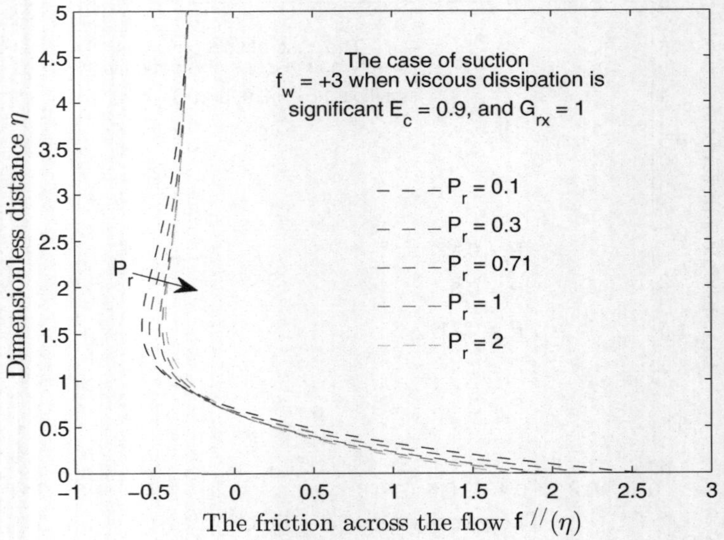

FIGURE 9.52
Variation in $f''(\eta)$ with P_r when $f_w = +3$ and $E_c = 0.9$.

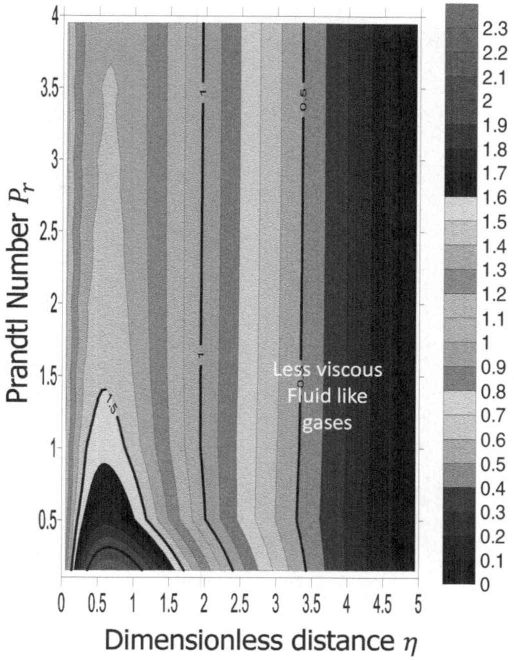

FIGURE 9.53
Variation in $\theta(\eta)$ with P_r when $f_w = +3$ and $E_c = 0.1$.

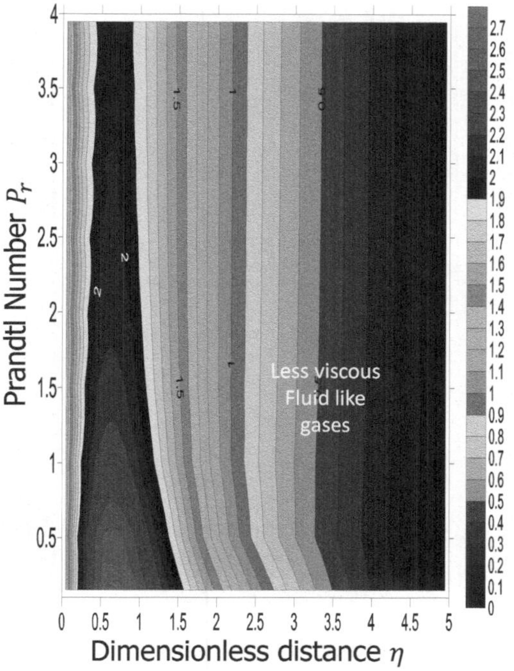

FIGURE 9.54
Variation in $\theta(\eta)$ with P_r when $f_w = +3$ and $E_c = 0.9$.

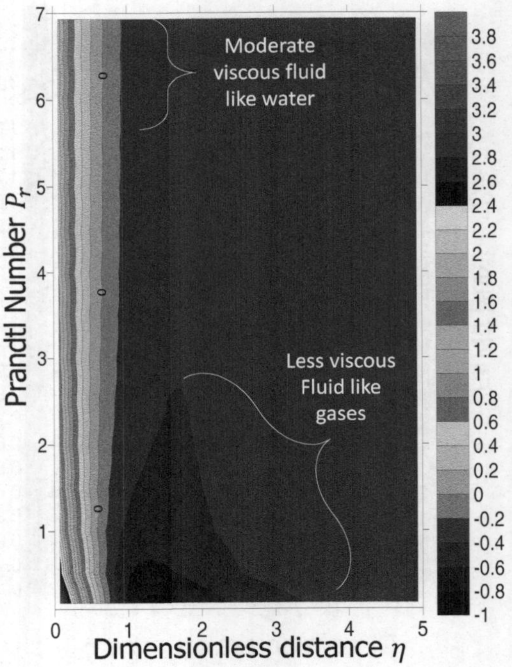

FIGURE 9.55
Variation in $\theta'(\eta)$ with P_r when $f_w = +3$ and $E_c = 0.1$.

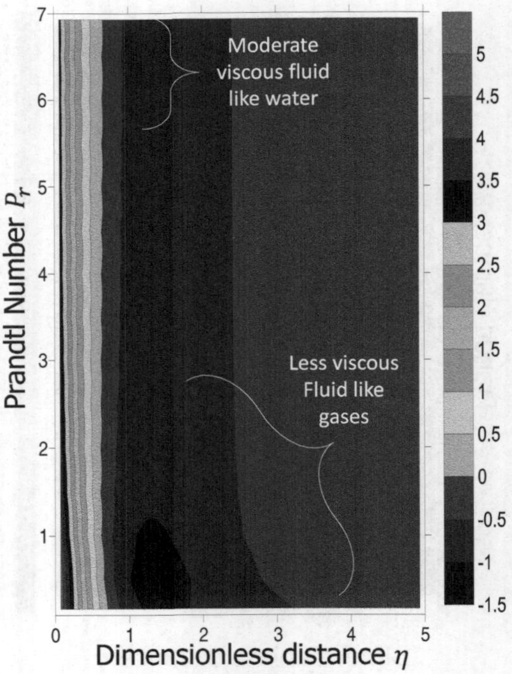

FIGURE 9.56
Variation in $\theta'(\eta)$ with P_r when $f_w = +3$ and $E_c = 0.9$.

9.5 Tutorial Questions

1. When viscous dissipation is minimal, how does the Prandtl number influence the velocity, friction throughout the domain, and temperature distribution of a flow generated by stretching at the wall?
2. When viscous dissipation is considerable in the presence of suction, what influence does the Prandtl number have on Sakiadis flow of a Newtonian fluid?
3. What impact does raising the Prandtl number have on the transport phenomena in the situation of mixed convection fluid flow when viscous dissipation is minimal and significant?
4. When viscous dissipation is minimal and significant, how does the Prandtl number cause a change in the local skin friction coefficients and heat transfer rate?

10

Analysis of Self-Similar Flows VI

10.1 Background Information

The growing impacts of the Prandtl number on the dynamics of certain fluids susceptible to thermal stratification are explored in this chapter.

10.2 Introduction: Thermal Stratification

At the wall, there are many patterns of fluctuation in heat energy. Thermal stratification is divided into three stages: epilimnion, hypolimnion, and metalimnion. The prescribed power-law surface temperature is modeled using

$$T = T_w = T_\infty + A\left(\frac{x}{l}\right)^2 \quad at \quad y = 0 \tag{10.1}$$

$$T \to T_\infty \quad as \quad y \to \infty. \tag{10.2}$$

The prescribed power-law heat flux on the wall is

$$-\kappa\frac{\partial T}{\partial y} = q_w = D\left(\frac{x}{l}\right)^2 \quad at \quad y = 0 \tag{10.3}$$

$$T \to T_\infty \quad as \quad y \to \infty. \tag{10.4}$$

The case of linear stratification of thermal energy is

$$T_w = T_o + m_1 x \quad at \quad y = 0 \tag{10.5}$$

$$T_\infty = T_o + m_2 x \quad as \quad y \to \infty \tag{10.6}$$

The effect of variation in the surface tension with temperature as the fluid flows along a horizontal surface is of the form

$$T = T_\infty + ax^2, \quad \mu\frac{\partial u}{\partial y} = \frac{\partial \sigma}{\partial T}\frac{\partial T}{\partial x} \quad at \quad y = 0 \tag{10.7}$$

where the dimensionless temperature and linear variation in surface tension with temperature $\sigma(T)$ are of the form

$$\theta(\eta) = \frac{T - T_\infty}{ax^2}, \quad \sigma(T) = \sigma_o[1 - \gamma(T - T_\infty)] \tag{10.8}$$

Here, γ is the rate of change of surface tension with temperature. At the epilimnion stage of thermal stratification, local skin friction coefficients increase with thermal stratification (Ajayi et al. [14]). Temperature distribution along the wall becomes minimal as the thermal

DOI: 10.1201/9781003217374-10

stratification increases where the wall temperature is zero. At the same time, the same property of the fluid decreases with the temperature at the free stream (Omowaye and Animasaun [219]). It should also be noted that when the wall temperature is zero, the temperature gradient at the wall reduces considerably; yet, the velocity of the flow rises negligibly with the thermal stratification parameter. In fact, in the case of two-dimensional convection of an excellent electrical conductor within a square domain, a higher Prandtl number was determined to be an agent suitable for distorting the flow due to viscous forces that dominate the emerging buoyancy forces, resulting in the formation of thermal stratification.

10.3 Fluid Flow Subject to Thermal Stratification

In this section, the motion of fluids on a horizontal surface subject to suction/injection and thermal stratification is examined in order to deduce the significance of thermal stratification and Prandtl number at epilimnion stage, metalimnion stage, and hypolimnion stage. As illustrated in Figure 10.1, the two-dimensional flow occurs on a horizontal surface where the velocity along the x-direction is $u(x, y)$ and the velocity along the y-direction is $v(x, y)$. The occurrence of linear variation form of thermal stratification at the wall temperature and at the free stream temperature where T_o is the reference temperature is defined as

$$T_\infty(x) = T_o + m_1 x, \qquad T_w(x) = T_o + m_2 x \tag{10.9}$$

The governing equation suitable to model the cases is written as

$$u_x + v_y = 0, \tag{10.10}$$

$$uu_x + vu_y = \frac{\mu}{\rho} u_{yy}, \tag{10.11}$$

$$uT_x + vT_y = \frac{\kappa}{\rho c_p} T_{yy}. \tag{10.12}$$

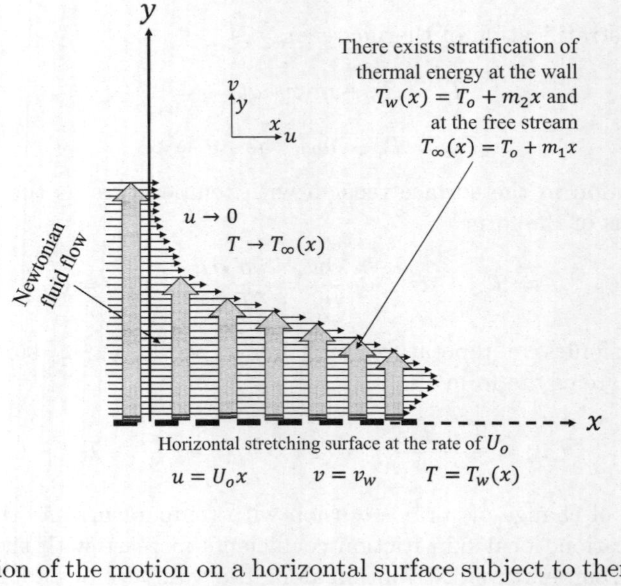

FIGURE 10.1
Graphical illustration of the motion on a horizontal surface subject to thermal stratification.

For this case, Eqs. (10.2)–(10.4) are subject to the boundary conditions

$$u = U_o x, \quad v = v_w, \quad T = T_w(x) \quad at \quad y = 0. \tag{10.13}$$

$$u \to 0, \quad T \to T_\infty(x), \quad as \quad y \to \infty \tag{10.14}$$

In order to obtain self-similar solution of Eqs. (10.2)–(10.14), the following similarity variables

$$\eta = y\sqrt{\frac{U_\infty}{\vartheta x}}, \quad \psi(x,y) = \sqrt{\vartheta x U_\infty} f(\eta), \quad u = \frac{\partial \psi}{\partial y}, \quad v = -\frac{\partial \psi}{\partial x},$$

$$\theta(\eta) = \frac{T - T_\infty}{T_w - T_o}, \quad P_r = \frac{\mu Cp}{\kappa} = \frac{\vartheta}{\alpha}, \quad S_t = \frac{m_1}{m_2}, \quad f_w = -\frac{v_w}{\sqrt{\vartheta U_o}}. \tag{10.15}$$

were used to obtain the dimensionless governing equation below:

$$\frac{d^3 f}{d\eta^3} + \frac{1}{2} f \frac{d^2 f}{d\eta^2} = 0, \tag{10.16}$$

$$\frac{d^2\theta}{d\eta^2} - S_t P_r \frac{df}{d\eta} - P_r \theta \frac{df}{d\eta} + \frac{1}{2} P_r f \frac{d\theta}{d\eta} = 0 \tag{10.17}$$

Dimensionless boundary conditions are

$$\frac{df}{d\eta} = 1, \quad f = f_w, \quad \theta = 1 - S_t \quad at \quad \eta = 0. \tag{10.18}$$

$$\frac{df}{d\eta} \to 0, \quad \theta \to 0 \quad as \quad \eta \to \infty. \tag{10.19}$$

10.3.1 Research Questions I

The aim of this subsection is to provide answers to the following research questions:

1. In the presence of either suction or injection at the horizontal surface, what is the effect of increasing Prandtl number on the local skin friction coefficients and Nusselt number when thermal stratification is at epilimnion stage, metalimnion stage, and hypolimnion stage?
2. When the thermal stratification is at the epilimnion stage, metalimnion stage, and hypolimnion stage, what are the effects of Prandtl number on the transport phenomenon and temperature distribution?

10.3.2 Analysis and Discussion of Results I

Equations (10.16) and (10.17) subject to Eqs. (10.18) and (10.19) were solved using the four-stage Lobatto IIIa formula as described in Chapter 1 using $\eta = 5$. It is evident from Tables (10.1) and (10.2) that the local skin friction coefficients $f''(0)$ proportional to the shear stress between the last layer of the fluid and the horizontal surface is a constant function of Prandtl number at various values of admissible thermal stratification. Such a result implies that at all the three stages of thermal stratification, for the case of either suction or injection at the wall, local skin friction coefficient $f''(0)$ does not change with increasing P_r. It is important to remark that the maximum local skin friction coefficient is achievable in the presence of injection as -0.0244; see Figures 10.2 and 10.3. Changes in the shear stress as $\eta \to 5$ for the case of suction and injection are illustrated in Figure F.1.

TABLE 10.1
Variation in $f''(0)$ and $-\theta'(0)$ with P_r When $f_w = +3$ (the Case of Suction) at Epilimnion Stage ($S_t = 0$), Metalimnion Stage ($S_t = 0.5$), and Hypolimnion Stage ($S_t = 1$)

P_r	$f''(0)$ $S_t = 0$	$-\theta'(0)$ $S_t = 0$	$f''(0)$ $S_t = 0.5$	$-\theta'(0)$ $S_t = 0.5$	$f''(0)$ $S_t = 1$	$-\theta'(0)$ $S_t = 1$	S_{lp}
0.1	−1.6477	0.3437	−1.6477	0.1981	−1.6477	0.0526	−0.2911
0.3	−1.6477	0.6780	−1.6477	0.4191	−1.6477	0.1601	
0.71	−1.6477	1.4143	−1.6477	0.8934	−1.6477	0.3724	
3	−1.6477	5.1546	−1.6477	3.3040	−1.6477	1.4535	
6	−1.6477	9.7782	−1.6477	6.2990	−1.6477	2.8198	−6.9584
S_{lp}	0	1.526511	0	1.032850	0	0.539178	

TABLE 10.2
Variation in $f''(0)$ and $-\theta'(0)$ with P_r When $f_w = -3$ (the Case of Injection) at Epilimnion Stage ($S_t = 0$), Metalimnion Stage ($S_t = 0.5$), and Hypolimnion Stage ($S_t = 1$)

P_r	$f''(0)$ $S_t = 0$	$-\theta'(0)$ $S_t = 0$	$f''(0)$ $S_t = 0.5$	$-\theta'(0)$ $S_t = 0.5$	$f''(0)$ $S_t = 1$	$-\theta'(0)$ $S_t = 1$	S_{lp}
0.1	−0.0244	0.2894	−0.0244	0.2255	−0.0244	0.1615	−0.1279
0.3	−0.0244	0.3972	−0.0244	0.3688	−0.0244	0.3403	
0.71	−0.0244	0.5008	−0.0244	0.4942	−0.0244	0.4877	
3	−0.0244	0.6170	−0.0244	0.6170	−0.0244	0.6170	
6	−0.0244	0.6410	−0.0244	0.6410	−0.0244	0.6410	0
S_{lp}	0	0.049919	0	0.057088	0	0.064266	

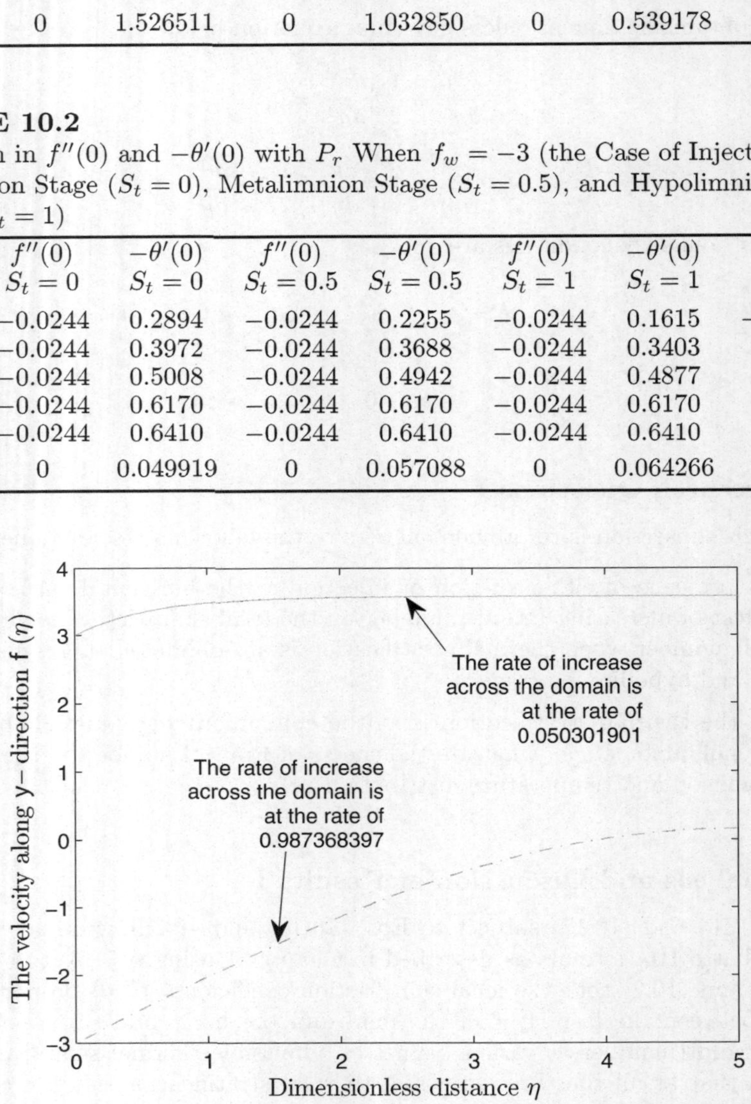

FIGURE 10.2
Variation in $f(\eta)$ due to P_r when $f_w = +3$ and $f_w = -3$.

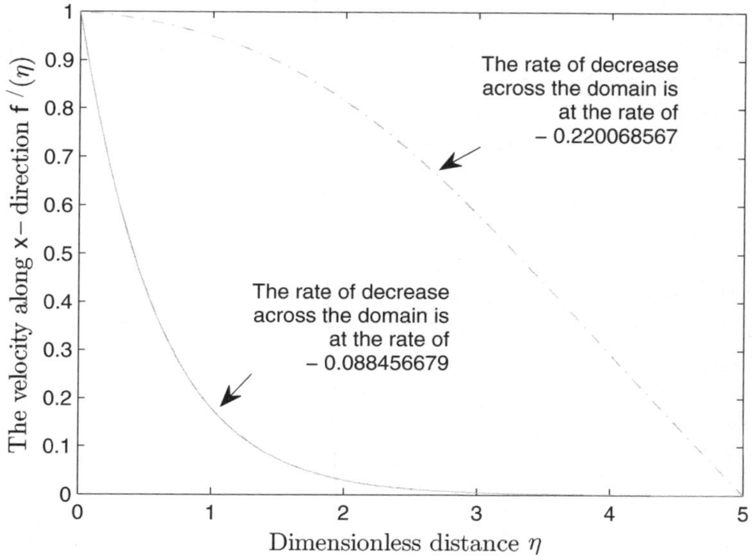

FIGURE 10.3
Variation in $f'(\eta)$ due to P_r when $f_w = +3$ and $f_w = -3$.

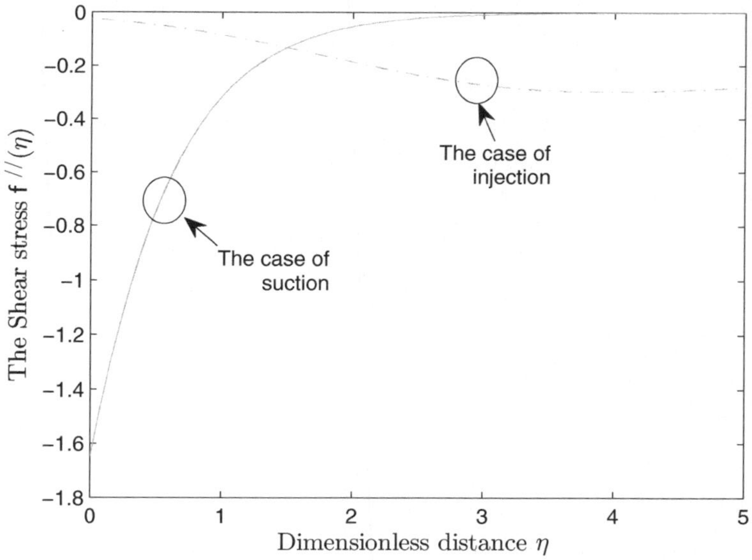

FIGURE 10.4
Variation in $f''(\eta)$ due to P_r when $f_w = +3$ and $f_w = -3$.

The observed constant local skin friction coefficient at all the values of P_r indicates that the horizontal and vertical velocities are also constant functions of Prandtl number and the wall temperature is not a factor in improving these velocities in a two-dimensional flow and shear stress across the dynamics. Table (10.1) shows that at all the three levels of thermal stratification in the presence of suction, the coordinates (η, f) to describe the observed variation in the motion along the x-direction are $(\eta = 0, f = 3)$ and $(\eta = 5, f = 3.5811)$

(Figure 10.4). In other words, the vertical velocity increases from the wall to the free stream at the rate of 0.050301901. Further investigation shows that the motion of the fluid along the x-direction decreases from the wall ($\eta = 0$) to the free stream ($\eta = 5$) at the rate of -0.088456679; see Figures 10.2 and 10.3.

When the Prandtl number is small ($P_r = 0.1$), the Nusselt number decreases with the thermal stratification-related parameter S_t at the rate of -0.2911; see Table (10.1). In the dynamics of a more viscous fluid such as water ($P_r = 6$), the Nusselt number is a decreasing property due to an increase in S_t at the rate of -6.9584. In addition, for the case of suction, as presented in Table 10.1, the Nusselt number increases with the Prandtl number at a larger rate in the case of epilimnion stage (i.e., $S_t = 0$ when the layers of fluid near the wall are warmest). Because the Nusselt number is inversely proportional to the exact heat transfer rate, as the Prandtl number increases, the rate of heat transfer across the fluid flow in the case of suction increases at the rate of 1.526511 when $S_t = 0$ and at the rate of 0.539178 when $S_t = 1$, as shown in Table 10.1. In the case of injection ($f_w = -3$), it is evident that the Prandtl number increases the Nusselt number at all the levels of thermal stratification within $0 \leq S_t \leq 1$. Surprisingly, it is clear that the gradual decrease in wall temperature, which is the corresponding impact of increasing S_t on the flow of more viscous fluids ($P_r = 3$ and $P_r = 6$), does not affect both the Nusselt number and the heat transfer rate. The coordinates (η, f) to describe the observed variation in the motion along the x-direction for the case of injection are ($\eta = 0, f = -3$) and ($\eta = 5, f = 0.1593$). In other words, the vertical velocity increases from the wall to the free stream at the rate of 0.987368397. The motion of the fluid along the x-direction decreases from the wall ($\eta = 0$) to the free stream ($\eta = 5$) at the rate of -0.220068567 (Figure 10.4).

The variation in the temperature distribution with the Prandtl number at the three levels of thermal stratification in the presence of suction and injection was investigated and is presented in Figures 10.5–10.10. In the first two cases of suction, the temperature distribution decreases with the Prandtl number at different patterns and rates; see Figures 10.5–10.9 presents another pattern of decrease in the temperature distribution due to a gradual increase in the Prandtl number at hypolimnion stage ($S_t = 1$). At the

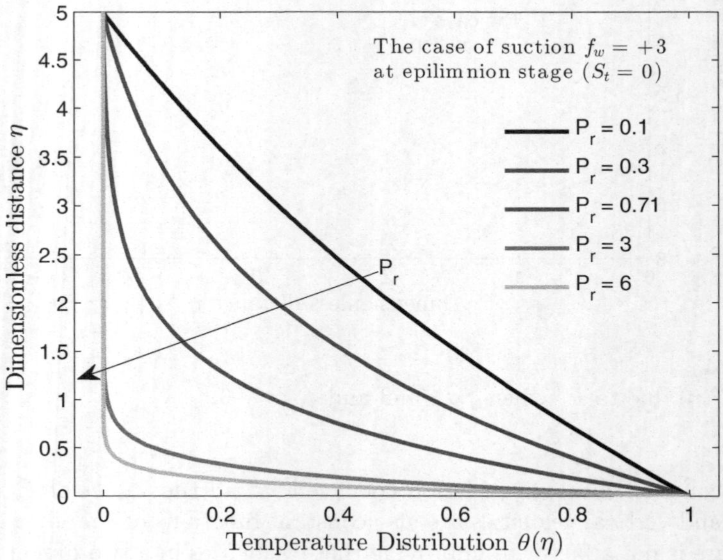

FIGURE 10.5
Variation in $\theta(\eta)$ due to P_r when $S_t = 0$ and $f_w = +3$.

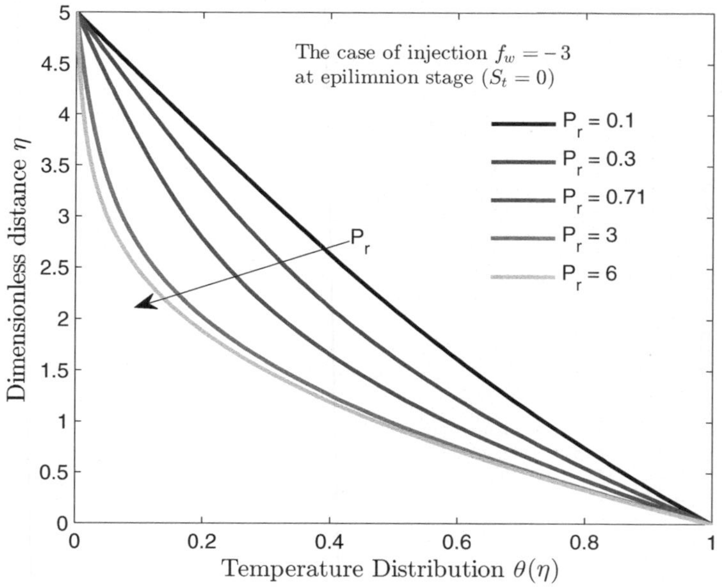

FIGURE 10.6
Variation in $\theta(\eta)$ due to P_r when $S_t = 0$ and $f_w = -3$.

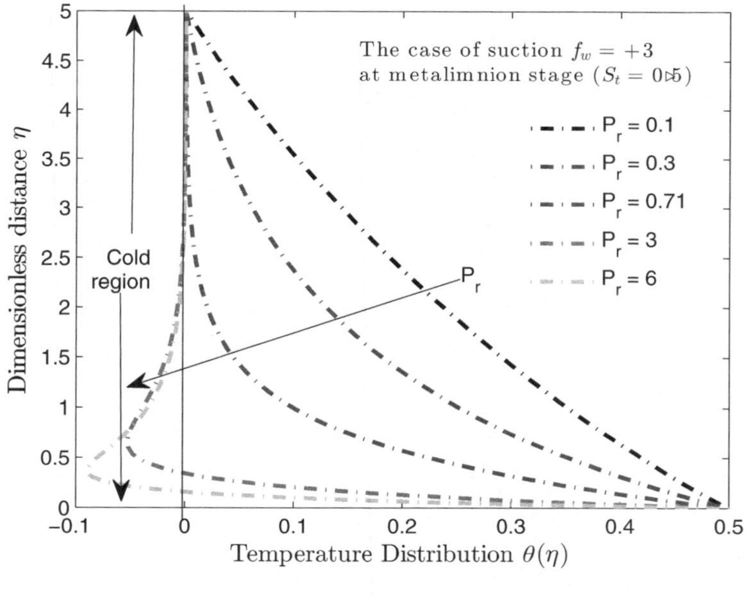

FIGURE 10.7
Variation in $\theta(\eta)$ due to P_r when $S_t = 0.5$ and $f_w = +3$.

epilimnion stage ($S_t = 0$), injection is found to subdue the nature of momentum diffusivity as the temperature distribution is found to be an increasing property of Prandtl number; see Figures 10.6–10.10 depict that at metalimnion stage ($S_t = 0.5$) and hypolimnion stage ($S_t = 1$), a decrease in the temperature distribution due to Prandtl number is achievable. As the magnitude of the Prandtl number increases, the momentum diffusivity increases, leading to a decrease in the temperature distribution. However, with an increase in the

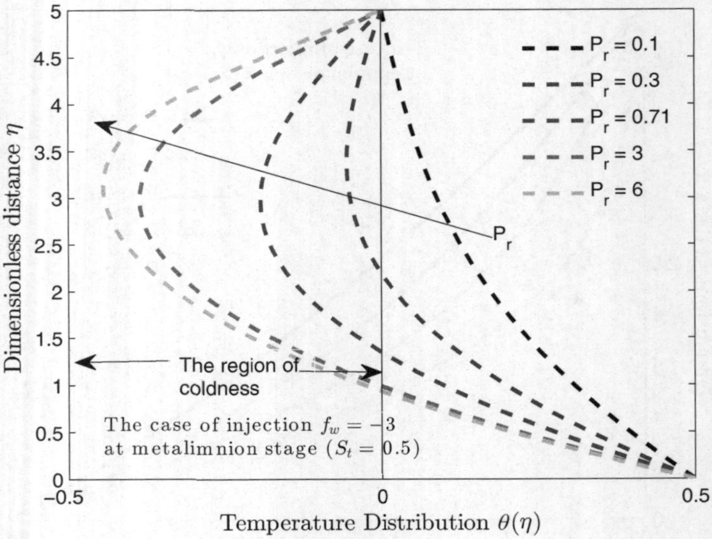

FIGURE 10.8
Variation in $\theta(\eta)$ due to P_r when $S_t = 0.5$ and $f_w = -3$.

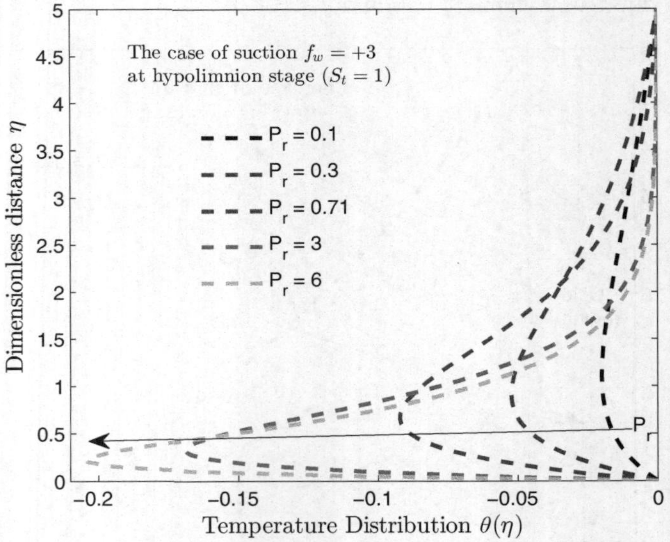

FIGURE 10.9
Variation in $\theta(\eta)$ due to P_r when $S_t = 1$ and $f_w = +3$.

thermal stratification S_t, the wall temperature $T_w(x) = T_o + m_2 x$ is bound to decrease. Such a result is found to be sufficient in reducing the overall temperature distribution across the domain. As the magnitude of dimensionless thermal stratification parameter S_t increases, the amount of wall temperature reduces. In other words, the warmest layer occurs within the fluid layers near the wall at the epilimnion stage ($S_t = 0$). At this stage, it is realistic to remark that the domain is much more heated.

The results illustrated in Figures 10.5 and 10.6 confirm that all the total heat energy is properly distributed due to suction that draws molecules of the liquid substance closer

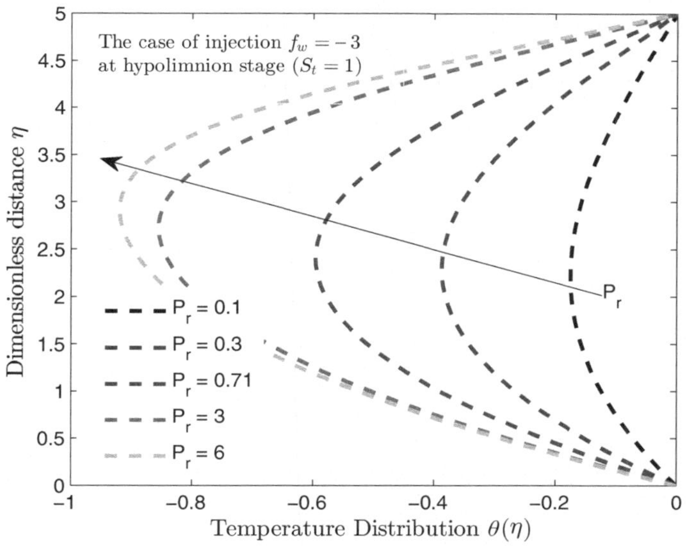

FIGURE 10.10
Variation in $\theta(\eta)$ due to P_r when $S_t = 1$ and $f_w = -3$.

TABLE 10.3
Variation in Heat Transfer Rate $-\theta'(\eta)$ Near the Horizontal Wall ($0 \leq \eta \leq 1$) for the Case of Water ($P_r = 0.71$)

η	$S_t = 0$ $f_w = +3$ $-\theta'(\eta)$	$S_t = 0$ $f_w = -3$ $-\theta'(\eta)$	$S_t = 0.5$ $f_w = +3$ $-\theta'(\eta)$	$S_t = 0.5$ $f_w = -3$ $-\theta'(\eta)$	$S_t = 1$ $f_w = +3$ $-\theta'(\eta)$	$S_t = 1$ $f_w = -3$ $-\theta'(\eta)$
0	1.4143	0.5008	0.8934	0.4942	0.3724	0.4877
0.2273	1.0000	0.4608	0.5881	0.4524	0.1761	0.4441
0.5051	0.6691	0.4086	0.3566	0.3967	0.0440	0.3849
0.7576	0.4715	0.3706	0.2281	0.3552	−0.0152	0.3398
1	0.3355	0.3223	0.1469	0.3005	−0.0418	0.2788
S_{lp}	−1.0604	−0.1767	−0.7314	−0.1915	−0.4023	−0.2063

to the wall (Figure 10.5). In the case of injection, the addition of molecules from the wall tends to push the distribution of heat energy away from the wall, as shown in Figure 10.6. The metalimnion stage ($S_t = 0.5$) implies that the wall temperature is reduced by 50%. The results illustrated in Figure 10.7 show that the remaining 50% of heat energy is drawn forcefully toward the wall. However, a highly viscous fluid reaches coldness even in the case of suction (see $P_r = 6$ in Figure 10.7). The injection is seen to drastically cool the fluid after a small distance away from the wall ($\eta = 1$); Figure 10.8. Scientifically, this is true because the heat energy near the wall remains 50% since $S_t = 0.5$. The case of the hypolimnion stage ($S_t = 1$) is similar to absolute zero of heat energy at the wall. Practically speaking, suction added molecules that are not sufficient to boost the heat energy at the wall; see Figure 10.9. Because of this, suction strongly affects the distribution of coolness at the wall, and injection spreads the coolness near the free stream (Figure 10.10). The corresponding effect of all these variations on the heat transfer rate can be deduced from Table 10.3. It is observed that the heat transfer rate reduces across the domain for suction and injection. However, the maximum heat transfer rate of 1.4143 is ascertained at the leading edge when $S_t = 0$ and $f_w = +3$.

10.4 Fluid Flow along a Vertical Thermally Stratified Surface

The following step is to investigate the same transport phenomena on a vertical surface caused by mixed convection, as shown in Figure 10.11. For this case, Eq. (10.3) is now replaced with the new momentum equation for the boundary layer flow:

$$uu_x + vu_y = -\frac{1}{\rho}\frac{dp}{dx} + \frac{\mu}{\rho}u_{yy} + g\beta(T - T_\infty). \tag{10.20}$$

For this case, the buoyancy parameter is defined as

$$Gr_x = \frac{gx\beta(T_w - T_o)}{U_\infty^2}$$

The boundary value ordinary differential equation that represents the dimensionless governing equation is

$$\frac{d^3 f}{d\eta^3} + \frac{1}{2}f\frac{d^2 f}{d\eta^2} + 1 + Gr_x\theta = 0, \tag{10.21}$$

$$\frac{d^2\theta}{d\eta^2} - S_t P_r\frac{df}{d\eta} - P_r\theta\frac{df}{d\eta} + \frac{1}{2}P_r f\frac{d\theta}{d\eta} = 0 \tag{10.22}$$

Dimensionless boundary conditions are

$$\frac{df}{d\eta} = 1, \quad f = f_w, \quad \theta = 1 - S_t \quad at \quad \eta = 0. \tag{10.23}$$

$$\frac{df}{d\eta} \to 0, \quad \theta \to 0 \quad as \quad \eta \to \infty. \tag{10.24}$$

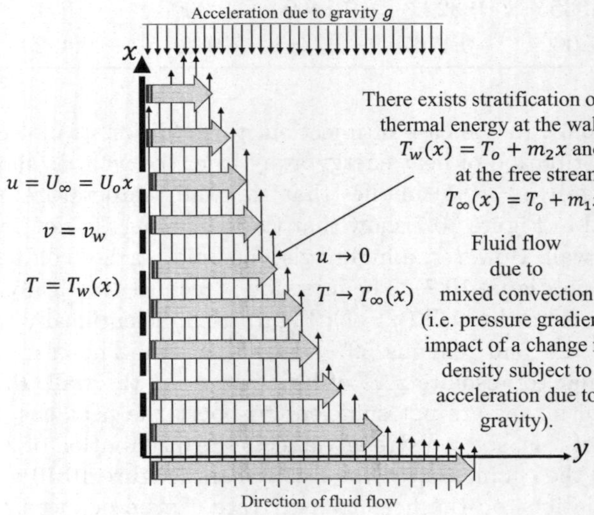

FIGURE 10.11
Graphical Illustration of the Motion on a Vertical Surface Subject to Thermal Stratification.

10.4.1 Research Questions II

The purpose of this subsection is to address the following research questions:

1. What impact does raising the Prandtl number have on the local skin friction coefficients and the Nusselt number when thermal stratification is at the epilimnion, metalimnion, and hypolimnion stages in the case of either suction or injection at the vertical surface?
2. What influence does the Prandtl number have on the transport phenomena and temperature distribution when the thermal stratification is at the epilimnion, metalimnion, and hypolimnion stages?

10.4.2 Analysis and Discussion of Results II

Equations (10.21) and (10.22) subject to Eqs. (10.23) and (10.24) were solved using the four-stage Lobatto IIIa formula as described in Chapter 1 using $\eta = 5$. The results show that the temperature gradient is a constant function of Prandtl number a small distance away from the wall when $S_t = 0$, $S_t = 0.5$, and $S_t = 1$ in the presence of either suction or injection. These results are not illustrated for brevity. As shown in Table 10.4, it is worth concluding that at all the levels of thermal stratification when buoyancy force is small and large in magnitude, the heat transfer rate $-\theta'(\eta)$ decreases within the thin layers of air near the wall ($0 \leq \eta \leq 1$) at different rates. Further exploration shows that in the case of suction, with an increase in the Grashof number by 400%, the percentage decrease in heat transfer rate is -5.2% when $S_t = 0$, but the percentage increase in heat transfer rate ascertained is 44.80% when $S_t = 1$. Such an occurrence is an indication that there exists a strong relationship between thermal stratification and thermal buoyancy. Scientifically,

TABLE 10.4
Variation in Heat Transfer Rate $-\theta'(\eta)$ Near the Vertical Wall ($0 \leq \eta \leq 1$) for the Case of a Less Viscous Fluid Like Air ($P_r = 0.71$) When $G_{rx} = 1$ and $G_{rx} = 5$

	$S_t = 0$	$S_t = 0$	$S_t = 0.5$	$S_t = 0.5$	$S_t = 1$	$S_t = 1$
$G_{rx} = 1$	$f_w = +3$	$f_w = -3$	$f_w = +3$	$f_w = -3$	$f_w = +3$	$f_w = -3$
η	$-\theta'(\eta)$	$-\theta'(\eta)$	$-\theta'(\eta)$	$-\theta'(\eta)$	$-\theta'(\eta)$	$-\theta'(\eta)$
0	1.7632	0.6705	1.4645	0.6222	1.1438	0.5677
0.25	1.1860	0.6527	0.9782	0.5973	0.7082	0.5072
0.5	0.7651	0.6055	0.5653	0.5499	0.4159	0.4750
0.75	0.4785	0.5229	0.3418	0.4855	0.2026	0.4261
1	0.2919	0.4436	0.1731	0.4101	0.0580	0.3460
S_{lp}	−1.4600	−0.2334	−1.2876	−0.2144	−1.0708	−0.2098
	$S_t = 0$	$S_t = 0$	$S_t = 0.5$	$S_t = 0.5$	$S_t = 1$	$S_t = 1$
$G_{rx} = 5$	$f_w = +3$	$f_w = -3$	$f_w = +3$	$f_w = -3$	$f_w = +3$	$f_w = -3$
η	$-\theta'(\eta)$	$-\theta'(\eta)$	$-\theta'(\eta)$	$-\theta'(\eta)$	$-\theta'(\eta)$	$-\theta'(\eta)$
0	1.8295	0.7804	1.3469	0.6177	0.6531	0.4202
0.25	1.2029	0.7621	0.8792	0.5689	0.3658	0.3475
0.5	0.7676	0.6782	0.4972	0.5128	0.1945	0.2706
0.75	0.4670	0.5748	0.3268	0.4415	0.1087	0.2079
1	0.2773	0.4278	0.1582	0.3425	0.0429	0.1320
S_{lp}	−1.5361	−0.357	−1.1719	−0.2711	−0.5910	−0.2864

this fact is true because both of them are dependent on reference temperature T_o and the direction of thermal stratification is perpendicular to the gravity g (Figure 10.11).

As it is expected, the motion of the fluid along the y-direction and the x-direction decreases with the enhancement in the Prandtl number due to the corresponding increase in the momentum diffusivity; see Figures 10.12–10.27. When $S_t = 0$, the maximum wall temperature is seen to also affect the vertical velocity as shown in Figures 10.12 and 10.14. Suction boosts the vertical velocity, while minimum vertical velocity is seen in the case of injection because the occurrence of both suction and injection is in the same direction with the flow (y-direction). The observed patterns when $S_t = 1$ imply that the wall temperature is zero; see Figures 10.13 and 10.15. In these cases, the influence of buoyancy is minimized

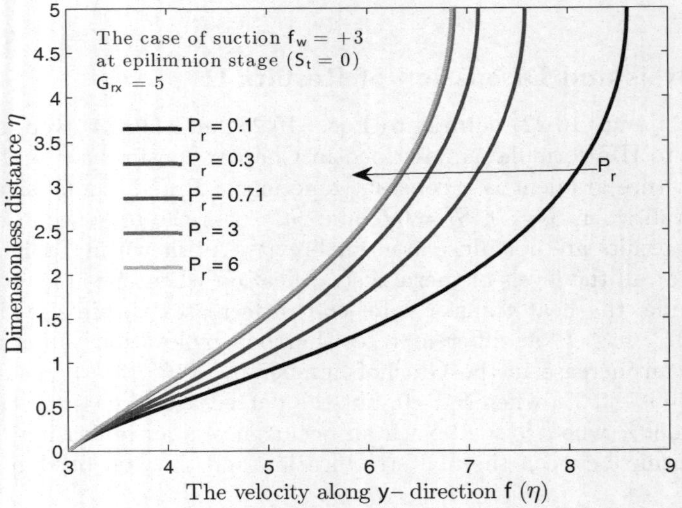

FIGURE 10.12
Variation in $f(\eta)$ due to P_r when $f_w = +3$ and $S_t = 0$.

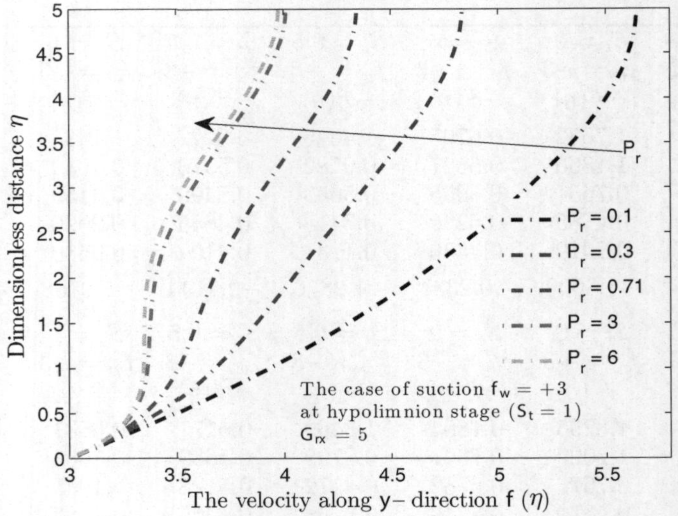

FIGURE 10.13
Variation in $f(\eta)$ due to P_r when $f_w = +3$ and $S_t = 1$.

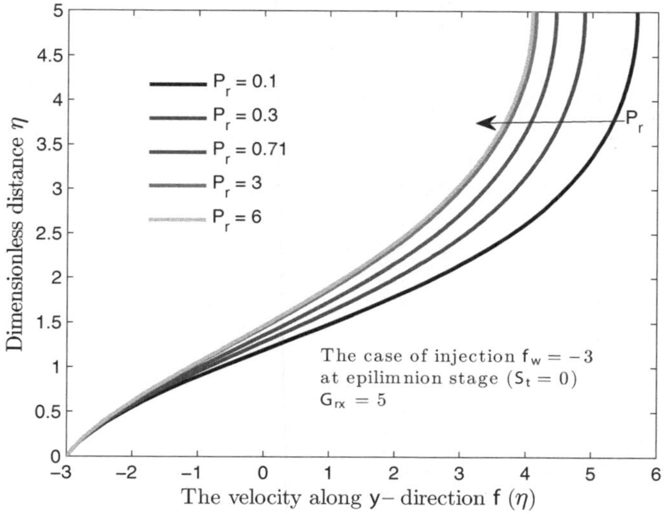

FIGURE 10.14
Variation in $f(\eta)$ due to P_r when $f_w = -3$ and $S_t = 0$.

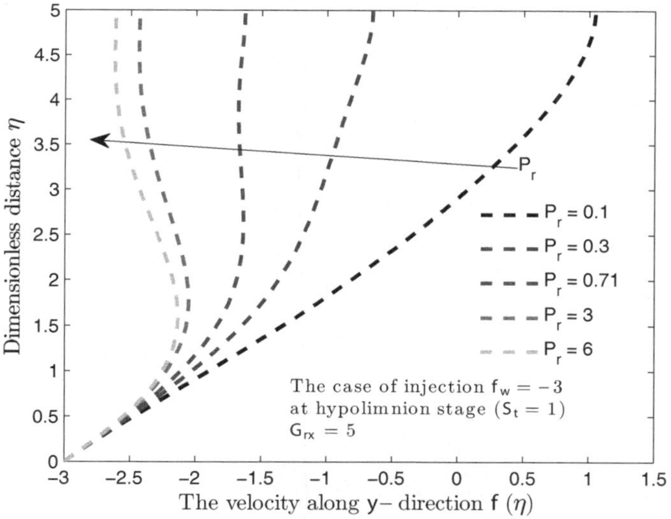

FIGURE 10.15
Variation in $f(\eta)$ due to P_r when $f_w = -3$ and $S_t = 1$.

because the process is dependent on the fact that the wall temperature is sufficiently larger than the free stream temperature or the fluid's temperature. Because of this, the overall vertical velocity for the hypolimnion stage ($S_t = 1$) is minimized compared to that of the epilimnion stage ($S_t = 0$), where the buoyancy force experiences no hindrance. When $S_t = 1$ in the case of injection (Figure 10.15), it is obvious that there exists a significant difference between the vertical velocities of more viscous fluids ($P_r = 3$ and $P_r = 6$) due to deceleration compared to those of extremely less viscous fluids ($P_r = 0.1$) that overcome the influence of injection after $\eta = 2.5$. This further implies that injection is capable of decelerating the motion along the y-direction, but this can be accelerated at some distance away from the wall $\eta = 1$ with the introduction of $S_t = 0$.

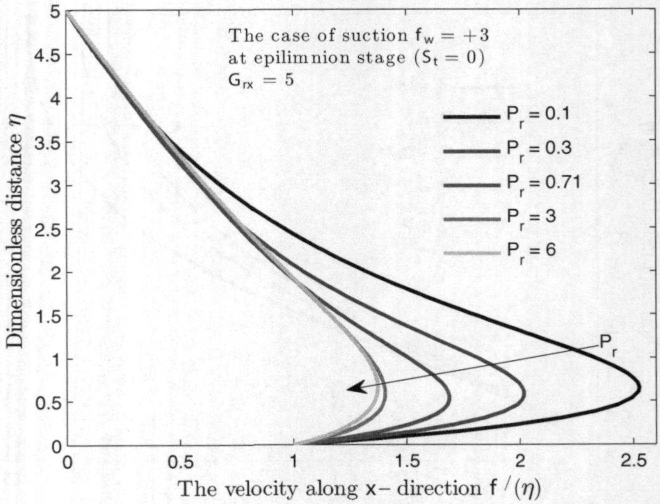

FIGURE 10.16
Variation in $f'(\eta)$ due to P_r when $f_w = +3$ and $S_t = 0$.

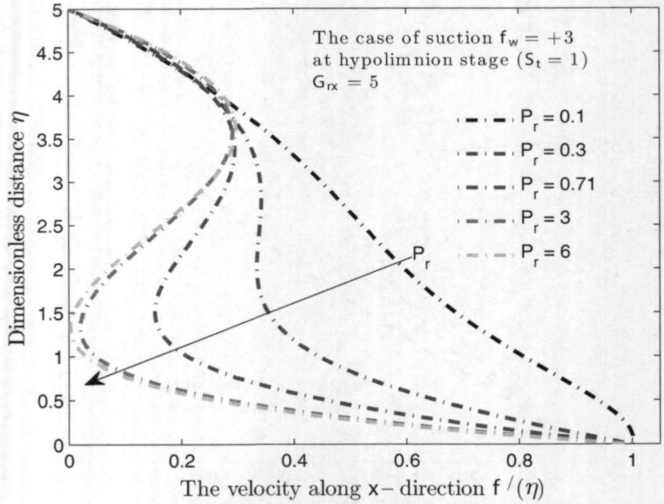

FIGURE 10.17
Variation in $f'(\eta)$ due to P_r when $f_w = +3$ and $S_t = 1$.

This is also applicable to the variations in the velocity along x-direction as illustrated in Figures 10.16–10.19. However, the stable solutions but of different patterns illustrated in Figures 10.17 and 10.19 could be traced to the fact that injection causes deceleration, setting the wall temperature to be zero ($S_t = 0$). Figures 10.20–10.23 present the combined effect of suction and injection on the shear stress proportional to friction across the flow at the epilimnion stage ($S_t = 0$) and hypolimnion stage ($S_t = 1$). The effects of Prandtl number on the temperature distribution when $S_t = 0$ and $S_t = 1$ on the transport phenomenon subject to suction and injection are illustrated in Figures 10.24–10.27. When the wall temperature is zero ($T_w(x) = 0$), both suction and injection are seen to drastically make the temperature distribution to vary at different levels of coldness.

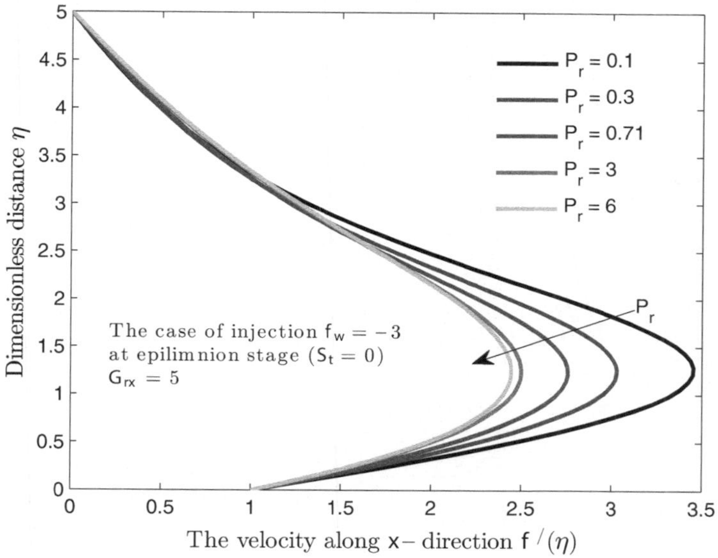

FIGURE 10.18
Variation in $f'(\eta)$ due to P_r when $f_w = -3$ and $S_t = 0$.

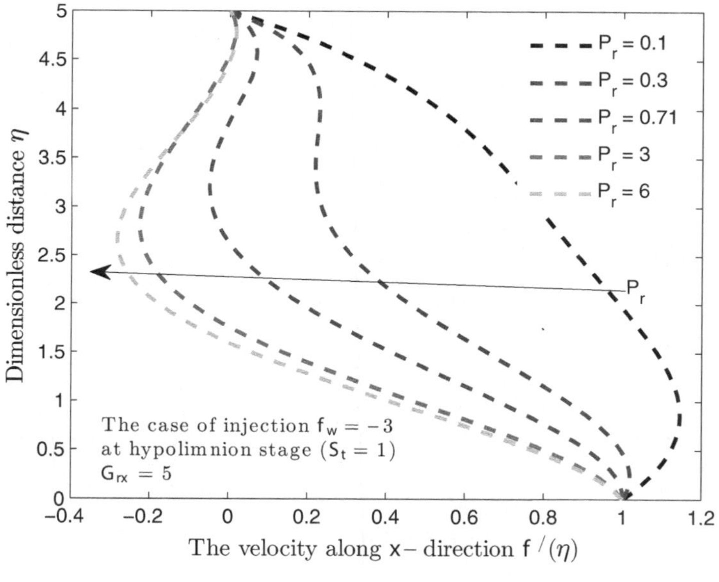

FIGURE 10.19
Variation in $f'(\eta)$ due to P_r when $f_w = -3$ and $S_t = 1$.

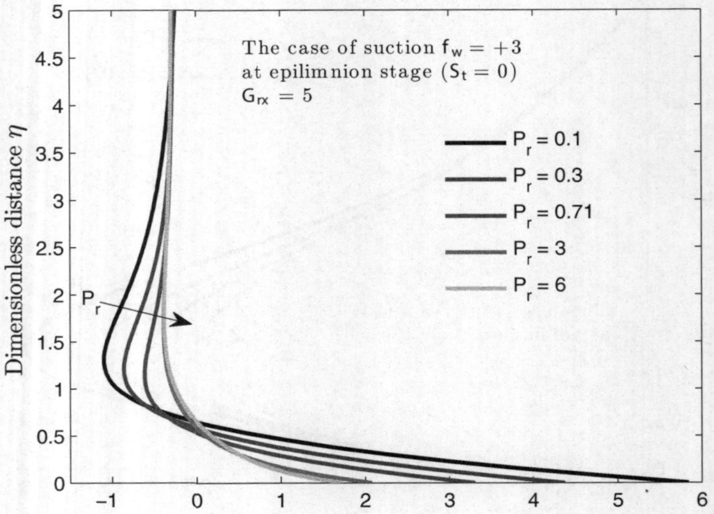

FIGURE 10.20
Variation in $f''(\eta)$ due to P_r when $f_w = +3$ and $S_t = 0$.

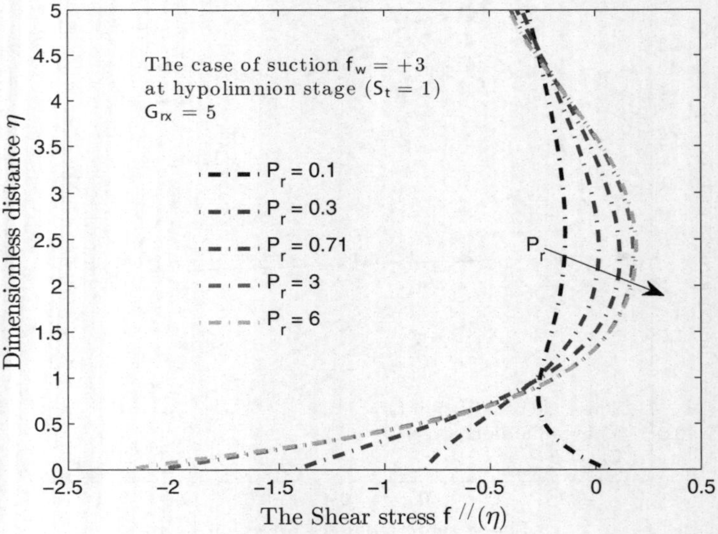

FIGURE 10.21
Variation in $f''(\eta)$ due to P_r when $f_w = +3$ and $S_t = 1$.

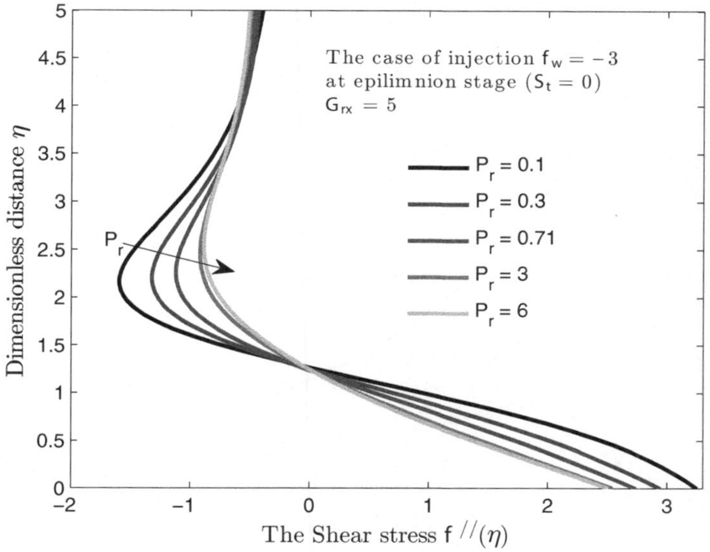

FIGURE 10.22
Variation in $f''(\eta)$ due to P_r when $f_w = -3$ and $S_t = 0$.

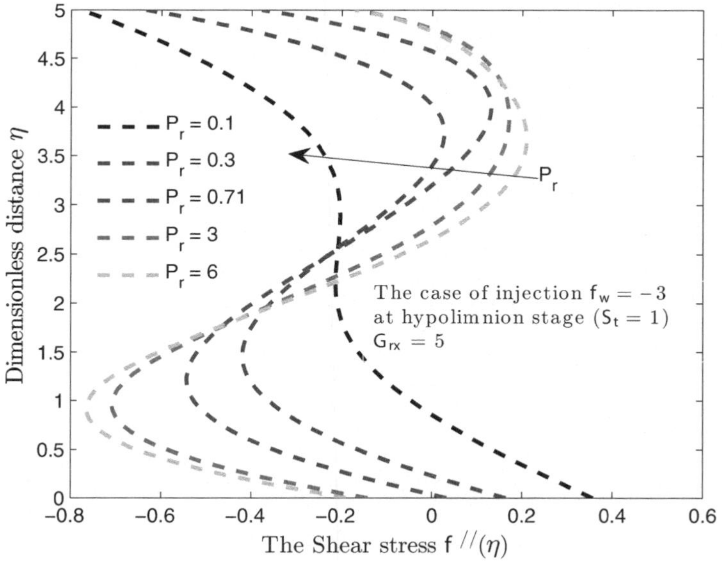

FIGURE 10.23
Variation in $f''(\eta)$ due to P_r when $f_w = -3$ and $S_t = 1$.

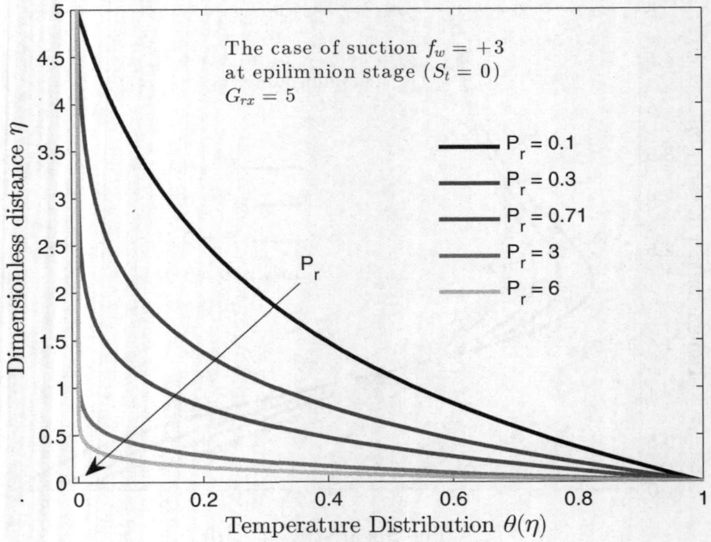

FIGURE 10.24
Variation in $\theta(\eta)$ due to P_r when $f_w = +3$ and $S_t = 0$.

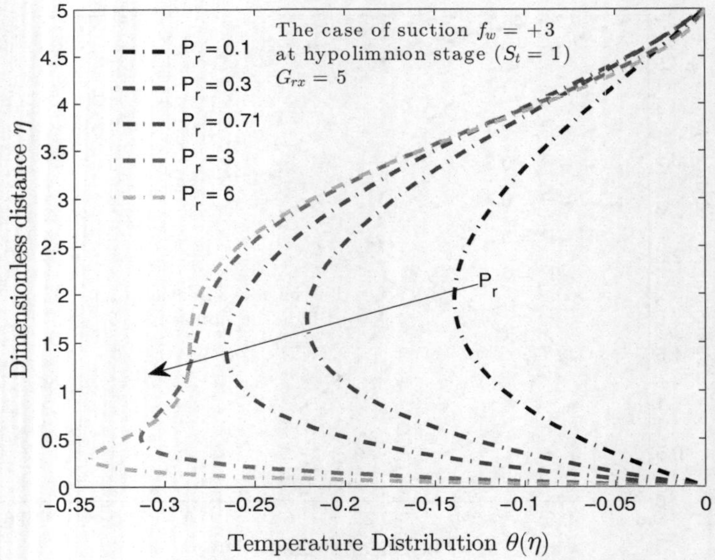

FIGURE 10.25
Variation in $\theta(\eta)$ due to P_r when $f_w = +3$ and $S_t = 1$.

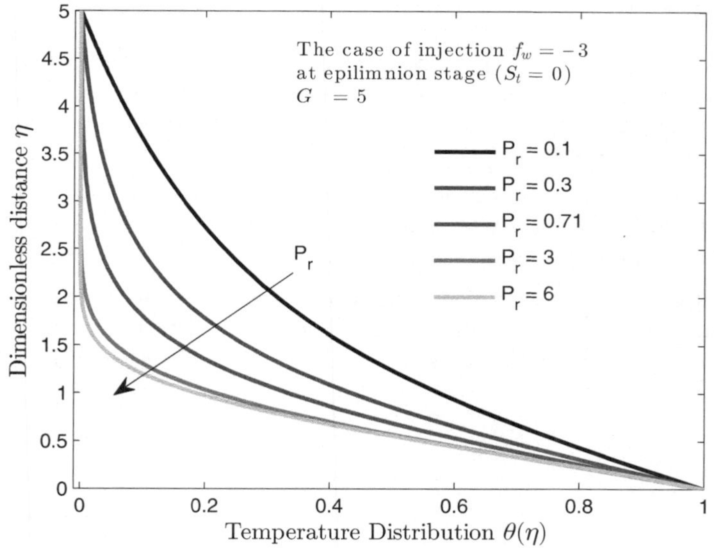

FIGURE 10.26
Variation in $\theta(\eta)$ due to P_r when $f_w = -3$ and $S_t = 0$.

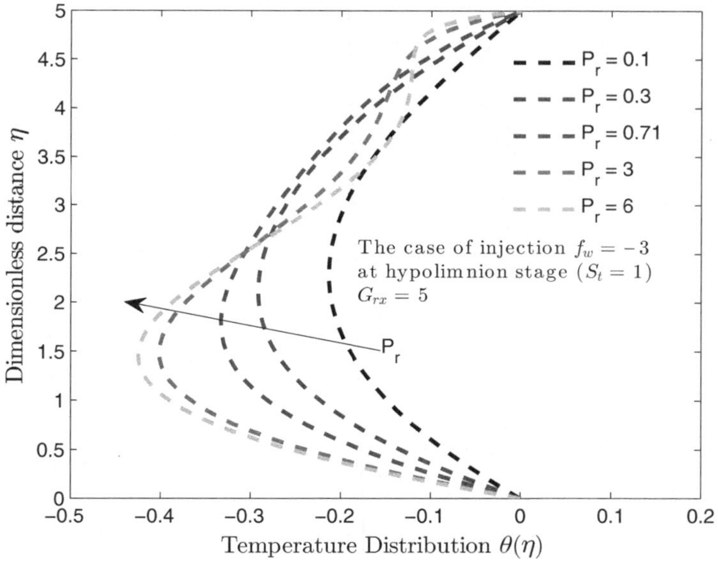

FIGURE 10.27
Variation in $\theta(\eta)$ due to P_r when $f_w = -3$ and $S_t = 1$.

10.5 Tutorial Questions

1. What is the impact of raising the Prandtl number on the local skin friction coefficients and Nusselt number when thermal stratification is at the epilimnion, metalimnion, and hypolimnion stages in the presence of suction or injection at the horizontal surface?
2. What are the implications of Prandtl number on the transport phenomena and temperature distribution when thermal stratification occurs at the epilimnion, metalimnion, and hypolimnion stages?
3. What impact does raising the Prandtl number have on the local skin friction coefficients and Nusselt number when thermal stratification is at the epilimnion, metalimnion, and hypolimnion stages in the case of either suction or injection at the vertical surface?

11

Analysis of Self-Similar Flows VII

11.1 Background Information

Thermo-migration of tiny/nanosized particles is known as thermophoresis. The movement of either tiny or nanoparticles from the region of a hot to a cold environment along the path of a temperature gradient is an unavoidable event during the colloidal crystal growth processes, deposition of aerosol particles across flowing fluids, determination of exhaust gas particle trajectories from combustion devices, removal of small particles from gas streams, and separation of micro-particles (Green and Lane [115]; Fuchs [102]). The dynamics of fluids through a porous medium is another focus of this chapter. This chapter presents the significance of the Prandtl number on the boundary layer flow when inertial effects due to additional friction for high-velocity flow are significant.

11.2 Introduction: Thermophoresis and Brownian Motion of Particles

When tiny particles in stagnant or flowing fluids possess a temperature gradient, a special force is exerted on the particles due to imbalanced forces related to molecular collisions from different regions with different temperatures. This force is common in cold regions and is often encountered with aero-colloidal particles rather than colloidal particles. The phenomenon tends to be small in liquids when compared to convection and other forces. The occurrence of this phenomenon can be traced to the collision of molecules on one side of the particles that possess different average velocities from those on the other side due to the temperature gradient (Sheikholeslami and Rokni [281], Grosan et al. [116], Eslamian et al. [97], Bahiraei and Hosseinalipour [50], Wakif et al. [309], and Qayyum et al. [238, 239]). The outcome of the meta-analysis by Wakif et al. [309] indicates that skin friction coefficients are a decreasing function of thermophoresis. Also, an increase in thermophoretic deposition is achievable due to an increase in thermophoresis. Thermal radiation, in particular, has a significant impact on the importance of small particle thermo-migration in fluid movement. The thermophoretic force was discovered by Ramachandran et al. [244] as a variable proportional to the ratio of dynamic viscosity to density of liquid substances and the temperature gradient. At more significant temperature gradients, atomic impulses in the hotter region of liquid substances are more vital.

The haphazard motion of tiny/nanosized particles is known as Brownian motion. According to Carus et al. [73], the mixing motion of dust particles in sunbeams has been linked to a substantial change in temperature and/or pressure. In 1785, the irregular movement of tiny coal dust on the surface of alcohol was observed by Dutch physiologist, biologist, and chemist Jan Ingenhousz, who also started the broad work on photosynthesis. In 1827, the Scottish botanist Robert Brown observed the random motion of tiny particles

DOI: 10.1201/9781003217374-11

suspended in gases or liquids. In the report of Einstein et al. [1], the kinetic theory was adopted to explain the Brownian motion, which is one of the phenomena that prove the existence of molecules. According to the kinetic theory of gases, gas molecules move randomly (which may be compared to disorder or zigzag motion), colliding with one another and the container walls. An internal pressure is generated due to this collision, and its amount is determined by the size and speed of the molecules.

11.3 Fluid Flow Subject to Brownian Motion and Thermophoresis of Tiny Particles due to Only Thermal Free Convection

Figure 11.1 depicts the transport phenomena of a typical Newtonian fluid over a vertical surface owing to free convection, where the thermo-migration of tiny/nanosized particles and random motion of tiny/nanosized particles are prominent. The two-dimensional flow along the x-direction with velocity u and the y-direction with velocity v was generated only by thermal buoyancy forces, as illustrated in Figure 11.1. Such an occurrence indicates that the buoyancy induced by concentration variations is believed to be minimal. In this example, mass diffusivity is denoted by D_a, while D_T denotes thermophoresis coefficient. The governing equation suitable to model the case is of the form

$$u_x + v_y = 0, \tag{11.1}$$

$$uu_x + vu_y = \frac{\mu}{\rho}u_{yy} + g\beta(T - T_\infty), \tag{11.2}$$

$$uT_x + vT_y = \frac{\kappa}{\rho c_p}T_{yy} + \tau\left[\frac{D_a}{\Delta C}C_yT_y + \frac{D_T}{T_\infty}T_yT_y\right], \tag{11.3}$$

$$uC_x + vC_y = D_aC_{yy} + \frac{D_T\Delta C}{T_\infty}T_{yy}. \tag{11.4}$$

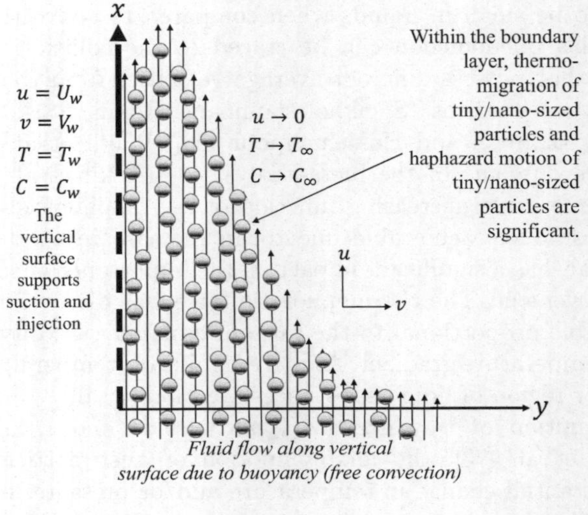

FIGURE 11.1
Fluid flow with Brownian motion and thermophoresis of nanoparticles owing to free convection.

subject to the boundary conditions

$$u = U_w, \quad v = v_w, \quad T = T_w, \quad C = C_w \quad at \quad y = 0. \tag{11.5}$$

$$u \to 0, \quad T \to T_\infty, \quad C \to C_\infty \quad as \quad y \to \infty. \tag{11.6}$$

The self-similar solution of Eqs. (11.1)–(11.4) subject to Eqs. (11.5) and (11.6) was obtained using

$$\eta = y\sqrt{\frac{U_\infty}{\vartheta x}}, \quad \psi(x,y) = \sqrt{\vartheta x U_\infty} f(\eta), \quad u = \frac{\partial \psi}{\partial y}, \quad v = -\frac{\partial \psi}{\partial x},$$

$$\theta(\eta) = \frac{T - T_\infty}{T_w - T_\infty}, \quad Gr_x = \frac{gx\beta(T_w - T_\infty)}{U_\infty^2}, \quad \phi(\eta) = \frac{C - C_\infty}{C_w - C_\infty}, \quad P_r = \frac{\mu c_p}{\kappa} = \frac{\vartheta}{\alpha},$$

$$S_c = \frac{\vartheta}{D_a}, \quad N_b = \frac{\tau D_a (C_w - C_\infty)}{\alpha \Delta C}, \quad N_t = \frac{\tau D_T}{T_\infty \alpha}(T_w - T_\infty), \quad f_w = -\frac{v_w}{\sqrt{\vartheta U_o}}. \tag{11.7}$$

The final dimensionless governing equation is

$$\frac{d^3 f}{d\eta^3} + \frac{1}{2} f \frac{d^2 f}{d\eta^2} + Gr_x \theta = 0, \tag{11.8}$$

$$\frac{d^2\theta}{d\eta^2} + \frac{1}{2} P_r f \frac{d\theta}{d\eta} + N_b \frac{d\theta}{d\eta}\frac{d\phi}{d\eta} + N_t \frac{d\theta}{d\eta}\frac{d\theta}{d\eta} = 0, \tag{11.9}$$

$$\frac{d^2\phi}{d\eta^2} + \frac{1}{2} S_c f \frac{d\phi}{d\eta} + \frac{N_t}{N_b}\frac{d^2\theta}{d\eta^2} = 0. \tag{11.10}$$

Dimensionless boundary conditions for Eqs. (11.5) and (11.6) are

$$\frac{df}{d\eta} = 1, \quad f = f_w, \quad \theta = 1, \quad \phi = 1 \quad at \quad \eta = 0. \tag{11.11}$$

$$\frac{df}{d\eta} \to 0, \quad \theta \to 0, \quad \phi \to 0, \quad as \quad \eta \to \infty. \tag{11.12}$$

11.3.1 Research Questions I

The goals of this case are to answer the following research questions:

1. What is the rising influence of Prandtl number on the transport phenomena in the presence of suction ($f_w = +3$) when the random motion of tiny/nanosized particles is insignificant and significant?
2. How does the Prandtl number impact the transport phenomena of a typical Newtonian fluid when thermo-migration of tiny/nanosized particles is minimal and substantial in the presence of suction ($f_w = +3$)?
3. What is the relevance of the Prandtl number on the transport phenomena in the situation of injection ($f_w = -3$) where the random motion of tiny/nanosized particles is both insignificant and significant?
4. What impact does the rising Prandtl number have on the transport phenomena of a typical Newtonian fluid where the thermo-migration of tiny/nanosized particles is minimal and substantial in the injection ($f_w = -3$)?

11.3.2 Analysis and Discussion of Results I

The numerical solution of dimensionless governing Eqs. (11.8)–(11.10) subject to Eqs. (11.11) and (11.12) was obtained using the four-stage Lobatto IIIa formula as described in Chapter 1 using $\eta = 5$. First, it is seen that increasing effects of the ratio of momentum diffusivity to thermal diffusivity on all the properties of the transport phenomenon can be influenced by the haphazard motion of tiny particles N_b, buoyancy forces G_{rx}, suction/injection $f_w = \pm 3$, and migration of suspended tiny particles through the fluid due to temperature gradient N_t. When $f_w = +3$, $S_c = 0.62$, $G_{rx} = 1$, $N_t = 0.1$, the haphazard motion of tiny particles is small and large in magnitude ($N_b = 0.1$ and $N_b = 10$). Figures 11.2 and 11.3 reveal that the velocity along the x-direction decreases significantly only for viscous fluid ($P_r = 0.1, 0.3$, and 0.71) with no effect on $f(\eta)$ when $N_b = 0.1$ and minor decreasing effect on $f(\eta)$ when $N_b = 10$. However, when the haphazard motion of tiny particles and migration of suspended tiny particles through the fluid due to temperature gradient are considerably large (i.e., $N_t = N_b = 5$), the observed decrease in high viscous fluids decreases slightly significantly, but moderately significantly when $G_{rx} = 5$ (Figures 11.14 and 11.15). It is worth deducing from Figures 11.4–11.17 that it may appear as if the Prandtl number has no effect on the horizontal velocity of highly viscous fluids when $N_b = 0.1$. When $N_b = 10$, the velocity decreases with the Prandtl number but with a slight difference in $f'(\eta)$ when $P_r = 6$ and $P_r = 10$.

As shown in Table (11.1), $S_{l}p$ for $f'(\eta)$ with η, a number that quantifies the rate of change in velocity along the x-direction within the domain $[0, 5]$, was estimated as -0.087815391 for $P_r = 6$ and -0.09877062 for $P_r = 10$. It is worth deducing from Figure 11.17 that the velocity of less viscous fluids ($P_r = 6$) could be made to be significantly higher than that of a more viscous fluid ($P_r = 10$) when buoyancy forces, migration of suspended tiny particles through the fluid due to temperature gradient, and haphazard motion of tiny particles are large. The results of this first case illustrated in Figures (11.6)–(11.19) further show that the shear stress which is a property directly proportional to the friction across the flow, increases with the Prandtl number at some distance away from the wall till the free

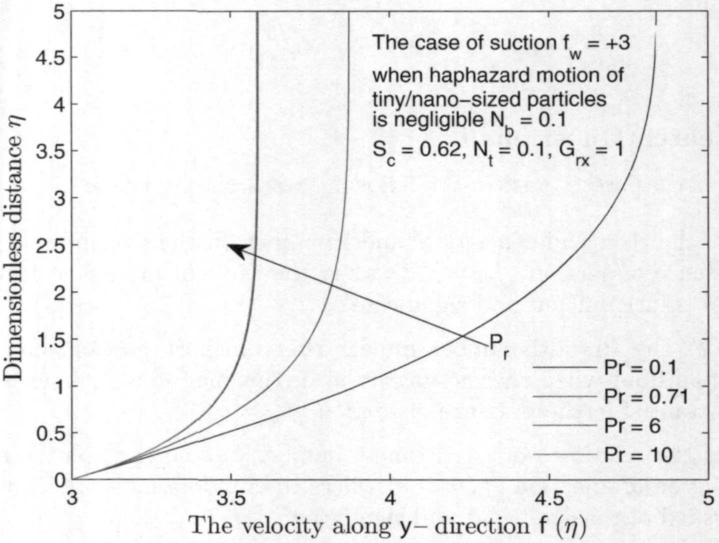

FIGURE 11.2
Variation in $f(\eta)$ due to P_r when $f_w = +3$ and $N_b = 0.1$.

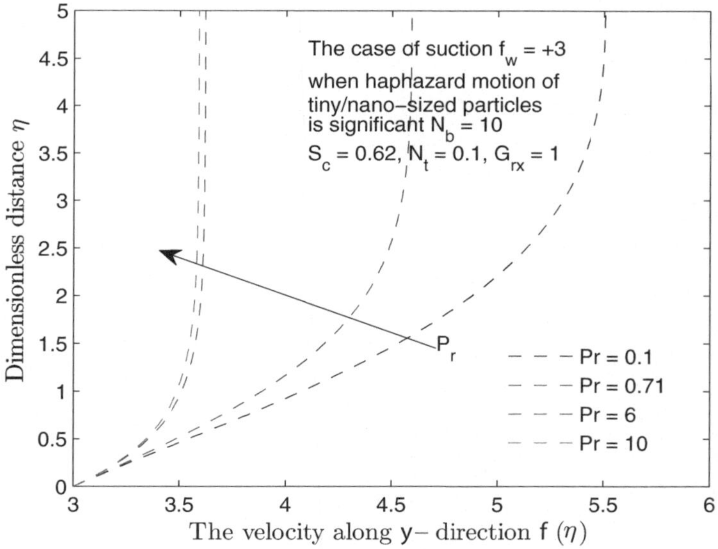

FIGURE 11.3
Variation in $f(\eta)$ due to P_r when $f_w = +3$ and $N_b = 10$.

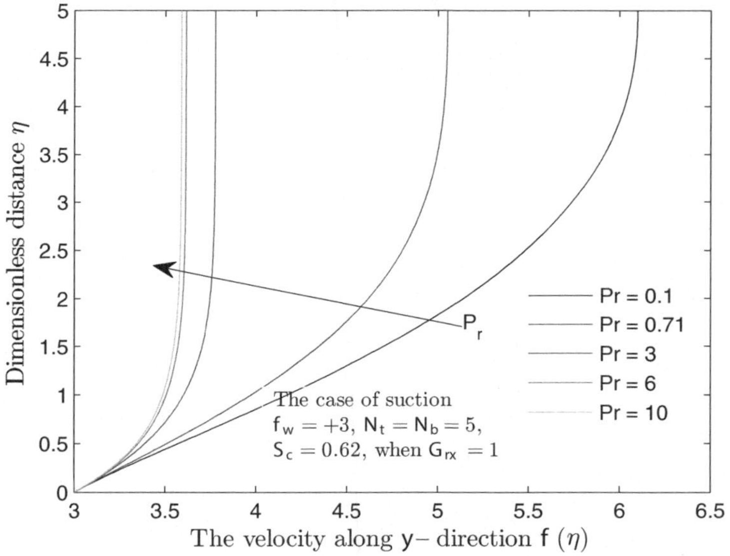

FIGURE 11.4
Variation in $f'(\eta)$ due to P_r when $f_w = +3$ and $N_b = 0.1$.

stream. Near the wall, most maximum and larger friction in the friction is achievable when $N_b = N_t = G_{rx} = 5$; see Figure 11.19. The shear stress proportional to friction across the domain is found to be a decreasing property of Prandtl number near the wall only. When the haphazard motion of tiny particles is large $N_b = 10$, such a decrease is influenced greatly for the dynamics of less viscous fluids; see Figures 11.8 and 11.9. As shown in Table 11.1, $S_l p$ for $f'(\eta)$ with η, a number that quantifies the rate of change in velocity along the x-direction

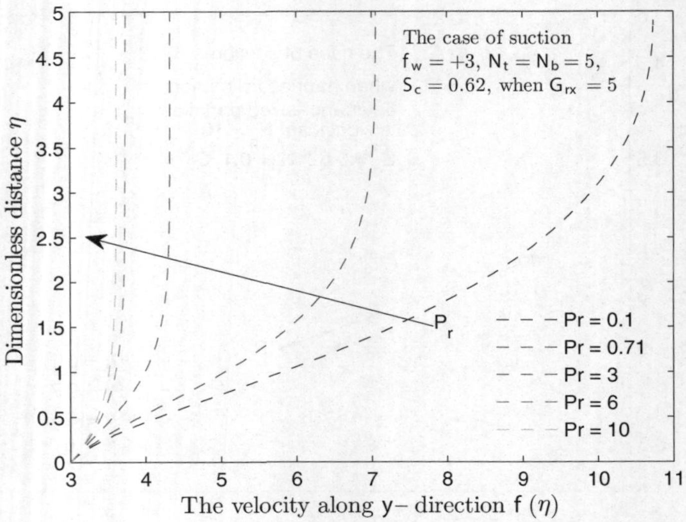

FIGURE 11.5
Variation in $f'(\eta)$ due to P_r when $f_w = +3$ and $N_b = 10$.

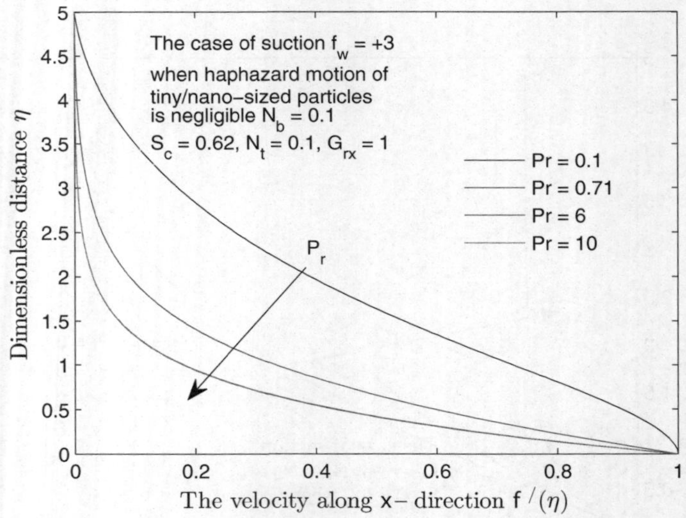

FIGURE 11.6
Variation in $f''(\eta)$ due to P_r when $f_w = +3$ and $N_b = 0.1$.

within the domain $\eta\varepsilon[0, 5]$, was estimated as -0.087815391 for $P_r = 6$ and -0.09877062 for $P_r = 10$. For further results on the combined effects of haphazard motion of tiny particles and migration of suspended tiny particles through the fluid due to temperature gradient, see Tables 11.2 and 11.3.

Figures 11.10–11.21 show that the temperature distribution decreases with the Prandtl number at different rates. It is obvious to note that an increase in the random mobility of small particles influences the pattern of temperature distribution reduction; see Figure 11.11. It is also observed in Figure 11.21 that when buoyancy forces, thermophoresis, and Brownian motion are high, the temperature distribution falls substantially. The variation in the

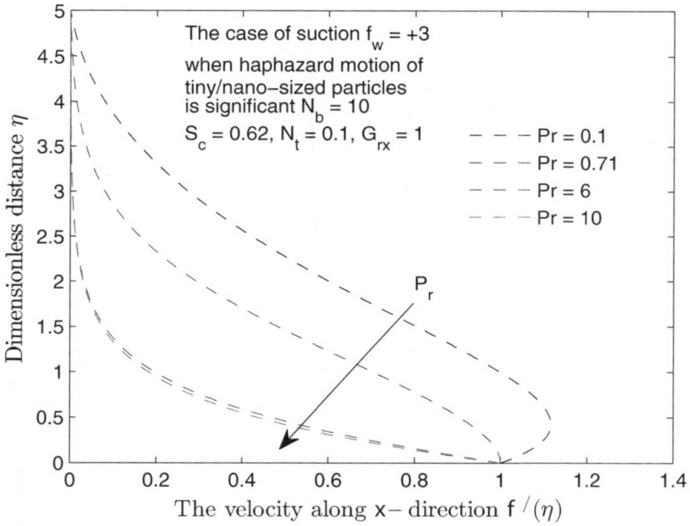

FIGURE 11.7
Variation in $f''(\eta)$ due to P_r when $f_w = +3$ and $N_b = 10$.

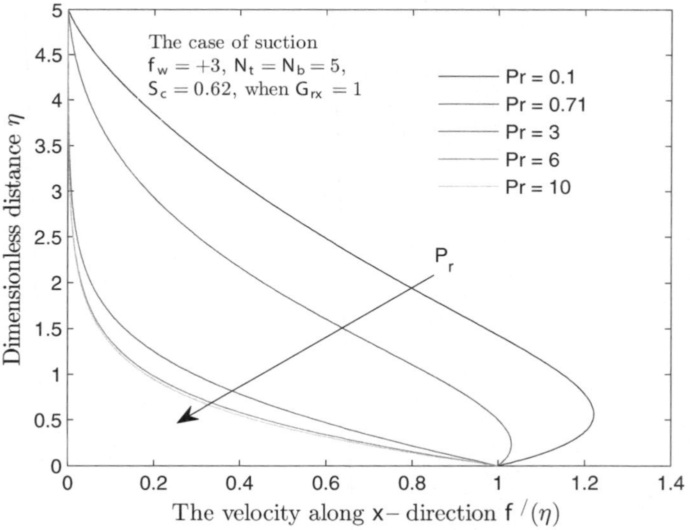

FIGURE 11.8
Variation in the friction across the flow at various levels of Prandtl number when $N_b = 0.1$: the case of Sakiadis flow and suction $f_w = +3$, $S_c = 0.62$, $N_t = 0.1$, $G_{rx} = 1$.

concentration of the fluid with the Prandtl number for less viscous fluids and moderate viscous fluids is presented in Figures 11.12–11.23. It is also observed that the concentration of a less viscous fluid ($P_r = 0.1$) is negative when thermophoresis, Brownian motion of tiny particles, and buoyancy forces are large (Figures 11.22–11.24). For the case of injection, analysis of Figures 11.25 and 11.26 shows that the velocity of the fluid along the y-direction is a constant function of Prandtl number near the wall, but decreases slightly near the free stream when buoyancy forces are small. When buoyancy force is large (i.e.,

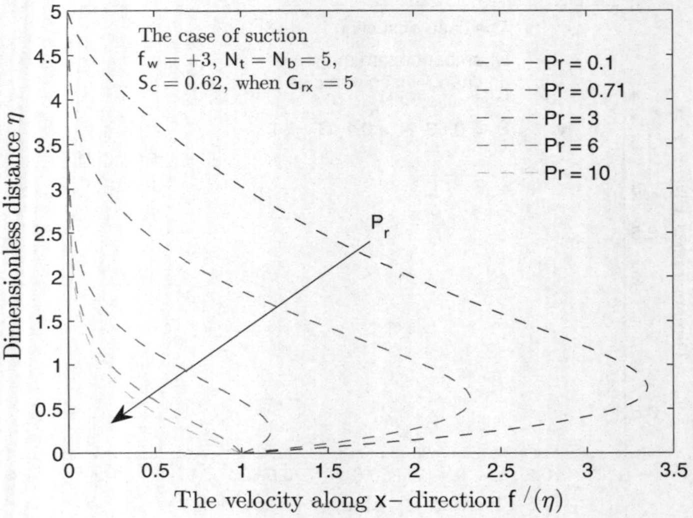

FIGURE 11.9
Variation in the friction across the flow at various levels of Prandtl number when $N_b = 10$: the case of Sakiadis flow and suction $f_w = +3$, $S_c = 0.62$, $N_t = 0.1$, $G_{rx} = 1$.

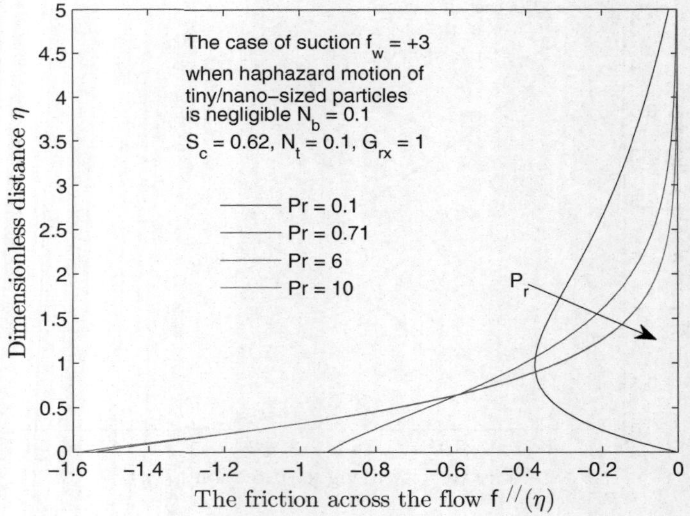

FIGURE 11.10
Variation in $\theta(\eta)$ due to P_r when $f_w = +3$ and $N_b = 0.1$.

$G_{rx} = 5$), the vertical velocity decreases more significantly with the Prandtl number. Figures 11.27 and 11.28 ascertain that the motion along the x-direction decreases with the Prandtl number. The observed decrease in the horizontal velocity with the Prandtl number is pronounced when buoyancy force is large (i.e., $G_{rx} = 5$). Maximum local skin friction coefficients are observed in the motion of a less fluid when buoyancy force, thermophoresis and Brownian motion of tiny particles are large ($G_{rx} = N_b = N_t = 5$); see Figures 11.29 and 11.30.

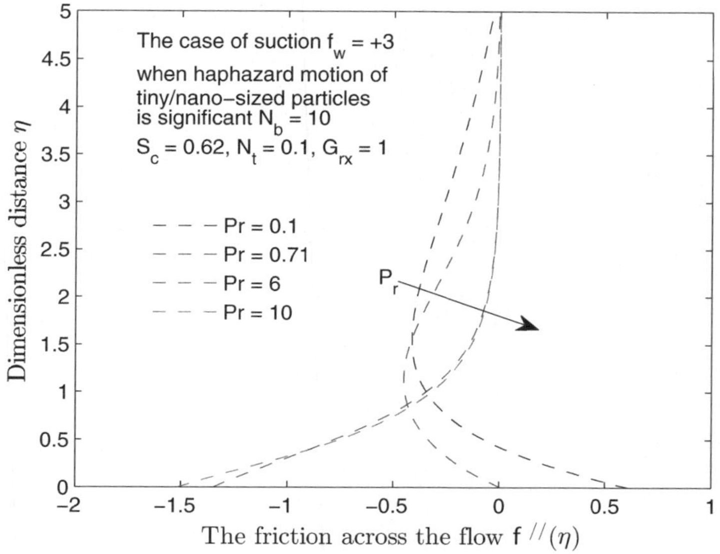

FIGURE 11.11
Variation in $\theta(\eta)$ due to P_r when $f_w = +3$ and $N_b = 10$.

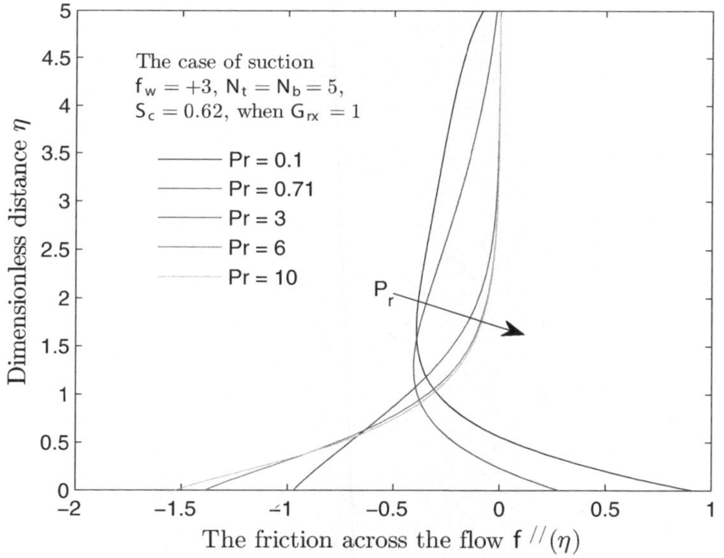

FIGURE 11.12
Variation in $\phi(\eta)$ due to P_r when $f_w = +3$ and $N_b = 0.1$.

For the case of injection, the temperature distribution diminishes when concentration enhances with the Prandtl number when buoyancy force is minimum (Figures 11.31 and 11.33). When buoyancy force is considerably large (i.e., $G_{rx} = 5$), the observed decrease in the temperature distribution and concentration occurs over a wider domain (Figures 11.32 and 11.34). When $N_b = 0.1$ and migration of suspended tiny particles through the fluid due to temperature gradient is more enhanced (i.e., $N_t = 5$), Table 11.4 reveals that the

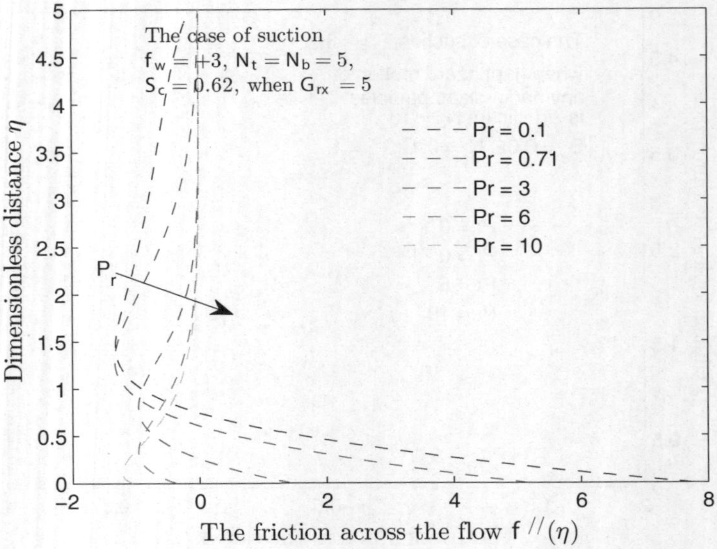

FIGURE 11.13
Variation in $\phi(\eta)$ due to P_r when $f_w = +3$ and $N_b = 10$.

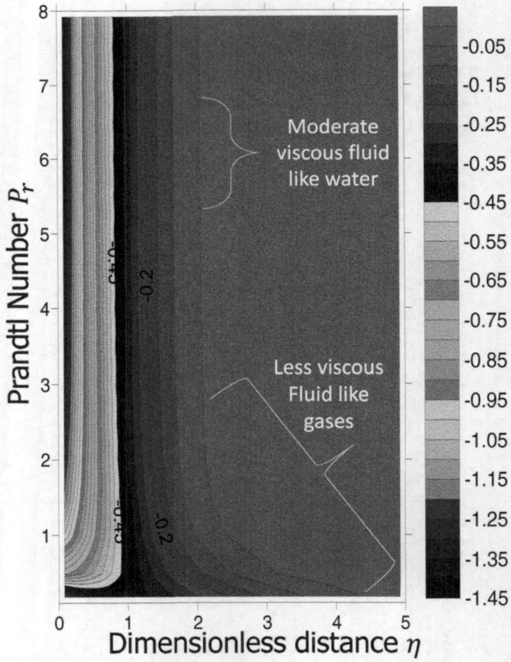

FIGURE 11.14
Variation in $f(\eta)$ due to P_r when $f_w = +3$ and $G_{rx} = 1$.

overall concentration of highly viscous fluids (i.e., $P_r = 6$ and $P_r = 10$) increases across the domain at the rate of 2.864676407 and 2.769319004, respectively, because of injection; see Figure 11.24. When $N_t = 5$ and $N_b = 5$, Table 11.5 shows that the local skin friction

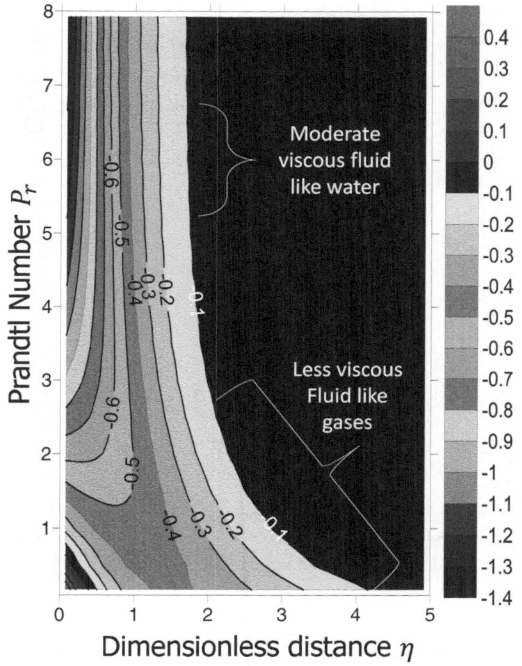

FIGURE 11.15
Variation in $f(\eta)$ due to P_r when $f_w = +3$ and $G_{rx} = 5$.

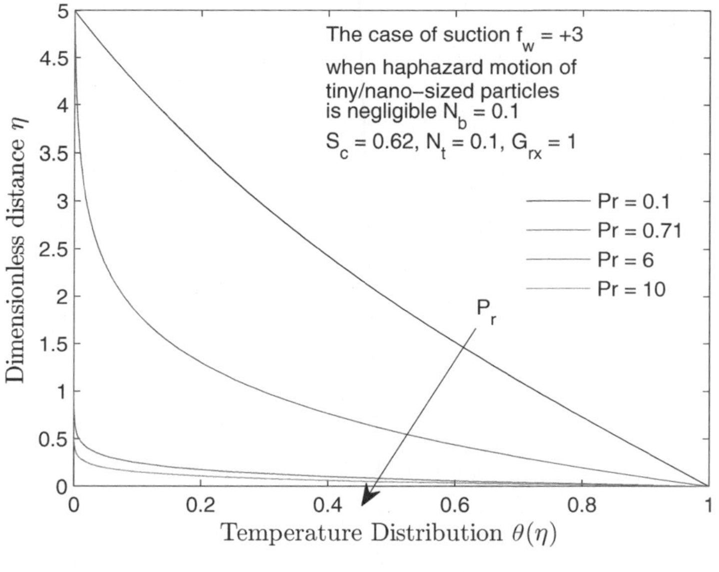

FIGURE 11.16
Variation in $f'(\eta)$ due to P_r when $f_w = +3$ and $G_{rx} = 1$.

coefficients decrease with the Prandtl number at the rate of -0.030966 for suction and at the rate of $-6.7E-05$ for injection when $G_{rx} = 0.1$. Reverse is the case when buoyancy is large as it is seen that the percentage decrease in local skin friction coefficient $f''(0)$ as f_w

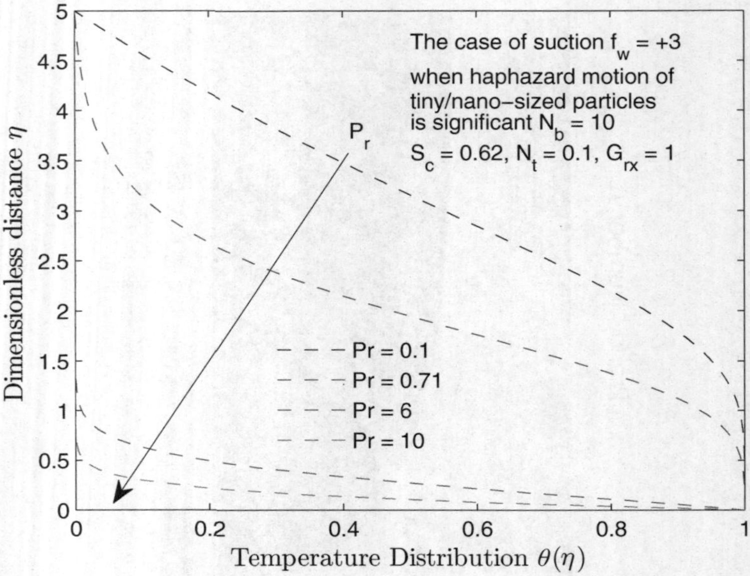

FIGURE 11.17
Variation in $f'(\eta)$ due to P_r when $f_w = +3$ and $G_{rx} = 5$.

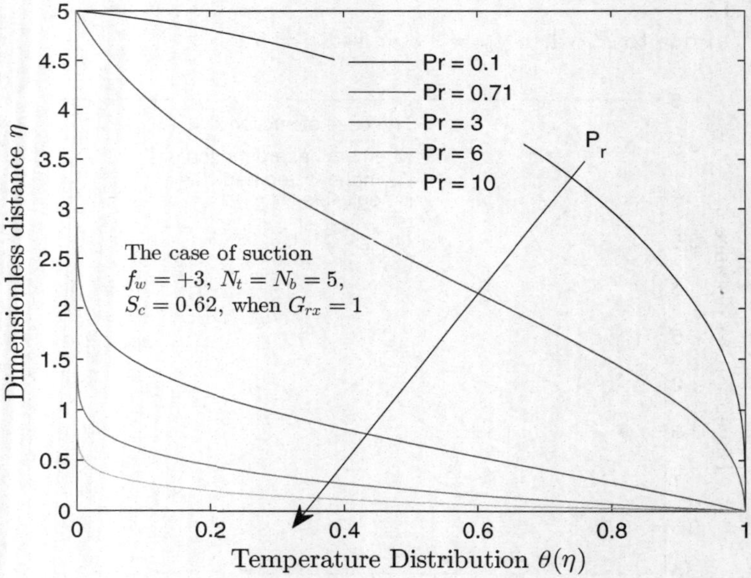

FIGURE 11.18
Variation in $f''(\eta)$ due to P_r when $f_w = +3$ and $G_{rx} = 1$.

moves from +3 to −3 is −89.71%. More so, when the haphazard motion of tiny particles and migration of suspended tiny particles through the fluid due to temperature gradient are more enhanced (i.e., $N_b = N_t = 5$), the variation in the observed mass transfer rate $-\phi'(0)$ with the Prandtl number is the most minimum when ($f_w = +3, G_{rx} = 10$) and maximum when ($f_w = -3, G_{rx} = 0.1$); see Table 11.5.

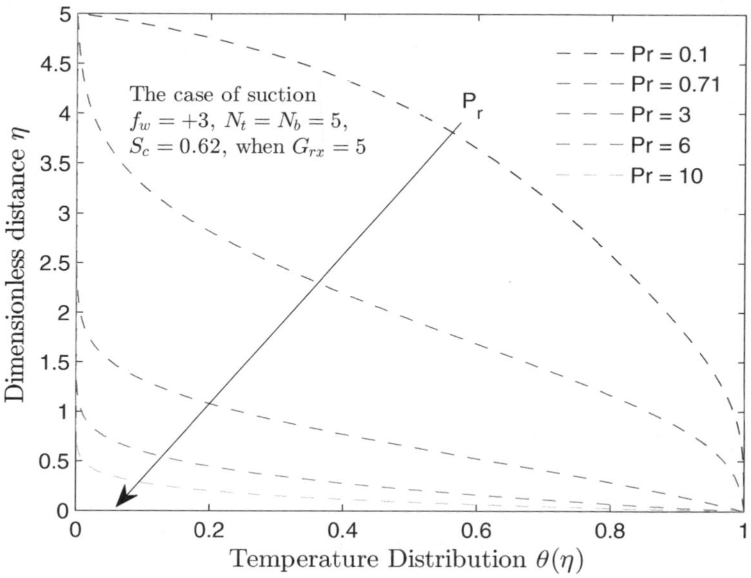

FIGURE 11.19
Variation in $f''(\eta)$ due to P_r when $f_w = +3$ and $G_{rx} = 5$.

TABLE 11.1
The Variation in $f'(\eta)$, $\theta(\eta)$, and $\phi(\eta)$ across the Domain η at Various Values of Prandtl Number When $f_w = +3$, $S_c = 0.62$, $N_t = 0.1$, and $G_{rx} = 1$

$N_b = 0.1$, $P_r \Rightarrow$	0.1	0.71	6	10
S_{lp} for $f'(\eta)$	−0.214838549	−0.112473639	−0.087815391	−0.09877062
S_{lp} for $\theta(\eta)$	−0.197829003	−0.103931279	−0.018706723	−0.013950748
S_{lp} for $\phi(\eta)$	−0.129288587	−0.189538426	−0.237366043	−0.262622126
$N_b = 10$, $P_r \Rightarrow$	0.1	0.71	6	10
S_{lp} for $f'(\eta)$	−0.262371999	−0.222389065	−0.091710238	−0.082916919
S_{lp} for $\theta(\eta)$	−0.228660738	−0.258265138	−0.05485638	−0.017205528
S_{lp} for $\phi(\eta)$	−0.132989614	−0.142255633	−0.121020374	−0.121161384

When there exists suction ($f_w = +3$), Table 11.6 indicates that the percentage decrease in the concentration of a highly viscous fluid ($P_r = 10$) within the domain under consideration as G_{rx} was increased from 0.1 to 10 is −10.15%. Meanwhile, in the case of injection ($f_w = -3$), Table 11.7 shows that when the Grashof number G_{rx} is raised from 0.1 to 10, the percentage reduction in the concentration of extremely viscous fluids ($P_r = 10$) from the wall to the free stream is −21.74%. These findings show a substantial difference in the fluctuation of extremely viscous fluid concentration within the region between suction and injection. For more information on the pattern of decrease in the velocity along the x-direction, temperature distribution, and concentration within the domain $\eta\varepsilon[0, 5]$ at various values of Prandtl number, see Table 11.7. The results obtained by Hayat et al. [123] indicate that the concentration field decreases with Brownian motion. The concentration of the fluid rises as the temperature rises in thermophoresis. The local Nusselt number was discovered to decrease with Brownian motion and thermophoresis. Brownian motion is observed to have a rising characteristic called the local Sherwood number. The link between the local Sherwood

TABLE 11.2
Effect of Prandtl Number P_r on Some of the Properties of the Flow When $S_c = 0.62$, $N_t = 0.1$, $G_{rx} = 1$

$f_w = -3$		$N_b = 0.1$			$N_b = 10$	
P_r	$f''(0)$	$-\theta'(0)$	$-\phi'(0)$	$f''(0)$	$-\theta'(0)$	$-\phi'(0)$
0.1	0.6177	0.1633	0.1253	0.7372	0.0000	0.1242
0.71	0.6262	0.0825	0.1838	0.7296	0.0001	0.1220
6	0.6500	0.0000	0.2092	0.6841	0.0000	0.1086
10	0.6515	0.0000	0.2043	0.6733	0.0000	0.1054
S_{lp}	0.003406	−0.014545	0.006141	−0.006670	$-5.3E-0.6$	−0.001964
$f_w = +3$		$N_b = 0.1$			$N_b = 10$	
P_r	$f''(0)$	$-\theta'(0)$	$-\phi'(0)$	$f''(0)$	$-\theta'(0)$	$-\phi'(0)$
0.1	−0.0030	0.2805	1.0141	0.6104	0.0000	1.1661
0.71	−0.9305	1.1333	0.0236	0.0084	0.0007	1.1281
6	−1.5433	9.1278	−7.9816	−1.3492	1.7329	1.0353
10	−1.5837	15.1416	−13.9903	−1.5174	6.2436	0.9860
S_{lp}	−0.131979	1.504201	−1.513562	−0.208832	0.596777	−0.017406

TABLE 11.3
Effect of Prandtl Number P_r on Some of the Properties of the Flow When $S_c = 0.62$, $N_b = 0.1$, and $G_{rx} = 1$

	$f_w = -3$	$N_t = 5$		$f_w = +3$	$N_t = 5$	
P_r	$f''(0)$	$-\theta'(0)$	$-\phi'(0)$	$f''(0)$	$-\theta'(0)$	$-\phi'(0)$
0.1	0.7245	0.0095	3.9905	0.7869	0.0228	19.1943
0.71	0.7081	0.0087	4.0040	−0.1774	0.3476	−10.1768
6	0.6667	0.0000	4.6427	−1.4966	6.8187	−335.1994
10	0.6617	0.0000	4.7565	−1.5663	12.6161	−624.6436
S_{lp}	−0.006235	−0.001053	0.084501	−0.218300	1.276618	−64.78046874

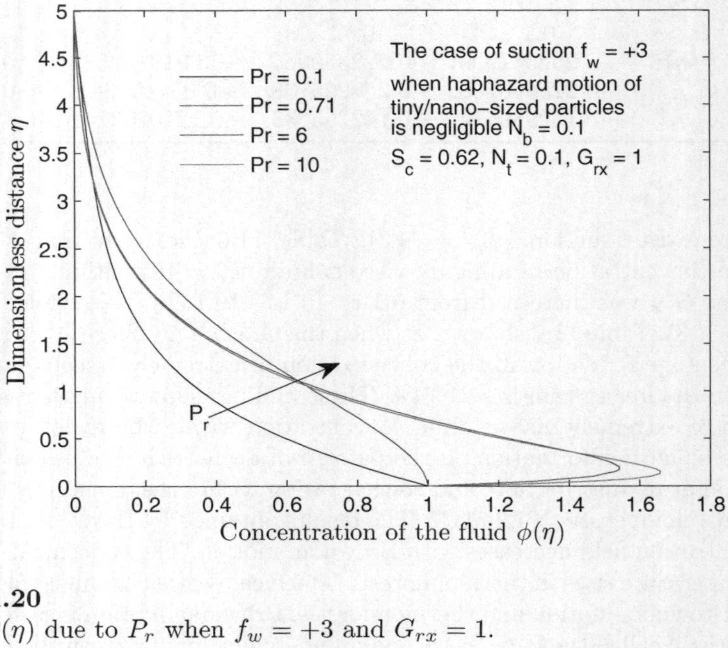

FIGURE 11.20
Variation in $\theta(\eta)$ due to P_r when $f_w = +3$ and $G_{rx} = 1$.

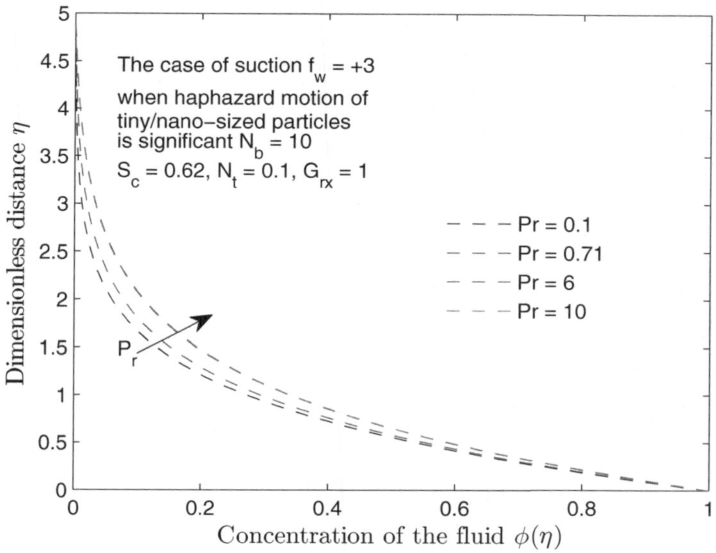

FIGURE 11.21
Variation in $\theta(\eta)$ due to P_r when $f_w = +3$ and $G_{rx} = 5$.

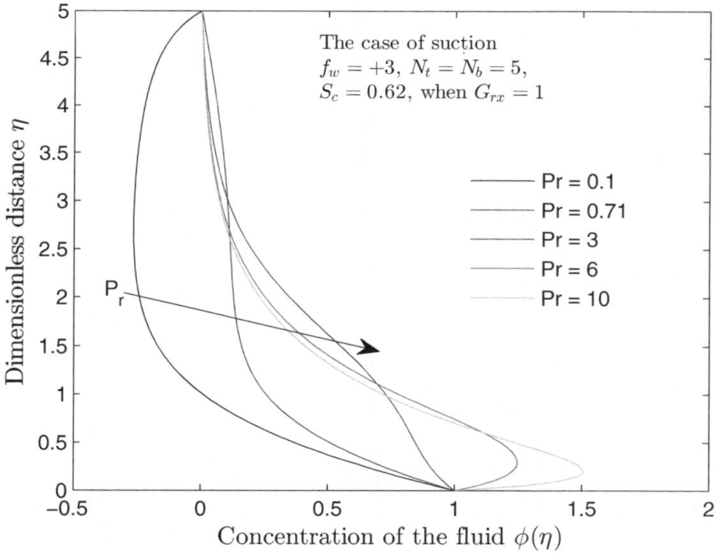

FIGURE 11.22
Variation in $\phi(\eta)$ due to P_r when $f_w = +3$ and $G_{rx} = 1$.

number and thermophoresis is the inverse since the same characteristic is a decreasing function of thermophoresis.

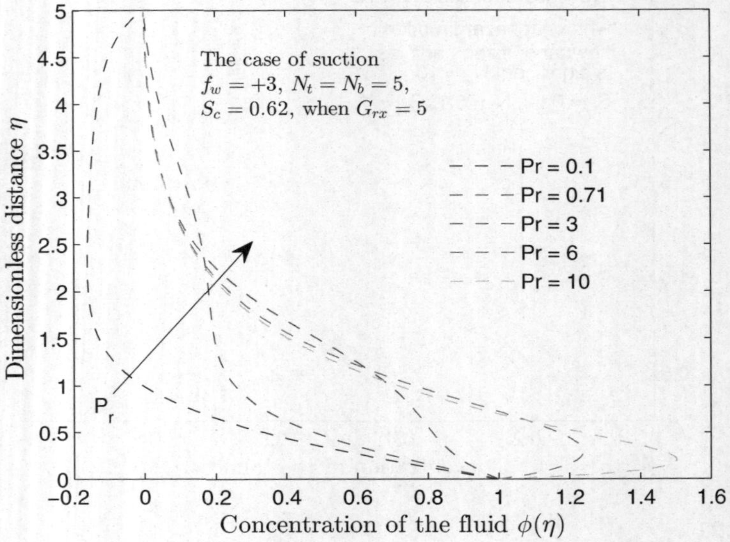

FIGURE 11.23
Variation in $\phi(\eta)$ due to P_r when $f_w = +3$ and $G_{rx} = 5$.

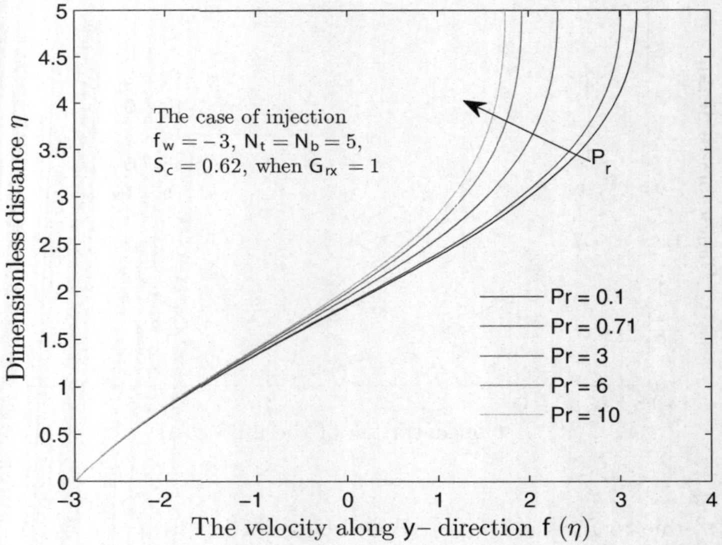

FIGURE 11.24
The effect of Prandtl number on the concentration: the case of injection $f_w = -3$, $N_t = 5$, $N_b = 0.1$, $S_c = 0.62$, and $G_{rx} = 1$.

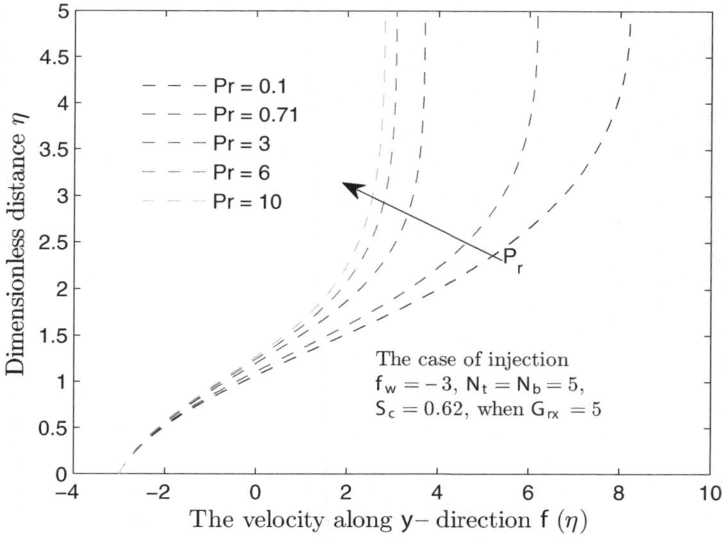

FIGURE 11.25
Variation in $f(\eta)$ due to P_r when $f_w = -3$ and $G_{rx} = 1$.

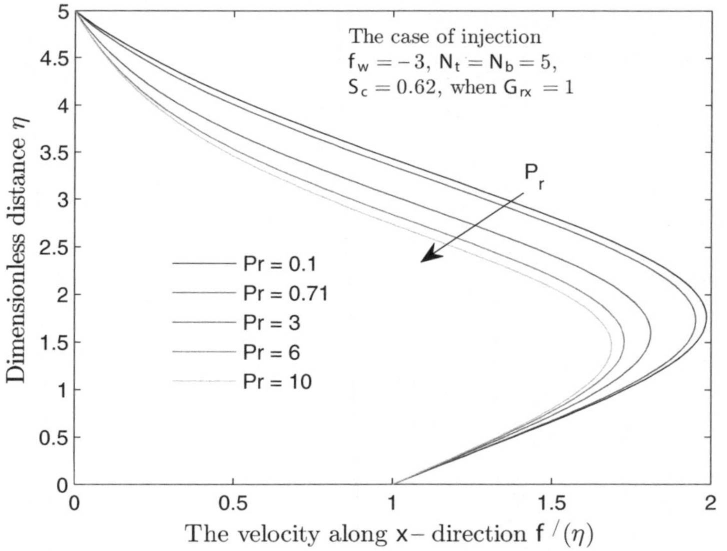

FIGURE 11.26
Variation in $f(\eta)$ due to P_r when $f_w = -3$ and $G_{rx} = 5$.

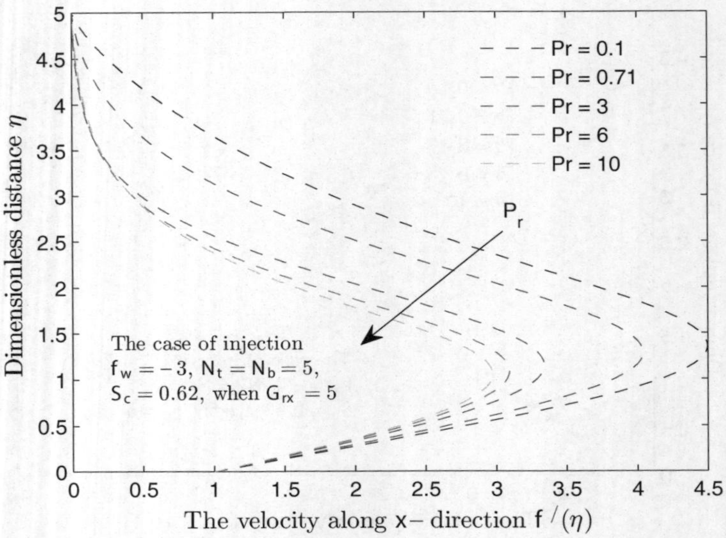

FIGURE 11.27
Variation in $f'(\eta)$ due to P_r when $f_w = -3$ and $G_{rx} = 1$.

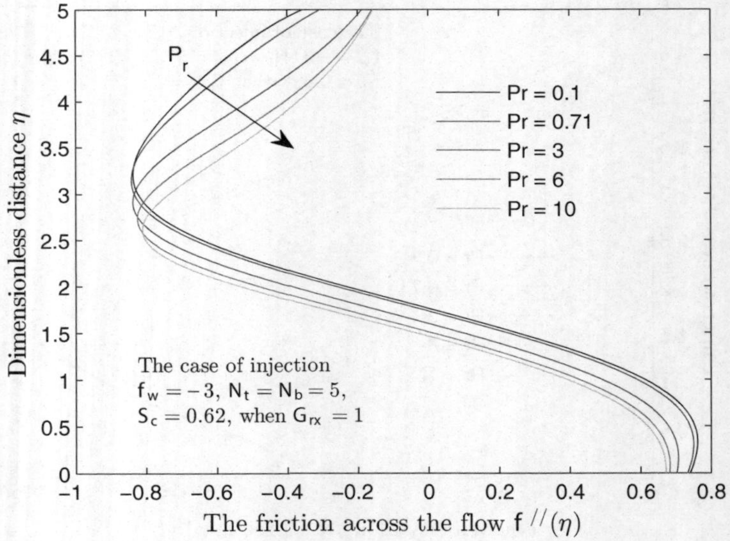

FIGURE 11.28
Variation in $f'(\eta)$ due to P_r when $f_w = -3$ and $G_{rx} = 5$.

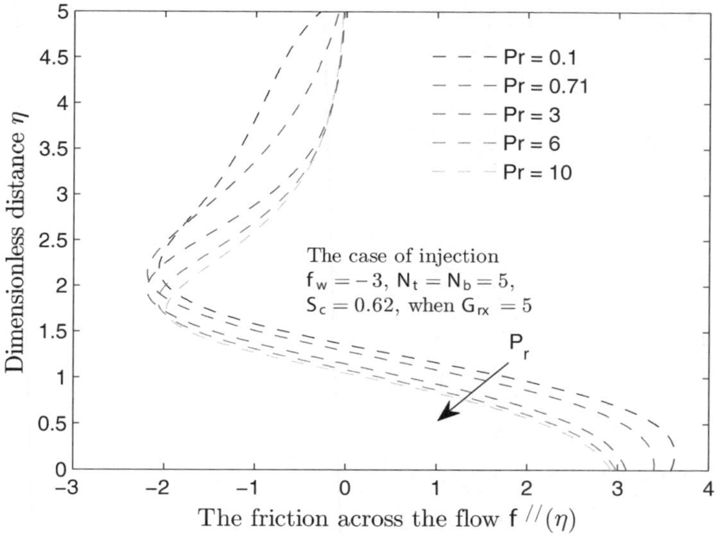

FIGURE 11.29
Variation in $f''(\eta)$ due to P_r when $f_w = -3$ and $G_{rx} = 1$.

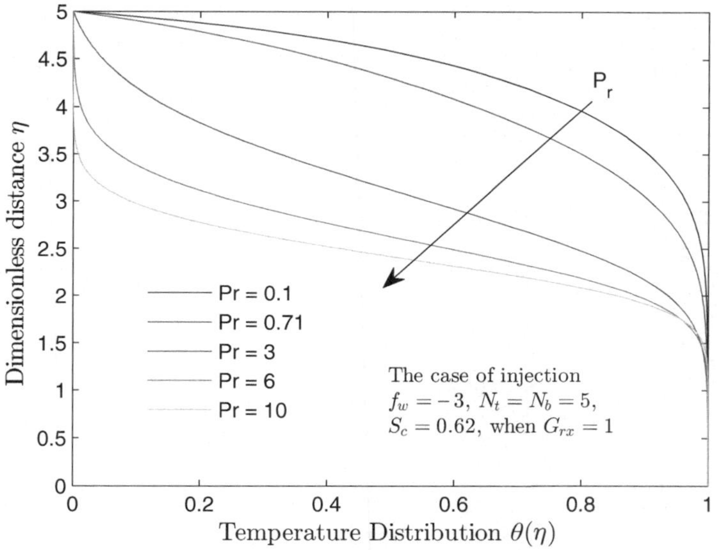

FIGURE 11.30
Variation in $f''(\eta)$ due to P_r when $f_w = -3$ and $G_{rx} = 5$.

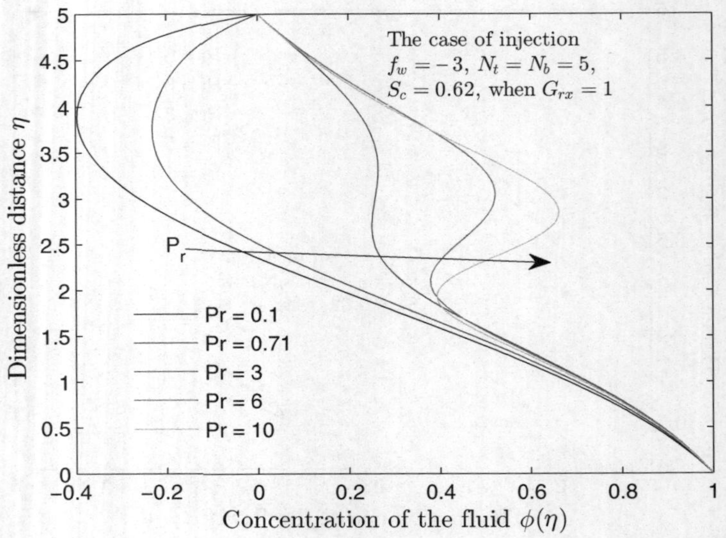

FIGURE 11.31
Variation in $\theta(\eta)$ due to P_r when $f_w = -3$ and $G_{rx} = 1$.

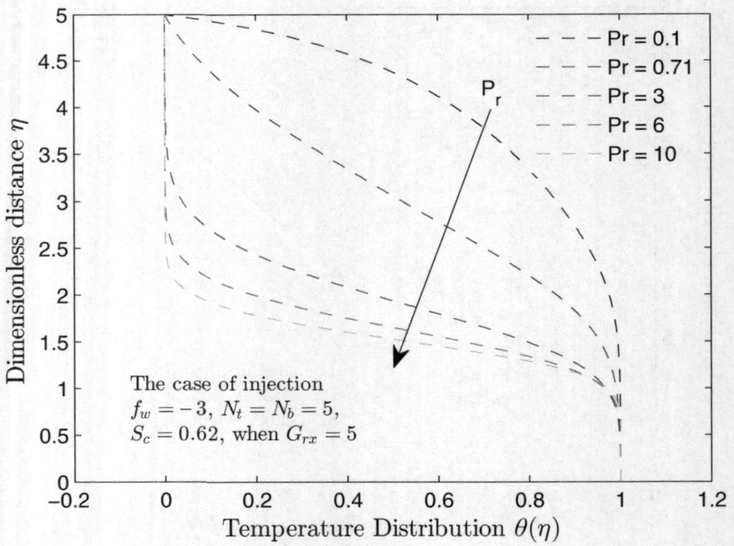

FIGURE 11.32
Variation in $\theta(\eta)$ due to P_r when $f_w = -3$ and $G_{rx} = 5$.

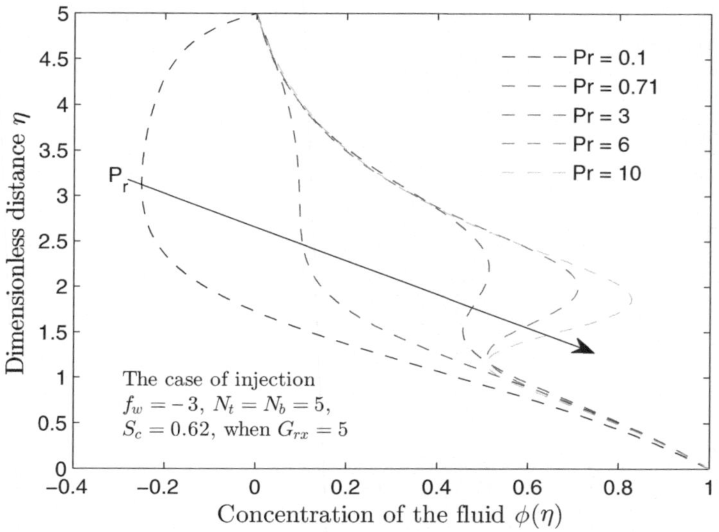

FIGURE 11.33
Variation in $\phi(\eta)$ due to P_r when $f_w = -3$ and $G_{rx} = 1$.

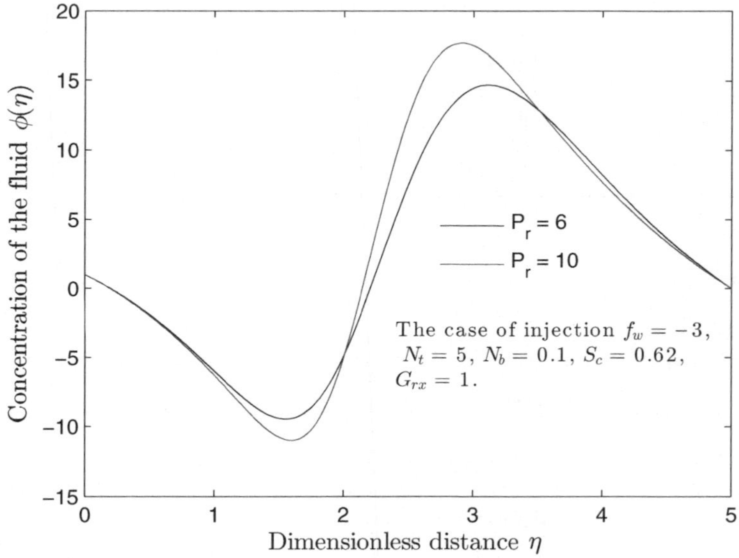

FIGURE 11.34
Variation in $\phi(\eta)$ due to P_r when $f_w = -3$ and $G_{rx} = 5$.

TABLE 11.4
Rate of Decrease in $\theta(\eta)$ and $\phi(\eta)$ with η each at Various Values of Prandtl Number When $N_t = +5$, $S_c = 0.62$, $N_b = 0.1$, and $G_{rx} = 1$

$f_w = -3$, $P_r \Rightarrow$	0.1	0.71	6	10
S_{lp} for $f'(\eta)$	−0.304740175	−0.321098774	−0.317320207	−0.338977219
S_{lp} for $\theta(\eta)$	−0.141774286	−0.190220475	−0.278675255	−0.280124296
S_{lp} for $\phi(\eta)$	−2.68727615	−0.569864164	2.864676407	2.769319004
$f_w = +3$, $P_r \Rightarrow$	0.1	0.71	6	10
S_{lp} for $f'(\eta)$	−0.271856939	−0.202258287	−0.101653467	−0.107101679
S_{lp} for $\theta(\eta)$	−0.161549329	−0.203845656	−0.030735406	−0.022043502
S_{lp} for $\phi(\eta)$	−0.177325563	−0.903409254	−6.366939882	−6.608096806

TABLE 11.5
Effect of Prandtl Number P_r on Some of the Properties of the Flow When $S_c = 0.62$, $N_t = 5$, and $N_b = 5$

$f_w = +3$		$G_{rx} = 0.1$			$G_{rx} = 10$	
P_r	$f''(0)$	$-\theta'(0)$	$-\phi'(0)$	$f''(0)$	$-\theta'(0)$	$-\phi'(0)$
0.1	−1.3314	0.0002	1.5518	14.2968	0.0006	1.7622
0.71	−1.4023	0.0069	1.2798	9.8232	0.0164	1.4794
6	−1.6218	3.2307	−2.1044	0.8989	3.0902	−1.8834
10	−1.6359	8.2641	−7.1215	−0.4731	8.2019	−7.0372
S_{lp}	−0.030966	0.815141	−0.848724	−1.428375	0.805304	−0.858961
$f_w = -3$		$G_{rx} = 0.1$			$G_{rx} = 10$	
P_r	$f''(0)$	$-\theta'(0)$	$-\phi'(0)$	$f''(0)$	$-\theta'(0)$	$-\phi'(0)$
0.1	0.0531	0.0001	0.1245	6.8698	0.0003	0.4876
0.71	0.0530	0.0000	0.1244	6.3169	0.0028	0.4071
6	0.0527	0.0000	0.1234	5.4337	0.0002	0.4225
10	0.0524	0.0000	0.1230	5.3122	0.0000	0.4290
S_{lp}	$-6.7E-05$	$-6.2E-06$	−0.000157	−0.146965	−0.000162	−0.002666

TABLE 11.6
Variation in $f'(\eta)$, $\theta(\eta)$, and $\phi(\eta)$ with η at Various Values of Prandtl Number When $f_w = +3$, $S_c = 0.62$, $N_t = 5$, and $N_b = 5$

$G_{rx} = 0.1$, $P_r \Rightarrow$	0.1	0.71	6	10
S_{lp} for $f'(\eta)$	−0.142650746	−0.135754849	−0.098068913	−0.115430339
S_{lp} for $\theta(\eta)$	−0.146130171	−0.217632484	−0.054902855	−0.033745957
S_{lp} for $\phi(\eta)$	−0.162061182	−0.131121035	−0.235765691	−0.279192697
$G_{rx} = 10$, $P_r \Rightarrow$	0.1	0.71	6	10
S_{lp} for $f'(\eta)$	−1.007822146	−0.695585275	−0.141401516	−0.109667944
S_{lp} for $\theta(\eta)$	−0.180268818	−0.232160978	−0.048433887	−0.025274084
S_{lp} for $\phi(\eta)$	−0.099590661	−0.111292333	−0.233877564	−0.250832998

TABLE 11.7
Variation in $f'(\eta)$, $\theta(\eta)$, and $\phi(\eta)$ with η at Various Values of Prandtl Number When $f_w = -3$, $S_c = 0.62$, $N_t = 5$, and $N_b = 5$

$G_{rx} = 0.1$, $P_r \Rightarrow$	0.1	0.71	6	10
S_{lp} for $f'(\eta)$	−0.205449749	−0.20892586	−0.216205251	−0.212031045
S_{lp} for $\theta(\eta)$	−0.079717356	−0.084325831	−0.128858697	−0.156449841
S_{lp} for $\phi(\eta)$	−0.337993418	−0.336179793	−0.29389094	−0.2621975
$G_{rx} = 10$, $P_r \Rightarrow$	0.1	0.71	6	10
S_{lp} for $f'(\eta)$	−1.0987571	−1.031199114	−0.804877503	−0.77176634
S_{lp} for $\theta(\eta)$	−0.155445076	−0.259875624	−0.230232409	−0.221918884
S_{lp} for $\phi(\eta)$	−0.191422762	−0.162927414	−0.200734437	−0.205175032

11.4 Introduction: Non-Darcy model for Dynamics through Porous Medium

Within the past 50 years, the analysis of transport phenomenon through a porous medium has been embraced to determine the velocity and friction. In most cases of fluid flow through a porous medium, a simple linear relationship exists between pressure drop and flow rate—Darcy's empirical flow model; see Darcy [86]. The well-known Darcy model discussed above is only valid and accurate to model the flow through porous media with minor porosity and small velocity (Hong et al. [134]). At a higher flow rate, the significance of the inertia effect is incorporated into the momentum equation through the addition of velocity squared (Forchheimer [101]). Non-Darcy flow is dynamics in which the inertial effects due to additional friction for high-velocity flow are appropriately considered. The relationship between buoyancy and the inertia effect is directly linked to thermal dispersion, which is inevitable in buoyancy-induced flow. Thermal dispersion is bound to be increasing when the inertia effect is increased; see Chen [77]. A comprehensive review of dynamics induced by convection and heat transfer in Darcy and non-Darcy porous media was provided by Nield and Bejan [214]. A simple procedure for measuring the pressure drop on fluid flow with superficial velocity along the height L of the porous media can be modeled using the expression of the Forchheimer equation for incompressible fluids

$$\frac{\Delta P}{L} = \frac{\mu V_s}{k_1} + \frac{\rho V_s^2}{k_2}$$

where P is the pressure drop, V_s is known as superficial velocity, and $k_1(m2)$ and $k_2(m)$ are the Darcy and non-Darcy permeability coefficients of the fluid with dynamic viscosity mu (Pa s) and density (kg/m^3). In the case of ceramic foam filters, Akbarnejad et al. [16] once remarked that it is imperative to know the Darcy and non-Darcy permeability coefficients to regulate the melt velocity and maintain the required pressure. With an increase in Darcy number, more porous are formed, and little resistance is experienced in such transport phenomenon. The reverse is the case when the Darcy number reduces as the dynamics experience higher resistance. Because of this, Behzadi et al. [56] concluded that (i) increase in the velocity is ascertained due to an increase in the Darcy number (ii) Nusselt number is an increasing property of the Darcy number while temperature decreases with the same dimensionless number.

11.5 Fluid Flow of Some Nanofluids through Porous Medium

The motion of water conveying different nanoparticles forming five nanofluids on a vertical surface through a porous medium is examined. The five nanoparticles under consideration are gold Ag, Copper Cu, Copper Oxide CuO, titania TiO_2, and alumina Al_2O_3 nanoparticles. The non-Darcy model was used to investigate the two-dimensional flow of incompressible fluid due to the nonlinear relationship between potential gradient (pressure drop) and flow rate when the velocity of the transport phenomenon due to mixed convection is very high. It is assumed that the two-dimensional flow occurs along x-direction with the velocity $u(x, y)$ along the vertical surface and the velocity $v(x, y)$ is along horizontal y-direction as shown in Figure 11.35. The governing equation suitable to model the cases is of the form

$$u_x + v_y = 0, \tag{11.13}$$

$$uu_x + vu_y = \frac{\mu_{nf}}{\rho_{nf}} u_{yy} + g\beta_{nf}(T - T_\infty) - \frac{\mu_{nf}}{\rho_{nf} K} u - \frac{b^*}{K} u^2 \tag{11.14}$$

$$uT_x + vT_y = \frac{\kappa_{nf}}{(\rho Cp)_{nf}} T_{yy}. \tag{11.15}$$

where the permeability of the porous medium is K and the inertial coefficient is b^*, Eqs. (B.1) and (11.15) are subject to the boundary conditions

$$u = U_o x, \quad v = 0, \quad T = T_w \quad at \;\; y = 0. \tag{11.16}$$

$$u \to 0, \quad T \to T_\infty, \quad as \;\; y \to \infty \tag{11.17}$$

Following Alloui [25], given that the solid volume fraction is denoted as ϕ, the effective density of the nanofluid ρ_{nf}, the density of each nanoparticle ρ_{sp}, the heat capacity of the nanofluid $(\rho C_p)_{nf}$, the heat capacity of the nanoparticles $(\rho C_p)_{sp}$, the volumetric thermal expansion of nanofluid β_{nf}, the thermal diffusivity of the nanofluid α_{nf}, the volumetric

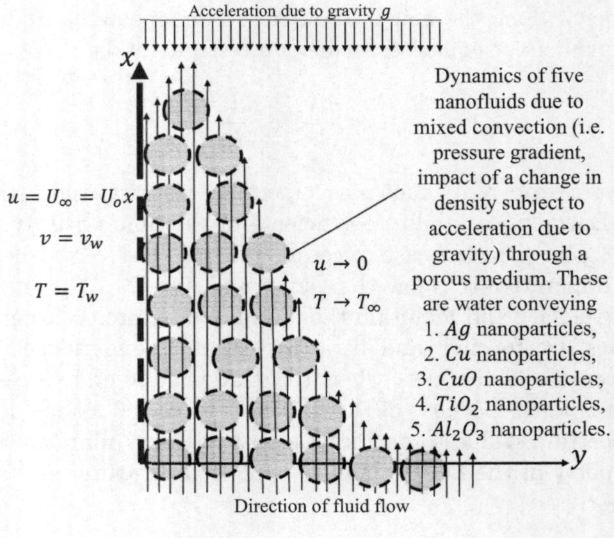

FIGURE 11.35
Illustration of nanofluids through a porous medium.

TABLE 11.8
Thermo-Physical Properties of Base Fluid (Water), Cupric-Oxide (CuO) Nanoparticles, and Alumina (Al_2O_3) Nanoparticles According to Alloui et al. [25]

Physical Properties	Water	CuO	Al_2O_3	Ag	Cu	TiO_2
$C_p(\text{JKg}^{-1}\text{K})$	4179	531.8	765	235	385	686.2
$\rho(\text{Kg/m}^3)$	997.1	6320	3970	10,500	8,933	4250
$\kappa(\text{W/mK})$	0.613	76.5	40	429	401	8.9538
$\beta\ (\text{K}^{-1})\ \times 10^5$	21	1.8	0.85	1.89	1.167	0.9

thermal expansion of nanoparticles β_{sp}, the dynamic viscosity of the nanofluid μ_{nf}, the dynamic viscosity of the base fluid μ_{bf}, the thermal conductivity of the nanofluid κ_{nf}, the thermal conductivity of the nanoparticles κ_{sp}, the thermal conductivity of the base fluid κ_{nf} are related mathematically as

$$\beta_{nf} = (1-\phi)\beta_{bf} + \phi\beta_{sp}, \quad \rho_{nf} = (1-\phi)\rho_{bf} + \phi\rho_{sp}, \quad (\rho C_p)_{nf} = (1-\phi)(\rho C_p)_{bf} + \phi(\rho C_p)_{sp},$$

$$\frac{\mu_{nf}}{\mu_{bf}} = \frac{1}{(1-\phi)^{2.5}}, \qquad \alpha_{nf} = \frac{\kappa_{nf}}{(\rho C_p)_{nf}}, \qquad \frac{\kappa_{nf}}{\kappa_{bf}} = \frac{\kappa_{sp} + 2\kappa_{bf} - 2\phi(\kappa_{bf} - \kappa_{sp})}{\kappa_{sp} + 2\kappa_{bf} + \phi(\kappa_{bf} - \kappa_{sp})}. \qquad (11.18)$$

Table 11.8 shows the thermo-physical characteristics of the base fluid (water), gold (Ag), copper (Cu), copper oxide (CuO), titania (TiO_2), and alumina (Al_2O_3) nanoparticles by Alloui et al. [25]. The heat capacity $(\rho C_p)_{bf}$ was estimated using the thermo-physical characteristics of the base fluid and each nanoparticle provided in Table 11.8.

In order to obtain self-similar solution of Eqs. (B.1), (B.2), (11.17), the following similarity variables were used:

$$\eta = y\sqrt{\frac{U_\infty}{\vartheta_{bf} x}}, \quad \psi(x,y) = \sqrt{\vartheta_{bf} x U_\infty} f(\eta), \quad u = \frac{\partial \psi}{\partial y}, \quad v = -\frac{\partial \psi}{\partial x}, \quad \theta(\eta) = \frac{T - T_\infty}{T_w - T_\infty},$$

$$F_s = \frac{b^*}{x}, \quad D_a = \frac{K}{x^2}, \quad Re = \frac{U_o x^2}{\vartheta_{bf}}, \quad G_{bf} = \frac{g\beta_{bf}(T_w - T_\infty)x}{U_\infty^2}, \quad G_{sp} = \frac{g\beta_{sp}(T_w - T_\infty)x}{U_\infty^2},$$

$$P_r = \frac{\mu_{bf} Cp_{bf}}{\kappa_{bf}} = \frac{\vartheta_{bf}}{\alpha_{bf}}, \quad A_1 = \frac{\frac{1}{(1-\phi)^{2.5}}}{1 - \phi + \phi\frac{\rho_{sp}}{\rho_{bf}}}, \quad A_2 = \frac{\frac{\kappa_{sp}+2\kappa_{bf}-2\phi(\kappa_{bf}-\kappa_{sp})}{\kappa_{sp}+2\kappa_{bf}+\phi(\kappa_{bf}-\kappa_{sp})}}{1 - \phi + \phi\frac{(\rho Cp)_{sp}}{(\rho Cp)_{bf}}}. \qquad (11.19)$$

to obtain

$$A_1 \frac{d^3 f}{d\eta^3} + \frac{1}{2} f \frac{d^2 f}{d\eta^2} + G_{bf}(1-\phi)\theta + G_{sp}\phi\theta - \frac{A_1}{D_a Re}\frac{df}{d\eta} - \frac{F_s}{D_a}\frac{df}{d\eta}\frac{df}{d\eta} = 0 \qquad (11.20)$$

$$A_2 \frac{d^2\theta}{d\eta^2} + \frac{1}{2} P_r f \frac{d\theta}{d\eta} = 0. \qquad (11.21)$$

subject to

$$\frac{df}{d\eta} = 1, \quad f = 0, \quad \theta = 0 \quad at \quad \eta = 0. \qquad (11.22)$$

$$\frac{df}{d\eta} \to 0, \quad \theta \to 0, \quad as \quad \eta \to \infty \qquad (11.23)$$

See the appendix for the derivation of Eq. (11.20).

11.5.1 Research Questions II

This subsection aims to provide answers to the following research questions:

1. What is the variational pattern of water conveying gold (Ag), Copper (Cu), Copper Oxide (CuO), titania (TiO_2), and alumina (Al_2O_3) nanoparticles?
2. How do the Reynolds number and the Prandtl number affect the transport phenomena of nanofluids through a porous medium?
3. What impact does rising (i) volume fraction, (ii) buoyancy-related parameter, (iii) Darcy number, (iv) Reynolds number, and (v) local Forchheimer number have on the transport phenomenon in this case?

11.5.2 Analysis and Discussion of Results II

The four-stage Lobatto IIIa formula described in Chapter 1 using $\eta = 5$ was used to obtain the solution of Eqs. (11.20) and (11.21) subject to Eqs. (11.22) and (11.23). Because the base fluid is water, $6 \leq P_r \leq 7$ was considered for the simulation. Examination of Tables 11.9–11.13 shows that maximum heat transfer rate $-\theta'(0)$ is seen in the case of water conveying titania (TiO_2) nanoparticles. The most minimum heat transfer rate is obtainable in the dynamics of water conveying gold (Ag) nanoparticles. Physically, the first and last positions in the heat transfer rate are associated with the minimum thermal conductivity of titania (TiO_2) nanoparticles and maximum thermal conductivity of gold (Ag) nanoparticles. Based on this, it is worthy to conclude that nanoparticles with minimum thermal conductivity possess a higher heat transfer rate at the wall. The outcome of this study indicates that the second position in the comparative analysis of the heat transfer rate is occupied by the dynamics of water conveying alumina (Al_2O_3) nanoparticles; see

TABLE 11.9
Variation in Some of the Properties of Water Conveying Gold (Ag) Nanoparticles When $\phi = 0.02$, $G_{rb} = G_{sp} = 3$, $F_s = 10$, $D_a = 0.2$, and $R_e = 0.5$ at Various Values of Prandtl Number

P_r	$f(\eta = 5)$ $H_2O - Ag$	$f'(\eta = 0.4798)$ $H_2O - Ag$	$\theta(\eta = 2)$ $H_2O - Ag$	$-\theta'(\eta = 0)$ $H_2O - Ag$
6	0.3760	0.1814	0.1632	0.6704
6.5	0.3661	0.1799	0.1473	0.7050
7	0.3572	0.1784	0.1332	0.7389
S_{lp}	−0.0188	−0.003	−0.03	0.0685

TABLE 11.10
Variation in Some of the Properties of Water Conveying Copper (Cu) Nanoparticles When $\phi = 0.02$, $G_{rb} = G_{sp} = 3$, $F_s = 10$, $D_a = 0.2$, and $R_e = 0.5$ at Various Values of Prandtl Number

P_r	$f(\eta = 5)$ $H_2O - Cu$	$f'(\eta = 0.4798)$ $H_2O - Cu$	$\theta(\eta = 2)$ $H_2O - Cu$	$-\theta'(\eta = 0)$ $H_2O - Cu$
6	0.3741	0.1814	0.1625	0.6728
6.5	0.3644	0.1799	0.1466	0.7076
7	0.3555	0.1784	0.1325	0.7417
S_{lp}	−0.0186	−0.003	−0.03	0.0689

TABLE 11.11
Variation in Some of the Properties of Water Conveying Copper Oxide (CuO) Nanoparticles When $\phi = 0.02$, $G_{rb} = G_{sp} = 3$, $F_s = 10$, $D_a = 0.2$, and $R_e = 0.5$ at Various Values of Prandtl Number

P_r	$f(\eta = 5)$ $H_2O - CuO$	$f'(\eta = 0.4545)$ $H_2O - CuO$	$\theta(\eta = 2)$ $H_2O - CuO$	$-\theta'(\eta = 0)$ $H_2O - CuO$
6	0.3719	0.1899	0.1628	0.6737
6.5	0.3623	0.1801	0.1469	0.7087
7	0.3536	0.1786	0.1327	0.7431
S_{lp}	−0.0183	−0.0113	−0.0301	0.0694

TABLE 11.12
Variation in Some of the Properties of Water Conveying Titania (TiO_2) Nanoparticles When $\phi = 0.02$, $G_{rb} = G_{sp} = 3$, $F_s = 10$, $D_a = 0.2$, and $R_e = 0.5$ at Various Values of Prandtl Number

P_r	$f(\eta = 5)$ H_2O-TiO_2	$f'(\eta = 0.4798)$ H_2O-TiO_2	$\theta(\eta = 2)$ H_2O-TiO_2	$-\theta'(\eta = 0)$ H_2O-TiO_2
6	0.3692	0.1906	0.1617	0.6773
6.5	0.3599	0.1802	0.1458	0.7126
7	0.3513	0.1787	0.1316	0.7474
S_{lp}	−0.0179	−0.0119	−0.0301	0.0701

TABLE 11.13
Variation in Some of the Properties of Water Conveying Alumina (Al_2O_3) Nanoparticles When $\phi = 0.02$, $G_{rb} = G_{sp} = 3$, $F_s = 10$, $D_a = 0.2$, and $R_e = 0.5$ at Various Values of Prandtl Number

P_r	$f(\eta = 5)$ $H_2O - Al_2O_3$	$f'(\eta = 0.4545)$ $H_2O - Al_2O_3$	$\theta(\eta = 2)$ $H_2O - Al_2O_3$	$-\theta'(\eta = 0)$ $H_2O - Al_2O_3$
6	0.3699	0.1908	0.1632	0.6742
6.5	0.3605	0.1803	0.1473	0.7094
7	0.3520	0.1789	0.1330	0.7439
S_{lp}	−0.0179	−0.0119	−0.0302	0.0697

Table 11.13. Although the thermal conductivity of Al_2O_3 nanoparticles is 40, this can be attributed to the nanoparticles' high specific heat, 765. It is fair to suppose that as the Prandtl number grows, the velocity and temperature distribution in the dynamics of these nanofluids diminish. Salman [266] once concluded that SiO_2 produced the highest Nusselt number proportional to heat transfer rate followed by alumina (Al_2O_3), Zinc Oxide (ZnO), and Copper Oxide (CuO) nanoparticles. When compared to the nanoparticles described above, water has the lowest Nusselt number (Table 11.14).

When $\phi = 0.02$, $G_{rb} = G_{sp} = 3$, $F_s = 10$, and $D_a = 0.2$, at various values of Prandtl number when $P_r = 6$ and $P_r = 7$, it is seen that due to an increase in the Reynolds number within the interval $0.1 \leq R_e \leq 0.5$, the motion of the TiO_2–water nanofluid along both the directions (x, y) increases, no significant effect on the shear stress exists, the temperature distribution decreases, and the temperature gradient proportional to heat transfer rate at the wall $-\theta'(0)$ increases; see Table 11.15. Moreover, Figures 11.36–11.39 further present the

TABLE 11.14
Variation in Some of the Properties of Water Conveying Titanium (TiO_2) Nanoparticles When $\phi = 0.02$, $G_{rb} = G_{sp} = 3$, $F_s = 10$, and $D_a = 0.2$ at Various Values of Reynold Number When $P_r = 6$

R_e **When** $P_r = 6$	$f(\eta = 5)$ H_2O-TiO_2	$f'(\eta = 0.4798)$ H_2O-TiO_2	$\theta(\eta = 2)$ H_2O-TiO_2	$-\theta'(\eta = 0)$ H_2O-TiO_2
0.1	0.2037	0.0664	0.2940	0.4838
0.2	0.2752	0.1250	0.2309	0.5766
0.3	0.3181	0.1567	0.1956	0.6262
0.4	0.3475	0.1768	0.1750	0.6567
0.5	0.3692	0.1906	0.1617	0.6773
S_{lp}	0.4033	0.3002	−0.3205	0.4671

TABLE 11.15
Variation in Some of the Properties of Water Conveying Titanium (TiO_2) Nanoparticles When $\phi = 0.02$, $G_{rb} = G_{sp} = 3$, $F_s = 10$, and $D_a = 0.2$ at Various Values of Reynolds Number When $P_r = 7$

R_e **When** $P_r = 7$	$f(\eta = 5)$ H_2O-TiO_2	$f'(\eta = 0.4798)$ H_2O-TiO_2	$\theta(\eta = 2)$ H_2O-TiO_2	$-\theta'(\eta = 0)$ H_2O-TiO_2
0.1	0.1976	0.0653	0.2604	0.5330
0.2	0.2643	0.1232	0.1970	0.6373
0.3	0.3040	0.1544	0.1631	0.6918
0.4	0.3312	0.1742	0.1439	0.7250
0.5	0.3513	0.1787	0.1316	0.7474
S_{lp}	0.018715	0.01389	−0.015535	0.025825

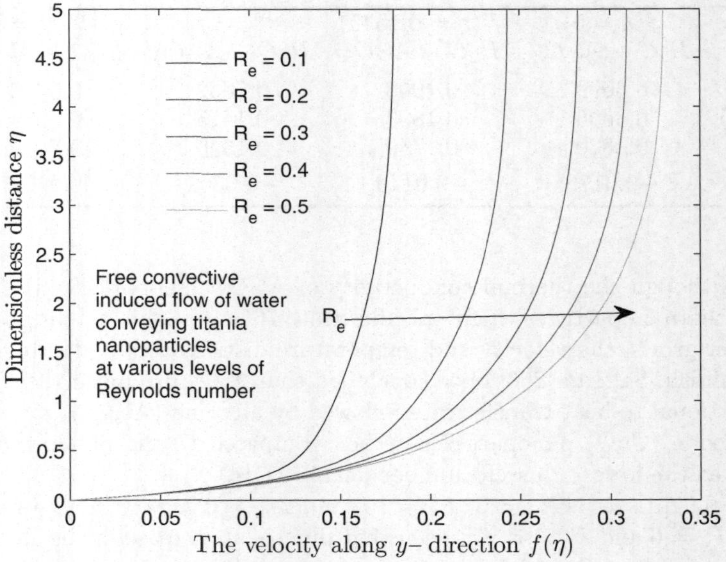

FIGURE 11.36
Variation in $f(\eta)$ due to R_e.

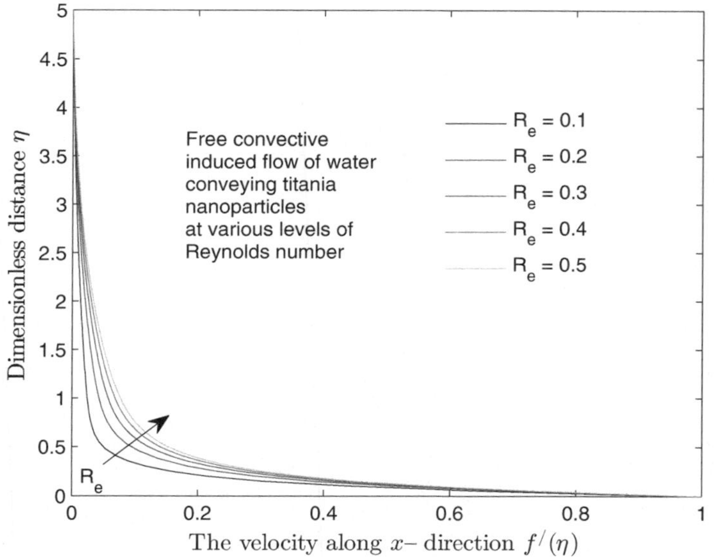

FIGURE 11.37
Variation in $f'(\eta)$ due to R_e.

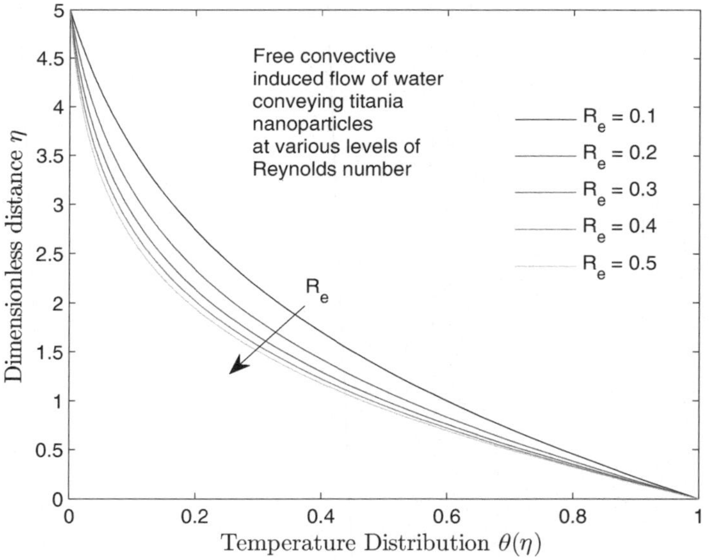

FIGURE 11.38
Variation in $\theta(\eta)$ due to R_e.

effect of increasing the ratio of inertial forces to viscous forces across the domain $0 \leq \eta \leq 5$. Physically, the Reynolds number corresponds to a direct increase in the inertial forces capable of laminar flow for a fixed viscous fluid $(1/\vartheta_{bf})$. This explains the observed increase in the velocity along both the directions (x, y). When velocity increases, the distribution of heat energy is also on a decreasing trend since the wall temperature is not increasing. This is in good agreement with the results published by Hayat et al. [123], which indicate

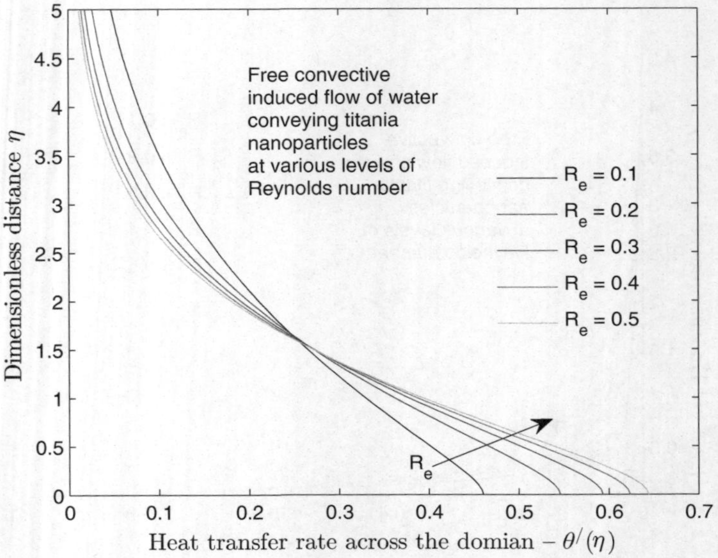

FIGURE 11.39
Variation in $-\theta'(\eta)$ due to R_e.

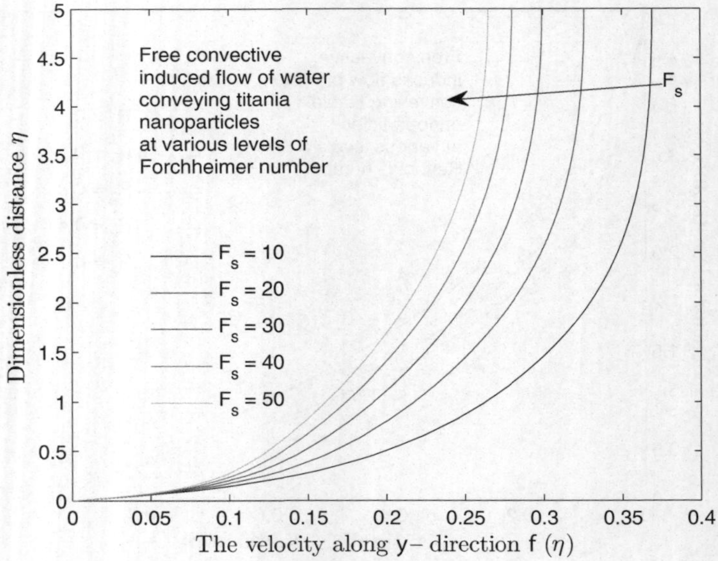

FIGURE 11.40
Variation in $f(\eta)$ due to F_s.

that a higher estimation of Reynolds number yields a lower temperature distribution. The increasing effects of Forchheimer number F_s on the motion of TiO_2–water nanofluid were captured using $\phi = 0.02$, $G_{rb} = G_{sp} = 3$, $P_r = 6$, $D_a = 0.2$, and $R_e = 0.5$ as illustrated in Figures 11.40–11.43.

The results indicate that the velocity along both the directions decreases due to an increase in F_s. Meanwhile, the temperature distribution is found to be an increasing property

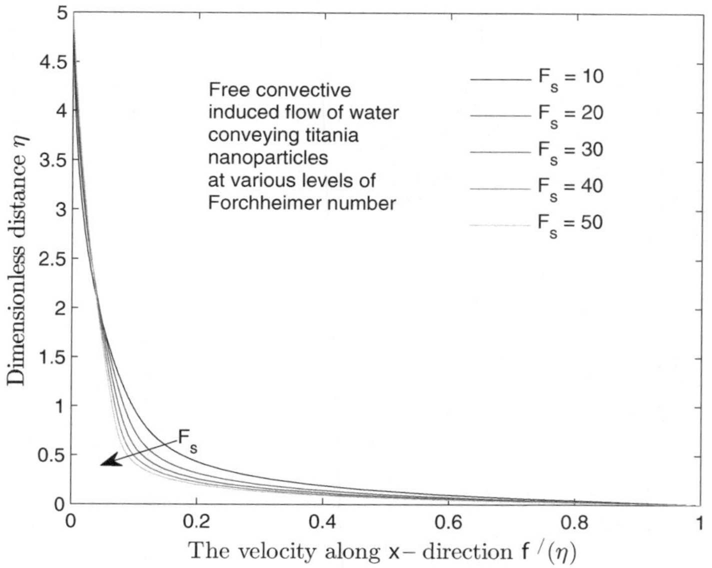

FIGURE 11.41
Variation in $f'(\eta)$ due to F_s.

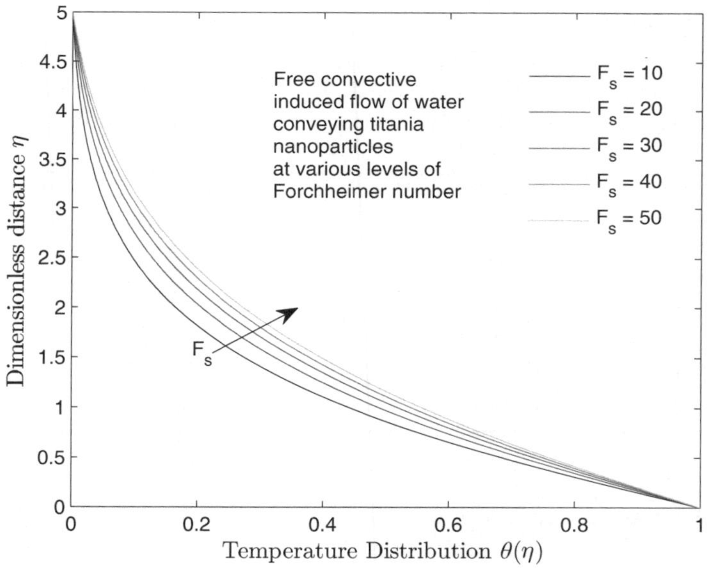

FIGURE 11.42
Variation in $\theta(\eta)$ due to F_s.

of F_s. The heat transfer rate $-\theta(\eta)$ decreases near the wall $(0 \leq \eta \leq 1)$ due to enlargement in the magnitude that represents F_s. The Reynolds number is the ratio of inertia force to viscous force. It is worthy of notice that for a more viscous fluid than air, it is inaccurate to remark that a higher Reynolds number implies a lower viscous force associated with the fluid's viscosity. Nevertheless, it seems correct and accurate to conclude that increasing the Reynolds number implies a higher inertia force when the viscous force is constant.

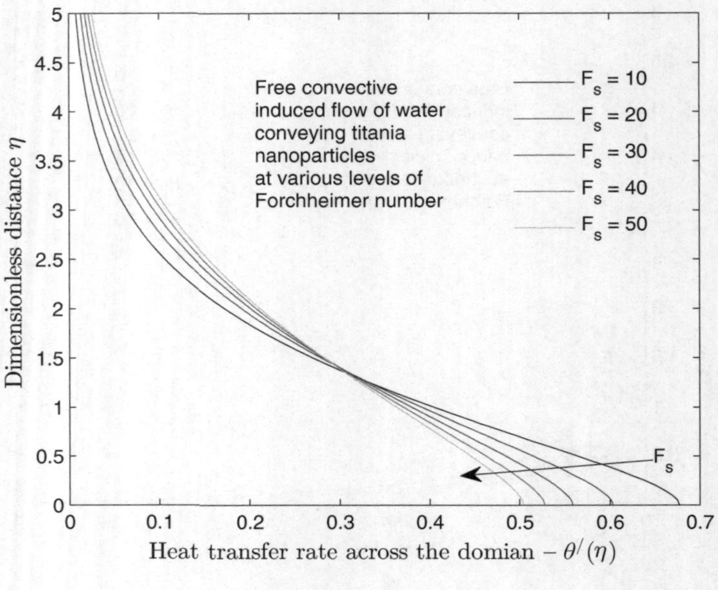

FIGURE 11.43
Variation in $-\theta'(\eta)$ due to F_s.

TABLE 11.16
Variation in Some of the Properties of Water Conveying Titanium (TiO_2) Nanoparticles When $\phi = 0.02$, $F_s = 10$, $D_a = 0.2$, and $R_e = 0.5$ at Various Values of Buoyancy-Related Parameters G_{rb} and G_{rs} when $P_r = 6$

G_{rb} & G_{rs} **When** $P_r = 6$	$f(\eta = 5)$ H_2O-TiO_2	$f'(\eta = 1.0101)$ H_2O-TiO_2	$\theta(\eta = 2)$ H_2O-TiO_2	$-\theta'(\eta = 0)$ H_2O-TiO_2
1	0.2786	0.0512	0.2199	0.6015
3	0.3692	0.0964	0.1617	0.6773
5	0.4297	0.1284	0.1293	0.7290
7	0.4762	0.1535	0.1079	0.7692
9	0.5144	0.1743	0.0925	0.8027
S_{lp}	0.02893	0.015165	−0.01543	0.024715

This is the major reason why larger values of Reynolds number cause higher velocity, as shown in Figures 11.36 and 11.37, and Table 11.15. First, the velocity and mass are directly proportional to the momentum. Second, more heat energy is consumed and transferred as the velocity increases due to the higher inertia force. The particles that form liquids vibrate and can move around heat transfer energy from the heated wall (region of higher temperature and causes temperature distribution to decline); see Figure 11.38. It is worth remarking that the observed results illustrated in Figure 11.38 corroborate with one of the results by Iqbal et al. [141], which says that the temperature distribution across the dynamics of Maxwell fluids is a decreasing property of Reynolds number (Table 11.16).

It is worth noticing that the inertial coefficients rise as the size of the Forchheimer number F_s grows. As a result, there are more impediments in the flow, lowering the velocity in both the directions. As b^* increases during deceleration, the fluid material heats up more, improving the temperature distribution. As demonstrated in Figure 11.43, the heat transmission rate is reducing in such a circumstance. For brevity, it is not illustrated that a larger magnitude of Forchheimer number F_s leads to a higher shear stress function $f''(\eta)$, but only within the interval $0.3 \leq \eta < 0.6$. Next is to examine the increasing effects of buoyancy-related parameter on the transport phenomenon of TiO_2–water nanofluid when $\phi = 0.02$, $F_s = 10$, $D_a = 0.2$, $P_r = 6$, and $R_e = 0.5$. First, an increment in the velocity along the (x, y) directions, a reduction in the temperature distribution, an increment in the shear stress function near the wall, and a reduction in the shear stress near the free stream are achievable due to an increase in buoyancy-related parameters ($G_{rb} = G_{sp}$); see Figures 11.44–11.47. At two different levels of Prandtl number ($P_r = 6$ and $P_r = 7$), as shown in Table 11.17, the heat transfer rate shows to be higher in magnitude when $P_r = 7$ at all values of buoyancy parameter ($G_{rb} = G_{sp}$). Exploration of the results presented as in Table 2 by Ghulam Rasool et al. [253] reveals that with an increase in the Forchheimer number defined as F_r, the Nusselt number decreases at the rate of -0.013211667, and the Sherwood number is also found to be a decreasing property of the Forchheimer number at the rate of -0.03629 (Table 11.18).

Ghulam Rasool et al. [253] concluded that the temperature distribution rises with Forchheimer number and porosity based on the data shown in their Figures 5 and 6. Figures 2 and 3 in their report show that the horizontal velocity diminishes with the Forchheimer number and porosity increases. The analysis presented in Figures 11.48–11.51 shows that a higher estimation of Darcy number D_a yields acceleration of the velocity along (x, y) directions, lower temperature, and higher heat transfer rate near the wall, but lower rate near the free stream. As demonstrated in Table 11.19, a decrease in the temperature distribution is found to be minimal at a higher Prandtl number (Table 11.20). The outcome of the analysis presented in Tables 11.21–11.29 shows that the velocity of the five nanofluids increases with greater volume fraction ϕ. The temperature distribution across the domain in the five nanofluids' flow is seen as an increasing volume fraction property. More so, a larger volume fraction ϕ leads to a lower heat transfer rate $-\theta'(0)$. The greatest rising rate in the temperature distribution with volume fraction was seen in the movement of water

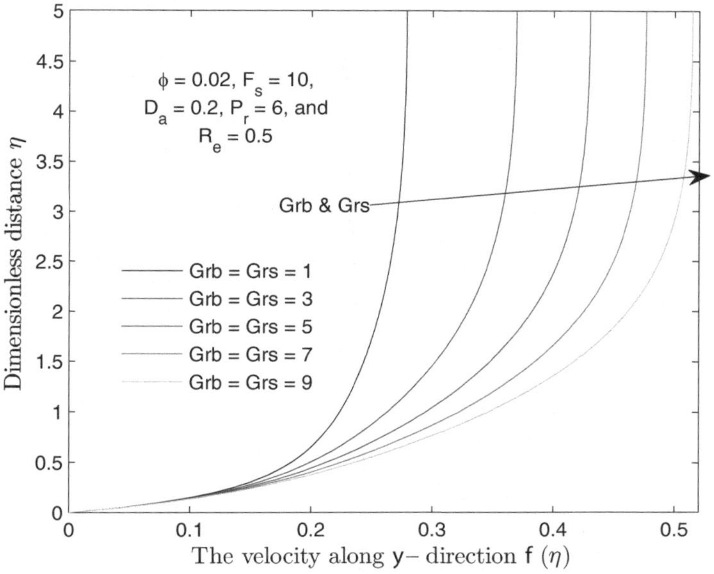

FIGURE 11.44
Variation in $f(\eta)$ with buoyancy-related parameter G_{rb} & G_{rs}.

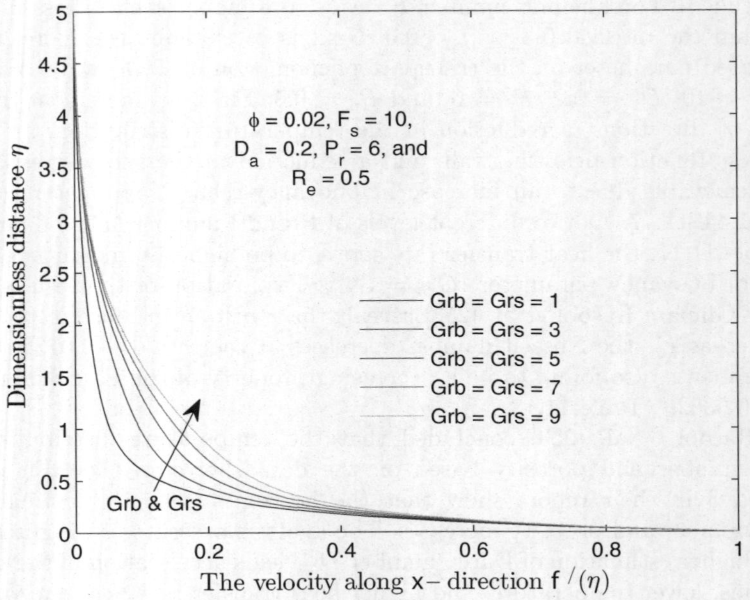

FIGURE 11.45
Variation in $f'(\eta)$ with buoyancy-related parameter G_{rb} & G_{rs}.

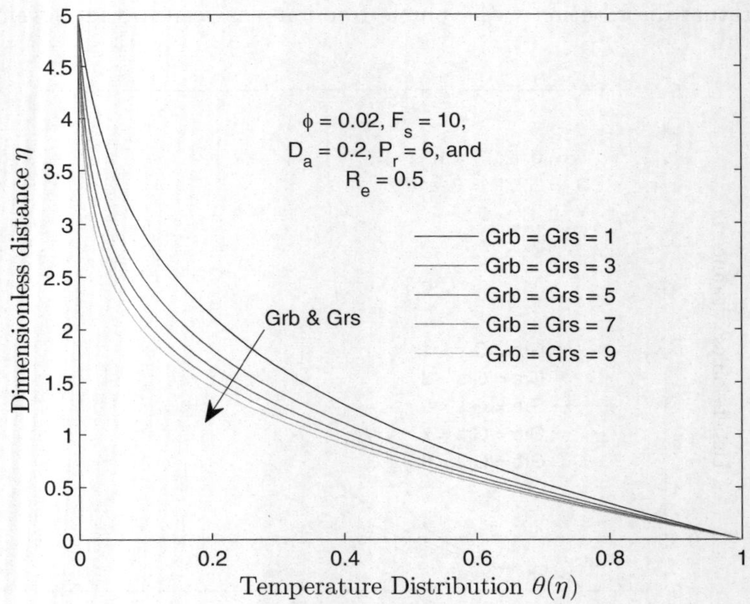

FIGURE 11.46
Variation in $\theta(\eta)$ with buoyancy-related parameter G_{rb} & G_{rs}.

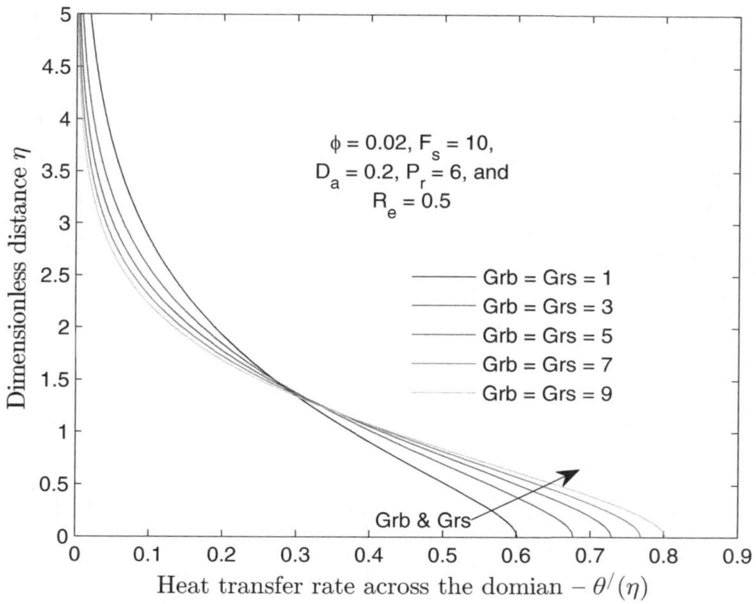

FIGURE 11.47
Variation in $-\theta'(\eta)$ with buoyancy-related parameter G_{rb} & G_{rs}.

TABLE 11.17
Variation in Some of the Properties of Water Conveying Titanium TiO_2 Nanoparticles When $\phi = 0.02$, $F_s = 10$, $D_a = 0.2$, and $R_e = 0.5$ at Various Values of Buoyancy Related Parameter G_{rb} & G_{rs} when $P_r = 7$

G_{rb} & G_{rs} **When** $P_r = 7$	$f(\eta = 5)$ H_2O-TiO_2	$f'(\eta = 1.0101)$ H_2O-TiO_2	$\theta(\eta = 2)$ H_2O-TiO_2	$-\theta'(\eta = 0)$ H_2O-TiO_2
1	0.2702	0.0487	0.1840	0.6688
3	0.3513	0.0906	0.1316	0.7474
5	0.4058	0.1200	0.1031	0.8010
7	0.4478	0.1431	0.0847	0.8429
9	0.4824	0.1621	0.0716	0.8778
S_{lp}	0.026045	0.013965	−0.013585	0.025675

TABLE 11.18
Variation in Some of the Properties of Water Conveying Titanium TiO_2 Nanoparticles When $\phi = 0.02$, $G_{rb} = G_{rs} = 3$, $F_s = 10$, $D_a = 0.2$, and $R_e = 0.5$ at Various Values of Darcy Number D_a when $P_r = 6$

D_a **When** $P_r = 6$	$f(\eta = 5)$ H_2O-TiO_2	$f'(\eta = 0.5)$ H_2O-TiO_2	$\theta(\eta = 1)$ H_2O-TiO_2	$-\theta'(\eta = 0)$ H_2O-TiO_2
0.2	0.3692	0.1735	0.4478	0.6773
1.2	0.7270	0.4580	0.2411	1.0102
2.2	0.8984	0.6043	0.1951	1.1033
3.2	1.0169	0.6438	0.1926	1.1530
4.2	1.1065	0.6933	0.1781	1.1851
S_{lp}	0.17645	0.12254	−0.05879	0.11584

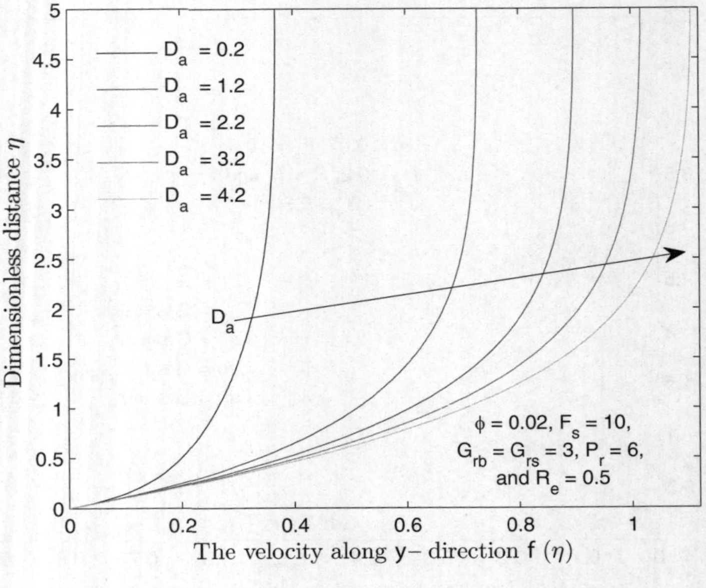

FIGURE 11.48
Variation in $f(\eta)$ with D_a.

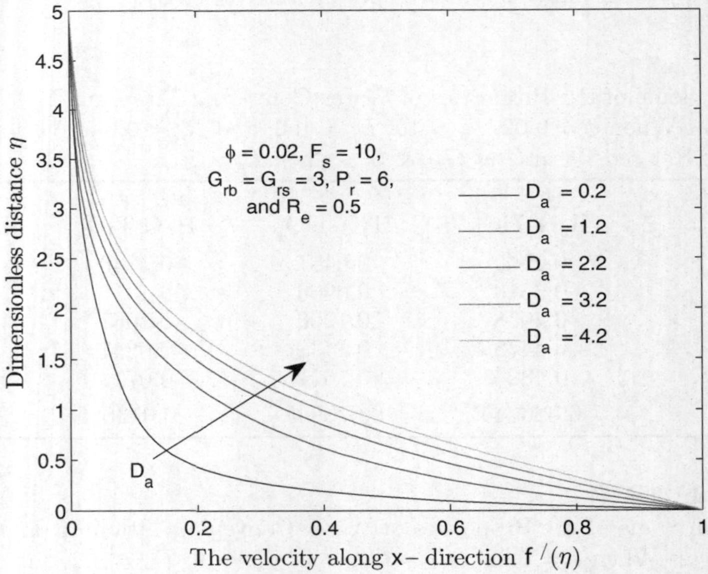

FIGURE 11.49
Variation in $f'(\eta)$ with D_a.

carrying Ag nanoparticles and was calculated to be 0.6348; see Table 11.27. Furthermore, when $Pr = 7$, the dynamics of H_2O–TiO_2 show a minimal estimation of the rising rate of temperature distribution with ϕ. Such a result was projected to be 0.538, as shown in Table 11.21.

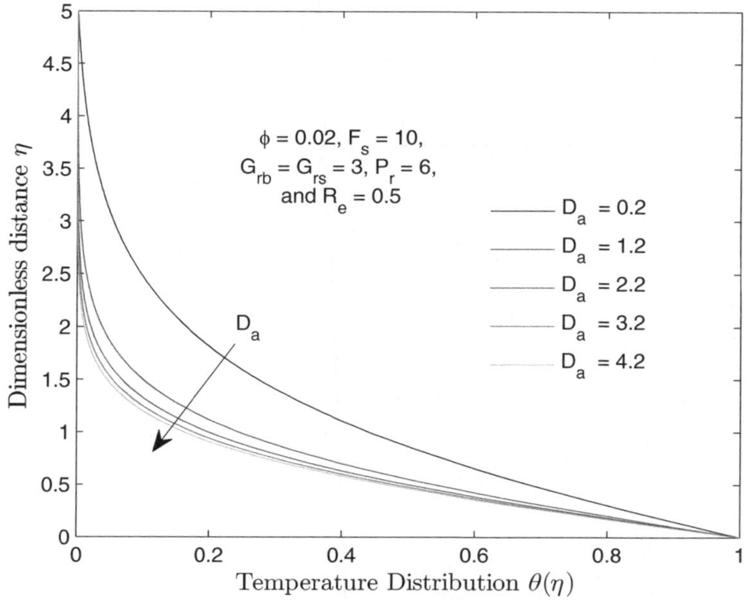

FIGURE 11.50
Variation in $\theta(\eta)$ with D_a.

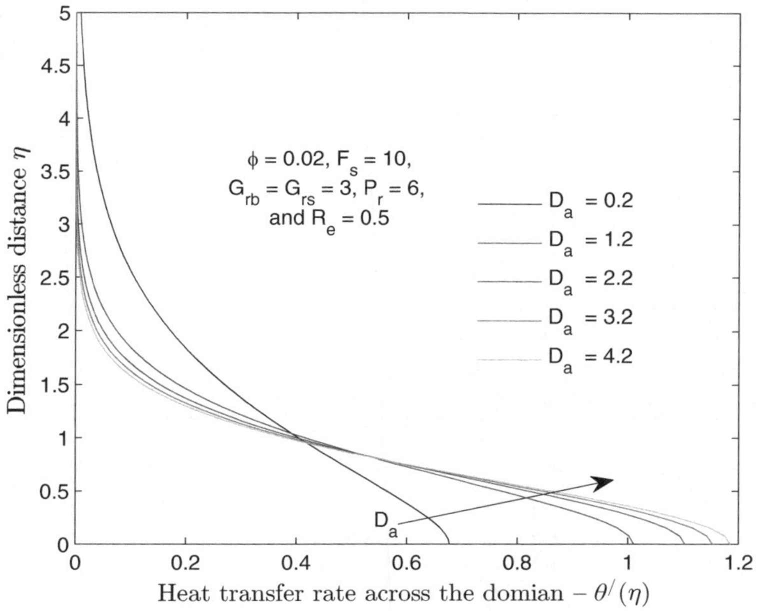

FIGURE 11.51
Variation in $-\theta'(\eta)$ with D_a.

TABLE 11.19
Variation in Some of the Properties of Water Conveying Titanium TiO_2 Nanoparticles When $\phi = 0.02$, $G_{rb} = G_{rs} = 3$, $F_s = 10$, $D_a = 0.2$, and $R_e = 0.5$ at Various Values of Darcy Number D_a when $P_r = 7$

D_a **When** $P_r = 7$	$f(\eta = 5)$ H_2O-TiO_2	$f'(\eta = 0.5)$ H_2O-TiO_2	$\theta(\eta = 1)$ H_2O-TiO_2	$-\theta'(\eta = 0)$ H_2O-TiO_2
0.2	0.3513	0.1787	0.3975	0.7474
1.2	0.6934	0.4482	0.2040	1.1040
2.2	0.8617	0.5925	0.1809	1.2009
3.2	0.9786	0.6291	0.1594	1.2520
4.2	1.0671	0.6772	0.1281	1.2850
S_{lp}	0.17168	0.11779	−0.05834	0.12232

TABLE 11.20
Variation in Some of the Properties of Water Conveying Titanium TiO_2 Nanoparticles When $D_a = 0.2$, $G_{rb} = G_{rs} = 3$, $F_s = 10$, $D_a = 0.2$, and $R_e = 0.5$ at Various Values of Volume Fraction ϕ when $P_r = 6$

ϕ **When** $P_r = 6$	$f(\eta = 5)$ H_2O-TiO_2	$f'(\eta = 2)$ H_2O-TiO_2	$\theta(\eta = 2)$ H_2O-TiO_2	$-\theta'(\eta = 0)$ H_2O-TiO_2
0	0.3623	0.0395	0.1509	0.7015
0.05	0.3793	0.0457	0.1781	0.6436
0.10	0.3949	0.0511	0.2059	0.5934
0.15	0.4087	0.0555	0.2340	0.5491
0.20	0.4202	0.0589	0.2621	0.5096
S_{lp}	0.2904	0.0972	0.5566	−0.9566

TABLE 11.21
Variation in Some of the Properties of Water Conveying Titanium TiO_2 Nanoparticles When $D_a = 0.2$, $G_{rb} = G_{rs} = 3$, $F_s = 10$, $D_a = 0.2$, and $R_e = 0.5$ at Various Values of Volume Fraction ϕ when $P_r = 7$

ϕ **When** $P_r = 7$	$f(\eta = 5)$ H_2O-TiO_2	$f'(\eta = 2)$ H_2O-TiO_2	$\theta(\eta = 2)$ H_2O-TiO_2	$-\theta'(\eta = 0)$ H_2O-TiO_2
0	0.3447	0.0333	0.1215	0.7746
0.05	0.3611	0.0393	0.1470	0.7094
0.10	0.3765	0.0447	0.1736	0.6527
0.15	0.3904	0.0493	0.2010	0.6026
0.20	0.4026	0.0530	0.2290	0.5580
S_{lp}	0.2902	0.0988	0.538	−1.08

TABLE 11.22
Variation in Some of the Properties of Water Conveying CuO Nanoparticles When $D_a = 0.2$, $G_{rb} = G_{rs} = 3$, $F_s = 10$, $D_a = 0.2$, and $R_e = 0.5$ at Various Values of Volume Fraction ϕ when $P_r = 6$

ϕ **When** $P_r = 6$	$f(\eta = 5)$ H$_2$O-CuO	$f'(\eta = 2)$ H$_2$O-CuO	$\theta(\eta = 2)$ H$_2$O-CuO	$-\theta'(\eta = 0)$ H$_2$O-CuO
0	0.3623	0.0395	0.1509	0.7015
0.05	0.3858	0.0488	0.1807	0.6359
0.10	0.4074	0.0568	0.2109	0.5810
0.15	0.4267	0.0635	0.2412	0.5337
0.20	0.4434	0.0689	0.2714	0.4924
S_{lp}	0.4062	0.147	0.603	−1.0408

TABLE 11.23
Variation in Some of the Properties of Water Conveying CuO Nanoparticles When $D_a = 0.2$, $G_{rb} = G_{rs} = 3$, $F_s = 10$, $D_a = 0.2$, and $R_e = 0.5$ at Various Values of Volume Fraction ϕ When $P_r = 7$

ϕ **When** $P_r = 7$	$f(\eta = 5)$ H$_2$O-CuO	$f'(\eta = 2)$ H$_2$O-CuO	$\theta(\eta = 2)$ H$_2$O-CuO	$-\theta'(\eta = 0)$ H$_2$O-CuO
0	0.3447	0.0333	0.1215	0.7746
0.05	0.3669	0.0421	0.1497	0.7002
0.10	0.3879	0.0500	0.1788	0.6379
0.15	0.4071	0.0568	0.2085	0.5844
0.20	0.4243	0.0625	0.2388	0.5376
S_{lp}	0.3988	0.1462	0.5868	−1.1796

TABLE 11.24
Variation in Some of the Properties of Water Conveying Al_2O_3 Nanoparticles When $D_a = 0.2$, $G_{rb} = G_{rs} = 3$, $F_s = 10$, $D_a = 0.2$, and $R_e = 0.5$ at Various Values of Volume Fraction ϕ When $P_r = 6$

ϕ **When** $P_r = 6$	$f(\eta = 5)$ H$_2$O-Al$_2$O$_3$	$f'(\eta = 2)$ H$_2$O-Al$_2$O$_3$	$\theta(\eta = 2)$ H$_2$O-Al$_2$O$_3$	$-\theta'(\eta = 0)$ H$_2$O-Al$_2$O$_3$
0	0.3623	0.0395	0.1509	0.7015
0.05	0.3809	0.0461	0.1820	0.6365
0.10	0.3979	0.0519	0.2137	0.5810
0.15	0.4128	0.0566	0.2455	0.5327
0.20	0.4252	0.0602	0.2770	0.4903
S_{lp}	0.3154	0.1038	0.6314	−1.0524

TABLE 11.25
Variation in Some of the Properties of Water Conveying Al_2O_3 Nanoparticles When $D_a = 0.2$, $G_{rb} = G_{rs} = 3$, $F_s = 10$, $D_a = 0.2$, and $R_e = 0.5$ at Various Values of Volume Fraction ϕ when $P_r = 7$

ϕ **When** $P_r = 7$	$f(\eta = 5)$ H_2O-Al_2O_3	$f'(\eta = 2)$ H_2O-Al_2O_3	$\theta(\eta = 2)$ H_2O-Al_2O_3	$-\theta'(\eta = 0)$ H_2O-Al_2O_3
0	0.3447	0.0333	0.1215	0.7746
0.05	0.3627	0.0397	0.1507	0.7014
0.10	0.3796	0.0455	0.1811	0.6387
0.15	0.3949	0.0505	0.2123	0.5842
0.20	0.4080	0.0545	0.2440	0.5361
S_{lp}	0.3176	0.1064	0.6132	−1.1884

TABLE 11.26
Variation in Some of the Properties of Water Conveying Ag Nanoparticles When $D_a = 0.2$, $G_{rb} = G_{rs} = 3$, $F_s = 10$, $D_a = 0.2$, and $R_e = 0.5$ at Various Values of Volume Fraction ϕ When $P_r = 6$

ϕ **When** $P_r = 6$	$f(\eta = 5)$ H_2O-Ag	$f'(\eta = 2)$ H_2O-A	$\theta(\eta = 2)$ H_2O-Ag	$-\theta'(\eta = 0)$ H_2O-Ag
0	0.3623	0.0395	0.1509	0.7015
0.05	0.3955	0.0536	0.1818	0.6291
0.10	0.4252	0.0651	0.2135	0.5704
0.15	0.4517	0.0745	0.2456	0.5208
0.20	0.4748	0.0822	0.2777	0.4779
S_{lp}	0.5624	0.2126	0.6348	−1.111

TABLE 11.27
Variation in Some of the Properties of Water Conveying Ag Nanoparticles When $D_a = 0.2$, $G_{rb} = G_{rs} = 3$, $F_s = 10$, $D_a = 0.2$, and $R_e = 0.5$ at Various Values of Volume Fraction ϕ when $P_r = 7$

ϕ **When** $P_r = 7$	$f(\eta = 5)$ H_2O-Ag	$f'(\eta = 2)$ H_2O-A	$\theta(\eta = 2)$ H_2O-Ag	$-\theta'(\eta = 0)$ H_2O-Ag
0	0.3447	0.0333	0.1215	0.7746
0.05	0.3754	0.0466	0.1511	0.6916
0.10	0.4040	0.0579	0.1817	0.6248
0.15	0.4302	0.0674	0.2134	0.5686
0.20	0.4537	0.0754	0.2457	0.5201
S_{lp}	0.5456	0.21	0.6214	−1.264

TABLE 11.28
Variation in Some of the Properties of Water Conveying Cu Nanoparticles When $D_a = 0.2$, $G_{rb} = G_{rs} = 3$, $F_s = 10$, $D_a = 0.2$, and $R_e = 0.5$ at Various Values of Volume Fraction ϕ when $P_r = 6$

ϕ **When** $P_r = 6$	$f(\eta = 5)$ H_2O-Cu	$f'(\eta = 2)$ H_2O-Cu	$\theta(\eta = 2)$ H_2O-Cu	$-\theta'(\eta = 0)$ H_2O-Cu
0	0.3623	0.0395	0.1509	0.7015
0.05	0.3911	0.0515	0.1800	0.6342
0.10	0.4171	0.0610	0.2071	0.5790
0.15	0.4401	0.0706	0.2428	0.5319
0.20	0.4602	0.0765	0.2691	0.4910
S_{lp}	0.4896	0.1862	0.5984	−1.0466

TABLE 11.29
Variation in Some of the Properties of Water Conveying Cu Nanoparticles When $D_a = 0.2$, $G_{rb} = G_{rs} = 3$, $F_s = 10$, $D_a = 0.2$, and $R_e = 0.5$ at Various Values of Volume Fraction ϕ when $P_r = 7$

ϕ **When** $P_r = 7$	$f(\eta = 5)$ H_2O-Cu	$f'(\eta = 2)$ H_2O-Cu	$\theta(\eta = 2)$ H_2O-Cu	$-\theta'(\eta = 0)$ H_2O-Cu
0	0.3447	0.0333	0.1215	0.7746
0.05	0.3714	0.0447	0.1492	0.6977
0.10	0.3964	0.0550	0.1800	0.6349
0.15	0.4192	0.0628	0.2070	0.5816
0.20	0.4396	0.0705	0.2402	0.5352
S_{lp}	0.4752	0.185	0.5904	−1.1898

11.6 Tutorial Questions

1. What is the rising influence of Prandtl number on the transport phenomena in the presence of suction ($f_w = +3$) when the random motion of tiny/nanosized particles is insignificant and significant?
2. When the thermo-migration of tiny/nanosized particles is insignificant and substantial in the presence of suction ($f_w = +3$), how does the Prandtl number impact the transport phenomena of a standard Newtonian fluid?
3. What is the relevance of the Prandtl number on the transport phenomena in the situation of injection ($f_w = -3$) when the random motion of tiny/nanosized particles is minimal and significant?
4. What impact does the rising Prandtl number have on the transport phenomena of a typical Newtonian fluid where the thermo-migration of tiny/nanosized particles is minimal and substantial in the injection ($f_w = -3$)?
5. What is the variational pattern of water conveying gold (Ag), Copper (Cu), Copper Oxide (CuO), titania (TiO_2), and alumina (Al_2O_3) nanoparticles?

6. What effect do the Reynolds number and the Prandtl number have on the transport of nanofluids through a porous medium?
7. What is the impact of rising (i) volume fraction, (ii) buoyancy-related parameter, (iii) Darcy number, (iv) Reynold number, and (v) local Forchheimer number.
8. When $P_r = 6$ and $P_r = 7$, how quickly does the vertical velocity of water-based titania TiO_2 nanofluid vary with the Reynolds number?

12

Conclusion and Recommendation

12.1 Background Information

This text provides a note on the significance of increasing the ratio of momentum diffusivity to thermal diffusivity. The first chapter defines the related thermo-physical characteristics, while the second chapter presents the study techniques. Chapters 3 and 4 offer theoretical and empirical reviews of more than one hundred screened published publications, focusing on published Prandtl number results. The fifth chapter examines the various outcomes of the slope linear regression via the data points provided in the previous two chapters. Chapters 5–11 examine similar flows, emphasizing the effects of increasing the Prandtl number. The purpose of this chapter was to lay forth the decisive conclusions and make some recommendations..

12.2 Conclusion

At the end of the inquiry, it is worth concluding that the fluid viscosity and heat conductivity have a basic yet proportional connection. The Prandtl number, connected to thermal and kinetic disturbances, substantially impacts buoyancy-opposed flows (Bera and Khalili [57]). Based on the scrutinization/meta-analysis, the following conclusions can be drawn:

1. Experimentally and theoretically, it is reasonable to conclude that the thermal diffusivity prevails for $P_r \ll 1$ as it is the fundamental cause for the low Prandtl number of gases in which heat diffuses quite easily and quickly. The opposite is true for $P_r \gg 1$, where momentum diffusivity dominates. This finding explains why the Prandtl number of motor oil is so high; it is because heat diffuses extremely slowly through such a medium. When $P_r = 1$, the rate of heat diffusion in the liquid is equal to momentum diffusivity.
2. An increase in the ratio of momentum diffusivity to thermal diffusivity correlates to an increase in the fluid's viscosity under consideration. According to the meta-analysis findings, a rise in the Prandtl number is equal to a reduction in thermal conductivity.
3. Velocity is a decreasing property of the Prandtl number. The melting heat transfer is one of the variables identified to enhance vertical and horizontal velocities with the Prandtl number. The Lewis number affects the vertical velocity in the two-dimensional flow when the Prandtl number is modest (gases) and substantial (air). When the Prandtl number is small (gases), the horizontal velocity changes rapidly with the Lewis number (air).
4. A larger Prandtl number is predicted to reduce temperature dispersion across fluid flows, although this should not be assumed in all circumstances. The results

DOI: 10.1201/9781003217374-12

mentioned above are based on the observed increase in temperature distribution as P_r increases (i) in the case of the dynamics of dusty fluid, (ii) when mass flux due to the temperature gradient and energy flux due to concentration gradient are significant, (iii) when there is melting heat transfer at both walls, (iv) when convective heating exists at both sides of the domain, and (v) when there is thermal jump at the surface.

5. Because of the direction of temperature changes and the pace of temperature changes, it is reasonable to conclude that the temperature gradient is an increasing feature of the Prandtl number. However, melting heat transfer at the wall, thermal leap between the wall and the free stream, and thermal stratification at the wall are all variables that can cause the temperature gradient to diminish as the Prandtl number increases. A considerable rise in temperature dispersion is observed at a significant heat source with low intensity due to a more significant Prandtl number. When both the heat source and the degree of the heat are magnificent, the opposite is true. The temperature distribution varies with the Lewis number when the Prandtl number is small (gases) and big (air).

6. when the intensity of the heat source/sink is modest or significant, the local skin friction coefficients in the case of forced convection Sakiadis flow maintain a constant function of Prandtl number. The result, as mentioned earlier, validates the difference in the internal heat source and buoyancy force. When the Prandtl and Dufour numbers are modest, an optimum rise in the local skin friction coefficients with mass flux owing to temperature gradient is achieved.

7. The Nusselt number is a characteristic of the increasing Prandtl number. Substantial phenomena that might induce a reduction in the Nusselt number owing to a higher Prandtl number include sinusoidal peristaltic waves at both walls, significant mass flux due to temperature gradient and energy flux due to concentration gradient, and melting heat transfer at the wall. The Nusselt number decreases with the Prandtl number for the heat source, but grows with the same dimensionless value for the heat sink. When the Dufour number is small, the Prandtl number is small (gases), and when the Prandtl number is large (air), the Nusselt number grows as the Soret number increases. However, whether the Dufour number is insignificant or significant, and the Prandtl number is minor (gases) or substantial (air), the Nusselt number falls as the Soret number increases.

8. Because mass/species transfer in a very viscous fluid is limited due to increased resistance and more essential intermolecular forces, the local Sherwood number is anticipated to decrease as the Prandtl number increases. When the Prandtl number is raised, two essential variables can induce a rise in mass transfer rates: mass flux owing to the temperature gradient and energy flux due to concentration gradient, and the lack of melting heat transfer. The Prandtl number is small (gases) and big (air) when minimal energy flows due to the concentration gradient. The Sherwood number decreases with a more significant Soret number. When the Dufour number is insignificant or significant, the Prandtl number is minor (gases) or substantial (air), the Soret number is large, and the Sherwood number grows with a more significant Soret number.

9. Increased viscosity as P_r rises indicates the inclusion of scattered chemicals, resulting in a minimal increase in relative molecular mass. As a result, it is capable of generating a modest boost in concentration.

10. Cases of decreasing concentration owing to a higher Prandtl number are linked with sinusoidal peristaltic waves at both walls, the presence of thermo-migration of small or nanosized particles, and random movement of tiny/nanosized particles. The concentration falls as the Lewis number increases. Nonetheless, variations in the concentration with the Lewis number when the Prandtl number is also low (gases) and big (air) are not implausible.

12.3 Recommendation

Based on the analysis of the results and the conclusion as mentioned above, it is worth making the following recommendations:

1. An increase in the Prandtl number is recommended to decrease the velocity of transport phenomenon of various fluids.
2. Introduction of melting heat transfer is recommended to establish increment in the velocities (vertical and horizontal) as the Prandtl number increases.
3. When the Prandtl number is small (gases) and high (air), a rapid change in horizontal velocity with the Lewis number is suggested.
4. An enhancement in the Prandtl number is recommended to diminish temperature distribution across various fluid flows.
5. (i) In the case of the dynamics of dusty fluid, (ii) when mass flux due to temperature gradient and energy flux due to concentration gradient are significant, (iii) when there is melting heat transfer at both walls, (iv) when convective heating exists at both sides of the domain, and (v) when there is thermal jump at the surface, an increase in the P_r is recommended to cause an increment in the temperature distribution.
6. Increasing property of the Prandtl number is recommended to cause an increment in the temperature gradient due to the direction of changes in temperature and the changes in the rate of temperature.
7. A rise in the ratio of momentum diffusivity to thermal diffusivity is recommended to diminish the temperature gradient due to melting heat transfer at the wall, thermal jump at the wall and free stream, and thermal stratification at the wall.
8. Increasing the Prandtl number is recommended to boost the Nusselt number. However, sinusoidal peristaltic waves at both walls, significant mass flux due to temperature gradient and energy flux due to concentration gradient, and melting heat transfer are recommended to reduce the Nusselt number.
9. Sequel to the transfer of mass/species in a highly viscous fluid minimal due to higher resistance and more vital intermolecular forces, an increase in the Prandtl number is recommended to reduce the local Sherwood number.
10. The presence of mass flux due to temperature gradient and energy flux due to concentration gradient, and the absence of melting heat transfer is recommended to cause a rise in the mass transfer rate when the Prandtl number is increased.
11. An increase in the viscosity as P_r increases denotes the addition of dispersed substances, consequently boosting the relative molecular mass negligible. Thus, this is capable of causing an increase in the concentration slightly.

12. A higher Prandtl number is recommended to cause a decrease in concentration due to sinusoidal peristaltic waves at both walls and the existence of thermo-migration of tiny/nanosized particles, and the haphazard motion tiny/nano-sized particles.

12.4 Tutorial Questions

1. Exclusively, outline the effects of increasing the Prandtl number on transport phenomenon experiencing viscous dissipation.
2. Under what circumstances does an increase in the Prandtl number become capable of producing a temperature gradient increase?
3. Remark on the relationship between the Prandtl number and temperature.
4. What is the effect of increasing the Prandtl number on the concentration of fluid flow?

A

Appendix I

Order of magnitude argument for the reduction of the Navier-Stoke Equation to the Boundary layer Equation

In compact form, the Navier-Stoke equation is of the form

$$\rho\frac{d\mathbf{V}}{dt} = \rho g - \nabla p + \nabla * \tau_{ij} \tag{A.1}$$

Equation (A.1) implies that density multiplied by the acceleration of the flow is equivalent to the gravity force per unit volume in the absence or removal of pressure force per unit volume and presence or addition of the viscous force per unit volume. Mathematically, for three-dimensional flow $\mathbf{V} = u(x, y, z), v(x, y, z), w(x, y, z)$; hence, Eq. (A.1) can be expressed as

$$\rho\left(\frac{\partial u}{\partial t} + u\frac{\partial u}{\partial x} + v\frac{\partial u}{\partial y} + w\frac{\partial u}{\partial z}\right) = \rho g_x - \frac{dp}{dx} + \frac{\partial \tau_{xx}}{\partial x} + \frac{\partial \tau_{yx}}{\partial y} + \frac{\partial \tau_{zx}}{\partial z}, \tag{A.2}$$

$$\rho\left(\frac{\partial v}{\partial t} + u\frac{\partial v}{\partial x} + v\frac{\partial v}{\partial y} + w\frac{\partial v}{\partial z}\right) = \rho g_y - \frac{dp}{dy} + \frac{\partial \tau_{xy}}{\partial x} + \frac{\partial \tau_{yy}}{\partial y} + \frac{\partial \tau_{zy}}{\partial z}, \tag{A.3}$$

$$\rho\left(\frac{\partial w}{\partial t} + u\frac{\partial w}{\partial x} + v\frac{\partial w}{\partial y} + w\frac{\partial w}{\partial z}\right) = \rho g_z - \frac{dp}{dz} + \frac{\partial \tau_{xz}}{\partial x} + \frac{\partial \tau_{yz}}{\partial y} + \frac{\partial \tau_{zz}}{\partial z}. \tag{A.4}$$

Sequel to the fact that the Navier–Stoke equation is based on the assumption, "the viscous stresses are proportional to the element strain rates and the coefficient of viscosity,"

$$\tau_{xx} = 2\mu\frac{\partial u}{\partial x}, \quad \tau_{yy} = 2\mu\frac{\partial v}{\partial y}, \quad \tau_{zz} = 2\mu\frac{\partial w}{\partial z}, \quad \tau_{xy} = \tau_{yx} = \mu\left(\frac{\partial u}{\partial y} + \frac{\partial v}{\partial x}\right)$$

$$\tau_{xz} = \tau_{zx} = \mu\left(\frac{\partial w}{\partial x} + \frac{\partial u}{\partial z}\right), \quad \tau_{yz} = \tau_{zy} = \mu\left(\frac{\partial w}{\partial y} + \frac{\partial v}{\partial z}\right). \tag{A.5}$$

Substituting Eq. (A.5) into Eqs. (A.2)–(A.4) and then differentiating, we get

$$\rho\left(\frac{\partial u}{\partial t} + u\frac{\partial u}{\partial x} + v\frac{\partial u}{\partial y} + w\frac{\partial u}{\partial z}\right) = \rho g_x - \frac{dp}{dx} + \mu\left(\frac{\partial^2 u}{\partial x^2} + \frac{\partial^2 u}{\partial y^2} + \frac{\partial^2 u}{\partial z^2}\right), \tag{A.6}$$

$$\rho\left(\frac{\partial v}{\partial t} + u\frac{\partial v}{\partial x} + v\frac{\partial v}{\partial y} + w\frac{\partial v}{\partial z}\right) = \rho g_y - \frac{dp}{dy} + \mu\left(\frac{\partial^2 v}{\partial x^2} + \frac{\partial^2 v}{\partial y^2} + \frac{\partial^2 v}{\partial z^2}\right), \tag{A.7}$$

$$\rho\left(\frac{\partial w}{\partial t} + u\frac{\partial w}{\partial x} + v\frac{\partial w}{\partial y} + w\frac{\partial w}{\partial z}\right) = \rho g_z - \frac{dp}{dz} + \mu\left(\frac{\partial^2 w}{\partial x^2} + \frac{\partial^2 w}{\partial y^2} + \frac{\partial^2 w}{\partial z^2}\right). \tag{A.8}$$

For two-dimensional unsteady flow, the order of $\frac{\partial u}{\partial t}$ is $O(1)$. For convection-induced flow, g_z is of order $O(1)$. For a steady case,

$$u\frac{\partial u}{\partial x} + v\frac{\partial u}{\partial y} = -\frac{1}{\rho}\frac{dp}{dx} + \frac{\mu}{\rho}\left(\frac{\partial^2 u}{\partial x^2} + \frac{\partial^2 u}{\partial y^2}\right), \tag{A.9}$$

$$u\frac{\partial v}{\partial x} + v\frac{\partial v}{\partial y} = -\frac{1}{\rho}\frac{dp}{dy} + \frac{\mu}{\rho}\left(\frac{\partial^2 v}{\partial x^2} + \frac{\partial^2 v}{\partial y^2}\right), \tag{A.10}$$

subject to

$$u = 0, \quad v = 0, \quad at \quad y = 0 \tag{A.11}$$

$$u = u_1(x) \quad at \quad x = 0 \tag{A.12}$$

$$u \to u_1(x) \quad as \quad \frac{y}{\delta} \to \infty \tag{A.13}$$

The order of magnitude argument looks somehow rough but approximate as it gives an idea of the relative importance of the individual term in Eqs. (A.9) and (A.10). However, for Reynold number $R_e \to \infty$ the argument becomes asymptotically exact. For a typical streamline, a representative change in velocity $\approx O(U_\infty)$ over a length $O(l)$ (i.e., change in velocity along the x-axis over distance l) $u = O(U_\infty)$ (Figure A.1). Next,

$$\frac{\partial u}{\partial x} = \frac{O(U_\infty)}{O(l)} = O\left(\frac{U_\infty}{l}\right) \tag{A.14}$$

$$\frac{\partial u}{\partial y} = \frac{O(U_\infty)}{O(\delta)} = O\left(\frac{U_\infty}{\delta}\right) \tag{A.15}$$

The maximum change in velocity in the y-direction for a streamline is $O(\delta)$. However,

$$\frac{v}{U_\infty} = \frac{O(\delta)}{O(l)} = O\left(\frac{\delta}{L}\right)$$

$$v = O\left(U_\infty \frac{\delta}{l}\right) \tag{A.16}$$

In fact, the continuity equation could also be used to obtain Eq. (A.16) as follows:

$$\frac{\partial u}{\partial x} + \frac{\partial v}{\partial y} = 0 \tag{A.17}$$

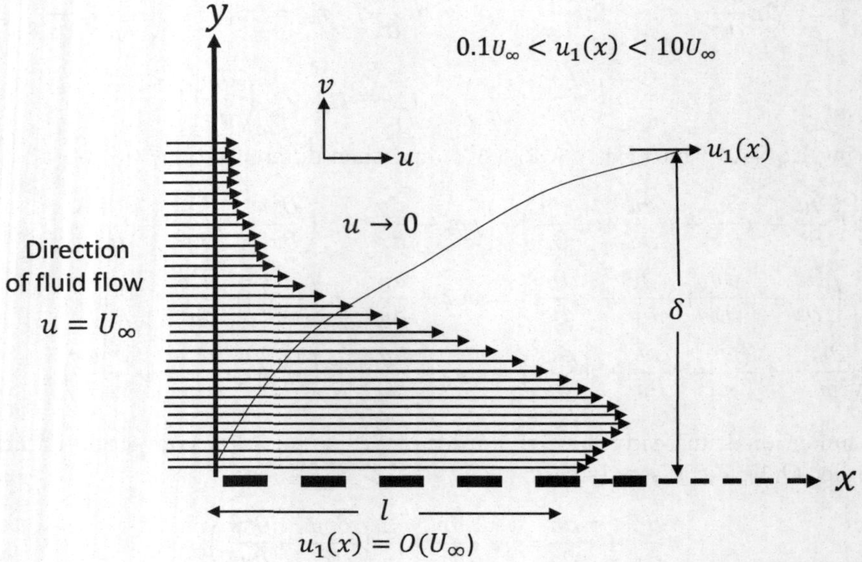

FIGURE A.1
Graphical illustration of a boundary layer flow

Substituting Eq. (A.14) into Eq. (A.17) and setting $\frac{\partial v}{\partial y} = \frac{O(v)}{O(\delta)} = O\left(\frac{v}{\delta}\right)$ we get

$$\frac{\partial u}{\partial x} + \frac{\partial v}{\partial y} = \underbrace{O\left(\frac{U_\infty}{l}\right) + O\left(\frac{v}{\delta}\right) = 0} \tag{A.18}$$

Thus, this leads to

$$O\left(\frac{U_\infty}{l}\right) = -O\left(\frac{v}{\delta}\right)$$

$$O(v) = \delta O\left(\frac{U_\infty}{l}\right)$$

Since we are interested in the magnitude, it is valid to ignore the sign and $O(v) = v$.

$$v = O\left(U_\infty \frac{\delta}{l}\right) \tag{A.19}$$

Then,

$$\frac{\partial v}{\partial x} = \frac{O(U_\infty \frac{\delta}{l})}{O(l)} = O\left(U_\infty \frac{\delta}{l^2}\right) \tag{A.20}$$

And

$$\frac{\partial v}{\partial y} = \frac{O(U_\infty \frac{\delta}{l})}{O(\delta)} = O\left(U_\infty \frac{1}{l}\right) \tag{A.21}$$

Next is to focus on the convective acceleration terms

$$u\frac{\partial u}{\partial x} = O(U_\infty) * O\left(\frac{U_\infty}{l}\right) = O\left(\frac{U_\infty^2}{l}\right) \tag{A.22}$$

$$v\frac{\partial u}{\partial y} = O\left(U_\infty \frac{\delta}{l}\right) * O\left(\frac{U_\infty}{\delta}\right) = O\left(\frac{U_\infty^2}{l}\right) \tag{A.23}$$

$$u\frac{\partial v}{\partial x} = O(U_\infty) * O\left(U_\infty \frac{\delta}{l^2}\right) = O\left(U_\infty^2 \frac{\delta}{l^2}\right) \tag{A.24}$$

$$v\frac{\partial v}{\partial y} = O\left(U_\infty \frac{\delta}{l}\right) * O\left(U_\infty \frac{1}{l}\right) = O\left(U_\infty^2 \frac{\delta}{l^2}\right) \tag{A.25}$$

Next is to consider the second derivatives

$$\frac{\partial^2 u}{\partial y^2} = \frac{\partial}{\partial y}\left(\frac{\partial u}{\partial y}\right) = \frac{\partial}{\partial y}\left[O\left(\frac{U_\infty}{\delta}\right)\right] = \frac{O\left(\frac{U_\infty}{\delta}\right)}{O(\delta)} = O\left(\frac{U_\infty}{\delta^2}\right) \tag{A.26}$$

And

$$\frac{\partial^2 u}{\partial x^2} = \frac{\partial}{\partial x}\left(\frac{\partial u}{\partial x}\right) = \frac{\partial}{\partial x}\left[O\left(\frac{U_\infty}{l}\right)\right] = \frac{O\left(\frac{U_\infty}{l}\right)}{O(l)} = O\left(\frac{U_\infty}{l^2}\right) \tag{A.27}$$

Next,

$$\frac{\partial^2 v}{\partial x^2} = \frac{\partial}{\partial x}\left(\frac{\partial v}{\partial x}\right) = \frac{\partial}{\partial x}\left[O\left(U_\infty \frac{\delta}{l^2}\right)\right] = O\left(U_\infty \frac{\delta}{l^3}\right) \tag{A.28}$$

And

$$\frac{\partial^2 v}{\partial y^2} = \frac{\partial}{\partial y}\left(\frac{\partial v}{\partial y}\right) = \frac{\partial}{\partial y}\left[O\left(U_\infty \frac{1}{l}\right)\right] = O\left(U_\infty \frac{1}{l\delta}\right) \tag{A.29}$$

Substituting Eqs. (A.14)–(A.29) into

$$u\frac{\partial u}{\partial x} + v\frac{\partial u}{\partial y} = -\frac{1}{\rho}\frac{dp}{dx} + \frac{\mu}{\rho}\left(\frac{\partial^2 u}{\partial x^2} + \frac{\partial^2 u}{\partial y^2}\right),$$

$$u\frac{\partial v}{\partial x} + v\frac{\partial v}{\partial y} = -\frac{1}{\rho}\frac{dp}{dy} + \frac{\mu}{\rho}\left(\frac{\partial^2 v}{\partial x^2} + \frac{\partial^2 v}{\partial y^2}\right).$$

we obtain

$$O\left(\frac{U_\infty^2}{l}\right) + O\left(\frac{U_\infty^2}{l}\right) = O\left(-\frac{1}{\rho}\frac{dp}{dx}\right) + \frac{\mu}{\rho}\left[O\left(\frac{U_\infty}{l^2}\right) + O\left(\frac{U_\infty}{\delta^2}\right)\right], \tag{A.30}$$

$$O\left(U_\infty^2\frac{\delta}{l^2}\right) + O\left(U_\infty^2\frac{\delta}{l^2}\right) = O\left(-\frac{1}{\rho}\frac{dp}{dx}\right) + \frac{\mu}{\rho}\left[O\left(U_\infty\frac{\delta}{l^3}\right) + O\left(U_\infty\frac{1}{l\delta}\right)\right]. \tag{A.31}$$

Multiply both sides of Eqs. (A.30) and (A.31) by $\frac{l}{U_\infty^2}$.

$$O(1) + O(1) = O\left(-\frac{l}{U_\infty^2}\frac{1}{\rho}\frac{dp}{dx}\right) + \left[O\left(\frac{\vartheta}{lU_\infty}\right) + O\left(\frac{l\vartheta}{U_\infty\delta^2}\right)\right], \tag{A.32}$$

$$O\left(\frac{\delta}{l}\right) + O\left(\frac{\delta}{l}\right) = O\left(-\frac{l}{U_\infty^2}\frac{1}{\rho}\frac{dp}{dy}\right) + \left[O\left(\frac{\vartheta\delta}{l^2U_\infty}\right) + O\left(\frac{\vartheta}{U_\infty\delta}\right)\right]. \tag{A.33}$$

Introduce the Reynolds number $R_e = \frac{lU_\infty}{\vartheta}$.

$$O(1) + O(1) = O\left(-\frac{l}{U_\infty^2}\frac{1}{\rho}\frac{dp}{dx}\right) + \left[O\left(\frac{1}{R_e}\right) + O\left(\frac{1}{R_e}\frac{l^2}{\delta^2}\right)\right], \tag{A.34}$$

$$O\left(\frac{\delta}{l}\right) + O\left(\frac{\delta}{l}\right) = O\left(-\frac{l}{U_\infty^2}\frac{1}{\rho}\frac{dp}{dy}\right) + \left[O\left(\frac{1}{R_e}\frac{\delta}{l}\right) + O\left(\frac{1}{R_e}\frac{l}{\delta}\right)\right]. \tag{A.35}$$

It is worth remarking that as $R_e \to \infty$, $\frac{1}{R_e} \to 0$; hence, $\frac{\partial^2 u}{\partial x^2}$ is insignificant to model boundary layer flow.

$$O(1) + O(1) = O\left(-\frac{l}{U_\infty^2}\frac{1}{\rho}\frac{dp}{dx}\right) + O\left(\frac{1}{R_e}\frac{l^2}{\delta^2}\right), \tag{A.36}$$

$$O\left(\frac{\delta}{l}\right) + O\left(\frac{\delta}{l}\right) = O\left(-\frac{l}{U_\infty^2}\frac{1}{\rho}\frac{dp}{dy}\right) + \left[O\left(\frac{1}{R_e}\frac{\delta}{l}\right) + O\left(\frac{1}{R_e}\frac{l}{\delta}\right)\right]. \tag{A.37}$$

Three different cases shall be considered (in order to conclude on the order of viscous term (i.e., $O\left(\frac{1}{R_e}\frac{l^2}{\delta^2}\right)$)) suitable to model the boundary layer flow.

Case A: $O\left(\frac{1}{R_e}\frac{l^2}{\delta^2}\right) \gg O(1)$

Here, the viscous terms dominate over inertial terms. For this case,

$$O\left(-\frac{l}{U_\infty^2}\frac{1}{\rho}\frac{dp}{dx}\right) = O\left(\frac{1}{R_e}\frac{l^2}{\delta^2}\right) \tag{A.38}$$

This implies the balance of pressure gradient and viscous forces. Such a situation would exist for fully developed flow in straight parallel pipes and ducts.

Case B: $O\left(\frac{1}{R_e}\frac{l^2}{\delta^2}\right) \ll O(1)$

This leaves a balance of inertia and pressure gradient forces (i.e., inviscid or Euler equation flow).

Case C: $O\left(\frac{1}{R_e}\frac{l^2}{\delta^2}\right) = O(1)$

In this case, inertia forces, pressure gradient, and viscous forces are all of the same importance. This case was considered by Ludwig Prandtl to be suitable to model boundary layer flow. However, this implies

$$O\left(\frac{1}{R_e}\frac{l^2}{\delta^2}\right) = O(1)$$

$$\frac{l^2}{\delta^2} = R_e, \quad \delta = \frac{l}{\sqrt{R_e}}$$

$$\delta = O\left(\frac{l}{\sqrt{R_e}}\right) \tag{A.39}$$

Substituting Eq. (A.39) into Eqs. (A.40) and (A.41)

$$O(1) + O(1) = O\left(-\frac{l}{U_\infty^2}\frac{1}{\rho}\frac{dp}{dx}\right) + O(1), \tag{A.40}$$

$$O\left(\frac{l}{R_e}\right) + O\left(\frac{l}{R_e}\right) = O\left(-\frac{l}{U_\infty^2}\frac{1}{\rho}\frac{dp}{dy}\right) + \left[O\left(\frac{1}{R_e}\frac{1}{\sqrt{R_e}}\right) + O\left(\frac{\sqrt{R_e}}{R_e}\right)\right]. \tag{A.41}$$

As $R_e \to \infty$, $\frac{1}{R_e} \to 0$. It is worth concluding that the boundary layer equation is of the form

$$\frac{\partial u}{\partial x} + \frac{\partial v}{\partial y} = 0, \tag{A.42}$$

$$u\frac{\partial u}{\partial x} + v\frac{\partial u}{\partial y} = -\frac{1}{\rho}\frac{dp}{dx} + \frac{\mu}{\rho}\frac{\partial^2 u}{\partial y^2}, \tag{A.43}$$

For Blasius flow, the boundary condition is

$$u = 0, \quad v = 0, \quad at \quad y = 0 \tag{A.44}$$

$$u \to u_1(x) \quad as \quad y \to \infty \tag{A.45}$$

where

$$u_1(x)\frac{du_1(x)}{dx} = -\frac{1}{\rho}\frac{dp}{dx}$$

[illegible]

[illegible]

[illegible]

[illegible]

[illegible]

B

Appendix II

Consider the momentum equation only for the boundary layer flow of nanofluid (Eq. (11.20))

$$\frac{\partial u}{\partial x} + \frac{\partial v}{\partial y} = 0, \tag{B.1}$$

$$u\frac{\partial u}{\partial x} + v\frac{\partial u}{\partial y} = \frac{\mu_{nf}}{\rho_{nf}}\frac{\partial^2 u}{\partial y^2} + g\beta_{nf}(T - T_\infty) - \frac{\mu_{nf}}{\rho_{nf}K}u - \frac{b^*}{K}u^2 \tag{B.2}$$

Given that

$$\beta_{nf} = (1-\phi)\beta_{bf} + \phi\beta_{sp}, \quad \rho_{nf} = (1-\phi)\rho_{bf} + \phi\rho_{sp}, \quad \frac{\mu_{nf}}{\mu_{bf}} = \frac{1}{(1-\phi)^{2.5}} \tag{B.3}$$

$$\eta = y\sqrt{\frac{U_\infty}{\vartheta_{bf}x}}, \quad \psi(x,y) = \sqrt{\vartheta_{bf}xU_\infty}f(\eta), \quad u = \frac{\partial\psi}{\partial y}, \quad v = -\frac{\partial\psi}{\partial x}, \quad \theta(\eta) = \frac{T - T_\infty}{T_w - T_\infty}$$

$$F_s = \frac{b^*}{x}, \quad D_a = \frac{K}{x^2}, \quad Re = \frac{U_o x^2}{\vartheta_{bf}}, \quad G_{bf} = \frac{g\beta_{bf}(T_w - T_\infty)x}{U_\infty^2}, \quad G_{sp} = \frac{g\beta_{sp}(T_w - T_\infty)x}{U_\infty^2}. \tag{B.4}$$

It is important to note that the continuity equation Eq. (B.1) is automatically satisfied. Next stage is to consider each of the equation components one after the other. In order to non-dimensionalize and parametrize the dimensional governing equation, Eq. (B.4) was substituted into Eq. (B.2) to obtain

$$u\frac{\partial u}{\partial x} = U_\infty^2\frac{df}{d\eta}\frac{d^2 f}{d\eta^2}y\sqrt{\frac{U_\infty}{\vartheta_{bf}}}\left(-\frac{1}{2}\right)x^{-3/2} \tag{B.5}$$

Next is to consider the second term of the convective acceleration term

$$v\frac{\partial u}{\partial y} = U_\infty^2\frac{df}{d\eta}\frac{d^2 f}{d\eta^2}y\sqrt{\frac{U_\infty}{\vartheta_{bf}}}\left(-\frac{1}{2}\right)x^{-3/2} - f(\eta)\sqrt{\vartheta_{bf}}\left(\frac{1}{2x}\right)\frac{U_\infty^2}{\sqrt{\vartheta_{bf}}}\frac{d^2 f}{d\eta^2} \tag{B.6}$$

The viscous term

$$\frac{\mu_{nf}}{\rho_{nf}}\frac{\partial^2 u}{\partial y^2} = \frac{\frac{\mu_{bf}}{(1-\phi)^{2.5}}}{(1-\phi)\rho_{bf} + \phi\rho_{sp}}U_\infty\frac{d^3 f}{d\eta^3}\frac{U_\infty}{\vartheta_{bf}x} = \frac{\vartheta_{bf}\frac{1}{(1-\phi)^{2.5}}}{1-\phi+\phi\frac{\rho_{sp}}{\rho_{bf}}}U_\infty\frac{d^3 f}{d\eta^3}\frac{U_\infty}{\vartheta_{bf}x} \tag{B.7}$$

Next is to consider the body force term using $T = T_\infty + \theta(\eta)(T_w - T_\infty)$

$$g\beta_{nf}(T - T_\infty) = g[(1-\phi)\beta_{bf} + \phi\beta_{sp}](T_\infty + \theta(\eta)(T_w - T_\infty) - T_\infty)$$

$$g\beta_{nf}(T - T_\infty) = g[(1-\phi)\beta_{bf} + \phi\beta_{sp}]\theta(\eta)(T_w - T_\infty) \tag{B.8}$$

We shall now consider the second pressure term

$$\frac{\mu_{nf}}{\rho_{nf}K}u = \frac{\mu_{bf}\frac{1}{(1-\phi)^{2.5}}}{[(1-\phi)\rho_{bf}+\phi\rho_{sp}]K}U_\infty\frac{df}{d\eta} = \frac{\vartheta_{bf}\frac{1}{(1-\phi)^{2.5}}}{[1-\phi+\phi\frac{\rho_{sp}}{\rho_{bf}}]K}U_\infty\frac{df}{d\eta} \tag{B.9}$$

Lastly,

$$\frac{b^*}{K}u^2 = \frac{b^*}{K}U_\infty^2\frac{df}{d\eta}\frac{df}{d\eta} \tag{B.10}$$

To obtain the dimensionless governing equation, we shall substitute Eqs. (B.5)–(B.11) into Eq. (B.2) to get

$$U_\infty^2\frac{df}{d\eta}\frac{d^2f}{d\eta^2}y\sqrt{\frac{U_\infty}{\vartheta_{bf}}}\left(-\frac{1}{2}\right)x^{-3/2} + U_\infty^2\frac{df}{d\eta}\frac{d^2f}{d\eta^2}y\sqrt{\frac{U_\infty}{\vartheta_{bf}}}\left(-\frac{1}{2}\right)x^{-3/2}$$

$$-f(\eta)\sqrt{\vartheta_{bf}}\left(\frac{1}{2x}\right)\frac{U_\infty^2}{\sqrt{\vartheta_{bf}}}\frac{d^2f}{d\eta^2} = \frac{\vartheta_{bf}\frac{1}{(1-\phi)^{2.5}}}{1-\phi+\phi\frac{\rho_{sp}}{\rho_{bf}}}U_\infty\frac{d^3f}{d\eta^3}\frac{U_\infty}{\vartheta_{bf}x}$$

$$+g[(1-\phi)\beta_{bf}+\phi\beta_{sp}]\theta(\eta)(T_w-T_\infty) - \frac{\vartheta_{bf}\frac{1}{(1-\phi)^{2.5}}}{[1-\phi+\phi\frac{\rho_{sp}}{\rho_{bf}}]K}U_\infty\frac{df}{d\eta} - \frac{b^*}{K}U_\infty^2\frac{df}{d\eta}\frac{df}{d\eta}$$

Upon simplification, we obtain

$$-f(\eta)\left(\frac{U_\infty^2}{2x}\right)\frac{d^2f}{d\eta^2} = \frac{\frac{1}{(1-\phi)^{2.5}}}{1-\phi+\phi\frac{\rho_{sp}}{\rho_{bf}}}U_\infty\frac{d^3f}{d\eta^3}\frac{U_\infty}{x} + g[(1-\phi)\beta_{bf}+\phi\beta_{sp}]\theta(\eta)(T_w-T_\infty)$$

$$-U_\infty\frac{\vartheta_{bf}\frac{1}{(1-\phi)^{2.5}}}{[1-\phi+\phi\frac{\rho_{sp}}{\rho_{bf}}]K}\frac{df}{d\eta} - \frac{b^*U_\infty^2}{K}\frac{df}{d\eta}\frac{df}{d\eta}$$

Divide both sides by U_∞^2.

$$-f(\eta)\left(\frac{1}{2x}\right)\frac{d^2f}{d\eta^2} = \frac{\frac{1}{(1-\phi)^{2.5}}}{1-\phi+\phi\frac{\rho_{sp}}{\rho_{bf}}}\frac{d^3f}{d\eta^3}\frac{1}{x} + g[(1-\phi)\beta_{bf}+\phi\beta_{sp}]\theta(\eta)\frac{(T_w-T_\infty)}{U_\infty^2}$$

$$-\frac{\vartheta_{bf}\frac{1}{(1-\phi)^{2.5}}}{[1-\phi+\phi\frac{\rho_{sp}}{\rho_{bf}}]U_\infty K}\frac{df}{d\eta} - \frac{b^*}{K}\frac{df}{d\eta}\frac{df}{d\eta}$$

Multiply both sides by x and rearrange.

$$\frac{\frac{1}{(1-\phi)^{2.5}}}{1-\phi+\phi\frac{\rho_{sp}}{\rho_{bf}}}\frac{d^3f}{d\eta^3} + \frac{1}{2}f\frac{d^2f}{d\eta^2} + g[(1-\phi)\beta_{bf}+\phi\beta_{sp}]\theta(\eta)\frac{(T_w-T_\infty)x}{U_\infty^2}$$

$$-x\frac{\vartheta_{bf}\frac{1}{(1-\phi)^{2.5}}}{[1-\phi+\phi\frac{\rho_{sp}}{\rho_{bf}}]U_\infty K}\frac{df}{d\eta} - \frac{xb^*}{K}\frac{df}{d\eta}\frac{df}{d\eta} = 0$$

Let us define the coefficients $A_1(\phi)$ as

$$A_1(\phi) = \frac{\frac{1}{(1-\phi)^{2.5}}}{1-\phi+\phi\frac{\rho_{sp}}{\rho_{bf}}}$$

Then, leads to

$$A_1\frac{d^3f}{d\eta^3}+\frac{1}{2}f\frac{d^2f}{d\eta^2}+g[(1-\phi)\beta_{bf}+\phi\beta_{sp}]\theta(\eta)\frac{(T_w-T_\infty)x}{U_\infty^2}-A_1\frac{x\vartheta_{bf}}{KU_\infty}\frac{df}{d\eta}-\frac{xb^*}{K}\frac{df}{d\eta}\frac{df}{d\eta}=0$$

It is very important to note that there are two buoyancy-related parameters. Buoyancy parameter is dependent on the thermal volumetric expansion of the base fluid G_{bf}, and the second one is dependent on the thermal volumetric expansion of the nanoparticles G_{sp}. These parameters are defined as

$$G_{bf}=\frac{g\beta_{bf}(T_w-T_\infty)x}{U_\infty^2},\quad G_{sp}=\frac{g\beta_{sp}(T_w-T_\infty)x}{U_\infty^2}$$

$$A_1\frac{d^3f}{d\eta^3}+\frac{1}{2}f\frac{d^2f}{d\eta^2}+G_{bf}(1-\phi)\theta+G_{sp}\phi\theta-A_1\frac{x\vartheta_{bf}}{KU_\infty}\frac{df}{d\eta}-\frac{xb^*}{K}\frac{df}{d\eta}\frac{df}{d\eta}=0$$

Next is to introduce the local Forchheimer number F_s, local Darcy number D_a and Reynold number R_e defined as

$$F_s=\frac{b^*}{x},\quad D_a=\frac{K}{x^2},\quad Re=\frac{U_ox^2}{\vartheta_{bf}}$$

Finally, obtained

$$A_1\frac{d^3f}{d\eta^3}+\frac{1}{2}f\frac{d^2f}{d\eta^2}+G_{bf}(1-\phi)\theta+G_{sp}\phi\theta-\frac{A_1}{D_aRe}\frac{df}{d\eta}-\frac{F_s}{D_a}\frac{df}{d\eta}\frac{df}{d\eta}=0 \tag{B.11}$$

C

Appendix III

Examine the dimensional governing equation presented by Ramzan et al. [250]. Remark on the governing equation and state if the transport phenomenon is forced convection, free convection, or mixed convection. Present the suitable governing equation for each mode of convection.

The governing equation for momentum adopted by Ramzan et al. [250] is written as

$$u\frac{\partial u}{\partial x} + v\frac{\partial u}{\partial y} = \vartheta\frac{\partial^2 u}{\partial y^2} + k_o\left(u\frac{\partial^3 u}{\partial x\partial y^2} + \frac{\partial u}{\partial x}\frac{\partial^2 u}{\partial y^2} + \frac{\partial u}{\partial y}\frac{\partial^2 v}{\partial y^2} + v\frac{\partial^3 u}{\partial y^3}\right)$$
$$- \frac{\vartheta}{K}u + g[\beta_T(T - T_\infty) + \beta_C(C - C_\infty)] \quad \text{(C.1)}$$

Without any doubt, two-dimensional governing Eq. (C.1) above for momentum is suitable to model the flow of viscoelastic fluid through a vertical porous medium due to free convection. The case of forced convection is of the form

$$u\frac{\partial u}{\partial x} + v\frac{\partial u}{\partial y} = -\frac{1}{\rho}\frac{dp}{dx} + \vartheta\frac{\partial^2 u}{\partial y^2} + k_o\left(u\frac{\partial^3 u}{\partial x\partial y^2} + \frac{\partial u}{\partial x}\frac{\partial^2 u}{\partial y^2} + \frac{\partial u}{\partial y}\frac{\partial^2 v}{\partial y^2} + v\frac{\partial^3 u}{\partial y^3}\right) - \frac{\vartheta}{K}u. \quad \text{(C.2)}$$

The pressure term in Eq. (C.2) is

$$-\frac{1}{\rho}\frac{dp}{dx} = u_e\frac{du_e}{dx} + \frac{\vartheta}{K}u_e. \quad \text{(C.3)}$$

For the case of mixed convection, the combination of pressure as written in Eq. (C.3) and buoyancy are incorporated into the governing equation as

$$u\frac{\partial u}{\partial x} + v\frac{\partial u}{\partial y} = -\frac{1}{\rho}\frac{dp}{dx} + \vartheta\frac{\partial^2 u}{\partial y^2} + k_o\left(u\frac{\partial^3 u}{\partial x\partial y^2} + \frac{\partial u}{\partial x}\frac{\partial^2 u}{\partial y^2} + \frac{\partial u}{\partial y}\frac{\partial^2 v}{\partial y^2} + v\frac{\partial^3 u}{\partial y^3}\right)$$
$$-\frac{\vartheta}{K}u + g[\beta_T(T - T_\infty) + \beta_C(C - C_\infty)]. \quad \text{(C.4)}$$

D

Appendix IV

(a) Remark on the relationship between the Prandtl number and the Schmidt number.

Conventionally, Prandtl number is P_r, Lewis number is L_e, and Schmidt number is S_c. Mathematically,

$$P_r \times L_e = \frac{\vartheta}{\alpha} \times \frac{\alpha}{D_m} = \frac{\vartheta}{D_m} = S_c.$$

where the momentum diffusivity ϑ, the thermal diffusivity is α, and the mass diffusivity is D_m. Hence, the ratio of the Schmidt number to the Prandtl number is the Lewis number. Whenever thermal diffusivity plays a significant role during mass transfer, the product of Prandtl number and Lewis number seems more appropriate to be incorporated into the concentration or species equation.

(b) Show that the emerged buoyancy parameter Gr_x defined below is dimensionless.

$$Gr_x = \frac{gx\beta(T_w - T_\infty)}{U_\infty^2} = \frac{g\beta(T_w - T_\infty)}{U_o^2 x}$$

Another name for the emerged buoyancy parameter is the Grashof number. The dimensionless number is a number depending on the acceleration due to gravity g ($\mathrm{ms^{-2}}$), volumetric thermal expansion β ($\mathrm{K^{-1}}$), temperature difference across the fluid flow of a certain domain $(T_w - T_\infty)$ (K), stretching rate U_o ($\mathrm{s^{-1}}$), characteristic length x (m), and stretching velocity $U_\infty = U_o x$ ($\mathrm{ms^{-1}}$). It is obvious that

$$Gr_x = \frac{\mathrm{ms^{-2} * K^{-1} * K}}{\mathrm{s^{-2} * m}} = 1$$

(c) Upon using the Buongiorno model to investigate the significance of Brownian motion and thermophoresis of very tiny particles, mention the two related parameters to quantify the phenomenon. Show that the two parameters are dimensionless.

Thermophoresis parameter N_t and Brownian motion parameter N_b are the two related parameters to quantify the significance of thermophoresis of very tiny particles and the haphazard motion of the tiny particles. These are defined as

$$N_b = \frac{\tau D_a(C_w - C_\infty)}{\alpha \Delta C}, \quad N_t = \frac{\tau D_T}{T_\infty \alpha}(T_w - T_\infty)$$

In this case, the dimensionless number τ is the ratio of heat capacity of the nanoparticles $(\rho C_p)_{sp}$ to the heat capacity of the base fluid $(\rho C_p)_{bf}$. For this case, Brownian diffusion coefficient is equivalent to mass diffusivity of nanoparticles D_a $(m^2 s^{-1})$, diffusivity of nanoparticles due to temperature gradient D_T $(m^2 s^{-1})$, and concentration gradient $(C_w - C_\infty)$; the change in concentration ΔC has the unit $mol * m^{-3}$; the thermal diffusivity α has the unit m^2s^{-1}; and the temperature difference $(T_w - T_\infty)$, and the free stream temperature T_∞ have the unit K. Hence,

$$N_b = \frac{\tau D_a(C_w - C_\infty)}{\alpha \Delta C} = \frac{1 * m^2s^{-1} * mol * m^{-3}}{m^2s^{-1} * mol * m^{-3}} = 1$$

$$N_t = \frac{\tau D_T}{T_\infty \alpha}(T_w - T_\infty) = \frac{1 * m^2s^{-1}}{K * m^2s^{-1}} K = 1$$

(d) Given that the dimensionless suction or injection velocity $f_w = -\frac{v_w}{\sqrt{\vartheta U_o}}$, show that it is indeed dimensionless.

It is worth mentioning that ϑ is the kinematic viscosity and its unit is m^2s^{-1}. The unit of the stretching rate U_o is s^{-1}. Also, the unit of suction/injection velocity v_w is ms^{-1}. Hence,

$$f_w = -\frac{v_w}{\sqrt{\vartheta U_o}} = \frac{ms^{-1}}{ms^{-1}} = -1$$

E

Appendix V

(a) The dynamics of an incompressible fluid within the thin boundary layer formed on a surface in the presence of either suction or injection on a Cartesian plane occurs at $(x, y = 0)$. Given that suction or injection velocity is $v = v_w$, show that the dimensionless quantity $f_w = -\frac{v_w}{\sqrt{\vartheta U_o}}$

The problem implies

$$v = v_w \quad at \quad y = 0, \tag{E.1}$$

whereas

$$\eta = y\sqrt{\frac{U_\infty}{\vartheta x}}, \quad \psi(x, y) = \sqrt{\vartheta x U_\infty} f(\eta), \quad v = -\frac{\partial \psi}{\partial x}, \quad U_\infty = U_o x.$$

In other words,

$$v = -\frac{\partial \psi}{\partial x} = -\frac{\partial}{\partial x}[\sqrt{\vartheta x U_\infty} f(\eta)]$$

Next is to consider the product rule of differentiation since $\eta \varepsilon (x, y)$

$$v = -\left[\sqrt{\vartheta x U_\infty}\frac{\partial f}{\partial x} + f(\eta)\frac{\partial}{\partial x}[\sqrt{\vartheta x U_\infty}]\right]$$

Whereas $U_\infty = U_o x$, then

$$v = -\left[\sqrt{\vartheta x U_\infty}\frac{df}{d\eta}\frac{\partial \eta}{\partial x} + f(\eta)[\sqrt{\vartheta U_o}]\right] \tag{E.2}$$

whereas, upon substituting $U_\infty = U_o x$ into $\eta = y\sqrt{\frac{U_o}{\vartheta}}$,

$$\frac{\partial \eta}{\partial x} = 0$$

Substituting this into Eq. (F.4), to get

$$v = -f(\eta)[\sqrt{\vartheta U_o}] \tag{E.3}$$

Hence, it is necessary to equate Eq. (F.6) to the suction or injection velocity $v = v_w$

$$-f(\eta)\sqrt{\vartheta U_o} = v_w$$

$$f = -\frac{v_w}{\sqrt{\vartheta U_o}} = f_w \tag{E.4}$$

(b) Determine the dimensionless suction or injection velocity f_w for the case of fluid flow due to Marangoni convection where surface tension plays a significant role.

In this case of fluid flow due to Marangoni convection, the shear stress at the wall τ is equivalent to $\frac{\partial \sigma}{\partial T}\frac{\partial T}{\partial x}$. In view of this, the appropriate similarity variables as suggested by Sastry et al. [273] are

$$\eta = c_2 y, \quad \psi(x,y) = c_1 x f(\eta), \quad v = -\frac{\partial \psi}{\partial x}, \quad c_1 = \left(\frac{\sigma_0 \gamma a \mu}{\rho_f^2}\right)^{\frac{1}{3}}, \quad c_2 = \left(\frac{\sigma_0 \gamma a \rho_f}{\mu_f}\right)^{\frac{1}{3}}.$$

starting with

$$v = -\frac{\partial \psi}{\partial x} = -\frac{\partial}{\partial x}[c_1 x f(\eta)]$$

In this case, η and c_2 are independent of x; hence,

$$v = -c_1 f(\eta)$$

Equating to the suction or injection velocity v_w

$$f(\eta) = -\frac{v_w}{c_1}$$

It is very important to remark that the unit of c_1 is ms^{-1}.

(c) Consider the case of fluid flow on a linear stratification of thermal energy on a horizontal surface. Given that the wall temperature $T_w(x) = T_o + m_2 x$ and the free stream temperature $T_\infty(x) = T_o + m_1 x$, obtain the dimensionless boundary conditions.

The question implies that as the fluid flows along an horizontal surface in the x-direction, T_o is the reference temperature and the wall temperature $T_w(x) = T_o + m_2 x$ implies that

$$At \;\; y = 0, \quad T_w(x) = T_o + m_2 x \tag{E.5}$$

and the free stream temperature $T_\infty(x) = T_o + m_1 x$ implies

$$As \;\; y \to \infty, \quad T_\infty(x) = T_o + m_1 x \tag{E.6}$$

The dimensionless temperature suitable to capture the thermal stratification is of the form

$$\theta(\eta) = \frac{T - T_\infty}{T_w - T_o}, \quad and \quad \eta = y\sqrt{\frac{U_\infty}{\vartheta x}} \tag{E.7}$$

Substituting Eqs. (E.5) and (E.6) into Eq. (E.7), to obtain

$$\theta(\eta) = \frac{T_o + m_2 x - T_o - m_1 x}{T_o + m_2 x - T_o} = \frac{m_2 x - m_1 x}{m_2 x}.$$

Given that the dimensionless thermal stratification parameter $S_t = \frac{m_1}{m_2}$, this leads to

$$\theta(\eta) = 1 - S_t$$

Since the condition holds at $y = 0$, substituting $y = 0$ into Eq. (E.7b) corresponds to $\eta = 0$. Likewise, substituting $y \to \infty$ into Eq. (E.7b) implies $\eta \to \infty$. Hence, the dimensionless conditions are

$$At \;\; \eta = 0, \;\; \theta(\eta) = 1 - S_t \tag{E.8}$$

$$As \;\; \eta \to \infty, \;\; \theta(\eta) \to 0 \tag{E.9}$$

F

Appendix VI

Given that the energy equation for two-dimensional flow within the region where the effect of viscosity is significant is Eq. (F.1), with the aid of the following similarity variables in Eq. (F.2), obtain the dimensionless governing equation

$$u\frac{\partial T}{\partial x} + v\frac{\partial T}{\partial y} = \frac{\kappa}{\rho Cp}\frac{\partial^2 T}{\partial y^2}. \tag{F.1}$$

$$\eta = y\sqrt{\frac{U_\infty}{\vartheta x}}, \quad \psi(x,y) = \sqrt{\vartheta x U_\infty} f(\eta), \quad u = \frac{\partial \psi}{\partial y}, \quad v = -\frac{\partial \psi}{\partial x}, \quad S_t = \frac{m_1}{m_2},$$
$$\theta(\eta) = \frac{T - T_\infty}{T_w - T_o}, \quad P_r = \frac{\mu Cp}{\kappa} = \frac{\vartheta}{\alpha}, \quad T_w(x) = T_o + m_2 x, \quad T_\infty(x) = T_o + m_1 x. \tag{F.2}$$

First, note that $U_\infty = U_o x$

$$u = \frac{\partial \psi}{\partial y} = \frac{\partial}{\partial y}(\sqrt{\vartheta x U_\infty} f(\eta)) = U_\infty \frac{df}{d\eta}$$

From Eq. (F.2), $T = T_\infty + \theta(T_w - T_o)$. Next is to consider

$$u\frac{\partial T}{\partial x} = U_\infty \frac{df}{d\eta}\frac{\partial}{\partial x}[T_\infty + \theta(T_w - T_o)]$$

$$u\frac{\partial T}{\partial x} = U_\infty \frac{df}{d\eta}\left[\frac{\partial T_\infty}{\partial x} + (T_w - T_o)\frac{\partial \theta}{\partial x} + \theta\frac{\partial}{\partial x}(T_w - T_o)\right] \tag{F.3}$$

This is necessary since at each point on y-axis, the temperature distribution varies along the x-axis. However,

$$\frac{\partial T_\infty}{\partial x} = \frac{\partial T_o}{\partial x} + \frac{\partial}{\partial x}(m_1 x) = m_1$$

$$(T_w - T_o)\frac{\partial \theta}{\partial x} = m_2 x \frac{d\theta}{d\eta}\frac{\partial \eta}{\partial x} = m_2 x \frac{d\theta}{d\eta}\left(-\frac{1}{2}\right) x^{-3/2} y \sqrt{\frac{U_\infty}{\vartheta}}$$

$$\theta\frac{\partial}{\partial x}(T_w - T_o) = \theta\frac{\partial}{\partial x}(m_2 x) = \theta m_2$$

It is very important that the reference temperature T_o is a constant. Hence, $\frac{\partial T_o}{\partial x} = 0$. Substituting these three equations into Eq. (F.3), to obtain

$$u\frac{\partial T}{\partial x} = U_\infty \frac{df}{d\eta}\left[m_1 + m_2 x \frac{d\theta}{d\eta}\left(-\frac{1}{2}\right) x^{-3/2} y \sqrt{\frac{U_\infty}{\vartheta}} + \theta m_2\right]$$

Expand to get

$$u\frac{\partial T}{\partial x} = m_1 U_\infty \frac{df}{d\eta} + U_\infty \frac{df}{d\eta} m_2 x \frac{d\theta}{d\eta}\left(-\frac{1}{2}\right) x^{-3/2} y \sqrt{\frac{U_\infty}{\vartheta}} + U_\infty \frac{df}{d\eta}\theta m_2 \tag{F.4}$$

Based on the fact that the stratification of thermal energy is only along x and not along y, it is valid to consider

$$v\frac{\partial T}{\partial y} = v\frac{\partial T}{\partial \theta}\frac{d\theta}{d\eta}\frac{\partial \eta}{\partial y} = vm_2x\frac{d\theta}{d\eta}\sqrt{\frac{U_\infty}{\vartheta x}} \tag{F.5}$$

whereas, it is easy to deduce that

$$v = -\frac{\partial \psi}{\partial x} = -x^{-1}U_\infty\frac{df}{d\eta}y\left(-\frac{1}{2}\right) - f\vartheta^{1/2}\frac{1}{2}x^{-1/2}U_\infty^{1/2}$$

Equation (F.6) becomes

$$v\frac{\partial T}{\partial y} = m_2x\frac{d\theta}{d\eta}\sqrt{\frac{U_\infty}{\vartheta x}}\left[-x^{-1}U_\infty\frac{df}{d\eta}y\left(-\frac{1}{2}\right) - f\vartheta^{1/2}\frac{1}{2}x^{-1/2}U_\infty^{1/2}\right]$$

Expand to get

$$v\frac{\partial T}{\partial y} = -x^{-1}U_\infty\frac{df}{d\eta}y\left(-\frac{1}{2}\right)m_2x\frac{d\theta}{d\eta}\sqrt{\frac{U_\infty}{\vartheta x}} - f\vartheta^{1/2}\frac{1}{2}x^{-1/2}U_\infty^{1/2}m_2x\frac{d\theta}{d\eta}\sqrt{\frac{U_\infty}{\vartheta x}} \tag{F.6}$$

Lastly,

$$\frac{\kappa}{\rho Cp}\frac{\partial^2 T}{\partial y^2} = \frac{\kappa}{\rho Cp}\frac{\partial}{\partial y}\left[\frac{\partial T}{\partial y}\right] = \frac{\kappa}{\rho Cp}\frac{d}{d\eta}\left[\frac{\partial T}{\partial y}\right]\frac{\partial \eta}{\partial y} = \frac{\kappa}{\rho Cp}\frac{d}{d\eta}\left[\frac{\partial T}{\partial \theta}\frac{d\theta}{d\eta}\frac{\partial \eta}{\partial y}\right]\frac{\partial \eta}{\partial y}$$

leads to

$$\frac{\kappa}{\rho Cp}\frac{\partial^2 T}{\partial y^2} = \frac{\kappa}{\rho Cp}\frac{d}{d\eta}\left[m_2x\frac{d\theta}{d\eta}\sqrt{\frac{U_\infty}{\vartheta x}}\right]\frac{\partial \eta}{\partial y}$$

$$\frac{\kappa}{\rho Cp}\frac{\partial^2 T}{\partial y^2} = \frac{\kappa}{\rho Cp}m_2x\frac{d^2\theta}{d\eta^2}\sqrt{\frac{U_\infty}{\vartheta x}}\sqrt{\frac{U_\infty}{\vartheta x}}$$

Simplify to get

$$\frac{\kappa}{\rho Cp}\frac{\partial^2 T}{\partial y^2} = \frac{\kappa}{\rho Cp}m_2x\frac{d^2\theta}{d\eta^2}\frac{U_\infty}{\vartheta x} \tag{F.7}$$

Substituting Eqs. (F.4)–(F.7) into Eq. (F.1)

$$m_1U_\infty\frac{df}{d\eta} + U_\infty\frac{df}{d\eta}m_2x\frac{d\theta}{d\eta}\left(-\frac{1}{2}\right)x^{-3/2}y\sqrt{\frac{U_\infty}{\vartheta}} + U_\infty\frac{df}{d\eta}\theta m_2$$

$$-x^{-1}U_\infty\frac{df}{d\eta}y\left(-\frac{1}{2}\right)m_2x\frac{d\theta}{d\eta}\sqrt{\frac{U_\infty}{\vartheta x}} - f\vartheta^{1/2}\frac{1}{2}x^{-1/2}U_\infty^{1/2}m_2x\frac{d\theta}{d\eta}\sqrt{\frac{U_\infty}{\vartheta x}} = \frac{\kappa}{\rho Cp}m_2x\frac{d^2\theta}{d\eta^2}\frac{U_\infty}{\vartheta x}$$

Divide both sides by $U_\infty m_2 x$.

$$\frac{m_1}{m_2x}\frac{df}{d\eta} + \underbrace{\frac{df}{d\eta}\frac{d\theta}{d\eta}\left(-\frac{1}{2}\right)x^{-3/2}y\sqrt{\frac{U_\infty}{\vartheta}}} + \frac{df}{d\eta}\theta\frac{m_2}{m_2x}$$

$$\underbrace{-x^{-1}\frac{df}{d\eta}y\left(-\frac{1}{2}\right)\frac{d\theta}{d\eta}\sqrt{\frac{U_\infty}{\vartheta x}}} - f\vartheta^{1/2}\frac{1}{2}x^{-1/2}\frac{d\theta}{d\eta}\sqrt{\frac{1}{\vartheta x}} = \frac{\kappa}{\rho Cp}\frac{d^2\theta}{d\eta^2}\frac{1}{\vartheta x}$$

It is worth noting that the terms with underbraces are struck out

$$\frac{m_1}{m_2x}\frac{df}{d\eta} + \frac{df}{d\eta}\theta\frac{m_2}{m_2x} - f\vartheta^{1/2}\frac{1}{2}x^{-1/2}\frac{d\theta}{d\eta}\sqrt{\frac{1}{\vartheta x}} = \frac{\kappa}{\rho Cp}\frac{d^2\theta}{d\eta^2}\frac{1}{\vartheta x}$$

In fact, the third term on the LHS reduces to

$$\frac{m_1}{m_2 x}\frac{df}{d\eta} + \frac{df}{d\eta}\theta\frac{m_2}{m_2 x} - f\frac{1}{2}\frac{d\theta}{d\eta}\frac{1}{x} = \frac{\kappa}{\rho Cp}\frac{d^2\theta}{d\eta^2}\frac{1}{\vartheta x}$$

Multiply both sides by x and introduce the thermal stratification parameter S_t

$$S_t\frac{df}{d\eta} + \theta\frac{df}{d\eta} - \frac{1}{2}f\frac{d\theta}{d\eta} = \frac{\kappa}{\rho Cp}\frac{d^2\theta}{d\eta^2}\frac{1}{\vartheta}$$

Finally, introduce Prandtl number P_r to get the final dimensionless equation as

$$\frac{d^2\theta}{d\eta^2} - S_t P_r\frac{df}{d\eta} - P_r\theta\frac{df}{d\eta} + \frac{1}{2}P_r f\frac{d\theta}{d\eta} = 0 \qquad \text{(F.8)}$$

Bibliography

[1] Eistein A, R. Furth, and A. D. Cowper. Investigation on the theory of the brownian movement. *The Scientific Monthly*, 83(6):314, 1956.

[2] S. O. Abah, B. E. Eletta, and S. O. Omale. The numerical analysis of the effect of free convection heat and mass transfer on the unsteady boundary layer flow past a vertical plate. *International Journal of Theoretical and Mathematical Physics*, 2(3):33 – 36, 2012. doi: 10.5923/j.ijtmp.20120203.02.

[3] N. Abbas, M. Y. Malik, S. Nadeem, and I. M. Alarifi. On extended version of Yamadaota and Xue models of hybrid nanofluid on moving needle. *The European Physical Journal Plus*, 135(2):145, 2020. doi:10.1140/epjp/s13360-020-00185-2.

[4] M. B. Abd-el Malek, N. A. Badran, H. S. Hassan, and H. H. Abbas. Lie group method for studying the heat generation effect on free convection laminar boundary-layer flow over a vertical flat plate. *Journal of Generalized Lie Theory and Applications*, 9(2):1000232, 2015. doi: 10.4172/1736-4337.1000232.

[5] A. K. Abdul Hakeem, P. Renuka, N. Vishnu Ganesh, R. Kalaivanan, and B. Ganga. Influence of inclined Lorentz forces on boundary layer flow of Casson fluid over an impermeable stretching sheet with heat transfer. *Journal of Magnetism and Magnetic Materials*, 401:354 – 361, 2016. doi:10.1016/j.jmmm.2015.10.026.

[6] M. S. Abel, P. L. Roa, and J. V Tawade. The effect of magnetic field on the boundary layer flow over a stretching sheet in a nanofluid with convective boundary condition. *International Journal of Mechanical Engineering and Technology*, 6(4):87 – 100, 2015.

[7] N Acharya, R. Bag, and P.K. Kundu. Influence of Hall current on radiative nanofluidflow over a spinning disk: A hybrid approach. *Physica E: Low-dimensional Systems and Nanostructures*, 111:103 – 112, 2019. doi:10.1016/j.physe.2019.03.006.

[8] D. J. Acheson and R. Hide. Hydromagnetics of rotating fluids. *Reports on Progress in Physics*, 36(2):159 – 221, 1973. doi:10.1088/0034-4885/36/2/002.

[9] K. S. Adegbie, A. J. Omowaye, A. B. Disu, and I. L. Animasaun. Heat and mass transfer of upper convected Maxwell fluid flow with variable thermo-physical properties over a horizontal melting surface. *Applied Mathematics*, 6(08):1362 – 1379, 2015. doi: 10.4236/am.2015.68129.

[10] W. Adrian. *English for Writing Research Papers*. Springer New York Dordrecht Heidelberg London, 2011. doi: 10.1007/978-1-4419-7922-3.

[11] I. Ahmad, M. Faisal, T. Javed, and I. L. Animasaun. Insight into the relationship between unsteady cattaneo-christov double diffusion, random motion and thermo-migration of tiny particles. Volume 13, Issue 1, January 2022, 101494. doi:10.1016/j.asej.2021.05.008.

[12] S. Ahmad, A. M. Rohni, and I. Pop. Blasius and sakiadis problems in nanofluids. *Acta Mechanica*, 218(3–4):195 – 204, 2010. doi:10.1007/s00707-010-0414-6.

[13] F. Ahsan, J. Razmi, and L. Ladani. Experimental measurement of thermal diffusivity, conductivity and specific heat capacity of metallic powders at room and high temperatures. *Powder Technology*, 374:648 – 657, 2020. doi:10.1016/j.powtec.2020.07.043.

[14] T. M. Ajayi, A. J. Omowaye, and I. L. Animasaun. Viscous dissipation effects on the motion of casson fluid over an upper horizontal thermally stratified melting surface of a paraboloid of revolution: boundary layer analysis. *Journal of Applied Mathematics*, 2017:13, 2017. doi: 10.1155/2017/1697135.

[15] S. Ajith Kumar, M. Mathur, A. Sameen, and S. Anil Lal. Effects of prandtl number on the laminar cross flow past a heated cylinder. *Physics of Fluids*, 28(11), 2016. doi:10.1063/1.4966937.

[16] S. Akbarnejad, M. Saffari Pour, L. T. I. Jonsson, and P. G. Jonsson. Effect of fluid bypassing on the experimentally obtained Darcy and non-darcy permeability parameters of ceramic foam filters. *Metallurgical and Materials Transactions B*, 48(1):197 – 207, 2016. doi:10.1007/s11663-016-0819-2.

[17] M. S. Alam, M. M. Rahman, and M. A. Sattar. Effects of variable suction and thermophoresis on steady MHD combined free-forced convective heat and mass transfer flow over a semi-infinite permeable inclined plate in the presence of thermal radiation. *International Journal of Thermal Sciences*, 47(6):758 – 765, 2008.

[18] H. Alfven. Existence of electromagnetic-hydrodynamic waves. *Nature*, 150(3805): 405 – 406, 1942.

[19] H. Alfven. Magneto-hydrodynamic waves and sunspots. *Arkiv for Astronomi*, 34: 1 – 20, 1948.

[20] H. Alfven. On the theory of sunspots. *Tellus*, 8(2):274 – 275, 1956.

[21] M. Alghamdi. On magnetohydrodynamic flow of viscoelastic nanofluids with homogeneous-heterogeneous reactions. *Coatings*, 10(1):55, 2020. doi:10.3390/coatings10010055.

[22] F. M. Ali, R. Nazar, N. M. Arifin, and I. Pop. Effect of Hall current on MHD mixed convection boundary layer flow over a stretched vertical flat plate. *Meccanica*, 46(5):1103 – 1112, 2011. doi: 10.1007/s11012-010-9371-3.

[23] M. E. Ali. The buoyancy effects on the boundary layers induced by continuous surfaces stretched with rapidly decreasing velocities. *Heat and Mass Transfer*, 40(3-4):285 – 291, 2004. doi: 10.1007/s00231-002-0405-9.

[24] N. Ali, J. A. Teixeira, and A. Addali. Aluminium nanofluids stability: A comparison between the conventional two-step fabrication approach and the controlled sonication bath temperature method. *Journal of Nanomaterials*, (1 - 9), 2019. doi:10.1155/2019/3930572.

[25] Z Alloui, P. Vasseur, and M. Reggio. Natural convection of nanofluids in a shallow cavity heated from below. *International Journal of Thermal Sciences*, 50:385 – 393, 2011. doi:10.1016/j.ijthermalsci.2010.04.006.

[26] F. E. Alsaadi, T. Hayat, S. A. Khan, F. E. Alsaadi, and M. I. Khan. Investigation of physical aspects of cubic autocatalytic chemically reactive flow of second grade nanomaterial with entropy optimization. *Computer Methods and Programs in Biomedicine*, 183, 2020. doi: 10.1016/j.cmpb.2019.105061.

[27] N. Anbuchezhian, K. Srinivasan, K. Chandrasekaran, and R. Kandasamy. Thermophoresis and brownian motion effects on boundary layer flow of nanofluid in presence of thermal stratification due to solar energy. *Applied Mathematics and Mechanics*, 33(6):765 – 780, 2012. doi: 10.1007/s10483-012-1585-8.

[28] E. N. D. C. Andrade. The viscosity of liquids. *Nature*, 125(3148):309 – 310, 1930. doi:10.1038/125309b0.

[29] I. L. Animasaun. *Effects of some thermo-physical parameters on free convective heat and mass transfer over vertical surface.* (Unpublished M.Tech Thesis). Federal University of Technology Akure, Nigeria, 2015.

[30] I. L. Animasaun. Melting heat and mass transfer in stagnation point micropolar fluid flow of temperature dependent fluid viscosity and thermal conductivity at constant vortex viscosity. *Journal of the Egyptian Mathematical Society*, 25(1):79 – 85, 2017. doi: 10.1016/j.joems.2016.06.007.

[31] I. L. Animasaun. *Significance of quartic autocatalysis on fluid flows conveying nanoparticles, dust-particles and microstructures over an object with non-uniform thickness.* (Unpublished doctoral dissertation). Federal University of Technology Akure, Nigeria, 2019.

[32] I. L. Animasaun. *Significance of Quartic Autocatalysis on the Flow of Some Fluids.* LAP LAMBERT Academic Publishing, Eastern Europe. ISBN: 9786200322845, 2019.

[33] I. L. Animasaun, E. A. Adebile, and A. I. Fagbade. Casson fluid flow with variable thermo-physical property along exponentially stretching sheet with suction and exponentially decaying internal heat generation using the homotopy analysis method. *Journal of the Nigerian Mathematical Society*, 35(1):1 – 17, 2016. doi:10.1016/j.jnnms.2015.02.001.

[34] I. L. Animasaun and O. B. Aluko. Analysis on variable fluid viscosity of non-Darcian flow over a moving vertical plate in a porous medium with suction and viscous dissipation. *IOSR Journal of Engineering*, 4(8):18 – 32, 2014. doi:10.9790/3021-04841832.

[35] I. L. Animasaun, R. O. Ibraheem, B. Mahanthesh, and H. A. Babatunde. A meta-analysis on the effects of haphazard motion of tiny/nano-sized particles on the dynamics and other physical properties of some fluids. *Chinese Journal of Physics*, 60(1):676 – 687, 2019. doi:10.1016/j.cjph.2019.06.007.

[36] I. L. Animasaun, O. K. Koriko, B. Mahanthesh, and A. S. Dogonchi. A note on the significance of quartic autocatalysis chemical reaction on the motion of air conveying dust particles. *Zeitschrift fur Naturforschung A*, 74(10):879 – 904, 2019. doi: 10.1515/zna-2019-0180.

[37] I. L. Animasaun and N. Sandeep. Buoyancy induced model for the flow of 36 nm alumina-water nanofluid along upper horizontal surface of a paraboloid of revolution with variable thermal conductivity and viscosity. *Powder Technology*, 301:858 – 867, 2016.

[38] S. R. Annasagaram, N N. M S. N. R. CH, A., and O. A. Beg. Hydromagnetic non-newtonian nanofluid transport phenomena from an isothermal vertical cone with partial slip: Aerospace nanomaterial enrobing simulation. *Heat Transfer-Asian Research*, 47(1):203 – 230, 2018. doi:10.1002/htj.21299.

[39] M. Archana, B. J. Gireesha, B. C. Prasannakumara, and R. S. R. Gorla. Numerical exploration of the combined effects of non-linear thermal radiation and variable thermo-physical properties on the flow of Casson nanofluid over a wedge. *Multidiscipline Modeling in Materials and Structures*, 13(4):628 – 647, 2017. doi:10.1108/mmms-05-2017-0037.

[40] M. Asma, W. A. M. Othman, T. Muhammad, F. Mallawi, and B. R. Wong. Numerical study for magnetohydrodynamic flow of nanofluid due to a rotating disk with binary chemical reaction and arrhenius activation energy. *Symmetry*, 11(10):1282, 2019. doi:10.3390/sym11101282.

[41] S. M. Atif, S. Hussain, and M. Sagheer. Numerical study of MHD micropolar Carreau nanofluid in the presence of induced magnetic field. *AIP Advances Symmetry*, 8(3):035219, 2018. doi:10.1063/1.5022681.

[42] K. Avinash and R. H. Reddy. Brownian moment and thermophoresis effects on unsteady liquid film flow of MHD Carreau fluid over a stretching sheet. *International Journal of Mathematics Trends and Technology*, 47(2):113 – 127, 2017. doi: 10.14445/22315373/ijmtt-v47p515.

[43] K. Avinash, R. H. Reddy, and A. O. Oyem. Analysis of MHD non-newtonian fluid over a stretching sheet with thermophoresis and brownian moment. *Advanced Engineering Forum*, 28:33 – 46, 2018. doi:10.4028/www.scientific.net/aef.28.33.

[44] K. Avinash, N. Sandeep, O. D. Makinde, and I. L. Animasaun. Aligned magnetic field effect on radiative bioconvection flow past a vertical plate with thermophoresis and brownian motion. *Defect and Diffusion Forum*, 377:127 – 140, 2017. doi: 10.4028/www.scientific.net/DDF.377.127.

[45] F. G. Awad, P. Sibanda, and A. A. Khidir. Thermodiffusion effects on magneto-nanofluid flow over a stretching sheet. *Boundary value problems*, 2013(1):136, 2013.

[46] R.U. Ayres. *Nuclear Power*. The History and Future of Technology. Springer, Cham., 2021.

[47] E. Azhar, Z. Iqbal, and E. N. Maraj. Impact of entropy generation on stagnation-point flow of sutterby nanofluid: a numerical analysis. *Zeitschrift fur Naturforschung A*, 71(9):837 – 848, 2016. doi: 10.1515/zna-2016-0188.

[48] H. Babar and H. M. Ali. Towards hybrid nanofluids: Preparation, thermophysical properties, applications, and challenges. *Journal of Molecular Liquids*, 281:598 – 633, 2019. doi:10.1016/j.molliq.2019.02.102.

[49] N. Bachok, A. Ishak, and I. Pop. Boundary layer stagnation-point flow toward a stretching/shrinking sheet in a nanofluid. *Journal of Heat Transfer*, 135(5):054501, 2013.

[50] M. Bahiraei and S. M. Hosseinalipour. Particle migration in nanofluids considering thermophoresis and its effect on convective heat transfer. *Thermochimica Acta*, 574:47–54, 2013. doi:10.1016/j.tca.2013.09.010.

[51] A. Y. Bakier and R. S. R. Gorla. Effects of thermophoresis and radiation on laminar flow along a semi-infinite vertical plate. *Heat and Mass Transfer*, 47(4):419 – 425, 2011. doi: 10.1007/s00231-010-0702-7.

[52] A. K. Baranwal and R. P. Chhabra. Free convection in confined power-law fluids from two side-by-side cylinders in a square enclosure. *Heat Transfer Engineering*, 37(18):1521 – 1537, 2016. doi: 10.1080/01457632.2016.1151296.

[53] H. T. Basha, R. Sivaraj, I. L. Animasaun, and O. D. Makinde. Influence of non-uniform heat source/sink on unsteady chemically reacting nanofluid flow over a cone and plate. *Defect and Diffusion Forum*, 389:50 – 59, 2018. doi:10.4028/www.scientific.net/ddf.389.50.

[54] R. C. Bataller. Radiation effects for the blasius and sakiadis flows with a convective surface boundary condition. *Applied Mathematics and Computation*, 206(2):832 – 840, 2008. doi:10.1016/j.amc.2008.10.001.

[55] O. A. Beg, A. S. Rao, N. Nagendra, C. H. Amanulla, and M. S. N. Reddy. Computational analysis of non-newtonian boundary layer flow of nanofluid past a vertical plate with partial slip. *Modelling, Measurement and Control B*, 86(1):271 – 295, 2017.

[56] T. Behzadi, K. M. Shirvan, S. Mirzakhanlari, and A. A. Sheikhrobat. Numerical simulation on effect of porous medium on mixed convection heat transfer in a ventilated square cavity. *Procedia Engineering*, 127:221 – 228, 2015. doi: 10.1016/j.proeng.2015.11.333.

[57] P. Bera and A. Khalili. Influence of prandtl number on stability of mixed convective flow in a vertical channel filled with a porous medium. *Physics of Fluids*, 18(12):124103, 2006. doi:10.1063/1.2405321.

[58] T. L. Bergman, F. P. Incropera, A. S. Lavine, and D. P. Dewitt. *Introduction to Heat Transfer*. Wiley, Hoboken, 2011.

[59] J. A. Berlin, M. P. Longnecker, and S. & Greenland. Meta-analysis of epidemiologic dose-response data. *Epidemiology*, 3(4):218 – 228, 1993. doi:10.1097/00001648-199305000-00005.

[60] M. H. U. Bhuiyan, R. Saidur, M. A. Amalina, R. M. Mostafizur, and A. Islam. Effect of nanoparticles concentration and their sizes on surface tension of nanofluids. *Procedia Engineering*, 105:431 – 437, 2015. doi:10.1016/j.proeng.2015.05.030.

[61] S. Bilal, M. Sohail, R. Naz, and M. Y. Malik. Dynamical and optimal procedure to analyse the exhibition of physical attribute imparted by sutterby magneto nano fluid in darcy medium yield by axially stretched cylinder. *Canadian Journal of Physics*, 98(1):1 – 10, 2020. doi:10.1139/cjp-2018-0581.

[62] S. Bilal, M. Sohail, R. Naz, M. Y. Malik, and M. Alghamdi. Upshot of ohmically dissipated darcy-forchheimer slip flow of magnetohydrodynamic sutterby fluid over radiating linearly stretched surface in view of cash and carp method. *Applied Mathematics and Mechanics*, 40(6):861 – 876, 2019. doi:10.1007/s10483-019-2486-9.

[63] P. R. H. Blasius. Grenzschichten in flussigkeiten mit kleiner reibung. *Zeitschrift fur Mathematik und Physik Band*, 56(Heft 1):1 – 58, 1908.

[64] A. Blass, P. Tabak, R. Verzicco, R. Stevens, and D. Lohse. The effect of Prandtl number on turbulent sheared thermal convection. *Journal of Fluid Mechanics*, 910:37, 2021.

[65] A. Bouras, M. Djezzar, and C. Ghernoug. Numerical simulation of natural convection between two elliptical cylinders: influence of rayleigh number and Prandtl number. *Energy Procedia*, 36:788 – 797, 2013.

[66] A. Boushehri and S. Afrashtehfar. Dufour effect in gases. *Bulletin of the Chemical Society of Japan*, 48(8):2372 – 2374, 1975. doi:10.1246/bcsj.48.2372.

[67] J. Boussinesq. *Theorie de l'ecoulement tourbillonnant et tumultueux des liquides dans les lits rectilignes a grande section,*. Gauthier-Villars Paris vol. 1. Open Library OL7070543M., 1897.

[68] J. Boussinesq. *Théorie Analytique de la Chaleur.* Gauthier-Villars., 1903.

[69] A. Brandenburg, N. E. L. Haugen, X.-Y. Li, and K. Subramanian. Varying the forcing scale in low prandtl number dynamos. *Monthly Notices of the Royal Astronomical Society*, 479(2):2827 – 2833, 2018. doi: 10.1093/mnras/sty1570.

[70] A.L. Braslow. *A History of Suction-type Laminar-flow Control with Emphasis on Flight Research, NASA History Division, Monographs in Aerospace History.* No. 13, Washington DC 20546., 1999.

[71] V. Buddakkagari and M. Kumar. Transient boundary layer laminar free convective flow of a nanofluid over a vertical cone/plate. *International Journal of Applied and Computational Mathematics*, 1(3):427 – 448, 2015. doi: 10.1007/s40819-015-0027-9.

[72] J. Campbell. The fracture of liquids. *The Mechanisms of Metallurgical Failure*, 1:1 – 165, 2020. doi:10.1016/b978-0-12-822411-3.00001-0.

[73] T. L. Carus, H. A. J. Munro, and G. Long. *On the nature of things.* Encyclopaedia Britannica, 1952.

[74] N. Chakraborty. Thermal transport regimes and effects of Prandtl number in molten pool transport in laser surface melting processes. *Numerical Heat Transfer, Part A: Applications*, 53(3):273 – 294, 2007. doi: 10.1080/10407780701557709.

[75] Sydney Chapman and T. G. Cowling. *The Mathematical Theory of Non-Uniform Gases (3rd ed.).* Cambridge University Press, 1970.

[76] A. Chaston Chapman and H. F. E. Hulton. Reviews. *The Analyst, Royal Society of Chemistry (RSC)*, 42(492):112, 1917. doi:10.1039/an9174200112.

[77] C. H. Chen. Non-Darcy mixed convection from a horizontal surface with variable surface heat flux in a porous medium. *Numerical Heat Transfer, Part A Applications*, 30(8):859 – 869, 1996.

[78] K. L. Chong, S. Wagner, M. Kaczorowski, O. Shishkina, and K. Q. Xia. Effect of prandtl number on heat transport enhancement in rayleigh-benard convection under geometrical confinement. *Physical Review Fluids*, 3(1):013501, 2018. doi: 10.1103/physrevfluids.3.013501.

[79] D. M. Christopher and B. Wang. Prandtl number effects for marangoni convection over a flat surface. *International Journal of Thermal Sciences*, 40(6):564 – 570, 2001. doi: 10.1016/s1290-0729(01)01244-3.

[80] K. Clusius and L. Waldmann. An overlooked kinetic effect in gases. *Naturwissenschaften*, 30:711 – 711, 1942.

[81] M. Combarnous. *Convection naturelle et convection mixte en milieu poroux.* Thesis, University of Paris, 1970.

[82] R. Cortell Bataller. Radiation effects in the Blasius flow. *Applied Mathematics and Computation*, 198(1):333 – 338, 2008. doi:10.1016/j.amc.2007.08.037.

[83] J. C. Crepeau and R. Clarksean. Similarity solutions of natural convection with internal heat generation. *Journal of Heat Transfer*, 119(1):183 – 185, 1997.

[84] Devaraj P. Daba, M. and S. V. Subhashini. Mixed convection boundary layer flow over a vertically stretching sheet with convective boundary condition and effect of partial slip. *Frontiers in Heat and Mass Transfer (FHMT)*, 6(1):3, 2015. doi: 10.5098/hmt.6.3.

[85] S. Dan, Y. Nandukumar, and P. Pal. Effect of Prandtl number on wavy rolls in Rayleigh-Benard convection. *Physica Scripta*, 90(3):035208, 2015. doi:10.1088/0031-8949/90/3/035208.

[86] H. Darcy. *Les fontainer publigues de la ville de Dijoin.* Dalmont., 1856.

[87] R. Das and A. K. Mohanty. Entrance effects on low Prandtl number convection through a circular duct. *Metallurgical Transactions B*, 24(1):197 – 200, 1993.

[88] S.K. Choi Das, W. S.U.S., Yu, and T. Predeep. *Nanofluids Science and Technology.* John Wiley and sons, Hoboken, New Jersey., 2008.

[89] G. B. Delancey and S. H. Chiang. Dufour effect in liquid systems. *AIChE Journal*, 14(4):664 – 665, 1968. doi:10.1002/aic.690140427.

[90] K. G. Denbigh. Note on a method of estimating the Prandtl number of liquids. *Journal of the Society of Chemical Industry*, 65(2):61 – 63, 1946.

[91] A. S. Dogonchi and D. D. Ganji. Analytical solution and heat transfer of two-phase nanofluid flow between non-parallel walls considering Joule heating effect. *Powder Technology*, 318:390 – 400, 2017.

[92] A. K. Doolittle. Studies in newtonian flow. II. the dependence of the viscosity of liquids on free-space. *Journal of Applied Physics*, 22(12):1471 – 1475, 1951. doi:10.1063/1.1699894.

[93] D. Eckert. *Heat and Mass Transfer.* McGraw - Hill, 1959.

[94] E. M. El-Omar. How to publish a scientific manuscript in a high-impact journal. *Advances in Digestive Medicine*, 1:105 – 109, 2014.

[95] A. H. A. Elnaby and I. I. E. Shamy. Effects of inclined magnetic field on a fluid with non-linear variable viscosity in a uniform tube with peristalsis. *Multidiscipline Modeling in Materials and Structures*, 5(4):395 – 409, 2009. doi:10.1163/157361109789808016.

[96] T. Elnaqeeb, I. L. Animasaun, and N. A. Shah. Ternary-hybrid nanofluids: significance of suction and dual-stretching on three-dimensional flow of water conveying nanoparticles with various shapes and densities. *Zeitschrift Fur Naturforschung A*, 76(3):231 – 243, 2021. doi:10.1515/zna-2020-0317.

[97] M. Eslamian, M. Ahmed, M. F. El-Dosoky, and M. Z. Saghir. Effect of thermophoresis on natural convection in a Rayleigh-Benard cell filled with a nanofluid. *International Journal of Heat and Mass Transfer*, 81:142 – 156, 2015. doi:10.1016/j.ijheatmasstransfer.2014.10.001.

[98] M. A. Ezzat and W. Khatan. A thermal shock-chemical reactive problem in flow of viscoelastic fluid with thermal relaxation. *Applied Mathematics*, 03(07):685 – 698, 2012. doi:10.4236/am.2012.37102.

[99] U. Farooq, T. Hayat, A. Alsaedi, and S. J. Liao. Series solutions of non-similarity boundary layer flows of nano-fluids over stretching surfaces. *Numerical Algorithms*, 70(1):43 – 59, 2015. doi: 10.1007/s11075-014-9934-9.

[100] U. Farooq, H. Waqas, M. I. Khan, S. U. Khan, Y.-M. Chu, and S. Kadry. Thermally radioactive bioconvection flow of Carreau nanofluid with modified cattaneo-christov expressions and exponential space-based heat source. *Alexandria Engineering Journal*, 60(3):3073 – 3086, 2021. doi:10.1016/j.aej.2021.01.050.

[101] P. Z. Forchheimer. Wasserbewegung durch boden. *Zeit Ver Deutsch Ing*, 45:1781 – 1788, 1901.

[102] N. A. Fuchs. *The Mechanics of Aerosols*. New York: The Macmillan Co., 1964.

[103] T. Gambaryan-Roisman. Marangoni convection, evaporation and interface deformation in liquid films on heated substrates with non-uniform thermal conductivity. *International Journal of Heat and Mass Transfer*, 53(1-3):390 – 402, 2010. doi:10.1016/j.ijheatmasstransfer.2009.09.017.

[104] J. A. Gbadeyan, E. O. Titiloye, and A. T. Adeosun. Effect of variable thermal conductivity and viscosity on Casson nanofluid flow with convective heating and velocity slip. *Heliyon*, 6(1):e03076, 2020. doi:10.1016/j.heliyon.2019.e03076.

[105] A. Ghadami Jadval Ghadam and A. Moradi. Effects of brownian motion and thermophoresis on MHD mixed convection stagnation-point flow of a nanofluid toward a stretching vertical sheet in porous medium. *Journal of Particle Science and Technology*, 1(4):225 – 240, 2015. doi: 10.22104/JPST.2015.271.

[106] M. Ghalambaz, E. Izadpanahi, A. Noghrehabadi, and A. Chamkha. Study of the boundary layer heat transfer of nanofluids over a stretching sheet: Passive control of nanoparticles at the surface. *Canadian Journal of Physics*, 93(7):725 – 733, 2014. doi: 10.1139/cjp-2014-0370.

[107] S. E. Ghasemi and M. Hatami. Solar radiation effects on MHD stagnation point flow and heat transfer of a nanofluid over a stretching sheet. *Case Studies in Thermal Engineering*, 25(7):100898, 2021. doi:10.1016/j.csite.2021.100898.

[108] M. Ghassemi and A. Shahidian. Solar radiation effects on MHD stagnation point flow and heat transfer of a nanofluid over a stretching sheet. *Fluid Mechanics. Nano and Bio Heat Transfer and Fluid Flow*, 25(7):57 – 87, 2017. doi:10.1016/b978-0-12-803779-9.00004-2.

[109] S.K. Ghosh, O.A. Beg, and A. Aziz. A mathematical model for magnetohydrodynamic convection flow in a rotating horizontal channel with inclined magnetic field, magnetic induction and Hall current effects. *World Journal of Mechanics*, 1(7):137 – 154, 2011. doi: 10.4236/wjm.2011.13019.

[110] A. E. Gill. The boundary-layer regime for convection in a rectangular cavity. *J. Fluid Mech*, 26:515 – 5536, 1962.

[111] G. V. Glass. Primary, secondary, and meta-analysis of research. *Educational Researcher*, 5(10):3 – 8, 1976. doi:10.3102/0013189x005010003.

[112] S. Gosukonda, V. P. N. S. Gorti, S. B. Baluguri, and S. R. Sakam. Particle spacing and chemical reaction effects on convective heat transfer through a nano-fluid in cylindrical annulus. *Procedia Engineering*, 127:263 – 270, 2015. doi:10.1016/j.proeng.2015.11.359.

[113] W. P. Graebel. The influence of prandtl number on free convection in a rectangular cavity. *International Journal of Heat and Mass Transfer*, 24(1):125 – 131, 1981. doi: 10.1016/0017-9310(81)90100-9.

[114] A. L. Graham. On the viscosity of suspensions of solid spheres. *Appl. Sci. Res*, 37:275 – 286, 1981. doi: 10.1016/0017-9310(81)90100-9.

[115] H. L. Green and W. R. Lane. *Particulate Clouds: Dusts, Smokes and Mists.* Van Nos-trand, New York., 1964.

[116] T. Grosan, M. A. Sheremet, I. Pop, and S. R. Pop. Double-diffusive natural convection in a differentially heated wavy cavity under thermophoresis effect. *Journal of Thermophysics and Heat Transfer*, 32(4):1045 – 1058, 2018. doi:10.2514/1.t5389.

[117] L. J. Grubka and K. M. Bobba. Heat transfer characteristics of a continuous, stretching surface with variable temperature. *Journal of Heat Transfer*, 107(1):248 – 250, 1985. doi:10.1115/1.3247387.

[118] F. M. Hady, A. Mahdy, R. A. Mohamed, and O. A. A. Zaid. Effects of viscous dissipation on unsteady mhd thermo bioconvection boundary layer flow of a nanofluid containing gyrotactic microorganisms along a stretching sheet. *World Journal of Mechanics*, 6(12):505, 2016.

[119] F. Hamba. Turbulent energy density and its transport equation in scale space. *Physics of Fluids*, 27(8):085108, 2015. doi:10.1063/1.4928698.

[120] H. A. Hamid and M. Khan. Unsteady mixed convective flow of williamson nanofluid with heat transfer in the presence of variable thermal conductivity and magnetic field. *Journal of Molecular Liquids*, 260:436 – 446, 2018. doi:10.1016/j.molliq.2018.03.079.

[121] R. Haq, S. Nadeem, Z. Khan, and T. Okedayo. Convective heat transfer and MHD effects on Casson nanofluid flow over a shrinking sheet. *Open Physics - Central European Journal of Physics*, 12(12):862 – 871, 2014. doi:10.2478/s11534-014-0522-3.

[122] S. J. Hawkes. The concept of density. *Journal of Chemical Education*, 81(1):14, 2004. doi:10.1021/ed081p14.

[123] T. Hayat, F. Hayat, T. Muhammad, and A. Alsaedi. Darcy-Forchheimer flow by rotating disk with partial slip. *Applied Mathematics and Mechanics*, 41(5):741 – 752, 2020. doi:10.1007/s10483-020-2608-9.

[124] T. Hayat, M. Imtiaz, and A. Alsaedi. Magnetohydrodynamic flow of nanofluid over permeable stretching sheet with convective boundary conditions. *Thermal science*, 29(6):1835 – 1845, 2016.

[125] T. Hayat, M. I. Khan, S. Qayyum, A. Alsaedi, and M. I. Khan. New thermodynamics of entropy generation minimization with nonlinear thermal radiation and nanomaterials. *Physics Letters A*, 382(11):749 – 760, 2018. doi:10.1016/j.physleta.2018.01.024.

[126] T. Hayat, F. Shah, A. Alsaedi, and M. Waqas. Numerical simulation for magneto nanofluid flow through a porous space with melting heat transfer. *Microgravity Science and Technology*, 30(3):265 – 275, 2018. doi: 10.1007/s12217-018-9595-8.

[127] T. Hayat, S. Shehzad, M. Qasim, and S. Obaidat. Flow of a second grade fluid with convective boundary conditions. *Thermal Science*, 15(2):253 – 261, 2011. doi:10.2298/tsci101014058h.

[128] T. Hayat, H. Zahir, M. Mustafa, and A. Alsaedi. Peristaltic flow of sutterby fluid in a vertical channel with radiative heat transfer and compliant walls: A numerical study. *Results in Physics*, 6(2):805 – 810, 2016. doi: 10.1016/j.rinp.2016.10.015.

[129] O. M. Heinrich. Thin accretion disks near the critical luminosity: the effect of turbulent energy transport. *Astronomy and Astrophysics*, 286:338 – 343, 1994.

[130] M. Hemmat Esfe, A. A. Abbasian Arani, M. Rezaie, W.-M. Yan, and A. Karimipour. Experimental determination of thermal conductivity and dynamic viscosity of Ag-MgO/water hybrid nanofluid. *International Communications in Heat and Mass Transfer*, 66:189 – 195, 2015. doi:10.1016/j.icheatmasstransfer.2015.06.003.

[131] M. Hemmat Esfe, M. Kiannejad Amiri, and A. Alirezaie. Thermal conductivity of a hybrid nanofluid. *Journal of Thermal Analysis and Calorimetry*, 134(2):1113 – 1122, 2018. doi:10.1007/s10973-017-6836-9.

[132] R.B. Hetnarski. *Encyclopedia of Thermal Stresses: Fourier Law of Heat Conduction (pp1736–1736)*. Springer Netherlands, 2014.

[133] J. P. Holman. *Heat Transfer (9th ed.)*. McGraw - Hill., 2002.

[134] J. T. Hong, Y. Yamada, and C. L. Tien. Effects of non-Darcian and non-uniform porosity on vertical-plate natural convection in porous media. *Journal of Heat Transfer*, 109(2):356 – 362, 1987. doi:10.1016/j.icheatmasstransfer.2015.05.014.

[135] W. Hort, S. J. Linz, and M. Lucke. Onset of convection in binary gas mixtures: Role of the Dufour effect. *Physical Review A*, 45(6):3737 – 3748, 1992. doi:10.1103/physreva.45.3737.

[136] J. R. Howell, M. P. Menguc, and R. Siegel. *Thermal Radiation Heat Transfer, 5th Edition.* Taylor and Francis Group, CRC press, Boca Raton London New York., 2010.

[137] A. Hussain, A. Rehman, S. Nadeem, M. R. Khan, and A. Issakhov. A computational model for the radiated kinetic molecular postulate of fluid-originated nanomaterial liquid flow in the induced magnetic flux regime. *Mathematical Problems in Engineering*, (6690366):1 – 17, 2021. doi:10.1155/2021/6690366.

[138] W. Ibrahim and B. Shanker. Boundary-layer flow and heat transfer of nanofluid over a vertical plate with convective surface boundary condition. *Journal of Fluids Engineering*, 134(8):081203, 2012. doi:10.1115/1.4007075.

[139] W. Ibrahim and A. Tulu. Magnetohydrodynamic (MHD) boundary layer flow past a wedge with heat transfer and viscous effects of nanofluid embedded in porous media. *Mathematical Problems in Engineering*, 2019:Article ID 4507852, 2019. doi:10.1155/2019/4507852.

[140] and Pendyala R. Ilyas, S. U. and N. Marneni. Settling characteristics of alumina nanoparticles in ethanol-water mixtures. *Applied Mechanics and Materials*, 372(143):143 – 148, 2013. doi:10.4028/www.scientific.net/amm.372.143.

[141] N. Iqbal, H. Yasmin, A. Bibi, and A. A. Attiya. Peristaltic motion of Maxwell fluid subject to convective heat and mass conditions. *Ain Shams Engineering Journal*, 12(3):3121 – 3131, 2021. doi:10.1016/j.asej.2021.01.015.

[142] A. Ishak. Mixed convection boundary layer flow over a horizontal plate with thermal radiation. *Heat and Mass Transfer*, 46(2):147, 2009. doi: 10.1007/s00231-009-0552-3.

[143] A. Ishak. Thermal boundary layer flow over a stretching sheet in a micropolar fluid with radiation effect. *Meccanica*, 45(3):367 – 373, 2009. doi:10.1007/s11012-009-9257-4.

[144] A. Ishak. Mhd boundary layer flow due to an exponentially stretching sheet with radiation effect. *Sains Malaysiana*, 40(4):391 – 395, 2011.

[145] S. Ittedi, D. Ramya, and S. Joga. Partial slip effect of mhd boundary layer flow of nanofluids and radiative heat transfer over a permeable stretching sheet. *Global Journal of Pure and Applied mathematics*, 13(7):3083 – 3103, 2017.

[146] M. Jahangiri, M. Saghafian, and M. R. Sadeghi. Numerical study of turbulent pulsatile blood flow through stenosed artery using fluid-solid interaction. *Computational and Mathematical Methods in Medicine*, 2015:1 – 10, 2015. doi:10.1155/2015/515613.

[147] Mohamoud Jama, Tejvir Singh, Seifelislam Mahmoud Gamaleldin, Muammer Koc, Ayman Samara, Rima J. Isaifan, and Muataz A. Atieh. Critical review on nanofluids: Preparation, characterization, and applications. *Journal of Nanomaterials*, 2016(Article ID 6717624):22 pages, 2016. doi:10.1155/2016/6717624.

[148] V. I. Kaljasin. Thermal diffusion. 2019. doi: 10.1615/AtoZ.t.thermal$_d$iffusion.

[149] R. Kandasamy, I. Muhaimin, and R. Mohamad. Thermophoresis and brownian motion effects on MHD boundary-layer flow of a nanofluid in the presence of thermal stratification due to solar radiation. *International Journal of Mechanical Sciences*, 70:146 – 154, 2013. doi: 10.1016/j.ijmecsci.2013.03.007.

[150] W. M. Kays. Turbulent prandtl number-where are we? *Journal of Heat Transfer*, 116(2):284 – 295, 1994.

[151] and Gorji M. and Sajjadi H. Kefayati, G. and D. Domiri Ganji. Investigation of prandtl number effect on natural convection MHD in an open cavity by lattice boltzmann method. *Engineering Computations*, 30(1):97 – 116, 2012. doi:10.1108/02644401311286035.

[152] R. Keimanesh and C. Aghanajafi. The effect of temperature-dependent viscosity and thermal conductivity on micropolar fluid over a stretching sheet. *Tehnicki Vjesnik - Technical Gazette*, 24(2):371 – 378, 2017. doi:10.17559/tv-20150517072748.

[153] and Haneef M. Khan, H., and Islam S. Shah, Z., W. Khan, and S. Muhammad. The combined magneto hydrodynamic and electric field effect on an unsteady Maxwell nanofluid flow over a stretching surface under the influence of variable heat and thermal radiation. *Applied Sciences*, 8(2):160, 2018. doi:10.3390/app8020160.

[154] J. A. Khan, M. Mustafa, T. Hayat, and A. Alsaedi. Numerical study on three-dimensional flow of nanofluid past a convectively heated exponentially stretching sheet. *Canadian Journal of Physics*, 93(10):1131 – 1137, 2015. doi:10.1139/cjp-2014-0433.

[155] M. I. Khan, T. Hayat, M. I. Khan, and A. Alsaedi. Activation energy impact in nonlinear radiative stagnation point flow of cross nanofluid. *International Communications in Heat and Mass Transfer*, 91:216 – 224, 2018. doi: 10.1016/j.icheatmasstransfer.2017.11.001.

[156] S. U. Khan, K. Al-Khaled, A. Aldabesh, M. Awais, and I. Tlili. Bioconvection flow in accelerated couple stress nanoparticles with activation energy: bio-fuel applications. *Scientific Reports*, 11(1):3331, 2021. doi:10.1038/s41598-021-82209-0.

[157] S. U. Khan, S. A. Shehzad, A. Rauf, and Z. Abbas. Thermally developed unsteady viscoelastic micropolar nanofluid with modified heat/mass fluxes: A generalized model. *Physica A: Statistical Mechanics and Its Applications*, 550:123986, 2020. doi:10.1016/j.physa.2019.123986.

[158] W. A. Khan and I. Pop. Boundary-layer flow of a nanofluid past a stretching sheet. *International Journal of Heat and Mass Transfer*, 53((11-12)):2477 – 2483, 2010. doi: 10.1016/j.ijheatmasstransfer.2010.01.032.

[159] J. Kierzenka and L. F. Shampine. A bvp solver based on residual control and the matlab pse. *ACM TOMS*, 27(3):299 – 316, 2001.

[160] M. Kogevinas and N. Pearce. Importance of meta-analyses for evaluating carcinogens. *Environmental health perspectives*, 120(4):a145 – a145, 2012.

[161] O. K. Koriko, K. S. Adegbie, A. S. Oke, and I. L. Animasaun. Exploration of coriolis force on motion of air over the upper horizontal surface of a paraboloid of revolution. *Physica Scripta*, 95(3):035210, 2020. doi:10.1088/1402-4896/ab4c50.

[162] O. K. Koriko, A. J. Omowaye, I. L. Animasaun, and M. E. Bamisaye. Melting heat transfer and induced-magnetic field effects on the micropolar fluid flow towards stagnation point: boundary layer analysis. *International Journal of Engineering Research in Africa*, 29:10 – 20, 2017. doi:10.4028/www.scientific.net/jera.29.10.

[163] O.K. Koriko, I.L. Animasaun, M. Gnaneswara Reddy, and N. Sandeep. Scrutinization of thermal stratification, nonlinear thermal radiation and quartic autocatalytic chemical reaction effects on the flow of three-dimensional eyring-powell alumina-water nanofluid. *Multidiscipline Modeling in Materials and Structures*, 14(2):261 – 283, 2018. doi: 10.1108/MMMS-08-2017-0077.

[164] O.K. Koriko, I.L. Animasaun, B. Mahanthesh, S. Saleem, G. Sarojamma, and R. Sivaraj. Heat transfer in the flow of blood-gold Carreau nanofluid induced by partial slip and buoyancy. *Heat Transfer - Asian Research*, 47(6):806 – 823, 2018. doi:10.1002/htj.21342.

[165] K. A. Kumar, V. Sugunamma, N. Sandeep, and J.V.R. Reddy. MHD stagnation point flow of williamson and casson fluids past an extended cylinder: A new heat flux model. *SN Applied Sciences*, 1(7):705, 2019. doi: 10.1007/s42452-019-0743-6.

[166] K. G. Kumar, M. R. Krishnamurthy, and R. NG. Boundary layer flow and melting heat transfer of Prandtl fluid over a stretching surface by considering Joule heating effect. *Multidiscipline Modeling in Materials and Structures*, 15(2):337 – 352, 2019.

[167] G. Kumaran, N. Sandeep, and I. L. Animasaun. Computational modeling of magnetohydrodynamic non-newtonian fluid flow past a paraboloid of revolution. *Alexandria Engineering Journal*, 57(3):1859 – 1865, 2018. doi:10.1016/j.aej.2017.03.019.

[168] J. Kunes. Interactive sensor network data retrieval and management using principal components analysis transform. *Technology and Mechanical Engineering*, ISBN: 978-0-12-416013-2:353 – 390, 2012. doi: 10.1016/B978-0-12-416013-2.

[169] T. Kunnegowda, B. Mahanthesh, G. Lorenzini, and I. L. Animasaun. Significance of induced magnetic field and exponential space dependent heat source on quadratic convective flow of Casson fluid in a micro-channel via hpm. *Mathematical Modelling of Engineering Problems*, 6(3):369 – 384, 2019. doi:10.18280/mmep.060308.

[170] R. Lakkaraju and M. Alam. Effects of prandtl number and a new instability mode in a plane thermal plume. *Journal of Fluid Mechanics*, 592:221 – 231, 2007. doi: 10.1017/s0022112007008610.

[171] S. Lam, X. Shang, S. Zhou, and K. Xia. Prandtl number dependence of the viscous boundary layer and the reynolds numbers in Rayleigh-Benard convection. *Physical Review E*, 65(6):066306, 2002. doi: 10.1103/physreve.65.066306.

[172] J. Langham, T. S. Eaves, and R. R. Kerswell. Stably stratified exact coherent structures in shear flow: the effect of Prandtl number. *Journal of Fluid Mechanics*, 882:A10, 2019. doi:10.1017/jfm.2019.811.

[173] K. Langhans, G. B. Scharmer, D. Kiselman, M. G. Lofdahl, and T. E. Berger. Inclination of magnetic fields and flows in sunspot penumbrae. *Astronomy and Astrophysics*, 436(3):1087 – 1101, 2005. doi:10.1051/0004-6361:20052678.

[174] A. I. Lare. Casson fluid flow with variable viscosity and thermal conductivity along exponentially stretching sheet embedded in a thermally stratified medium with exponentially heat generation. *Journal of Heat and Mass Transfer Research*, 2(2):63 – 78, 2015. doi: 10.22075/jhmtr.2015.346.

[175] S. Lee, S. U.-S. Choi, S. Li, and J. A. Eastman. Measuring thermal conductivity of fluids containing oxide nano particles. *Journal of Heat Transfer*, 121(2):280 – 289, 1999.

[176] S. W. Lee, S. D. Park, S. Kang, I. C. Bang, and J. H. Kim. Investigation of viscosity and thermal conductivity of sic nanofluids for heat transfer applications. *International Journal of Heat and Mass Transfer*, 54(1 – 3):433 – 438, 2011.

[177] X. Li, J. He, Y. Tian, P. Hao, and S. Huang. Effects of prandtl number in quasi-two-dimensional Rayleigh-Benard convection. *Journal of Fluid Mechanics*, 915:A60, 2021. doi:10.1017/jfm.2021.21.

[178] N. Liron and H. E. Wilhelm. Integration of the magnetohydrodynamic boundary-layer equations by meksyn's method. *ZAMM-Journal of Applied Mathematics and Mechanics/Zeitschrift fur Angewandte Mathematik und Mechanik*, 54(1):27 – 37, 1974.

[179] I. S. Liu. On fourier's law of heat conduction. *Continuum mechanics and Thermodynamics*, 2(4):301 – 305, 1990.

[180] D. Lu, M. Ramzan, N. ul Huda, J. D. Chung, and U. Farooq. Nonlinear radiation effect on mhd carreau nanofluid flow over a radially stretching surface with zero mass flux at the surface. *Scientific Reports*, 8(1):3709, 2018. doi:10.1038/s41598-018-22000-w.

[181] L. A. Lund, Z. Omar, and I. Khan. Analysis of dual solution for mhd flow of Williamson fluid with slippage. *Heliyon*, 5(3):e01345, 2019. doi:10.1016/j.heliyon.2019.e01345.

[182] Hady F. M., Mohamed R., M.R. Eid, Abd-Elsalam, and Mostafa A. Ahmed. The Blasius and Sakiadis flow in a nanofluid through a porous medium in the presence of thermal radiation under a convective surface boundary condition. *International Journal of Engineering and Innovative Technology*, 3(3):225 – 234, 2013. doi:10.1115/1.3247387.

[183] F. Mabood, S. M. Ibrahim, and W. A. Khan. Framing the features of Brownian motion and thermophoresis on radiative nanofluid flow past a rotating stretching sheet with magnetohydrodynamics. *Results in Physics*, 6:1015 – 1023, 2016. doi:10.1016/j.rinp.2016.11.046.

[184] B. Mahanthesh, B. J. Gireesha, I. L. Animasaun, T. Muhammad, and N. S. Shashikumar. Mhd flow of swcnt and mwcnt nanoliquids past a rotating stretchable disk with thermal and exponential space dependent heat source. *Physica Scripta*, 94(8):085214, 2019. doi:10.1088/1402-4896/ab18ba.

[185] B. Mahanthesh, O. D. Makinde, B. J. Gireesha, K. L. Krupalakshmi, and I. L. Animasaun. Two-phase flow of dusty casson fluid with Cattaneo-Christov heat flux and heat source past a cone, wedge and plate. *Defect and Diffusion Forum*, 387:625 – 639, 2018. doi:10.4028/www.scientific.net/ddf.387.625.

[186] N. C. Mahanti and P. Gaur. Effects of varying viscosity and thermal conductivity on steady free convective flow and heat transfer along an isothermal vertical plate in the presence of heat sink. *Journal of Applied Fluid Mechanics*, 2(1):23 – 28, 2009. doi:10.36884/jafm.2.01.11852.

[187] I. M. Mahbubul, R. Saidur, and M. A. Amalina. Heat transfer and pressure drop characteristics of Al_2O_3-r141b nanorefrigerant in horizontal smooth circular tube. *Procedia Engineering*, 56:323 – 329, 2013. doi:10.1016/j.proeng.2013.03.126.

[188] O. D. Makinde and I. L. Animasaun. Bioconvection in mhd nanofluid flow with nonlinear thermal radiation and quartic autocatalysis chemical reaction past an upper surface of a paraboloid of revolution. *International Journal of Thermal Sciences*, 109:159 – 171, 2016. doi: 10.1016/j.ijthermalsci.2016.06.003.

[189] O. D. Makinde and P. O. Olanrewaju. Buoyancy effects on thermal boundary layer over a vertical plate with a convective surface boundary condition. *Journal of Fluids Engineering*, 132(4):044502, 2010. doi: 10.1115/1.4001386.

[190] O. D. Makinde, N. Sandeep, T. M. Ajayi, and I. L. Animasaun. Numerical exploration of heat transfer and lorentz force effects on the flow of MHD casson fluid over an upper horizontal surface of a thermally stratified melting surface of a paraboloid of revolution. *International Journal of Nonlinear Sciences and Numerical Simulation*, 19(2):93 – 106, 2018. doi:10.1515/ijnsns-2016-0087.

[191] O.D. Makinde and I. L. Animasaun. Thermophoresis and Brownian motion effects on mhd bioconvection of nanofluid with nonlinear thermal radiation and quartic chemical reaction past an upper horizontal surface of a paraboloid of revolution. *Journal of Molecular Liquids*, 221:733 – 743, 2016. doi: 10.1016/j.molliq.2016.06.047.

[192] H. Masuda, A. Ebata, K. Teramae, and N. Hishinuma. Alteration of thermal conductivity and viscosity of liquid by dispersing ultra-fine particles dispersion of Al_2O_3, sio_2 and tio_2 ultra-fine particles. *Netsu Bussei*, 7(4):227 – 233, 1993.

[193] J. C. Maxwell. *A Treatise on Electricity and Magnetism*. Clarendon, Oxford, UK., 1873.

[194] P. Megaraju, S. R. Sheri, and M. R. Shekar. Transient MHD flows through an exponentially accelerated isothermal vertical plate with Hall effect and chemical reaction effect: Fem. *Partial Differential Equations in Applied Mathematics*, 4:100047, 2021. doi:10.1016/j.padiff.2021.100047.

[195] S. Mishra, D. Pal, H. Mondal, and P. Sibanda. On radiative-magnetoconvective heat and mass transfer of a nanofluid past a non-linear stretching surface with ohmic heating and convective surface boundary condition. *Propulsion and Power Research*, 5(4):326 – 337, 2016. doi:10.1016/j.jppr.2016.11.007.

[196] M. B. K. Moorthy and K. Senthilvadivu. Soret and Dufour effects on natural convection flow past a vertical surface in a porous medium with variable viscosity. *Journal of Applied Mathematics*, 2012:1 – 15, 2012. doi: 10.1155/2012/634806.

[197] K. Morimoto, S. Sawai, and K. Hisanno. Measurement of specific heat capacity from 150 to 310 k by thermal radiation calorimetry. *International Journal of Thermophysics*, 20:721 – 731, 1999. doi:10.1109/icmmt.1998.768272.

[198] S. S. Motsa and I. L. Animasaun. Bivariate spectral quasi-linearisation exploration of heat transfer in the boundary layer flow of micropolar fluid with strongly concentrated particles over a surface at absolute zero due to impulsive. *International Journal of Computing Science and Mathematics*, 9(5):455 – 473, 2018. doi: 10.18869/acadpub.jafm.68.236.25597.

[199] T. Muhammad, A. Alsaedi, S. A. Shehzad, and T. Hayat. A revised model for Darcy-forchheimer flow of Maxwell nanofluid subject to convective boundary condition. *Chinese Journal of Physics*, 55(3):963 – 976, 2017. doi: 10.1016/j.cjph.2017.03.006.

[200] M. Mustafa, T. Hayat, and A. Alsaedi. Unsteady boundary layer flow of nanofluid past an impulsively stretching sheet. *Journal of Mechanics*, 29(3):423 – 432, 2013. doi:10.1017/jmech.2013.9.

[201] M. Mustafa, M. Nawaz, T. Hayat, and A. & Alsaedi. MHD boundary layer flow of second-grade nanofluid over a stretching sheet with convective boundary conditions. *Journal of Aerospace Engineering*, 27(4):04014006, 2012. doi: 10.1061/(asce)as.1943-5525.0000314.

[202] Sandeep N., Chamkha A. J., and I. L. Animasaun. Numerical exploration of magnetohydrodynamic nanofluid flow suspended with magnetite nanoparticles. *The Journal of Brazilian Society of Mechanical Sciences and Engineering*, 39(9):3635 – 3644, 2018. doi:10.1007/s40430-017-0866-x.

[203] S. Nadeem, R. U. Haq, N. S. Akbar, C. Lee, and Z. H. Khan. Numerical study of boundary layer flow and heat transfer of Oldroyd-b nanofluid towards a stretching sheet. *PloS one*, 8(8):e69811, 2013. doi:10.1371/journal.pone.0069811.

[204] F. Nasirzadehroshenin, A. pourmozafari, H. Maddah, and H. Sakhaeinia. Experimental and theoretical investigation of thermophysical properties of synthesized hybrid nanofluid developed by modeling approaches. *Arabian Journal for Science and Engineering*, 45(9):7205 – 7218, 2020. doi:10.1007/s13369-020-04352-6.

[205] R. Nasrin, S. Parvin, and M. A. Alim. Effect of Prandtl number on free convection in a solar collector filled with nanofluid. *Procedia Engineering*, 56:54 – 62, 2013. doi: 10.1016/j.proeng.2013.03.088.

[206] M. Naveed, Z. Abbas, and M. Sajid. Nonlinear radiative heat transfer in blasius and sakiadis flows over a curved surface. *International Journal of Thermophysics*, 38(1):14, 2016. doi:10.1007/s10765-016-2154-x.

[207] M. K. Nayak, V. S. Pandey, D. Tripathi, N. S. Akbar, and O. D. Makinde. 3D MHD cross flow over an exponential stretching porous surface. *Heat Transfer - Asian Research*, 49(3):1256 – 1280, 2020. doi:10.1002/htj.21661.

[208] M. K. Nayak, S. Shaw, and A. J. Chamkha. 3D MHD free convective stretched flow of a radiative nanofluid inspired by variable magnetic field. *Arabian Journal for Science and Engineering*, 44(2):1269 – 1282, 2019. doi:10.1007/s13369-018-3473-y.

[209] M. K. Nayak, A. Wakif, I. L. Animasaun, and M. S. H. Alaoui. Numerical differential quadrature examination of steady mixed convection nanofluid flows over an isothermal thin needle conveying metallic and metallic oxide nanomaterials: A comparative investigation. *Arabian Journal for Science and Engineering*, 45(7):5331 – 5346, 2020. doi:10.1007/s13369-020-04420-x.

[210] A. S. Nehad, I. L. Animasaun, Jae Dong Chung, Abderrahim Wakif, F. I. Alao, and C.S.K. Raju. Significance of nanoparticle's radius, heat flux due to concentration gradient, and mass flux due to temperature gradient: The case of water conveying copper nanoparticles. *Scientific Reports*, 11:1882, 2021. doi:10.1038/s41598-021-81417-y.

[211] A. S. Nehad, I. L. Animasaun, A. Wakif, O. K. Koriko, R. Sivaraj, K. S. Adegbie, Zahra Abdelmalek, Hanumesh Vaidya, A. F. Ijirimoye, and K. V. Prasad. Significance of suction and dual stretching on the dynamics of various hybrid nanofluids: Comparative analysis between type I and type II models. *Physica Scripta*, 95(9):095205, 2020. doi:10.1088/1402-4896/aba8c6.

[212] J. L. Neyeloff, S. C. Fuchs, and L. B. Moreira. Meta-analyses and forest plots using a microsoft excel spreadsheet:step-by-step guide focusing on descriptive data analysis. *BMC Research Notes*, 5(1):52, 2012.

[213] C.T. Nguyen, F. Desgranges, and N. Galanisetal. Viscosity data for Al_2O_3-water nanofluid - hysteresis: Is heat transfer enhancement using nanofluids reliable? *International Journal of Thermal Sciences*, 47(2):103 – 111, 2008.

[214] D. A. Nield and A. Bejan. *Convection in Porous Media.* Berlin Heidelberg New York Tokyo: Springer-Verlag, New York., 1992.

[215] A. Nordmann, B. Kasenda, and M. Briel. Meta-analyses: What they can and cannot do? *Swiss Medical Weekly*, 142:w13518, 2012. doi:10.4414/smw.2012.13518.

[216] A. S. Oke, I. L. Animasaun, W. N. Mutuku, M. Kimathi, N. A. Shah, and S. Saleem. Significance of coriolis force, volume fraction, and heat source/sink on the dynamics of water conveying 47 nm alumina nanoparticles over a uniform surface. *Chinese Journal of Physics*, 71:716 – 727, 2021. doi:10.1016/j.cjph.2021.02.005.

[217] M. A. Olanrewaju, J. A. Gbadeyan, and A. S. Idowu. Flow and heat transfer analysis of a second grade fluid with newtonian heating in the presence of elastic deformation in a porous medium. *The Pacific Journal of Science and Technology*, 17(1):30 – 47, 2016.

[218] P. O. Olanrewaju, J. A. Gbadeyan, O. O. Agboola, and S. O. Abah. Radiation and viscous dissipation effects for the Blasius and Sakiadis flows with a convective surface boundary condition. *International Journal of Advances in science and Technology*, 2(4):102 – 115, 2011.

[219] A. J. Omowaye and I. L. Animasaun. Upper-convected Maxwell fluid flow with variable thermo-physical properties over a melting surface situated in hot environment subject to thermal stratification. *Journal of Applied Fluid Mechanics*, 9(4):1777 – 1790, 2016.

[220] K. O'Rourke and S. Greenland. *Meta-Analysis.* Lippincott Williams and Wilkins, Philadelphia, USA, 652 - 682, 2008.

[221] B. C. Pak and Y. I. Cho. Hydrodynamic and heat transfer study of dispersed fluids with submicron metallic oxide particles. *Experimental Heat Transfer*, 1(2):151 – 170, 1998.

[222] D. Pal and H. Mondal. Influence of soret-dufour and thermophoresis on hydromagnetic mixed convection heat and mass transfer over an inclined flat plate with non-uniform heat source/sink and chemical reaction. *International Journal for Computational Methods in Engineering Science and Mechanics*, 19(2):49 – 60, 2018.

[223] G. Palmer. Thermal conductivity of liquids. *Industrial and Engineering Chemistry*, 40(1):89 – 92, 1948.

[224] A. Pantokratoras. The blasius, sakiadis and Blasius Sakiadis flow with uniform suction: a unified approach. *Progress in Computational Fluid Dynamics, An International Journal*, 7(8):482, 2007. doi:10.1504/pcfd.2007.015781.

[225] A. Pantokratoras. Non-similar Blasius and Sakiadis flow of a non-Newtonian Carreau fluid. *Journal of the Taiwan Institute of Chemical Engineers*, 56:1 – 5, 2015. doi:10.1016/j.jtice.2015.03.021.

[226] J. Parker, C. Caulfield, and R. Kerswell. The effects of Prandtl number on the nonlinear dynamics of kelvin-helmholtz instability in two dimensions. *Journal of Fluid Mechanics*, 915:A37, 2021. doi:10.1017/jfm.2021.125.

[227] S. Parvin, R. Nasrin, M. A. Alim, and N. F. Hossain. Effect of Prandtl number on forced convection in a two sided open enclosure using nanofluid. *Journal of Scientific Research*, 5(1):67 – 75, 2013. doi: 10.3329/jsr.v5i1.9641.

[228] H. R. Patel. Effects of cross diffusion and heat generation on mixed convective mhd flow of Casson fluid through porous medium with non-linear thermal radiation. *Heliyon*, 5(4):e01555, 2019. doi:10.1016/j.heliyon.2019.e01555.

[229] K. K. Patra and S. Panda. Formation of the capillary ridge on the free surface dynamics of second-grade fluid over an inclined locally heated plate. *Zeitschrift fur Naturforschung A*, 74(12):1099 – 1108, 2019. doi: 10.1515/zna-2019-0126.

[230] G. M. Pavithra and B. J. Gireesha. Effect of internal heat generation/absorption on dusty fluid flow over an exponentially stretching sheet with viscous dissipation. *Journal of Mathematics*, 2013:1 – 10, 2013. doi:10.1155/2013/583615.

[231] R. Pfister, K. Schwarz, R. Carson, and M. Jancyzk. Easy methods for extracting individual regression slopes: Comparing spss, r, and excel. *Tutorials in Quantitative Methods for Psychology*, 9(2):72 – 78, 2013.

[232] F. B. Pidduck, S. Chapman, and T. G. Cowling. The mathematical theory of non-uniform gases. *The Mathematical Gazette*, 23(257):488, 1939. doi:10.2307/3607024.

[233] R. L. Plackett. Studies in the history of probability and statistics: Vii. the principle of the arithmetic mean. *Biometrika*, 45(1/2):130, 1958. doi: 10.2307/2333051.

[234] D. R. Poirier and G. H. Geiger. Fick's law and diffusivity of materials. *In Transport Phenomena in Materials Processing*, (2):419–461, 2016.

[235] T. Poornima and N. B. Reddy. Radiation effects on MHD free convective boundary layer flow of nanofluids over a nonlinear stretching sheet. *Advances in Applied Science Research*, 4(2):190 – 202, 2013.

[236] S. Pramanik. Casson fluid flow and heat transfer past an exponentially porous stretching surface in presence of thermal radiation. *Ain Shams Engineering Journal*, 5(1):205 – 212, 2014. doi: 10.1016/j.asej.2013.05.003.

[237] S. Qayyum. *Nonlinear models for flows of non-Newtonian fluids with heat transfer.* (Unpublished doctoral dissertation). Quaid-I-Azam University Islamabad, Pakistan., 2019.

[238] S. Qayyum, T. Hayat, S. A. Shehzad, and A. Alsaedi. Mixed convection and heat generation/absorption aspects in MHD flow of tangent-hyperbolic nanoliquid with Newtonian heat/mass transfer. *Radiation Physics and Chemistry*, 144(1):396 – 404, 2018. doi:10.1016/j.radphyschem.2017.10.002.

[239] S. Qayyum, M. Imtiaz, A. Alsaedi, and T. Hayat. Analysis of radiation in a suspension of nanoparticles and gyrotactic microorganism for rotating disk of variable thickness. *Chinese Journal of Physics*, 56(5):2404 – 2423, 2018. doi:10.1016/j.cjph.2018.06.020.

[240] M. M. Rahman, A. A. Mamun, M. A. Azim, and M. A. Alim. Effects of temperature dependent thermal conductivity on magnetohydrodynamic (MHD) free convection flow along a vertical flat plate with heat conduction. *Nonlinear Analysis: Modeling and Control*, 13(4):513 – 524, 2008. doi: 10.15388/na.2008.13.4.14555.

[241] M. M. Rahman, S. Parvin, N. A. Rahim, M. R. Islam, R. Saidur, and M. Hasanuzzaman. Effects of Reynolds and prandtl number on mixed convection in a ventilated cavity with a heat-generating solid circular block. *Applied Mathematical Modelling*, 36(5):2056 – 2066, 2012. doi: 10.1016/j.apm.2011.08.014.

[242] A. Rai, K. Park, L. Zhou, and M. R. Zachariah. Understanding the mechanism of aluminium nanoparticle oxidation. *Combustion Theory and Modelling*, 10(5):843 – 859, 2006. doi:10.1080/13647830600800686.

[243] R. K. Rajput. *A Textbook of fluid mechanics and hydraulic machines*. S. Chand and Company Ltd., New Delhi., 2004.

[244] S. Ramachandran, C.B. Sobhan, and G.P. Peterson. Thermophoresis of nanoparticles in liquids. *International Journal of Heat and Mass Transfer*, 147:118925, 2020. doi: 10.1016/j.ijheatmasstransfer.2019.118925.

[245] B. Ramadevi, K. A. Kumar, V. Sugunamma, and N. Sandeep. Influence of non-uniform heat source / sink on the three-dimensional magnetohydrodynamic Carreau fluid flow past a stretching surface with modified Fourier's law. *Pramana*, 93(6):86, 2019. doi:10.1007/s12043-019-1847-7.

[246] G. K. Ramesh and B. J. Gireesha. Flow over a stretching sheet in a dusty fluid with radiation effect. *Journal of Heat Transfer*, 135(10):102702, 2013. doi:10.1115/1.4024587.

[247] K. Ramesh and J. Prakash. Thermal analysis for heat transfer enhancement in electroosmosis-modulated peristaltic transport of sutterby nanofluids in a microfluidic vessel. *Journal of Thermal Analysis and Calorimetry*, 138(2):1311 – 1326, 2019. doi:10.1007/s10973-018-7939-7.

[248] M. L. V . Ramires, C. A. Nieto de Castr o, Y . Nagasaka, A. Nagashima, M. J. Assael, and W. A. Wakeham. *Standard reference data for the thermal conductivity of water*. https:www.nist.govsitesdefaultfilesdocumentssrdjpcrd493.pdf NIST. Retrieved 25 May 2017., 1994.

[249] M. Ramzan, J. D. Chung, and N. Ullah. Radiative magnetohydrodynamic nanofluid flow due to gyrotactic microorganisms with chemical reaction and non-linear thermal radiation. *International Journal of Mechanical Sciences*, 130:31 – 40, 2017. doi:10.1016/j.ijmecsci.2017.06.009.

[250] M. Ramzan, F. Yousaf, M. Farooq, and J. D. Chung. Mixed convective viscoelastic nanofluid flow past a porous media with Soret-Dufour effects. *Communications in Theoretical Physics*, 66(1):133 – 142, 2016. doi:10.1088/0253-6102/66/1/133.

[251] B. E. Rapp. Chapter 20: Surface tension. microfluidics: Modelling, mechanics and mathematics. pages 421 – 444, 2017. doi:10.1016/b978-1-4557-3141-1.50003-4.

[252] A. Rasekh, and Tavakoli S. Ganji, D.D., and S. Naeejee. Melting effect on steady laminar flow of a micropolar fluid over a stagnation point on a vertical surface. *J Eng Phy Thermophys*, 86:210 – 1216, 2013. doi:10.1016/b978-1-4557-3141-1.50003-4.

[253] Ghulam Rasool, Anum Shafiq, Ilyas Khan, Dumitru Baleanu, Kottakkaran Sooppy Nisar, and Gullnaz Shahzadi. Entropy generation and consequences of MHD in Darcy-Forchheimer nanofluid flow bounded by non-linearly stretching surface. *Symmetry*, 12:652, 2020. doi:10.3390/sym12040652.

[254] S. P. Rayapole and A. R. Jakkula. Effects of brownian motion and thermophoresis on magneto hydrodynamics stagnation point of a nanofluid boundary layer flow on a stretching surface with variable thickness. *International Journal of Current Research and Review*, 9(22):5 – 12, 2017. doi: 10.7324/ijcrr.2017.9222.

[255] J. Raza, A. M. Rohni, Z. Omar, and M. Awais. Heat and mass transfer analysis of MHD nanofluid flow in a rotating channel with slip effects. *Journal of Molecular Liquids*, 219:703 – 708, 2016. doi: 10.1016/j.molliq.2016.04.003.

[256] G. J. Reddy, B. Kethireddy, M. Kumar, H. P. Rani, and R. S. R. Gorla. Effect of prandtl number for Casson fluid flow over a vertical cylinder: Heatline approach. *International Journal of Applied and Computational Mathematics*, 4(3):85, 2018. doi:10.1007/s40819-018-0516-8.

[257] C. Revnic, T. Grosan, M. Sheremet, and I. Pop. Numerical simulation of MHD natural convection flow in a wavy cavity filled by a hybrid $Cu - Al_2O_3$-water nanofluid with discrete heating. *Applied Mathematics and Mechanics*, 41:1345 – 1358, 2020.

[258] V. J. Rossow. On flow of electrically conducting fluids over a flat plate in the presence of a transverse magnetic field. *NACA Tech*, page 3071, 1958.

[259] K. J. Rothman, S. Greenland, and T. L. Lash. Chapter 3 - epidemiologic study designs. *Epidemiology and Medical Statistics*, pages 64 – 108, 2007.

[260] Saleem S., Rafiq H., Al-Qahtani A., Abd El-Aziz M., Malik M. Y., and Animasaun I. L. Magneto Jeffrey nanofluid bioconvection over a rotating vertical cone due to gyrotactic microorganism. *Mathematical Problems in Engineering*, 2019(Article ID 3478037):11, 2019. doi: 10.1155/2019/3478037.

[261] R. K. Sahu and S. S. Hiremath. Synthesis of aluminium nanoparticles in a water/polyethylene glycol mixed solvent using $\mu-$edm. *IOP Conference Series: Materials Science and Engineering*, 225:012257, 2017. doi:10.1088/1757-899x/225/1/012257.

[262] Z. Said, R. Saidur, M. A. Sabiha, N. A. Rahim, and M. R. Anisur. Thermo-physical properties of single wall carbon nanotubes and its effect on exergy efficiency of a flat plate solar collector. *Solar Energy*, 115:757 – 769, 2015.

[263] M. Saidi and G. Karimi. Free convection cooling in modified l-shape enclosures using copper-water nanofluid. *Energy*, 70:251 – 271, 2014. doi:10.1016/j.energy.2014.03.121.

[264] B. C. Sakiadis. Boundary layer behavior on continuous solid surfaces: The boundary layer on a continuous flat surface. *American Institute of Chemical Engineers*, 7(2):221 – 225, 1961.

[265] T. Salahuddin, M. Khan, T. Saeed, M. Ibrahim, and Y.-M. Chu. Induced MHD impact on exponentially varying viscosity of williamson fluid flow with variable conductivity and diffusivity. *Case Studies in Thermal Engineering*, 25:100895, 2021. doi:10.1016/j.csite.2021.100895.

[266] S. D. Salman. Comparative study on heat transfer enhancement of nanofluids flow in ribs tube using CFD simulation. *Heat Transfer-Asian Research*, 48(1):148 – 163, 2018. doi:10.1002/htj.21376.

[267] M. Samee AD, A. Afzal, A. Razak RK, and MK. Ramis. Effect of Prandtl number on the average exit temperature of coolant in a heat-generating vertical parallel plate channel: A conjugate analysis. *Heat Transfer-Asian Research*, 47(4):603 – 619, 2018. doi:10.1002/htj.21330.

[268] R. L. Sani and J. N. Koster. Low-gravity fluid dynamics and transport phenomena (chapter - analysis of convective situations with the soret effect). *American Institute of Aeronautics and Astronautics*, pages 437 – 485, 1990. doi:10.2514/5.9781600866036.0437.0485.

[269] S. Saravanan and P. Kandaswamy. Buoyancy convection in low Prandtl number liquids with large temperature variation. *Meccanica*, 37(6):599 – 608, 2002. doi:10.1023/A:1020959623479.

[270] G. S. Sariyerli, O. Sakarya, and U. Y. Akcadag. Comparison tests for the determination of the viscosity values of reference liquids by capillary viscometers and stabinger viscometer svm 3001. *International Journal of Metrology and Quality Engineering*, 9:7, 2018. doi:10.1051/ijmqe/2018004.

[271] G. Sarojamma, R. V. Lakshmi, P. S. Narayana, and I. L. Animasaun. Exploration of the significance of autocatalytic chemical reaction and Cattaneo-Christov heat flux on the dynamics of a micropolar fluid. *Journal of Applied and Computational Mechanics*, 6(1):77 – 89, 2020. doi: 10.22055/JACM.2019.28742.1501.

[272] G. Sarojamma and K. Vendabai. Boundary layer flow of a Casson nanofluid past a vertical exponentially stretching cylinder in the presence of a transverse magnetic field with internal heat generation/absorption. *World Academy of Science, Engineering and Technology, International Journal of Mechanical, Aerospace, Industrial, Mechatronic and Manufacturing Engineering*, 9(1):138 – 143, 2015. doi.org/10.5281/zenodo.1099346.

[273] D. R. V. S. R. K. Sastry, A. S. N. Murti, and T. P. Kantha. The effect of heat transfer on MHD Marangoni boundary layer flow past a flat plate in nanofluid. *International Journal of Engineering Mathematics*, page 581507, 2013. doi: 10.1155/2013/581507.

[274] W. Schneider. A similarity solution for combined forced and free convection flow over a horizontal plate. *International Journal of Heat and Mass Transfer*, 22(10):1401 – 1406, 1979. doi: 10.1016/0017-9310(79)90202-3.

[275] F. Selimefendigil, H. F. Oztop, and N. Abu-Hamdeh. Mixed convection due to rotating cylinder in an internally heated and flexible walled cavity filled with SiO_2-water nanofluids: Effect of nanoparticle shape. *International Communications in Heat and Mass Transfer*, 71(10):9 – 19, 2016. doi:10.1016/j.icheatmasstransfer.2015.12.007.

[276] N. A. Shah, I. L. Animasaun, R. O. Ibraheem, H. A. Babatunde, N. Sandeep, and I. Pop. Scrutinization of the effects of Grashof number on the flow of different fluids driven by convection over various surfaces. *Journal of Molecular Liquids*, 249(10):980 – 990, 2018. doi:10.1016/j.molliq.2017.11.042.

[277] Z. Shah, S. Islam, T. Gul, E. Bonyah, and M. Altaf Khan. The electrical MHD and hall current impact on micropolar nanofluid flow between rotating parallel plates. *Results in Physics*, 9:1201 – 1214, 2018. doi: 10.1016/j.rinp.2018.01.064.

[278] V. Sharma and A. K. Dhiman. Heat transfer from a rotating circular cylinder in the steady regime: Effects of Prandtl number. *Thermal Science*, 16(1):79 – 91, 2012. doi: 10.2298/tsci100914057s.

[279] S. Shaw, A. S. Dogonchi, M. K. Nayak, and O. D. Makinde. Impact of entropy generation and nonlinear thermal radiation on Darcy-Forchheimer flow of $MnFe_2O_4$-casson/water nanofluid due to a rotating disk: Application to brain dynamics. *Arabian*

Journal for Science and Engineering, 45(7):5471 – 5490, 2020. doi:10.1007/s13369-020-04453-2.

[280] S. A. Shehzad, A. Alsaedi, and T. Hayat. Influence of thermophoresis and Joule heating on the radiative flow of Jeffrey fluid with mixed convection. *Brazilian Journal of Chemical Engineering*, 30(4):897 – 908, 2013. doi: 10.1590/s0104-66322013000400021.

[281] M. Sheikholeslami and H. B. Rokni. Effect of melting heat transfer on nanofluid flow in existence of magnetic field considering buongiorno model. *Chinese Journal of Physics*, 55(4):1115 – 1126, 2017. doi:10.1016/j.cjph.2017.04.019.

[282] Shell Chemical LP. Houston Texas 77210, USA.

[283] H. Sithole, H. Mondal, and P. Sibanda. Entropy generation in a second grade magnetohydrodynamic nanofluid flow over a convectively heated stretching sheet with nonlinear thermal radiation and viscous dissipation. *Results in Physics*, 9:1077 – 1085, 2018.

[284] R. Sivaraj, A. J. Benazir, S. Srinivas, and A. J. Chamkha. Investigation of cross-diffusion effects on casson fluid flow in existence of variable fluid properties. *The European Physical Journal Special Topics*, 228(1):35 – 53, 2019. doi:10.1140/epjst/e2019-800187-3.

[285] S. R. Smith, R. Rafati, A. Sharifi Haddad, A. Cooper, and H. Hamidi. Application of aluminium oxide nanoparticles to enhance rheological and filtration properties of water based muds at hpht conditions. *Colloids and Surfaces A: Physicochemical and Engineering Aspects*, 537:361 – 371, 2018. doi:10.1016/j.colsurfa.2017.10.050.

[286] C. W. Somerton. The Prandtl number effect in porous layer convection. *Applied Scientific Research*, 40(4):333 – 344, 1983.

[287] Y.-Q. Song, B. D. Obideyi, N. A. Shah, I. L. Animasaun, Y. M. Mahrous, and J. D. Chung. Significance of haphazard motion and thermal migration of alumina and copper nanoparticles across the dynamics of water and ethylene glycol on a convectively heated surface. *Case Studies in Thermal Engineering*, 26:101050, 2021. doi:10.1016/j.csite.2021.101050.

[288] C. Sulochana and G. P. Ashwinkumar. Impact of brownian moment and thermophoresis on magnetohydrodynamic flow of magnetic nanofluid past an elongated sheet in the presence of thermal diffusion. *Multidiscipline Modeling in Materials and Structures*, 14(4):744 – 755, 2018. doi: 10.1108/MMMS-12-2017-0168.

[289] F. Sultan, W. A. Khan, M. Ali, M. Shahzad, H. Sun, and M. Irfan. Importance of entropy generation and infinite shear rate viscosity for non-Newtonian nanofluid. *Journal of the Brazilian Society of Mechanical Sciences and Engineering*, 41(10):439, 2019. doi:10.1007/s40430-019-1950-1.

[290] H. Tamao. Changes of specific heat capacity and heat capacity during weathering. *Chemical Geology*, 60(1987):131 – 136, 1986. doi:10.1007/s40430-019-1950-1.

[291] S. Tanvir and L. Qiao. Surface tension of nanofluid-type fuels containing suspended nanomaterials. *Nanoscale research letters*, 7(1):1 – 10, 2012. doi:10.1007/s40430-019-1950-1.

[292] S. Tanweer, A. Dewan, and S. Sanghi. Study on effects of Prandtl number on cross buoyancy flow past a square cylinder using open foam. *Journal of Applied Fluid Mechanics*, 12(1):257 – 269, 2019. doi:10.29252/jafm.75.253.29020.

[293] H. Tariq, A. A. Khan, and A. Zaman. Peristaltically wavy motion on dusty Walter's b fluid with inclined magnetic field and heat transfer. *Arabian Journal for Science and Engineering*, 44(9):7799 – 7808, 2019. doi:10.1007/s13369-019-03909-4.

[294] G. M. Tawfik, K. A. S. Dila, M. Y. F. Mohamed, D. N. H. Tam, N. D. Kien, A. M. Ahmed, and N. T. Huy. A step by step guide for conducting a systematic review and meta-analysis with simulation data. *Tropical Medicine and Health*, 47(1):46, 2019. doi:10.1186/s41182-019-0165-6.

[295] E. V. Timofeeva, J. L. Routbort, and D. Singh. Particle shape effects on thermophysical properties of alumina nanofluids. *Journal of Applied Physics*, 106(1):014304, 2009. doi:10.1063/1.3155999.

[296] I. Tlili, H. A. Nabwey, S. P. Samrat, and N. Sandeep. 3d MHD nonlinear radiative flow of cuo-mgo/methanol hybrid nanofluid beyond an irregular dimension surface with slip effect. *Scientific Reports*, 10(1):9181, 2020. doi:10.1038/s41598-020-66102-w.

[297] M. Trivedi, O. Otegbeye, M. S. Ansari, and S. S. Motsa. A paired quasi-linearization on magnetohydrodynamic flow and heat transfer of Casson nanofluid with Hall effects. *Journal of Applied and Computational Mechanics*, 5(5):849 – 860, 2019. doi:10.1038/s41598-020-66102-w.

[298] M. J. Uddin, M. Ferdows, and O. A. Beg. Group analysis and numerical computation of magneto-convective non-Newtonian nanofluid slip flow from a permeable stretching sheet. *Applied Nanoscience*, 4(7):897 – 910, 2013. doi:10.1007/s13204-013-0274-1.

[299] M. J. Uddin, M. A. A. Hamad, and A. M. Ismail. Thermophoresis and brownian motion effect on chemically reacting MHD boundary layer slips flow of a nanofluid. *AIP Conference Proceedings*, 1450(1):183 – 189, 2012. doi: 10.1063/1.4724137.

[300] M. J. Uddin, W. A. Khan, F.-T. Zohra, and A. I. M. Ismail. Blasius and Sakiadis slip flows of nanofluid with radiation effects. *Journal of Aerospace Engineering*, 29(4):04015080, 2016. doi:10.1061/(asce)as.1943-5525.0000575.

[301] H. Ur Rasheed, A. AL-Zubaidi, S. Islam, S. Saleem, Z. Khan, and W. Khan. Effects of Joule heating and viscous dissipation on magnetohydrodynamic boundary layer flow of Jeffrey nanofluid over a vertically stretching cylinder. *Coatings*, 11(3):353, 2021. doi:10.3390/coatings11030353.

[302] K. Ur Rehman, N. Ul Saba, M. Y. Malik, and I. Zehra. Nanoparticles individualities in both Newtonian and Casson fluid models by way of stratified media: A numerical analysis. *The European Physical Journal E*, 41:37, 2018. doi: 10.1140/epje/i2018-11641-8.

[303] M. Usman, F. A. Soomro, R. Ul Haq, W. Wang, and O. Defterli. Thermal and velocity slip effects on Casson nanofluid flow over an inclined permeable stretching cylinder via collocation method. *International Journal of Heat and Mass Transfer*, 122:1255 – 1263, 2018. doi: 10.1016/j.ijheatmasstransfer.2018.02.045.

[304] S. Vafaei, D. Wen, and T. Borca-Tasciuc. Nanofluid surface wettability through asymptotic contact angle. *Langmuir*, 27(6):2211 – 2218, 2011. doi: 10.1016/j.ijheatmasstransfer.2018.02.045.

[305] R. S. Vajjha and D. K. Das. Measurements of specific heat and density of Al_2O_3 nanofluid. *AIP Conference Proceedings*, 1063(1):361, 2008. doi:10.1063/1.3027181.

[306] R. Verzicco and R. Camussi. Prandtl number effects in convective turbulence. *Journal of Fluid Mechanics*, 383(1):55 – 73, 1999. doi: 10.1017/s0022112098003619.

[307] D. J. Vitello, R. M. Ripper, M. R. Fettiplace, G. L. Weinberg, and J. M. Vitello. Blood density is nearly equal to water density: A validation study of the gravimetric method of measuring intraoperative blood loss. *Journal of Veterinary Medicine*, 2015(1):1 – 4, 2015. doi:10.1155/2015/152730.

[308] A. Wakif, I. L. Animasaun, P. V. Satya Narayana, and G. Sarojamma. Meta-analysis on thermo-migration of tiny/nano-sized particles in the motion of various fluids. *Chinese Journal of Physics*, 68(1):293 – 307, 2020. doi: 10.1016/j.cjph.2019.12.002.

[309] A. Wakif, Z. Boulahia, A. Amine, I. L. Animasaun, M. I. Afridi, M. Qasim, and R. Sehaqui. Magneto-convection of alumina - water nanofluid within thin horizontal layers using the revised generalized buongiorno's model. *Frontiers in Heat and Mass Transfer (FHMT)*, 12(3), 2019. doi:10.5098/hmt.12.3

[310] A. Wakif, Z. Boulahia, S.R. Mishra, M.M. Rashidi, and R. Sehaqui. Influence of a uniform transverse magnetic field on the thermohydrodynamic stability in water-based nanofluids with metallic nanoparticles using the generalized buongiorno's mathematical model. *The European Physical Journal Plus*, 133:181, 2018. doi: 10.1140/epjp/i2018-12037-7.

[311] B. Wang, L. Zhou, and X. Peng. Viscosity, thermal diffusivity and Prandtl number of nanoparticle suspensions. *Progress in Natural Science*, 14(10):922 – 926, 2004. doi: 10.1080/10020070412331344571.

[312] H. Wang, M. Liang, Z. He, X. Kang, Y. Zhao, and Y. Zhao. Usefulness of adding magnetic nano-powders to water: Introducing novel criteria for assessing heat transfer applications. *Powder Technology*, 133(181):1 – 16, 2021. doi:10.1016/j.powtec.2021.03.052.

[313] M. Waqas, M. I. Khan, T. Hayat, M. M. Gulzar, and A. Alsaedi. Transportation of radiative energy in viscoelastic nanofluid considering buoyancy forces and convective conditions. *Chaos Solitons & Fractals*, 130:109415, 2020. doi: 10.1016/j.chaos.2019.109415.

[314] P. S. Wei, J. S. Yeh, C. N. Ting, T. DebRoy, F. K. Chung, and C. L. Lin. The effects of Prandtl number on wavy weld boundary. *International Journal of Heat and Mass Transfer*, 52(15-16):3790 – 3798, 2009. doi: 10.1016/j.ijheatmasstransfer.2009.02.020.

[315] S. Wei and C. Ming-Hsiung. Effect of Prandtl number on buoyancy-induced transport processes with and without solidification. *International Journal of Heat and Mass Transfer*, 33(11):2565 – 2578, 1990. doi: 10.1016/j.ijheatmasstransfer.2009.02.020.

[316] H. F. Wong, M. Sohail, Z. Siri, and N. F. M. Noor. Numerical solutions for heat transfer of an unsteady cavity with viscous heating. *Computers, Materials & Continua*, 68(1):319 – 336, 2021. doi:10.32604/cmc.2021.015710.

[317] R. C. Wrede and M. Spiegel. *Theory and problems of advanced calculus.* Schaum's Outline Series McGraw-Hill Companies, Inc, New York., 2002. doi: 10.1036/0071398341.

[318] W. Wuest and C.-H. Chun. Thermal Marangoni convection. *COSPAR: Space Research*, 68:559 – 562, 1979. doi:10.1016/b978-0-08-023417-5.50104-6.

[319] K. Q. Xia, S. Lam, and S. Q. Zhou. Heat-flux measurement in high-Prandtl-number turbulent Rayleigh-Benard convection. *Physical review letters*, 88(6):064501, 2002. doi: 10.1103/PhysRevLett.88.064501.

[320] R. C. Xin and M. A. Ebadian. The effects of Prandtl numbers on local and average convective heat transfer characteristics in helical pipes. *Journal of Heat Transfer*, 119(3):467 – 473, 1997.

[321] X. Xu and S. Chen. Cattaneo-christov heat flux model for heat transfer of Marangoni boundary layer flow in a copper-water nanofluid. *Heat Transfer-Asian Research*, 46(8):1281 – 1293, 2017.

[322] Y. Xuan and W. Roetzel. Conceptions for heat transfer correlation of nanofluids. *International Journal of Heat and Mass Transfer*, 43(19):3701 – 3707, 2000.

[323] Hari Krishna Y, Ramana Murthy MV, Bhikshu NL, and Venkata Ramana G. Effects of radiation and chemical reaction on mhd flow past an oscillating inclined porous plate with variable temperature and mass diffusion. *International Journal of Chemical Sciences*, 15(3):1 – 12, 2017.

[324] N Yacob, A Ishak, and I. Pop. Melting heat transfer in boundary layer stagnation-point flow towards a stretching/shrinking sheet in a micropolar fluid. *Computers and Fluids*, 47:16 – 21, 2011. doi:10.1016/j.compfluid.2011.01.040.

[325] Y. Yang, A. Oztekin, S. Neti, and S. Mohapatra. Particle agglomeration and properties of nanofluids. *Journal of Nanoparticle Research*, 14(5):852, 2012. doi:10.1007/s11051-012-0852-2.

[326] A. D. Young. The effects of small changes of Prandtl number and the viscosity-temperature index on skin-friction. *Aeronautical Quarterly*, 15(04):392 – 406, 1964.

[327] P. Zhang, Z. N. Meng, H. Zhu, Y. L. Wang, and S. P. Peng. Melting heat transfer characteristics of a composite phase change material fabricated by paraffin and metal foam. *Applied Energy*, 185:1971 – 1983, 2017. doi:10.1016/j.apenergy.2015.10.075.

[328] D. W. Zhao, G. H. Su, Tian W. X., K. Sugiyama, and Qiu S. Z. Experimental and theoretical study on transition boiling along a downward-facing horizontal surface in confined space. *Nuclear Engineering and Design*, 238(9):2460 – 2467, 2008. doi: 10.1016/j.nucengdes.2008.02.015.

[329] E. J. Zita. Melting heat transfer characteristics of a composite phase change material fabricated by paraffin and metal foam. *American Journal of Physics*, 74:462, 2006. doi: 10.1119/1.2174055.

[330] Jang, S.P. Choi SUS. Role of Brownian motion in the enhanced thermal conductivity of nanofluids. *Appl. Phys. Lett.*, 84(21):4316, 2004. doi:10.1063/1.1756684.

[331] Cao, W., Animasaun, I. L., Yook, S. J., Oladipupo, V. A., and Ji, X. Simulation of the dynamics of colloidal mixture of water with various nanoparticles at different levels of partial slip: Ternary-hybrid nanofluid. *International Communications in Heat and Mass Transfer*, 135:106069, 2022. doi:10.1016/j.icheatmasstransfer.2022.106069

[illegible]

[illegible]

[illegible]

[illegible]

[illegible]

[illegible]

[illegible]

[illegible]

[illegible]

[illegible]

[illegible]

[illegible]

[illegible]

[illegible]

Index

Note: **Bold** page numbers refer to tables and *italic* page numbers refer to figures.